T0360794

FRACTIONAL
PARTIAL
DIFFERENTIAL
EQUATIONS

FRACTIONAL PARTIAL DIFFERENTIAL EQUATIONS

Yong Zhou

Xiangtan University, China

Macau University of Science and Technology, China

World Scientific

NEW JERSEY · LONDON · SINGAPORE · BEIJING · SHANGHAI · HONG KONG · TAIPEI · CHENNAI · TOKYO

Published by

World Scientific Publishing Co. Pte. Ltd.

5 Toh Tuck Link, Singapore 596224

USA office: 27 Warren Street, Suite 401-402, Hackensack, NJ 07601

UK office: 57 Shelton Street, Covent Garden, London WC2H 9HE

Library of Congress Control Number: 2024007633

British Library Cataloguing-in-Publication Data
A catalogue record for this book is available from the British Library.

FRACTIONAL PARTIAL DIFFERENTIAL EQUATIONS

ISBN 978-981-12-9040-4 (hardcover)
ISBN 978-981-12-9041-1 (ebook for institutions)
ISBN 978-981-12-9042-8 (ebook for individuals)

For any available supplementary material, please visit
https://www.worldscientific.com/worldscibooks/10.1142/13764#t=suppl

Desk Editors: Nandha Kumar/Kwong Lai Fun

Typeset by Stallion Press
Email: enquiries@stallionpress.com

Printed in Singapore

Preface

The fractional calculus started more than three centuries ago. In the last years, the fractional calculus is playing a very important role in various scientific fields. In fact, it has been recognized as one of the best tools to describe long-memory processes. Fractional-order models are interesting not only for engineers and physicists, but also for mathematicians. Among such models those described by partial differential equations (PDEs) containing fractional derivatives are of utmost importance. Their evolution was more complex than for the classical integer-order counterpart. Nonetheless, classical PDEs' methods are hardly applicable directly to fractional PDEs. Therefore, new theories and methods are required, with concepts and algorithms specifically developed for fractional PDEs. In the recent years, the theory of fractional PDEs has been highly developed and constitutes an important branch of differential equations.

This monograph gives a presentation of the theory for time-fractional partial differential equations. Many of the basic techniques and results recently developed about this theory are presented, including the well-posedness, regularity, approximation and optimal control. The contents of the book cover fractional Navier-Stokes equations, fractional Rayleigh-Stokes equations, fractional Fokker-Planck equations and fractional Schrödinger equations. Several examples of applications relating to these equations are also discussed in detail. The materials in this monograph are based on the research work carried out by the author and some experts during the past five years. It provides the necessary background material required to go further into the subject and explore the rich research literature. It is useful for researchers, graduate or PhD students dealing with differential equations, applied analysis, and related areas of research.

I would like to thank Professors B. Ahmad, M. Fečkan, P. Górka, M. Kirane, V. Kiryakova, K.N. Le, W. McLean, S.K. Ntouyas, G.M. N'Guerekata, H. Prado, M. Stynes, J.J. Trujillo and N.H. Tuan for their support. I also wish to express my appreciation to doctoral students J.W. He, L. Peng and J.N. Wang for their help. Finally, I thank the editorial assistance of World Scientific Publishing Co., especially Ms. L.F. Kwong.

I acknowledge with gratitude the support of National Natural Science Foundation of China (12071396) and the Macau Science and Technology Development Fund (0092/2022/A).

Yong Zhou
November 2023, Macau, China

About the Author

Dr.**Yong Zhou** is a Full Professor at Xiangtan University and a Distinguished Guest Professor at Macau University of Science and Technology. His research fields include fractional differential equations, functional differential equations, evolution equations and control theory. Zhou has published seven monographs in Springer, Elsevier, De Gruyter, World Scientific and Science Press respectively, and more than three hundred research papers in international journals including *Mathematische Annalen, Journal of Functional Analysis, Nonlinearity, Inverse Problems, Proceedings of the Royal Society of Edinburgh, Bulletin des Sciences Mathématiques, Comptes rendus Mathematique, International Journal of Bifurcation and Chaos, Zeitschrift für Angewandte Mathematik und Physik,* and so on. He was included in Highly Cited Researchers list from Thompson Reuters (2014) and Clarivate Analytics (2015–2021). Zhou has undertaken five projects from National Natural Science Foundation of China, and two projects from the Macau Science and Technology Development Fund. He won the second prize of Chinese University Natural Science Award in 2000, and the second prize of Natural Science Award of Hunan Province, China in 2017 and 2021. Zhou had worked as an Associate Editor for *IEEE Transactions on Fuzzy Systems, Journal of Applied Mathematics & Computing, Mathematical Inequalities & Applications,* and an Editorial Board Member of *Fractional Calculus and Applied Analysis.*

Introduction

Fractional calculus has been attracting the attention of mathematicians and engineers from long time ago. The concept of fractional (or, more precisely, noninteger) differentiation appeared for the first time in a famous correspondence between L'Hospital and Leibniz, in 1695. Many mathematicians have further developed this area and we can mention the studies of Euler, Laplace, Abel, Liouville and Riemann. However, the fractional calculus remained for centuries a purely theoretical topic, with little if any connections to practical problems of physics and engineering. In the past thirty years, the fractional calculus has been recognized as an effective modelling methodology for researchers. Fractional differential equations are generalizations of classical differential equations to an arbitrary (noninteger) order. Based on the wide applications in engineering and sciences such as physics, mechanics, electricity, chemistry, biology, economics, and many others, research on fractional differential equations is active and extensive around the world.

In the recent years, there has been a significant development in ordinary and partial differential equations involving fractional derivatives, see the monographs of Miller et al. [268], Podlubny [303], Hilfer [171], Kilbas et al. [197], Diethelm [101], Zhou [386, 387] and the references therein. A strong motivation for investigating this class of equations comes mainly from a compelling reasons: the fractional order models of real systems are often more adequate than the classical integer order models, since the description of some systems is more accurate when the fractional derivative is used. As an example, considering anomalous diffusion phenomena, the classical diffusion equation fails to describe diffusion phenomenon in heterogeneous porous media that exhibits fractal characteristics. How is the classical diffusion equation modified to make it appropriate to depict anomalous diffusion phenomena? This problem is interesting for researchers. Fractional calculus has been found effective in modelling anomalous diffusion processes since it has been recognized as one of the best tools to characterize the long memory processes. Consequently, it is reasonable and significative to propose the generalized diffusion equations with fractional derivative operators, which can be used to simulate anomalous diffusion in fractal media. Its evolutionary behave in a much more complex way than in classical integer order case and the corresponding investigation becomes more challenging.

Such models are interesting not only for engineers and physicists but also for pure mathematicians.

This monograph gives a presentation of the theory for time-fractional partial differential equations. The contents of the book cover fractional Navier-Stokes equations, fractional Rayleigh-Stokes equations, fractional Fokker-Planck equations and fractional Schrödinger equations. Many of the basic techniques and results recently developed about this theory are presented, including well-posedness, regularity, approximation and optimal control for fractional partial differential equations. Several examples of applications relating to these equations are also discussed in detail. Such basic theory should be the starting point for further research concerning the dynamics, numerical analysis and applications of fractional partial differential equations.

This monograph is arranged and organized as follows.

In order to make the book self-contained, we devote the first chapter to a description of general information on fractional calculus, Mittag-Leffler functions, integral transforms and semigroups.

The second chapter deals with the Navier-Stokes equations with time-fractional derivative of order $\alpha \in (0,1)$. The second section is concerned with the existence and uniqueness of local and global mild solutions. Meanwhile, we also give local mild solutions in $\overline{C_\sigma^\infty(\Omega)}^{|\cdot|_q}$. Moreover, we prove the existence and regularity of classical solutions for such equations in $\overline{C_\sigma^\infty(\Omega)}^{|\cdot|_q}$. The third section is devoted to investigation of existence, uniqueness and Hölder continuity of the local mild solutions. In the fourth section, we obtain the existence and uniqueness of the solutions to approximate equations, as well as the convergence of the approximate solutions. Furthermore, we present some convergence results for the Faedo-Galerkin approximations of the given problems. In the fifth section, we firstly give the concept of the weak solutions and establish the existence criterion of weak solutions by means of the Galerkin approximations in the case that the dimension $n \leq 4$. Moreover, a complete proof of the uniqueness is given when $n = 2$. At last we give a sufficient condition of optimal control pairs. In the sixth section, we make use of energy methods to study the time-fractional Navier-Stokes equations. In the first part, we construct a regularized equation by using a smoothing process to transform unbounded differential operators into bounded operators and then obtain the approximate solutions. The second part describes a procedure to take a limit in the approximation program to present a global solution to the objective equation. In the final section of this chapter, we use the tools from harmonic analysis to study the Cauchy problem for time-fractional Navier-Stokes equations. Two results concerning the local existence of solutions for the given problem in Sobolev spaces are addressed.

In the third chapter, we are interested in discussing the nonlinear Rayleigh-Stokes problem for a generalized second grade fluid with the Riemann-Liouville fractional derivative. In the first section, we firstly show that the solution operator of the problem is compact, and continuous in the uniform operator topology. Furthermore, we give an existence result of mild solutions for the nonlinear problem. In the second section, we are devoted to the global/local well-posedness of mild solutions for a semilinear time-fractional Rayleigh-Stokes problem on $\mathbb{R}^N$. We are concerned with, the approaches rely on the Gagliardo-Nirenberg's inequalities, operator theory, standard fixed point technique and harmonic analysis methods. We also present several results on the continuation, a blow-up alternative with the blow-up rate and the integrability in Lebesgue spaces. In the third section, we consider the fractional Rayleigh-Stokes problem with the nonlinearity term satisfies certain critical conditions. The local existence, uniqueness and continuous dependence upon the initial data of ϵ-regular mild solutions are obtained. Furthermore, a unique continuation result and a blow-up alternative result of ϵ-regular mild solutions are given. In the fourth section, we are devoted to the study of the existence, uniqueness and regularity of weak solutions in $L^\infty(0, b; L^2(\Omega)) \bigcap L^2(0, b; H_0^1(\Omega))$ of the Rayleigh-Stokes problem. Furthermore, we prove an improved regularity result of weak solutions. In the fifth section, we prove a long time existence result for fractional Rayleigh-Stokes equations. More precisely, we discuss the existence, uniqueness, continuous dependence on initial value and asymptotic behavior of global solutions in Besov-Morrey spaces. The results are formulated that allows for a larger class in initial value than the previous works and the approach is also suitable for fractional diffusion cases. In the final section, we study a fractional nonlinear Rayleigh-Stokes problem with final value condition. By means of the finite dimensional approximation, we first obtain the compactness of solution operators. Moreover, we handle the problem in weighted continuous function spaces, and then the existence result of solutions is established. Finally, because of the ill-posedness of backward problem in the sense of Hadamard, the quasi-boundary value method is utilized to get the regularized solutions, and the corresponding convergence rate is obtained.

In the fourth chapter, we study the time-fractional Fokker-Planck equations which can be used to describe the subdiffusion in an external time and space-dependent force field $F(t, x)$. In the first section, we present some results on existence and uniqueness of mild solutions allowing the "working space" that may have low regularity. Secondly, we analyze the relationship between "working space" and the value range of α when investigating the problem of classical solutions. Finally, by constructing a suitable weighted Hölder continuous function space, the existence of classical solutions is derived without the restriction on $\alpha \in (\frac{1}{2}, 1)$. In the second section, a time-fractional Fokker–Planck initial-boundary value problem is considered, the spatial domain $\Omega \subset \mathbb{R}^d$, where $d \geq 1$, has a smooth boundary. Existence, uniqueness and regularity of a mild solution u are proved under the hypothesis that the initial data u_0 lies in $L^2(\Omega)$. For $1/2 < \alpha < 1$ and $u_0 \in H^2(\Omega) \cap H_0^1(\Omega)$, it is shown that u becomes a classical solution of the problem. Estimates of time derivatives of the classical solution are derived.

The final chapter is concerned with time-fractional Schrödinger equation. In the first section of this chapter, we study the linear fractional Schrödinger equation on a Hilbert space, with a time-fractional derivative of order $0 < \alpha < 1$, and a self-adjoint generator A. We show that A generates a family of bounded operators $\{U_\alpha(t)\}_{t \geq 0}$ which are defined by the functional calculus of A via the Mittag-Leffler function when evaluated at A. Using the spectral theorem we prove existence and uniqueness of strong solutions. Moreover, we prove that the solution family $U_\alpha(t)$ converges strongly to the family of unitary operators e^{-itA}, as α approaches to 1. In the second section of this chapter, we apply the tools of harmonic analysis to study the Cauchy problem for nonlinear fractional Schrödinger equation. Some fundamental properties of two solution operators are estimated. The existence and a sharp decay estimate for solutions of the given problem in two different spaces are addressed.

The materials in this monograph are based on the research work carried out by the author and some experts during the past five years. The contents in Section 2.3 are taken from Zhou, Peng and Huang [404]. The material in Section 2.4 is taken from Peng, Debbouche and Zhou [293]. The results in Section 2.5 are adopted from Zhou and Peng [403]. The contents of Section 2.6 due to Zhou and Peng [406]. Section 2.7 is taken from Zhou, Peng, Ahmad et al. [298]. The results in Section 3.1 due to Zhou and Wang [409]. The main results in Section 3.2 are adopted from He, Zhou and Peng [163]. Section 3.3 is taken from Wang, Alsaedi, Ahmad and Zhou [346]. The contents in Section 3.4 are taken from Wang, Zhou, Alsaedi and Ahmad [348]. The results in Section 3.5 due to Peng and Zhou [296]. Section 3.6 is from Wang, Zhou and He [347]. The contents in Section 4.1 are taken from Peng and Zhou [294]. The results of Section 4.2 are due to Le, McLean and Stynes [219]. The results in Section 5.1 due to Górka, Prado and Trujillo [152]. The contents in Section 5.2 are taken from Peng, Zhou and Ahmad [299].

Contents

Chapter 1

Preliminaries

In this chapter, we introduce some notations and basic facts on fractional calculus, Mittag-Leffler functions, the Wright-type function, integral transforms and semi-groups, which are needed throughout this monograph.

1.1 Fractional Calculus

1.1.1 *Definitions*

A number of definitions for the fractional derivative has emerged over the years, we refer the reader to Diethelm [101], Hilfer [171], Kilbas, Srivastava and Trujillo [197], Miller and Ross [268], Podlubny [303]. In this book, we restrict our attention to the use of the Riemann-Liouville and Caputo fractional derivatives. In this section, we introduce some basic definitions and properties of the fractional integrals and fractional derivatives which are used further in this book. The materials in this section are taken from [197].

As usual $\mathbb{Z}$ denotes the set of integer numbers, $\mathbb{N}^+$ denotes the set of positive integer numbers and $\mathbb{N}_0$ denotes the set of nonnegative integer numbers. $\mathbb{R}$ denotes the set of real numbers, $\mathbb{R}_+$ denotes the set of nonnegative reals and $\mathbb{R}^+$ the set of positive reals, $\mathbb{R}^-$ denotes the set of nonpositive reals. Let $\mathbb{C}$ be the set of complex numbers.

Let $J = [a, b]$ $(-\infty < a < b < \infty)$ be a finite interval of $\mathbb{R}$. We assume that X is a Banach space with the norm $|\cdot|$. Denote $C(J, X)$ be the Banach space of all continuous functions from J into X with the norm $\|x\| = \sup_{t \in J} |x(t)|$, where $x \in C(J, X)$. $C^n(J, X)$ $(n \in \mathbb{N}_0)$ denotes the set of mappings having n times continuously differentiable on J, $AC(J, X)$ is the space of functions which are absolutely continuous on J and $AC^n(J, X)$ $(n \in \mathbb{N}^+)$ is the space of functions f such that $f \in C^{n-1}(J, X)$ and $f^{(n-1)} \in AC(J, X)$. In particular, $AC^1(J, X) = AC(J, X)$.

Let $1 \leq p \leq \infty$. $L^p(J, X)$ denotes the Banach space of all measurable functions $f : J \to X$. $L^p(J, X)$ is normed by

$$
\|f\|_{L^p J} = \begin{cases} \left(\int_J |f(t)|^p dt \right)^{\frac{1}{p}}, & 1 \leq p < \infty, \\ \inf_{\mu(\bar{J})=0} \left\{ \sup_{t \in J \setminus \bar{J}} |f(t)| \right\}, & p = \infty. \end{cases}
$$

In particular, $L^1(J, X)$ is the Banach space of measurable functions $f : J \to X$ with the norm $\|f\|_{LJ} = \int_J |f(t)| dt$, and $L^\infty(J, X)$ is the Banach space of measurable functions $f : J \to X$ which are bounded, equipped with the norm $\|f\|_{L^\infty J} = \inf\{c > 0| \ |f(t)| \leq c, \text{ a.e. } t \in J\}$.

The Gamma function $\Gamma(z)$ is defined by

$$
\Gamma(z) = \int_0^\infty t^{z-1} e^{-t} dt \ \ (Re(z) > 0),
$$

where $t^{z-1} = e^{(z-1)\ln(t)}$. This integral is convergent for all complex $z \in \mathbb{C}$ $(Re(z) > 0)$.

For this function the reduction formula

$$
\Gamma(z+1) = z\Gamma(z) \ \ (Re(z) > 0)
$$

holds. In particular, if $z = n \in \mathbb{N}_0$, then

$$
\Gamma(n+1) = n! \ \ (n \in \mathbb{N}_0)
$$

with (as usual) $0! = 1$.

Let us consider some of the starting points for a discussion of fractional calculus. One development begins with a generalization of repeated integration. Thus if f is locally integrable on (c, ∞), then the n-fold iterated integral is given by

$$
{}_cD_t^{-n} f(t) = \int_c^t ds_1 \int_c^{s_1} ds_2 \cdots \int_c^{s_{n-1}} f(s_n) ds_n
$$

$$
= \frac{1}{(n-1)!} \int_c^t (t-s)^{n-1} f(s) ds
$$

for almost all t with $-\infty \leq c < t < \infty$ and $n \in \mathbb{N}^+$. Writing $(n-1)! = \Gamma(n)$, an immediate generalization is the integral of f of fractional order $\alpha > 0$,

$$
{}_cD_t^{-\alpha} f(t) = \frac{1}{\Gamma(\alpha)} \int_c^t (t-s)^{\alpha-1} f(s) ds \ \ \text{(left hand)}
$$

and similarly for $-\infty < t < d \leq \infty$

$$
{}_tD_d^{-\alpha} f(t) = \frac{1}{\Gamma(\alpha)} \int_t^d (s-t)^{\alpha-1} f(s) ds \ \ \text{(right hand)}
$$

both being defined for suitable f.

Definition 1.1. (Left and right Riemann-Liouville fractional integrals) Let $J = [a, b]$ $(-\infty < a < b < \infty)$ be a finite interval of $\mathbb{R}$. The left and right Riemann-Liouville fractional integrals ${}_aD_t^{-\alpha} f(t)$ and ${}_tD_b^{-\alpha} f(t)$ of order $\alpha \in \mathbb{R}^+$, are defined by

$$
{}_aD_t^{-\alpha} f(t) = \frac{1}{\Gamma(\alpha)} \int_a^t (t-s)^{\alpha-1} f(s) ds, \ \ t > a, \ \alpha > 0 \tag{1.1}
$$

and

$$\,_t D_b^{-\alpha} f(t) = \frac{1}{\Gamma(\alpha)} \int_t^b (s-t)^{\alpha-1} f(s) ds, \quad t < b, \quad \alpha > 0, \tag{1.2}$$

respectively, provided that the right-hand sides are pointwise defined on $[a, b]$. When $\alpha = n \in \mathbb{N}^+$, the definitions (1.1) and (1.2) coincide with the n-th integrals of the form

$$\,_a D_t^{-n} f(t) = \frac{1}{(n-1)!} \int_a^t (t-s)^{n-1} f(s) ds$$

and

$$\,_t D_b^{-n} f(t) = \frac{1}{(n-1)!} \int_t^b (s-t)^{n-1} f(s) ds.$$

Definition 1.2. (Left and right Riemann-Liouville fractional derivatives) The left and right Riemann-Liouville fractional derivatives $\,_a D_t^{\alpha} f(t)$ and $\,_t D_b^{\alpha} f(t)$ of order $\alpha \in \mathbb{R}_+$, are defined by

$$\,_a D_t^{\alpha} f(t) = \frac{d^n}{dt^n} \,_a D_t^{-(n-\alpha)} f(t)$$

$$= \frac{1}{\Gamma(n-\alpha)} \frac{d^n}{dt^n} \left(\int_a^t (t-s)^{n-\alpha-1} f(s) ds \right), \quad t > a$$

and

$$\,_t D_b^{\alpha} f(t) = (-1)^n \frac{d^n}{dt^n} \,_t D_b^{-(n-\alpha)} f(t)$$

$$= \frac{1}{\Gamma(n-\alpha)} (-1)^n \frac{d^n}{dt^n} \left(\int_t^b (s-t)^{n-\alpha-1} f(s) ds \right), \quad t < b,$$

respectively, where $n = [\alpha] + 1$, $[\alpha]$ means the integer part of α. In particular, when $\alpha = n \in \mathbb{N}_0$, then

$$\,_a D_t^0 f(t) = \,_t D_b^0 f(t) = f(t),$$

$$\,_a D_t^n f(t) = f^{(n)}(t) \quad \text{and} \quad \,_t D_b^n f(t) = (-1)^n f^{(n)}(t),$$

where $f^{(n)}(t)$ is the usual derivative of $f(t)$ of order n. If $0 < \alpha < 1$, then

$$\,_a D_t^{\alpha} f(t) = \frac{1}{\Gamma(1-\alpha)} \frac{d}{dt} \left(\int_a^t (t-s)^{-\alpha} f(s) ds \right), \quad t > a$$

and

$$\,_t D_b^{\alpha} f(t) = -\frac{1}{\Gamma(1-\alpha)} \frac{d}{dt} \left(\int_t^b (s-t)^{-\alpha} f(s) ds \right), \quad t < b.$$

Remark 1.1. If $f \in C([a, b]; \mathbb{R}^N)$, it is obvious that the Riemann-Liouville fractional integral of order $\alpha > 0$ exists on $[a, b]$. On the other hand, following Lemma 2.2 in [197], we know that the Riemann-Liouville fractional derivative of order $\alpha \in [n-1, n)$ exists almost everywhere on $[a, b]$ if $f \in AC^n([a, b]; \mathbb{R}^N)$.

The left and right Caputo fractional derivatives are defined via above the Riemann-Liouville fractional derivatives .

Definition 1.3. (Left and right Caputo fractional derivatives) The left and right Caputo fractional derivatives $^{C}_{a}D^{\alpha}_{t}f(t)$ and $^{C}_{t}D^{\alpha}_{b}f(t)$ of order $\alpha \in \mathbb{R}_{+}$ are defined by

$$^{C}_{a}D^{\alpha}_{t}f(t) = {}_{a}D^{\alpha}_{t}\left(f(t) - \sum_{k=0}^{n-1} \frac{f^{(k)}(a)}{k!}(t-a)^{k}\right)$$

and

$$^{C}_{t}D^{\alpha}_{b}f(t) = {}_{t}D^{\alpha}_{b}\left(f(t) - \sum_{k=0}^{n-1} \frac{f^{(k)}(b)}{k!}(b-t)^{k}\right),$$

respectively, where

$$n = [\alpha] + 1 \text{ for } \alpha \notin \mathbb{N}_{0}; \; n = \alpha \text{ for } \alpha \in \mathbb{N}_{0}. \tag{1.3}$$

In particular, when $0 < \alpha < 1$, then

$$^{C}_{a}D^{\alpha}_{t}f(t) = {}_{a}D^{\alpha}_{t}(f(t) - f(a))$$

and

$$^{C}_{t}D^{\alpha}_{b}f(t) = {}_{t}D^{\alpha}_{b}(f(t) - f(b)).$$

The Riemann-Liouville fractional derivative and the Caputo fractional derivative are connected with each other by the following relations.

Proposition 1.1.

(i) *If $\alpha \notin \mathbb{N}_{0}$ and $f(t)$ is a function for which the Caputo fractional derivatives $^{C}_{a}D^{\alpha}_{t}f(t)$ and $^{C}_{t}D^{\alpha}_{b}f(t)$ of order $\alpha \in \mathbb{R}^{+}$ exist together with the Riemann-Liouville fractional derivatives $_{a}D^{\alpha}_{t}f(t)$ and $_{t}D^{\alpha}_{b}f(t)$, then*

$$^{C}_{a}D^{\alpha}_{t}f(t) = {}_{a}D^{\alpha}_{t}f(t) - \sum_{k=0}^{n-1} \frac{f^{(k)}(a)}{\Gamma(k-\alpha+1)}(t-a)^{k-\alpha}$$

and

$$^{C}_{t}D^{\alpha}_{b}f(t) = {}_{t}D^{\alpha}_{b}f(t) - \sum_{k=0}^{n-1} \frac{f^{(k)}(b)}{\Gamma(k-\alpha+1)}(b-t)^{k-\alpha},$$

where $n = [\alpha] + 1$. In particular, when $0 < \alpha < 1$, we have

$$^{C}_{a}D^{\alpha}_{t}f(t) = {}_{a}D^{\alpha}_{t}f(t) - \frac{f(a)}{\Gamma(1-\alpha)}(t-a)^{-\alpha}$$

and

$$^{C}_{t}D^{\alpha}_{b}f(t) = {}_{t}D^{\alpha}_{b}f(t) - \frac{f(b)}{\Gamma(1-\alpha)}(b-t)^{-\alpha}.$$

(ii) *If $\alpha = n \in \mathbb{N}_0$ and the usual derivative $f^{(n)}(t)$ of order n exists, then ${}^{C}_{a}D^n_t f(t)$ and ${}^{C}_{t}D^n_b f(t)$ are represented by*

$$
{}^{C}_{a}D^n_t f(t) = f^{(n)}(t) \quad \text{and} \quad {}^{C}_{t}D^n_b f(t) = (-1)^n f^{(n)}(t). \tag{1.4}
$$

Proposition 1.2. *Let $\alpha \in \mathbb{R}_+$ and let n be given by (1.3). If $f \in AC^n([a,b]; \mathbb{R}^N)$, then the Caputo fractional derivatives ${}^{C}_{a}D^\alpha_t f(t)$ and ${}^{C}_{t}D^\alpha_b f(t)$ exist almost everywhere on $[a,b]$.*

(i) *If $\alpha \notin \mathbb{N}_0$, ${}^{C}_{a}D^\alpha_t f(t)$ and ${}^{C}_{t}D^\alpha_b f(t)$ are represented by*

$$
{}^{C}_{a}D^\alpha_t f(t) = \frac{1}{\Gamma(n-\alpha)} \left(\int_a^t (t-s)^{n-\alpha-1} f^{(n)}(s) ds \right)
$$

and

$$
{}^{C}_{t}D^\alpha_b f(t) = \frac{(-1)^n}{\Gamma(n-\alpha)} \left(\int_t^b (s-t)^{n-\alpha-1} f^{(n)}(s) ds \right),
$$

respectively, where $n = [\alpha] + 1$. In particular, when $0 < \alpha < 1$, $f \in AC([a,b]; \mathbb{R}^N)$,

$$
{}^{C}_{a}D^\alpha_t f(t) = \frac{1}{\Gamma(1-\alpha)} \left(\int_a^t (t-s)^{-\alpha} f'(s) ds \right) \tag{1.5}
$$

and

$$
{}^{C}_{t}D^\alpha_b f(t) = -\frac{1}{\Gamma(1-\alpha)} \left(\int_t^b (s-t)^{-\alpha} f'(s) ds \right). \tag{1.6}
$$

(ii) *If $\alpha = n \in \mathbb{N}_0$, then ${}^{C}_{a}D^\alpha_t f(t)$ and ${}^{C}_{t}D^\alpha_b f(t)$ are represented by (1.4). In particular,*

$$
{}^{C}_{a}D^0_t f(t) = {}^{C}_{t}D^0_b f(t) = f(t).
$$

Remark 1.2. If f is an abstract function with values in Banach space X, then integrals which appear in above definitions are taken in Bochner's sense.

The fractional integrals and derivatives, defined on a finite interval $[a,b]$ of $\mathbb{R}$, are naturally extended to whole axis $\mathbb{R}$.

Definition 1.4. (Left and right Liouville-Weyl fractional integrals on the real axis) The left and right Liouville-Weyl fractional integrals ${}_{-\infty}D^{-\alpha}_t f(t)$ and ${}_{t}D^{-\alpha}_{+\infty} f(t)$ of order $\alpha > 0$ on the whole axis $\mathbb{R}$ are defined by

$$
{}_{-\infty}D^{-\alpha}_t f(t) = \frac{1}{\Gamma(\alpha)} \int_{-\infty}^t (t-s)^{\alpha-1} f(s) ds \tag{1.7}
$$

and

$$
{}_{t}D^{-\alpha}_{+\infty} f(t) = \frac{1}{\Gamma(\alpha)} \int_t^\infty (s-t)^{\alpha-1} f(s) ds,
$$

respectively, where $t \in \mathbb{R}$ and $\alpha > 0$.

Definition 1.5. (Left and right Liouville-Weyl fractional derivatives on the real axis) The left and right Liouville-Weyl fractional derivatives $_{-\infty}D_t^\alpha f(t)$ and $_tD_{+\infty}^\alpha f(t)$ of order α on the whole axis $\mathbb{R}$ are defined by

$$_{-\infty}D_t^\alpha f(t) = \frac{d^n}{dt^n}(_{-\infty}D_t^{-(n-\alpha)}f(t))$$

$$= \frac{1}{\Gamma(n-\alpha)}\frac{d^n}{dt^n}\left(\int_{-\infty}^t (t-s)^{n-\alpha-1}f(s)ds\right)$$

and

$$_tD_{+\infty}^\alpha f(t) = (-1)^n \frac{d^n}{dt^n}(_tD_{+\infty}^{-(n-\alpha)}f(t))$$

$$= \frac{1}{\Gamma(n-\alpha)}(-1)^n \frac{d^n}{dt^n}\left(\int_t^\infty (s-t)^{n-\alpha-1}f(s)ds\right),$$

respectively, where $n = [\alpha]+1$, $\alpha \geq 0$ and $t \in \mathbb{R}$.

In particular, when $\alpha = n \subset \mathbb{N}_0$, then

$$_{-\infty}D_t^0 f(t) = {}_tD_{+\infty}^0 f(t) = f(t),$$

$$_{-\infty}D_t^n f(t) = f^{(n)}(t) \quad \text{and} \quad {}_tD_{+\infty}^n f(t) = (-1)^n f^{(n)}(t),$$

where $f^{(n)}(t)$ is the usual derivative of $f(t)$ of order n. If $0 < \alpha < 1$ and $t \in \mathbb{R}$, then

$$_{-\infty}D_t^\alpha f(t) = \frac{1}{\Gamma(1-\alpha)}\frac{d}{dt}\left(\int_{-\infty}^t (t-s)^{-\alpha}f(s)ds\right)$$

$$= \frac{\alpha}{\Gamma(1-\alpha)}\int_0^\infty \frac{f(t)-f(t-s)}{s^{\alpha+1}}ds$$

and

$$_tD_{+\infty}^\alpha f(t) = -\frac{1}{\Gamma(1-\alpha)}\frac{d}{dt}\left(\int_t^\infty (s-t)^{-\alpha}f(s)ds\right)$$

$$= \frac{\alpha}{\Gamma(1-\alpha)}\int_0^\infty \frac{f(t)-f(t+s)}{s^{\alpha+1}}ds.$$

Formulas (1.5) and (1.6) can be used for the definition of the Caputo fractional derivatives on the whole axis $\mathbb{R}$.

Definition 1.6. (Left and right Caputo fractional derivatives on the real axis) The left and right Caputo fractional derivatives $_{-\infty}^C D_t^\alpha f(t)$ and $_t^C D_{+\infty}^\alpha f(t)$ of order α (with $\alpha > 0$ and $\alpha \notin \mathbb{N}^+$) on the whole axis $\mathbb{R}$ are defined by

$$_{-\infty}^C D_t^\alpha f(t) = \frac{1}{\Gamma(n-\alpha)}\left(\int_{-\infty}^t (t-s)^{n-\alpha-1}f^{(n)}(s)ds\right) \tag{1.8}$$

and

$$_t^C D_{+\infty}^\alpha f(t) = \frac{(-1)^n}{\Gamma(n-\alpha)}\left(\int_t^\infty (s-t)^{n-\alpha-1}f^{(n)}(s)ds\right), \tag{1.9}$$

respectively.

When $0 < \alpha < 1$, the relations (1.8) and (1.9) take the following forms

$$_{-\infty}^C D_t^\alpha f(t) = \frac{1}{\Gamma(1-\alpha)}\left(\int_{-\infty}^t (t-s)^{-\alpha}f'(s)ds\right)$$

and

$$_t^C D_{+\infty}^\alpha f(t) = -\frac{1}{\Gamma(1-\alpha)}\left(\int_t^\infty (s-t)^{-\alpha}f'(s)ds\right).$$

1.1.2 *Properties*

We present here some properties of the fractional integral and fractional derivative operators that will be useful throughout this book.

Proposition 1.3. *If $\beta > 0$, then*

$$_aD_t^{-\alpha}(t-a)^{\beta-1} = \frac{\Gamma(\beta)}{\Gamma(\beta+\alpha)}(t-a)^{\beta+\alpha-1} \quad (\alpha > 0),$$

$$_aD_t^{\alpha}(t-a)^{\beta-1} = \frac{\Gamma(\beta)}{\Gamma(\beta-\alpha)}(t-a)^{\beta-\alpha-1} \quad (\alpha \geq 0)$$

and

$$_tD_b^{-\alpha}(b-t)^{\beta-1} = \frac{\Gamma(\beta)}{\Gamma(\beta+\alpha)}(b-t)^{\beta+\alpha-1} \quad (\alpha > 0),$$

$$_tD_b^{\alpha}(b-t)^{\beta-1} = \frac{\Gamma(\beta)}{\Gamma(\beta-\alpha)}(b-t)^{\beta-\alpha-1} \quad (\alpha \geq 0).$$

In particular, if $\beta = 1$ and $\alpha \geq 0$, then the Riemann-Liouville fractional derivatives of a constant are, in general, not equal to zero:

$$_aD_t^{\alpha}1 = \frac{(t-a)^{-\alpha}}{\Gamma(1-\alpha)}, \quad _tD_b^{\alpha}1 = \frac{(b-t)^{-\alpha}}{\Gamma(1-\alpha)}.$$

On the other hand, for $j = 1, 2, ..., [\alpha] + 1$,

$$_aD_t^{\alpha}(t-a)^{\alpha-j} = 0, \quad _tD_b^{\alpha}(b-t)^{\alpha-j} = 0.$$

The semigroup property of the fractional integral operators $_aD_t^{-\alpha}$ and $_tD_b^{-\alpha}$ are given by the following results.

Proposition 1.4. *If $\alpha > 0$ and $\beta > 0$, then the equations*

$$_aD_t^{-\alpha}\left(_aD_t^{-\beta}f(t)\right) = _aD_t^{-\alpha-\beta}f(t) \text{ and } _tD_b^{-\alpha}\left(_tD_b^{-\beta}f(t)\right) = _tD_b^{-\alpha-\beta}f(t) \quad (1.10)$$

are satisfied at almost every point $t \in [a, b]$ for $f \in L^p(a, b; \mathbb{R}^N)$ $(1 \leq p < \infty)$. If $\alpha + \beta > 1$, then the relations in (1.10) hold at any point of $[a, b]$.

Proposition 1.5.

(i) *If $\alpha > 0$ and $f \in L^p(a, b; \mathbb{R}^N)$ $(1 \leq p \leq \infty)$, then the following equalities*

$$_aD_t^{\alpha}\left(_aD_t^{-\alpha}f(t)\right) = f(t) \text{ and } _tD_b^{\alpha}\left(_tD_b^{-\alpha}f(t)\right) = f(t)$$

hold almost everywhere on $[a, b]$.

(ii) *If $\alpha > \beta > 0$, then, for $f \in L^p(a, b; \mathbb{R}^N)$ $(1 \leq p \leq \infty)$, the relations*

$$_aD_t^{\beta}\left(_aD_t^{-\alpha}f(t)\right) = _aD_t^{-\alpha+\beta}f(t) \text{ and } _tD_b^{\beta}\left(_tD_b^{-\alpha}f(t)\right) = _tD_b^{-\alpha+\beta}f(t)$$

hold almost everywhere on $[a, b]$.

In particular, when $\beta = k \in \mathbb{N}^+$ and $\alpha > k$, then

$$_aD_t^{k}\left(_aD_t^{-\alpha}f(t)\right) = _aD_t^{-\alpha+k}f(t) \text{ and } _tD_b^{k}\left(_tD_b^{-\alpha}f(t)\right) = (-1)^k{}_tD_b^{-\alpha+k}f(t).$$

To present the next property, we use the spaces of functions $_aD_t^{-\alpha}(L^p)$ and $_tD_b^{-\alpha}(L^p)$ defined for $\alpha > 0$ and $1 \leq p \leq \infty$ by

$$_aD_t^{-\alpha}(L^p) = \{f : f = {_aD_t^{-\alpha}}\varphi, \ \varphi \in L^p(a,b;\mathbb{R}^N)\}$$

and

$$_tD_b^{-\alpha}(L^p) = \{f : f = {_tD_b^{-\alpha}}\phi, \ \phi \in L^p(a,b;\mathbb{R}^N)\},$$

respectively. The composition of the fractional integral operator $_aD_t^{-\alpha}$ with the fractional derivative operator $_aD_t^\alpha$ is given by the following results.

Proposition 1.6. *Let $\alpha > 0$, $n = [\alpha] + 1$ and let $f_{n-\alpha}(t) = {_aD_t^{-(n-\alpha)}}f(t)$ be the fractional integral (1.1) of order $n - \alpha$.*

(i) *If $1 \leq p \leq \infty$ and $f \in {_aD_t^{-\alpha}}(L^p)$, then*

$$_aD_t^{-\alpha}\left({_aD_t^\alpha}f(t)\right) = f(t).$$

(ii) *If $f \in L^1(a,b;\mathbb{R}^N)$ and $f_{n-\alpha} \in AC^n([a,b];\mathbb{R}^N)$, then the equality*

$$_aD_t^{-\alpha}\left({_aD_t^\alpha}f(t)\right) = f(t) - \sum_{j=1}^n \frac{f_{n-\alpha}^{(n-j)}(a)}{\Gamma(\alpha-j+1)}(t-a)^{\alpha-j}$$

holds almost everywhere on $[a,b]$.

Proposition 1.7. *Let $\alpha > 0$ and $n = [\alpha] + 1$. Also let $g_{n-\alpha}(t) = {_tD_b^{-(n-\alpha)}}g(t)$ be the fractional integral (1.2) of order $n - \alpha$.*

(i) *If $1 \leq p \leq \infty$ and $g \in {_tD_b^{-\alpha}}(L^p)$, then*

$$_tD_b^{-\alpha}\left({_tD_b^\alpha}g(t)\right) = g(t).$$

(ii) *If $g \in L^1(a,b;\mathbb{R}^N)$ and $g_{n-\alpha} \in AC^n([a,b];\mathbb{R}^N)$, then the equality*

$$_tD_b^{-\alpha}\left({_tD_b^\alpha}g(t)\right) = g(t) - \sum_{j=1}^n \frac{(-1)^{n-j}g_{n-\alpha}^{(n-j)}(a)}{\Gamma(\alpha-j+1)}(b-t)^{\alpha-j}$$

holds almost everywhere on $[a,b]$.

In particular, if $0 < \alpha < 1$, then

$$_tD_b^{-\alpha}\left({_tD_b^\alpha}g(t)\right) = g(t) - \frac{g_{1-\alpha}(a)}{\Gamma(\alpha)}(b-t)^{\alpha-1},$$

where $g_{1-\alpha}(t) = {_tD_b^{\alpha-1}}g(t)$ while for $\alpha = n \in \mathbb{N}^+$, the following equality holds:

$$_tD_b^{-n}\left({_tD_b^n}g(t)\right) = g(t) - \sum_{k=0}^{n-1} \frac{(-1)^k g^{(k)}(a)}{k!}(b-t)^k.$$

Proposition 1.8. *Let $\alpha > 0$ and let $y \in L^\infty(a,b;\mathbb{R}^N)$ or $y \in C([a,b];\mathbb{R}^N)$. Then*

$$_a^C D_t^\alpha\left({_aD_t^{-\alpha}}y(t)\right) = y(t) \quad \text{and} \quad {_t^C D_b^\alpha}\left({_tD_b^{-\alpha}}y(t)\right) = y(t).$$

Proposition 1.9. *Let $\alpha > 0$ and let n be given by (1.3). If $y \in AC^n([a,b];\mathbb{R}^N)$ or $y \in C^n([a,b];\mathbb{R}^N)$, then*

$$_aD_t^{-\alpha}\left(_a^C D_t^\alpha y(t)\right) = y(t) - \sum_{k=0}^{n-1} \frac{y^{(k)}(a)}{k!}(t-a)^k$$

and

$$_tD_b^{-\alpha}\left(_t^C D_b^\alpha y(t)\right) = y(t) - \sum_{k=0}^{n-1} \frac{(-1)^k y^{(k)}(b)}{k!}(b-t)^k.$$

In particular, if $0 < \alpha \leq 1$ and $y \in AC([a,b];\mathbb{R}^N)$ or $y \in C([a,b];\mathbb{R}^N)$, then

$$_aD_t^{-\alpha}\left(_a^C D_t^\alpha y(t)\right) = y(t) - y(a) \quad\text{and}\quad _tD_b^{-\alpha}\left(_t^C D_b^\alpha y(t)\right) = y(t) - y(b). \qquad (1.11)$$

1.2 Some Results from Analysis

1.2.1 Mittag-Leffler Function

Definition 1.7. [268, 303] The Mittag-Leffler function $E_{\alpha,\beta}$ is defined by

$$E_{\alpha,\beta}(z) := \sum_{k=0}^{\infty} \frac{z^k}{\Gamma(\alpha k + \beta)} = \frac{1}{2\pi i} \int_\Upsilon \frac{\lambda^{\alpha-\beta} e^\lambda}{\lambda^\alpha - z} d\lambda, \quad \alpha > 0, \beta \in \mathbb{R}, \ z \in \mathbb{C},$$

where Υ is a contour which starts and ends as $-\infty$ and encircles the disc $|\lambda| \leq |z|^{1/\alpha}$ counter-clockwise.

The function $E_{\alpha,\beta}(z)$ is an entire function, and so it is real analytic when restricted to the real line. Moreover, the approximation form of Mittag-Leffler function is given by

$$E_{\alpha,\beta}(z) = -\sum_{k=1}^{N} \frac{1}{\Gamma(\beta - \alpha k)} \frac{1}{z^k} + O\left(\frac{1}{z^{N+1}}\right),$$

with $|z| \to \infty$, $\mu \leq |\arg(z)| \leq \pi$ for $\mu > 0$, and $N \in \mathbb{N}^+$. In particular,

$$E_{\alpha,1}(z) = -\frac{1}{\Gamma(1-\alpha)} \frac{1}{z} + O\left(\frac{1}{z^2}\right), \qquad (1.12)$$

with $|z| \to \infty$, $\mu \leq |\arg(z)| \leq \pi$ for $\mu > 0$.

For short, set

$$E_\alpha(z) := E_{\alpha,1}(z), \quad e_\alpha(z) := E_{\alpha,\alpha}(z).$$

Then Mittag-Leffler functions have the following properties.

Proposition 1.10. *[268, 303] For $\alpha \in (0,1)$ and $t \in \mathbb{R}$,*

(i) $E_\alpha(t), e_\alpha(t) > 0$;

(ii) $(E_\alpha(t))' = \frac{1}{\alpha}e_\alpha(t)$;

(iii) $\lim\limits_{t\to-\infty} E_\alpha(t) = \lim\limits_{t\to-\infty} e_\alpha(t) = 0$;

(iv) $_0^C D_t^\alpha E_\alpha(\omega t^\alpha) = \omega E_\alpha(\omega t^\alpha)$, $_0 D_t^{\alpha-1}(t^{\alpha-1}e_\alpha(\omega t^\alpha)) = E_\alpha(\omega t^\alpha)$, $\omega \in \mathbb{C}$.

Proposition 1.11. *[303] Let $0 < \alpha < 1$ and $\lambda > 0$. Then*

(i) $\frac{d}{dt}E_\alpha(-\lambda t^\alpha) = -\lambda t^{\alpha-1}e_\alpha(-\lambda t^\alpha)$, *for $t > 0$;*

(ii) $\frac{d}{dt}\left(t^{\alpha-1}e_\alpha(-\lambda t^\alpha)\right) = t^{\alpha-2}E_{\alpha,\alpha-1}(-\lambda t^\alpha)$, *for $t > 0$;*

(iii) $\int_0^\infty e^{-st}E_\alpha(-\lambda t^\alpha)dt = \frac{s^{\alpha-1}}{s^\alpha+\lambda}$, *for $Re(s) > \lambda^{1/\alpha}$.*

It is well known that $E_{\alpha,1}(-t)$ is a positive and completely monotonic function for $\alpha \in (0,1)$, $t > 0$, that is, for all $t > 0$, $k \in \mathbb{N}_0$, we have

$$(-1)^k \left(\frac{d}{dt}\right)^k E_{\alpha,1}(-t) \geq 0.$$

Additionally, one can find that $w(t) := E_{\alpha,1}(\lambda t^\alpha)$ is a solution of equation $_0^C D_t^\alpha w(t) = \lambda w(t)$, $\lambda \in \mathbb{R}$, $\alpha \in (0,2)$. We use the notation $a \lesssim b$ that stands for $a \leq Cb$, with a positive constant C that does not depend on a, b. The following propositions will be frequently used and can be found in [303].

Proposition 1.12. *For $\lambda > 0$, $\alpha > 0$, $\beta \in \mathbb{R}$ and any arbitrary positive number m, we have*

$$\left(\frac{d}{dt}\right)^m \left(t^{\beta-1}E_{\alpha,\beta}(-\lambda t^\alpha)\right) = t^{\beta-m-1}E_{\alpha,\beta-m}(-\lambda t^\alpha), \quad t > 0.$$

In particular,

$$\frac{d}{dt}E_{\alpha,1}(-\lambda t^\alpha) = -\lambda t^{\alpha-1}E_{\alpha,\alpha}(-\lambda t^\alpha), \quad t > 0.$$

Proposition 1.13. *If $0 < \alpha < 2$, $\beta \in \mathbb{R}$, $\pi\alpha/2 < \theta < \min\{\pi, \pi\alpha\}$, then*

$$|E_{\alpha,\beta}(z)| \lesssim \frac{1}{1+|z|}, \quad z \in \mathbb{C}, \quad \theta \leq |\arg z| \leq \pi.$$

Proposition 1.14. *If $0 < \alpha < 2$, $\beta \in \mathbb{R}$, θ is such that $\pi\alpha/2 < \theta < \min\{\pi, \pi\alpha\}$, then*

$$|E_{\alpha,\beta}(z)| \lesssim (1+|z|)^{(1-\beta)/\alpha} \exp\left(Re(z^{1/\alpha})\right) + \frac{1}{1+|z|}, \quad z \in \mathbb{C}, \quad |\arg z| \leq \theta.$$

By the fractional order term-by-term integration of the series, there is a more general relationship obtained as follows

$$\frac{1}{\Gamma(\vartheta)} \int_0^t (t-s)^{\vartheta-1}s^{\beta-1}E_{\alpha,\beta}(\lambda s^\beta)ds = t^{\beta+\vartheta-1}E_{\alpha,\beta+\vartheta}(\lambda t^\beta), \quad \vartheta > 0, \ \beta > 0, \ t > 0.$$

$$(1.13)$$

Proposition 1.15. *[14] Let $1 < \beta < 2$, $\beta' \in \mathbb{R}$ and $\lambda > 0$. Then the following estimates hold.*

(i) *Let* $0 \le \alpha \le 1, 0 < \beta' < \beta$. *Then* $\left| \lambda^\alpha t^{\beta'} E_{\beta,\beta'}(-\lambda t^\beta) \right| \lesssim t^{\beta'-\beta\alpha}, \quad t > 0$.

(ii) *Let* $0 \le \beta' \le 1$. *Then* $\left| \lambda^{1-\beta'} t^{\beta-2} E_{\beta,\beta'}(-\lambda t^\beta) \right| \lesssim t^{\beta\beta'-2}, \quad t > 0$.

In what follows, let us state the definition and some properties of a function $\mathcal{M}_\alpha(\cdot)$ which is also called the Wright-type function. This function is a special case of the Wright function that plays an important role in different areas of fractional calculus and it is introduced by Mainardi to characterize the solution of initial value problem for fractional diffusion-wave equations.

Definition 1.8. [254] The Wright-type function $\mathcal{M}_\alpha$ is defined by

$$\mathcal{M}_\alpha(z) := \sum_{n=0}^\infty \frac{(-z)^n}{n! \Gamma(-\alpha n + 1 - \alpha)}$$

$$= \frac{1}{\pi} \sum_{n=1}^\infty \frac{(-z)^n}{(n-1)!} \Gamma(n\alpha) \sin(n\pi\alpha), \quad \text{for} \quad 0 < \alpha < 1, z \in \mathbb{C}.$$

For $-1 < r < \infty, \lambda > 0$, the Wright-type function has the properties.

Proposition 1.16.

(W1) $\mathcal{M}_\alpha(t) \ge 0, t > 0$;

(W2) $\displaystyle\int_0^\infty \frac{\alpha}{t^{\alpha+1}} \mathcal{M}_\alpha(\frac{1}{t^\alpha}) e^{-\lambda t} dt = e^{-\lambda^\alpha}$;

(W3) $\displaystyle\int_0^\infty \mathcal{M}_\alpha(t) t^r dt = \frac{\Gamma(1+r)}{\Gamma(1+\alpha r)}$;

(W4) $\displaystyle\int_0^\infty \mathcal{M}_\alpha(t) e^{-zt} dt = E_\alpha(-z), \quad z \in \mathbb{C}$;

(W5) $\displaystyle\int_0^\infty \alpha t \mathcal{M}_\alpha(t) e^{-zt} dt = e_\alpha(-z), \quad z \in \mathbb{C}$.

1.2.2 *Laplace and Fourier Transforms*

In this subsection we present definitions and some properties of Laplace and Fourier transforms.

Definition 1.9. The Laplace transform of a function $f(t)$ of a real variable $t \in \mathbb{R}^+$ is defined by

$$(\mathcal{L}f)(s) = \mathcal{L}[f(t)](s) = \bar{f}(s) := \int_0^\infty e^{-st} f(t) dt \quad (s \in \mathbb{C}). \tag{1.14}$$

The inverse Laplace transform is given for $x \in \mathbb{R}^+$ by the formula

$$(\mathcal{L}^{-1}f)(x) = \mathcal{L}^{-1}[f(s)](x) := \frac{1}{2\pi i} \int_{\gamma-i\infty}^{\gamma+i\infty} e^{sx} f(s) ds \quad (\gamma = Re(s)). \tag{1.15}$$

Proposition 1.17. *Let $f(t)$ be defined on $(0, \infty)$ and $0 < \alpha < 1$. Then the Laplace transform of fractional integral and fractional differential operators satisfy*

(i) $\overline{{}_0D_t^{-\alpha}f}(s) = s^{-\alpha}\bar{f}(s)$;

(ii) $\overline{{}_0D_t^{\alpha}f}(s) = s^{\alpha}\bar{f}(s) - ({}_0D_t^{\alpha-1}f)(0)$;

(iii) $\overline{{}_0^CD_t^{-\alpha}f}(s) = s^{\alpha}\bar{f}(s) - s^{\alpha-1}f(0)$.

Definition 1.10. The Fourier transform of a function $f(t)$ of a real variable $t \in \mathbb{R}$ is defined by

$$(\mathcal{F}f)(w) = \mathcal{F}[f(t)](w) = \hat{f}(w) := \int_{-\infty}^{\infty} e^{-it\cdot w}f(t)dt \quad (w \in \mathbb{R}). \tag{1.16}$$

The inverse Fourier transform is given by the formula

$$(\mathcal{F}^{-1}g)(w) = \mathcal{F}^{-1}[g(t)](w) = \frac{1}{2\pi}\hat{g}(-w) := \frac{1}{2\pi}\int_{-\infty}^{\infty} e^{it\cdot w}g(t)dt \quad (w \in \mathbb{R}). \tag{1.17}$$

The integrals in (1.16) and (1.17) converge absolutely for functions $f, g \in L^1(\mathbb{R})$ and in the norm of the space $L^2(\mathbb{R})$ for $f, g \in L^2(\mathbb{R})$.

Proposition 1.18. *Let $f(t)$ be defined on $(-\infty, \infty)$ and $0 < \alpha < 1$. Then Fourier transform of Liouville-Weyl fractional integral and fractional differential operators satisfy*

(i) $\widehat{{}_{-\infty}D_t^{-\alpha}f}(w) = (iw)^{-\alpha}\hat{f}(w)$;

(ii) $\widehat{{}_tD_\infty^{-\alpha}f}(w) = (-iw)^{-\alpha}\hat{f}(w)$;

(iii) $\widehat{{}_{-\infty}D_t^{\alpha}f}(w) = (iw)^{\alpha}\hat{f}(w)$;

(iv) $\widehat{{}_tD_\infty^{\alpha}f}(w) = (-iw)^{\alpha}\hat{f}(w)$.

1.3   Semigroups

1.3.1 C_0-*Semigroup*

Let us recall the definitions and properties of operator semigroups, for details see Pazy [291]. Let X be a Banach space and $\mathfrak{L}(X)$ be the Banach space of linear bounded operators with the norm $\|\cdot\|$.

Definition 1.11. A one parameter family $\{T(t)\}_{t\geq 0} \subset \mathfrak{L}(X)$ is a semigroup of bounded linear operators on X if

(i) $T(t)T(s) = T(t+s)$, for $t, s \geq 0$;
(ii) $T(0) = I$; here, I denotes the identity operator in X.

Definition 1.12. A semigroup of bounded linear operators $\{T(t)\}_{t\geq 0}$ is uniformly continuous if

$$\lim_{t\to 0^+} \|T(t) - I\| = 0.$$

From the definition it is clear that if $\{T(t)\}_{t\geq 0}$ is a uniformly continuous semigroup of bounded linear operators, then

$$\lim_{s\to t} \|T(s) - T(t)\| = 0.$$

Definition 1.13. We say that the semigroup $\{T(t)\}_{t\geq 0}$ is strongly continuous (or a C_0-semigroup) if the mapping $t \to T(t)u$ is strongly continuous, for each $u \in X$, i.e.,

$$\lim_{t\to 0^+} T(t)u = u, \quad \forall\, u \in X.$$

Definition 1.14. Let $\{T(t)\}_{t\geq 0}$ be a C_0-semigroup defined on X. The linear operator A is the infinitesimal generator of $\{T(t)\}_{t\geq 0}$ defined by

$$Au = \lim_{t\to 0^+} \frac{T(t)u - u}{t}, \quad \text{for } u \in D(A),$$

where $D(A) = \left\{u \in X : \lim_{t\to 0^+} \frac{T(t)u-u}{t} \text{ exists in } X\right\}$.

Definition 1.15. The family $R(\lambda, A) = (\lambda I - A)^{-1}$, $\lambda \in \rho(A)$ of bound linear operator is called of the resolvent of A, where $\rho(A)$ is the set of the all complex number λ for which $\lambda I - A$ is invertible.

If there are $M \geq 0$ and $\nu \in \mathbb{R}$ such that $\|T(t)\| \leq Me^{\nu t}$, then

$$(\lambda I - A)^{-1}u = \int_0^\infty e^{-\lambda t}T(t)u\,dt, \quad \text{Re}(\lambda) > \nu, \ u \in X. \tag{1.18}$$

A C_0-semigroup $\{T(t)\}_{t\geq 0}$ is called exponentially stable if there exist constants $M > 0$ and $\delta > 0$ such that

$$\|T(t)\| \leq Me^{-\delta t}, \quad t \geq 0. \tag{1.19}$$

The growth bound ν_0 of $\{T(t)\}_{t\geq 0}$ is defined by

$$\nu_0 = \inf\{\delta \in \mathbb{R} : \text{ there exists } M_\delta > 0 \text{ such that } \|T(t)\| \leq M_\delta e^{\delta t}, \ \forall\, t \geq 0\}. \tag{1.20}$$

Furthermore, ν_0 can also be obtained by the following formula:

$$\nu_0 = \limsup_{t\to +\infty} \frac{\ln \|T(t)\|}{t}. \tag{1.21}$$

Definition 1.16. A C_0-semigroup $\{T(t)\}_{t\geq 0}$ is called uniformly bounded if there exists a constant $M > 0$ such that

$$\|T(t)\| \leq M, \quad t \geq 0. \tag{1.22}$$

Definition 1.17. A C_0-semigroup $\{T(t)\}_{t\geq 0}$ is called compact if $T(t)$ is compact for $t > 0$.

Proposition 1.19. *If $\{T(t)\}_{t>0}$ is compact, then $\{T(t)\}_{t>0}$ is equicontinuous for $t > 0$.*

Definition 1.18. A C_0-semigroup $\{T(t)\}_{t\geq 0}$ is called positive if $T(t)u \geq \theta$ for all $u \geq \theta$ and $t \geq 0$, where θ is the zero element.

1.3.2 Analytic Semigroup

Definition 1.19. Let $\Delta := \{z : \varphi_1 < \arg z < \varphi_2, \ \varphi_1 < 0 < \varphi_2\}$. The family $\{T(z)\}_{z\in\Delta} \subset \mathcal{L}(X)$ is called an analytic semigroup in Δ if

(i) $z \mapsto T(z)$ is analytic in Δ;
(ii) $T(0) = I$ and $\lim_{z\in\Delta, z\to 0} T(z)x = x$ for every $x \in X$;
(iii) $T(z_1 + z_2) = T(z_1)T(z_2)$ for $z_1,\ z_2 \in \Delta$.

A semigroup $T(t)$ is called analytic if it is analytic in some sector Δ containing the nonnegative real axis.

Theorem 1.1. *Let $\{T(t)\}_{t\geq 0}$ be a uniformly bounded C_0-semigroup. Let A be the infinitesimal generator of $\{T(t)\}_{t\geq 0}$ and assume $0 \in \rho(A)$. The following statements are equivalent:*

(i) *$T(t)$ can be extended to an analytic semigroup in a sector $\Sigma_\delta := \{z \in \mathbb{C} : |\arg z| < \delta\}$ and $\|T(z)\|$ is uniformly bounded in every closed subsector $\overline{\Sigma}_{\delta'}$, $\delta' < \delta$, of Σ_δ;*
(ii) *there exists a positive constant C such that for every $\sigma > 0$, $\tau \neq 0$,*

$$\|R(\sigma + i\tau, A)\| \leq \frac{C}{|\tau|};$$

(iii) *there exist $0 < \delta < \frac{\pi}{2}$ and $M > 0$ such that*

$$\rho(A) \supset \Sigma := \{\lambda \in \mathbb{C} : |\arg \lambda| < \frac{\pi}{2} + \delta\} \cup \{0\}$$

and

$$\|R(\lambda, A)\| \leq \frac{M}{|\lambda|}, \quad \text{for } \lambda \in \Sigma, \ \lambda \neq 0;$$

(iv) *$T(t)$ is differentiable for $t > 0$ and there is a constant C such that*

$$\|AT(t)\| \leq \frac{C}{t}, \quad \text{for } t > 0.$$

Chapter 2

Fractional Navier-Stokes Equations

2.1 Introduction

Navier-Stokes equations have been investigated by many researchers in view of their crucial role in fluid mechanics and turbulence problems. For more details, we refer the reader to the monographs by Ben-Artzi et al. [43], Cannone [59], Lemarié-Rieusset [221], Raugel and Sell [308], Temam [330] and Varnhorn [340]. There are extensive literatures on the well-posedness of initial value problems for these equations, see, e.g., Dubois [109], Giga [138], Hieber and Sawada [169], Hieber and Shibata [170], Ibrahim and Keraani [178], Koch et al. [202], Koch and Tataru [203], Lemarié-Rieusset [221], Miura [269], Von Wahl [343], Yamazaki [370] and references therein.

It is worth mentioning that Leray carried out an initial study that a boundary-value problem for the time-dependent Navier-Stokes equations possesses a unique smooth solution on some intervals of time provided the data are sufficiently smooth. Since then many results on the existence for weak, mild and strong solutions for the Navier-Stokes equations have been investigated intensively by many authors, see, e.g., Almeida and Ferreira [12], Heck et al. [164], Iwabuchi and Takada [182], Koch et al. [202], Masmoudi and Wong [249], and Weissler [362]. Moreover, one can find results on regularity of weak and strong solutions from Amrouche and Rejaiba [16], Chemin and Gallagher [71], Chemin et al. [72], Choe [80], Danchin [93], Giga [140], Kozono [206], Raugel and Sell [308] and the references therein. In the last few years considerable process has been made in the existence, uniqueness and smoothness properties of weak solutions related to the Navier-Stokes equations, see, e.g., Barbu [33], Duchon and Robert [110], Feireisl et al. [120], Jia and Šverák [184], Jungel [185], Vasseur and Yu [341] and the references therein.

The topic of the global existence of weak, mild and strong solutions supplemented with small initial data received considerable attention. For example, Leray [223] carried out a pioneering study on the existence of global weak solutions in the energy space and the uniqueness of such solutions in $\mathbb{R}^2$. Lemarié-Rieusset [222] established the existence of global mild solutions in different types of frameworks in Morrey-Campanato spaces. Later, Iwabuchi and Takada [183] discussed the same problem in function spaces of Besov type. A similar result was established by Lei and Lin [220] in the space $\mathfrak{X}^{-1}$. Similar results were studied by Kato [186] in $L^n(\mathbb{R}^n)$, Giga and Miyakawa [139], and Taylor [328] in Morrey spaces, Cannone [60] and Planchon [302] in the Besov spaces $B_{p,\infty}^{-1+n/p}(\mathbb{R}^n)$, $1 < p < \infty$.

It is worth mentioning that Lions [231] was the first to carry out the study that the Navier-Stokes equations have a weak solution with time-fractional derivative of order less than $\frac{1}{4}$ provided the space dimension is not further than four. After that, only a few research results on this subject have been achieved; for example, Zhang [382] proved that this type of equations has a weak solution whose time-fractional derivative of order is no more than $\frac{1}{2}$.

A strong motivation for investigating the time-fractional Navier-Stokes equations arises from the fact that the phenomena of anomalous diffusion in fractal media can be simulated by Navier-Stokes equations with the time-fractional derivative. Also, fractional partial differential equations play a significant role in modelling many practical situations such as fluid flow, diffusive transport akin to diffusion and so on.

Comparing with theory of the classical Navier-Stokes equations, the researches on time-fractional Navier-Stokes equations are only on their initial stage of development. The main effort on time-fractional Navier-Stokes equations has been put into attempts to derive numerical solutions and analytical solutions, see El-Shahed et al. [113], Ganji et al. [134], and Moman and Zaid [270]. However, to the best of our knowledge, there are few results on time-fractional Navier-Stokes equations. For the first time, Carvalho-Neto [64] dealt with the existence and uniqueness of global and local mild solutions for the time-fractional Navier-Stokes equations.

This chapter deals with the Navier-Stokes equations with time-fractional derivative of order $\alpha \in (0,1)$. Section 2.2 is concerned with the existence and uniqueness of local and global mild solutions in $H^{\beta,q}$. Meanwhile, we also give local mild solutions in J_q. Moreover, we prove the existence and regularity of classical solutions for such equations in J_q. Section 2.3 is devoted to investigation of existence, uniqueness and Hölder continuity of the local mild solutions. In Section 2.4, we obtain the existence and uniqueness of the solutions to approximate equations, as well as the convergence of the approximate solutions. Furthermore, we present some convergence results for the Faedo-Galerkin approximations of the given problems. In Section 2.5, we firstly give the concept of the weak solutions and establish the existence criterion of weak solutions by means of the Galerkin approximations in the case that the dimension $n \leq 4$. Moreover, a complete proof of the uniqueness is given when $n = 2$. At last we give a sufficient condition of optimal control pairs. In Section 2.6, we make use of energy methods to study the time-fractional Navier-Stokes equations. In the first step, we construct a regularized equation by using a smoothing process to transform unbounded differential operators into bounded operators and then obtain the approximate solutions. The second part describes a procedure to take a limit in the approximation program to present a global solution to the objective equation. In Section 2.7, we use the tools from harmonic analysis to study the Cauchy problem for time-fractional Navier-Stokes equations. Two results concerning the local existence of solutions for the given problem in Sobolev spaces are addressed.

2.2 Cauchy Problem in $\mathbb{R}^n$

In this section, we study the following time-fractional Navier-Stokes equations in an open set $\Omega \subset \mathbb{R}^n$ ($n \geq 3$):

$$\begin{cases} \partial_t^\alpha u - \nu \Delta u + (u \cdot \nabla)u = -\nabla p + f, & t > 0, \\ \nabla \cdot u = 0, \\ u|_{\partial\Omega} = 0, \\ u(0, x) = a, \end{cases} \tag{2.1}$$

where ∂_t^α is the Caputo fractional derivative of order $\alpha \in (0, 1)$, $u = (u_1(t, x), u_2(t, x), \ldots, u_n(t, x))$ represents the velocity field at a point $x \in \Omega$ and time $t > 0$, $p = p(t, x)$ is the pressure, ν the viscosity, $f = f(t, x)$ is the external force and $a = a(x)$ is the initial velocity. From now on, we assume that Ω has a smooth boundary.

Firstly, we get rid of the pressure term by applying Helmholtz projector P to equation (2.1), which converts equation (2.1) to

$$\begin{cases} \partial_t^\alpha u - \nu P \Delta u + P(u \cdot \nabla)u = Pf, & t > 0, \\ \nabla \cdot u = 0, \\ u|_{\partial\Omega} = 0, \\ u(0, x) = a. \end{cases}$$

The operator $-\nu P \Delta$ with Dirichlet's boundary conditions is, basically, the Stokes operator A in the divergence-free function space under consideration. Then we rewrite (2.1) as the following abstract form

$$\begin{cases} {}^C_0 D_t^\alpha u = -Au + F(u, u) + Pf, & t > 0, \\ u(0) = a, \end{cases} \tag{2.2}$$

where $F(u, v) = -P(u \cdot \nabla)v$. If one can give sense to the Helmholtz projection P and the Stokes operator A, then the solution of equation (2.2) is also the solution of equation (2.1).

In this section, we establish the existence and uniqueness of global and local mild solutions of problem (2.2) in $H^{\beta,q}$. Further, we prove the regularity results which state essentially that if Pf is Hölder continuous then there is a unique classical solution $u(t)$ such that Au and ${}^C_0 D_t^\alpha u(t)$ are Hölder continuous in J_q.

The section is organized as follows. In Subsection 2.2.1 we recall some notations, definitions, and preliminary facts. Subsection 2.2.2 is devoted to the existence and uniqueness of global mild solution in $H^{\beta,q}$ of problem (2.2), then proceed to study the local mild solution in $H^{\beta,q}$. In Subsection 2.2.3, we use the iteration method to obtain the existence and uniqueness of local mild solution in J_q of problem (2.2). Finally, Subsection 2.2.4 is concerned with the existence and regularity of classical solution in J_q of problem (2.2).

2.2.1   Definitions and Lemmas

In this subsection, we introduce notations, definitions, and preliminary facts which are used throughout this section.

Let $\Omega = \{(x_1,\ldots,x_n) : x_n > 0\}$ be an open subset of $\mathbb{R}^n$, where $n \geq 3$. Let $1 < q < \infty$. Then there is a bounded projection P called the Hodge projection on $(L^q(\Omega))^n$, whose range is the closure of

$$C_\sigma^\infty(\Omega) := \{u \in (C^\infty(\Omega))^n : \nabla \cdot u = 0, \ u \text{ has compact support in } \Omega\},$$

and whose null space is the closure of

$$\{u \in (C^\infty(\Omega))^n : u = \nabla\phi, \ \phi \in C^\infty(\Omega)\}.$$

For notational convenience, let $J_q := \overline{C_\sigma^\infty(\Omega)}^{|\cdot|_q}$, which is a closed subspace of $(L^q(\Omega))^n$. $(W^{m,q}(\Omega))^n$ is a Sobolev space with the norm $|\cdot|_{m,q}$.

$A = -\nu P\Delta$ denotes the Stokes operator in J_q whose domain is $D(A) = D(\Delta) \cap J_q$, here,

$$D(\Delta) = \{u \in (W^{2,q}(\Omega))^n : u|_{\partial\Omega} = 0\}.$$

It is known that $-A$ is a closed linear operator and generates the bounded analytic semigroup $\{e^{-tA}\}$ on J_q, and we have for $\theta \in (\pi/2, \pi)$,

$$\|(z + A)^{-1}\|_{\mathcal{L}(X)} \leq M/|z| \quad \text{for} \quad z \in \Sigma_\theta = \{z \in \mathbb{C} : z \neq 0, \ |\arg z| < \theta\},$$

where $\mathcal{L}(X)$ be the Banach space of linear bounded operators with the norm $\|\cdot\|_{\mathcal{L}(X)}$.

So as to state our results, we need to introduce the definitions of the fractional power spaces associated with $-A$. For $\beta > 0$ and $u \in J_q$, define

$$A^{-\beta}u = \frac{1}{\Gamma(\beta)} \int_0^\infty t^{\beta-1}e^{-tA}u\,dt.$$

Then $A^{-\beta}$ is a bounded, one-to-one operator on J_q. Let A^β be the inverse of $A^{-\beta}$. For $\beta > 0$, we denote the space $H^{\beta,q}$ by the range of $A^{-\beta}$ with the norm

$$|u|_{H^{\beta,q}} = |A^\beta u|_q.$$

It is easy to check that e^{-tA} extends (or restricts) to a bounded analytic semigroup on $H^{\beta,q}$. For more details, we refer to Von Wahl [343].

Let X be a Banach space and J be an interval of $\mathbb{R}$. $C(J, X)$ denotes the set of all continuous X-valued functions. For $0 < \vartheta < 1$, $C^\vartheta(J, X)$ stands for the set of all functions which are Hölder continuous with the exponent ϑ.

Let X be a Banach space, $\alpha \in (0, 1]$ and $v : [0, T] \to X$. The fractional integral of order α with the lower limit zero for the function v is defined as

$$_0D_t^{-\alpha}v(t) = (g_\alpha * v)(t), \ v \in L^1(0, T; X), \ t \in [0, T],$$

where $g_\alpha = \frac{t^{\alpha-1}}{\Gamma(\alpha)}, t > 0$ and $*$ denotes the convolution. Further, the Caputo fractional derivative operator of order α for the function v is defined by

$$_0^C D_t^\alpha v(t) = \frac{d}{dt} \left[_0 D_t^{-(1-\alpha)} (v(t) - v(0)) \right].$$

In general, for $u : [0, T] \times \mathbb{R}^n \to \mathbb{R}^n$, the Caputo fractional derivative with respect to time of the function u can be written as

$$\partial_t^\alpha u(t, x) = \partial_t \left[_0 D_t^{-(1-\alpha)} (u(t, x) - u(0, x)) \right].$$

Let us introduce the generalized Mittag-Leffler special functions:

$$E_\alpha(-t^\alpha A) = \int_0^\infty \mathcal{M}_\alpha(s) e^{-st^\alpha A} ds, \quad E_{\alpha,\alpha}(-t^\alpha A) = \int_0^\infty \alpha s \mathcal{M}_\alpha(s) e^{-st^\alpha A} ds,$$

where $\mathcal{M}_\alpha(\theta)$ is the Wright-type function defined in Definition 1.8.

Proposition 2.1.

(i) $E_{\alpha,\alpha}(-t^\alpha A) = \dfrac{1}{2\pi i} \displaystyle\int_{\Gamma_\theta} E_{\alpha,\alpha}(-\mu t^\alpha)(\mu I + A)^{-1} d\mu$, where

$$\Gamma_\theta = \{re^{-i\theta} : r \geq \delta\} \cup \{\delta e^{i\psi} : |\psi| \leq \theta\} \cup \{re^{i\theta} : r \geq \delta\} \quad \text{for } \delta > 0.$$

(ii) $A^\gamma E_{\alpha,\alpha}(-t^\alpha A) = \dfrac{1}{2\pi i} \displaystyle\int_{\Gamma_\theta} \mu^\gamma E_{\alpha,\alpha}(-\mu t^\alpha)(\mu I + A)^{-1} d\mu.$

Proof. (i) In view of $\int_0^\infty \alpha s \mathcal{M}_\alpha(s) e^{-st} ds = E_{\alpha,\alpha}(-t)$ and the Fubini theorem, we get

$$E_{\alpha,\alpha}(-t^\alpha A) = \int_0^\infty \alpha s \mathcal{M}_\alpha(s) e^{-st^\alpha A} ds$$

$$= \frac{1}{2\pi i} \int_0^\infty \alpha s \mathcal{M}_\alpha(s) \int_{\Gamma_\theta} e^{-\mu st^\alpha} (\mu I + A)^{-1} d\mu ds$$

$$= \frac{1}{2\pi i} \int_{\Gamma_\theta} E_{\alpha,\alpha}(-\mu t^\alpha)(\mu I + A)^{-1} d\mu,$$

where Γ_θ is a suitable integral path.

(ii) A similar argument shows that

$$A^\gamma E_{\alpha,\alpha}(-t^\alpha A) = \int_0^\infty \alpha s \mathcal{M}_\alpha(s) A^\gamma e^{-st^\alpha A} ds$$

$$= \frac{1}{2\pi i} \int_0^\infty \alpha s \mathcal{M}_\alpha(s) \int_{\Gamma_\theta} \mu^\gamma e^{-\mu st^\alpha} (\mu I + A)^{-1} d\mu ds$$

$$= \frac{1}{2\pi i} \int_{\Gamma_\theta} \mu^\gamma E_{\alpha,\alpha}(-\mu t^\alpha)(\mu I + A)^{-1} d\mu.$$

$\square$

Moreover, we have the following results.

Lemma 2.1. *[352] For $t > 0$, $E_\alpha(-t^\alpha A)$ and $E_{\alpha,\alpha}(-t^\alpha A)$ are continuous in the uniform operator topology. Moreover, for every $r > 0$, the continuity is uniform on $[r, \infty)$.*

Lemma 2.2. *[352] Let $0 < \alpha < 1$. Then*

(i) *for all $u \in X$, $\lim_{t \to 0^+} E_\alpha(-t^\alpha A)u = u$;*

(ii) *for all $u \in D(A)$ and $t > 0$, ${}_0^C D_t^\alpha E_\alpha(-t^\alpha A)u = -AE_\alpha(-t^\alpha A)u$;*

(iii) *for all $u \in X$, $E_\alpha'(-t^\alpha A)u = -t^{\alpha-1}AE_{\alpha,\alpha}(-t^\alpha A)u$;*

(iv) *for all $u \in X$ and $t > 0$, $E_\alpha(-t^\alpha A)u = {}_0D_t^{-(1-\alpha)}\left(t^{\alpha-1}E_{\alpha,\alpha}(-t^\alpha A)u\right)$.*

Before presenting the definition of mild solution of problem (2.2), we give the following lemma for a given function $h : [0, \infty) \to X$. For more details we refer to Zhou [386, 387].

Lemma 2.3. *If*

$$u(t) = a + \frac{1}{\Gamma(\alpha)} \int_0^t (t-s)^{\alpha-1}\left(-Au(s) + h(s)\right)ds, \quad for\ t \geq 0 \qquad (2.3)$$

holds, then we have

$$u(t) = E_\alpha(-t^\alpha A)a + \int_0^t (t-s)^{\alpha-1}E_{\alpha,\alpha}(-(t-s)^\alpha A)h(s)ds.$$

We rewrite (2.2) as

$$u(t) = a + \frac{1}{\Gamma(\alpha)} \int_0^t (t-s)^{\alpha-1}\left(-Au(s) + F(u(s), u(s)) + Pf(s)\right)ds, \quad for\ t \geq 0.$$

Inspired by above discussion, we adopt the following concepts of mild solution to problem (2.2).

Definition 2.1. A function $u : [0, \infty) \to H^{\beta,q}$ is called a global mild solution of the problem (2.2) in $H^{\beta,q}$, if $u \in C([0, \infty); H^{\beta,q})$ and for $t \in [0, \infty)$,

$$u(t) = E_\alpha(-t^\alpha A)a + \int_0^t (t-s)^{\alpha-1}E_{\alpha,\alpha}(-(t-s)^\alpha A)F(u(s), u(s))ds$$
$$+ \int_0^t (t-s)^{\alpha-1}E_{\alpha,\alpha}(-(t-s)^\alpha A)Pf(s)ds. \qquad (2.4)$$

Definition 2.2. Let $0 < T < \infty$. A function $u : [0, T] \to H^{\beta,q}$ (or J_q) is called a local mild solution of the problem (2.2) in $H^{\beta,q}$ (or J_q), if $u \in C([0, T]; H^{\beta,q})$ (or $C([0, T]; J_q)$) and u satisfies (2.4) for $t \in [0, T]$.

For convenience, we define two operators Φ and $\mathcal{G}$ as follows:

$$\Phi(t) = \int_0^t (t-s)^{\alpha-1}E_{\alpha,\alpha}(-(t-s)^\alpha A)Pf(s)ds,$$

$$\mathcal{G}(u,v)(t) = \int_0^t (t-s)^{\alpha-1}E_{\alpha,\alpha}(-(t-s)^\alpha A)F(u(s), v(s))ds.$$

In subsequent proof we use the following fixed point result.

Lemma 2.4. *[59] Let $(X, |\cdot|)$ be a Banach space, $G : X \times X \to X$ a bilinear operator and L a positive real number such that*

$$|G(u,v)| \le L|u||v|, \ \forall \ u, \ v \in X.$$

Then for any $u_0 \in X$ with $|u_0| < \frac{1}{4L}$, the equation $u = u_0 + G(u,u)$ has a unique solution $u \in X$.

2.2.2   Global/Local Existence

Our main purpose in this subsection is to establish sufficient conditions for existence and uniqueness of mild solution to the problem (2.2) in $H^{\beta,q}$. To this end we assume that:

(f) Pf is continuous for $t > 0$ and $|Pf(t)|_q = o(t^{-\alpha(1-\beta)})$ as $t \to 0$ for $0 < \beta < 1$.

Lemma 2.5. *[133, 362] Let $1 < q < \infty$ and $\beta_1 \le \beta_2$. Then there is a constant $C = C(\beta_1, \beta_2)$ such that*

$$|e^{-tA}v|_{H^{\beta_2,q}} \le Ct^{-(\beta_2-\beta_1)}|v|_{H^{\beta_1,q}}, \quad t > 0,$$

for $v \in H^{\beta_1,q}$. Furthermore, $\lim_{t\to 0} t^{(\beta_2-\beta_1)}|e^{-tA}v|_{H^{\beta_2,q}} = 0$.

Now, we study an important technical lemma, which helps us to prove the final main theorems of this subsection.

Lemma 2.6. *Let $1 < q < \infty$ and $\beta_1 \le \beta_2 < \beta_1 + 1$. Then for any $T > 0$, there exists a constant $C_1 = C_1(\alpha, \beta_1, \beta_2) > 0$ such that*

$$|E_\alpha(-t^\alpha A)v|_{H^{\beta_2,q}} \le C_1 t^{-\alpha(\beta_2-\beta_1)}|v|_{H^{\beta_1,q}},$$
$$|E_{\alpha,\alpha}(-t^\alpha A)v|_{H^{\beta_2,q}} \le C_1 t^{-\alpha(\beta_2-\beta_1)}|v|_{H^{\beta_1,q}},$$

for all $v \in H^{\beta_1,q}$ and $t \in (0,T]$. Furthermore,

$$\lim_{t\to 0} t^{\alpha(\beta_2-\beta_1)}|E_\alpha(-t^\alpha A)v|_{H^{\beta_2,q}} = 0.$$

Proof. Let $v \in H^{\beta_1,q}$. By Lemma 2.5, we estimate

$$|E_\alpha(-t^\alpha A)v|_{H^{\beta_2,q}} \le \int_0^\infty M_\alpha(s)|e^{-st^\alpha A}v|_{H^{\beta_2,q}}ds$$
$$\le \left(C \int_0^\infty M_\alpha(s)s^{-(\beta_2-\beta_1)}ds \right) t^{-\alpha(\beta_2-\beta_1)}|v|_{H^{\beta_1,q}}$$
$$\le C_1 t^{-\alpha(\beta_2-\beta_1)}|v|_{H^{\beta_1,q}}, \quad \text{for} \ \beta_1 - \beta_2 > -1.$$

More precisely, the Lebesgue's dominated convergence theorem shows

$$\lim_{t\to 0} t^{\alpha(\beta_2-\beta_1)}|E_\alpha(-t^\alpha A)v|_{H^{\beta_2,q}} \le \int_0^\infty M_\alpha(s) \lim_{t\to 0} t^{\alpha(\beta_2-\beta_1)}|e^{-st^\alpha A}v|_{H^{\beta_2,q}}ds = 0.$$

Similarly,

$$|E_{\alpha,\alpha}(-t^{\alpha}A)v|_{H^{\beta_2,q}} \leq \int_0^{\infty} \alpha s \mathcal{M}_{\alpha}(s)|e^{-st^{\alpha}A}v|_{H^{\beta_2,q}}ds$$

$$\leq \left(\alpha C \int_0^{\infty} \mathcal{M}_{\alpha}(s)s^{1-(\beta_2-\beta_1)}ds \right)t^{-\alpha(\beta_2-\beta_1)}|v|_{H^{\beta_1,q}}$$

$$\leq C_1 t^{-\alpha(\beta_2-\beta_1)}|v|_{H^{\beta_1,q}},$$

where the constant $C_1 = C_1(\alpha, \beta_1, \beta_2)$ is such that

$$C_1 \geq C \max\left\{ \frac{\Gamma(1-\beta_2+\beta_1)}{\Gamma(1+\alpha(\beta_1-\beta_2))}, \frac{\alpha\Gamma(2-\beta_2+\beta_1)}{\Gamma(1+\alpha(1+\beta_1-\beta_2))} \right\}.$$

$\square$

In the following, we study the global mild solution of problem (2.2) in $H^{\beta,q}$. For convenience, we denote

$$M(t) = \sup_{s\in(0,t]} \{s^{\alpha(1-\beta)}|Pf(s)|_q\},$$

$$B_1 = C_1 \max\{B(\alpha(1-\beta), 1-\alpha(1-\beta)), B(\alpha(1-\gamma), 1-\alpha(1-\beta))\},$$

$$L \geq MC_1 \max\{B(\alpha(1-\beta), 1-2\alpha(\gamma-\beta)), B(\alpha(1-\gamma), 1-2\alpha(\gamma-\beta))\},$$

where M is given later.

Theorem 2.1. *Let $1 < q < \infty$, $0 < \beta < 1$ and (f) hold. For every $a \in H^{\beta,q}$, suppose that*

$$C_1|a|_{H^{\beta,q}} + B_1 M_{\infty} < \frac{1}{4L}, \tag{2.5}$$

where $M_{\infty} := \sup_{s\in(0,\infty)}\{s^{\alpha(1-\beta)}|Pf(s)|_q\}$. If $\frac{n}{2q} - \frac{1}{2} < \beta$, then there is a $\gamma > \max\{\beta, \frac{1}{2}\}$ and a unique function $u : [0, \infty) \to H^{\beta,q}$ satisfying:

(a) *$u : [0, \infty) \to H^{\beta,q}$ is continuous and $u(0) = a$;*
(b) *$u : (0, \infty) \to H^{\gamma,q}$ is continuous and $\lim_{t\to 0} t^{\alpha(\gamma-\beta)}|u(t)|_{H^{\gamma,q}} = 0$;*
(c) *u satisfies (2.4) for $t \in [0, \infty)$.*

Proof. Let $\gamma = \frac{(1+\beta)}{2}$. Define $X_{\infty} = X[\infty]$ as the space of all curves $u : (0, \infty) \to H^{\beta,q}$ such that:

(i) *$u : [0, \infty) \to H^{\beta,q}$ is bounded and continuous;*
(ii) *$u : (0, \infty) \to H^{\gamma,q}$ is bounded and continuous, moreover,*
  $\lim_{t\to 0} t^{\alpha(\gamma-\beta)}|u(t)|_{H^{\gamma,q}} = 0$

with its natural norm

$$\|u\|_{X_{\infty}} = \max\left\{ \sup_{t\geq 0}|u(t)|_{H^{\beta,q}}, \sup_{t\geq 0} t^{\alpha(\gamma-\beta)}|u(t)|_{H^{\gamma,q}} \right\}.$$

It is obvious that X_{∞} is a non-empty complete metric space.

From an argument of Weissler [362], we know that $F : H^{\gamma,q} \times H^{\gamma,q} \to J_q$ is a bounded bilinear map, then there exists M such that for $u, v \in H^{\gamma,q}$,

$$|F(u,v)|_q \leq M|u|_{H^{\gamma,q}}|v|_{H^{\gamma,q}},$$

$$|F(u,u) - F(v,v)|_q \leq M(|u|_{H^{\gamma,q}} + |v|_{H^{\gamma,q}})|u - v|_{H^{\gamma,q}}. \tag{2.6}$$

Step 1. Let $u, v \in X_\infty$. We show that the operator $\mathcal{G}(u(t), v(t))$ belongs to $C([0,\infty); H^{\beta,q})$ as well as $C((0,\infty); H^{\gamma,q})$.

For arbitrary $t_0 \geq 0$ fixed and $\varepsilon > 0$ enough small, consider $t > t_0$ (the case $t < t_0$ follows analogously), we have

$$|\mathcal{G}(u(t), v(t)) - \mathcal{G}(u(t_0), v(t_0))|_{H^{\beta,q}}$$

$$\leq \int_{t_0}^t (t-s)^{\alpha-1}|E_{\alpha,\alpha}(-(t-s)^\alpha A)F(u(s),v(s))|_{H^{\beta,q}} ds$$

$$+ \int_0^{t_0} |((t-s)^{\alpha-1} - (t_0-s)^{\alpha-1})E_{\alpha,\alpha}(-(t-s)^\alpha A)F(u(s),v(s))|_{H^{\beta,q}} ds$$

$$+ \int_0^{t_0-\varepsilon} (t_0-s)^{\alpha-1}|(E_{\alpha,\alpha}(-(t-s)^\alpha A)$$

$$- E_{\alpha,\alpha}(-(t_0-s)^\alpha A))F(u(s),v(s))|_{H^{\beta,q}} ds$$

$$+ \int_{t_0-\varepsilon}^{t_0} (t_0-s)^{\alpha-1}|(E_{\alpha,\alpha}(-(t-s)^\alpha A)$$

$$- E_{\alpha,\alpha}(-(t_0-s)^\alpha A))F(u(s),v(s))|_{H^{\beta,q}} ds$$

$$=: I_{11}(t) + I_{12}(t) + I_{13}(t) + I_{14}(t).$$

We estimate each of the four terms separately. For $I_{11}(t)$, in view of Lemma 2.6, we obtain

$$I_{11}(t) \leq C_1 \int_{t_0}^t (t-s)^{\alpha(1-\beta)-1}|F(u(s),v(s))|_q ds$$

$$\leq MC_1 \int_{t_0}^t (t-s)^{\alpha(1-\beta)-1}|u(s)|_{H^{\gamma,q}}|v(s)|_{H^{\gamma,q}} ds$$

$$\leq MC_1 \int_{t_0}^t (t-s)^{\alpha(1-\beta)-1}s^{-2\alpha(\gamma-\beta)} ds \sup_{s\in[0,t]} \{s^{2\alpha(\gamma-\beta)}|u(s)|_{H^{\gamma,q}}|v(s)|_{H^{\gamma,q}}\}$$

$$= MC_1 \int_{t_0/t}^1 (1-s)^{\alpha(1-\beta)-1}s^{-2\alpha(\gamma-\beta)} ds \sup_{s\in[0,t]} \{s^{2\alpha(\gamma-\beta)}|u(s)|_{H^{\gamma,q}}|v(s)|_{H^{\gamma,q}}\}.$$

By the properties of the Beta function, there exists $\delta > 0$ small enough such that for $0 < t - t_0 < \delta$,

$$\int_{t_0/t}^1 (1-s)^{\alpha(1-\beta)-1}s^{-2\alpha(\gamma-\beta)} ds \to 0,$$

which follows that $I_{11}(t)$ tends to 0 as $t - t_0 \to 0$. For $I_{12}(t)$, since

$$I_{12}(t) \leq C_1 \int_0^{t_0} ((t_0-s)^{\alpha-1} - (t-s)^{\alpha-1})(t-s)^{-\alpha\beta}|F(u(s),v(s))|_q ds$$

$$\leq MC_1 \int_0^{t_0} \left((t_0-s)^{\alpha-1} - (t-s)^{\alpha-1}\right)(t-s)^{-\alpha\beta} s^{-2\alpha(\gamma-\beta)} ds$$

$$\times \sup_{s\in[0,t_0]} \left\{s^{2\alpha(\gamma-\beta)} |u(s)|_{H^{\gamma,q}} |v(s)|_{H^{\gamma,q}}\right\}.$$

Noting that

$$\int_0^{t_0} |(t_0-s)^{\alpha-1} - (t-s)^{\alpha-1}|(t-s)^{-\alpha\beta} s^{-2\alpha(\gamma-\beta)} ds$$

$$\leq \int_0^{t_0} (t-s)^{\alpha-1}(t-s)^{-\alpha\beta} s^{-2\alpha(\gamma-\beta)} ds + \int_0^{t_0} (t_0-s)^{\alpha-1}(t-s)^{-\alpha\beta} s^{-2\alpha(\gamma-\beta)} ds$$

$$\leq 2\int_0^{t_0} (t_0-s)^{\alpha(1-\beta)-1} s^{-2\alpha(\gamma-\beta)} ds$$

$$= 2B(\alpha(1-\beta), 1-2\alpha(\gamma-\beta)),$$

then by the Lebesgue's dominated convergence theorem, we have

$$\int_0^{t_0} \left((t_0-s)^{\alpha-1} - (t-s)^{\alpha-1}\right)(t-s)^{-\alpha\beta} s^{-2\alpha(\gamma-\beta)} ds \to 0, \quad \text{as } t\to t_0,$$

one deduces that $\lim_{t\to t_0} I_{12}(t) = 0$. For $I_{13}(t)$, since

$$I_{13}(t) \leq \int_0^{t_0-\varepsilon} (t_0-s)^{\alpha-1} \big|\big(E_{\alpha,\alpha}(-(t-s)^\alpha A)$$

$$+ E_{\alpha,\alpha}(-(t_0-s)^\alpha A)\big)F(u(s),v(s))\big|_{H^{\beta,q}} ds$$

$$\leq C_1 \int_0^{t_0-\varepsilon} (t_0-s)^{\alpha-1}\left((t-s)^{-\alpha\beta} + (t_0-s)^{-\alpha\beta}\right)|F(u(s),v(s))|_q ds$$

$$\leq 2MC_1 \int_0^{t_0-\varepsilon} (t_0-s)^{\alpha(1-\beta)-1} s^{-2\alpha(\gamma-\beta)} ds$$

$$\times \sup_{s\in[0,t_0]} \left\{s^{2\alpha(\gamma-\beta)} |u(s)|_{H^{\gamma,q}} |v(s)|_{H^{\gamma,q}}\right\},$$

using the Lebesgue's dominated convergence theorem again, the fact from the uniform continuity of the operator $E_{\alpha,\alpha}(-t^\alpha A)$ due to Lemma 2.1 shows

$$\lim_{t\to t_0} I_{13}(t) = \int_0^{t_0-\varepsilon} (t_0-s)^{\alpha-1} \lim_{t\to t_0} \big|\big(E_{\alpha,\alpha}(-(t-s)^\alpha A)$$

$$- E_{\alpha,\alpha}(-(t_0-s)^\alpha A)\big)F(u(s),v(s))\big|_{H^{\beta,q}} ds$$

$$= 0.$$

For $I_{14}(t)$, by immediate calculation, we estimate

$$I_{14}(t) \leq C_1 \int_{t_0-\varepsilon}^{t_0} (t_0-s)^{\alpha-1}\left((t-s)^{-\alpha\beta} + (t_0-s)^{-\alpha\beta}\right)|F(u(s),v(s))|_q ds$$

$$\leq 2MC_1 \int_{t_0-\varepsilon}^{t_0} (t_0-s)^{\alpha(1-\beta)-1} s^{-2\alpha(\gamma-\beta)} ds$$

$$\times \sup_{s\in[t_0-\varepsilon,t_0]} \left\{s^{2\alpha(\gamma-\beta)} |u(s)|_{H^{\gamma,q}} |v(s)|_{H^{\gamma,q}}\right\}$$

$$\to 0, \quad \text{as } \varepsilon \to 0$$

according to the properties of the Beta function. Thenceforth, it follows

$$|\mathcal{G}(u(t), v(t)) - \mathcal{G}(u(t_0), v(t_0))|_{H^{\beta,q}} \to 0, \quad \text{as } t \to t_0.$$

The continuity of the operator $\mathcal{G}(u, v)$ evaluated in $C((0, \infty); H^{\gamma,q})$ follows by the similar discussion as above. So, we omit the details.

Step 2. We show that the operator $\mathcal{G} : X_\infty \times X_\infty \to X_\infty$ is a continuous bilinear operator.

By Lemma 2.6, we have

$$\begin{aligned}
|\mathcal{G}(u(t), v(t))|_{H^{\beta,q}} &\leq \left| \int_0^t (t-s)^{\alpha-1} E_{\alpha,\alpha}(-(t-s)^\alpha A) F(u(s), v(s)) ds \right|_{H^{\beta,q}} \\
&\leq C_1 \int_0^t (t-s)^{\alpha(1-\beta)-1} |F(u(s), v(s))|_q ds \\
&\leq M C_1 \int_0^t (t-s)^{\alpha(1-\beta)-1} s^{-2\alpha(\gamma-\beta)} ds \\
&\quad \times \sup_{s \in [0,t]} \{ s^{2\alpha(\gamma-\beta)} |u(s)|_{H^{\gamma,q}} |v(s)|_{H^{\gamma,q}} \} \\
&\leq M C_1 B(\alpha(1-\beta), 1 - 2\alpha(\gamma-\beta)) \|u\|_{X_\infty} \|v\|_{X_\infty}
\end{aligned}$$

and

$$\begin{aligned}
|\mathcal{G}(u(t), v(t))|_{H^{\gamma,q}} &\leq \left| \int_0^t (t-s)^{\alpha-1} E_{\alpha,\alpha}(-(t-s)^\alpha A) F(u(s), v(s)) ds \right|_{H^{\gamma,q}} \\
&\leq C_1 \int_0^t (t-s)^{\alpha(1-\gamma)-1} |F(u(s), v(s))|_q ds \\
&\leq M C_1 \int_0^t (t-s)^{\alpha(1-\gamma)-1} s^{-2\alpha(\gamma-\beta)} ds \\
&\quad \times \sup_{s \in [0,t]} \{ s^{2\alpha(\gamma-\beta)} |u(s)|_{H^{\gamma,q}} |v(s)|_{H^{\gamma,q}} \} \\
&\leq M C_1 t^{-\alpha(\gamma-\beta)} B(\alpha(1-\gamma), 1 - 2\alpha(\gamma-\beta)) \|u\|_{X_\infty} \|v\|_{X_\infty},
\end{aligned}$$

it follows that

$$\sup_{t \in [0,\infty)} t^{\alpha(\gamma-\beta)} |\mathcal{G}(u(t), v(t))|_{H^{\gamma,q}} \leq M C_1 B(\alpha(1-\gamma), 1 - 2\alpha(\gamma-\beta)) \|u\|_{X_\infty} \|v\|_{X_\infty}.$$

More precisely,

$$\lim_{t \to 0} t^{\alpha(\gamma-\beta)} |\mathcal{G}(u(t), v(t))|_{H^{\gamma,q}} = 0.$$

Hence, $\mathcal{G}(u, v) \in X_\infty$ and $\|\mathcal{G}(u, v)\|_{X_\infty} \leq L \|u\|_{X_\infty} \|v\|_{X_\infty}$.

Step 3. We verify that (c) holds.

Let $0 < t_0 < t$. Since

$$|\Phi(t) - \Phi(t_0)|_{H^{\beta,q}}$$

$$\leq \int_{t_0}^t (t-s)^{\alpha-1} |E_{\alpha,\alpha}(-(t-s)^\alpha A) P f(s)|_{H^{\beta,q}} ds$$

$$+ \int_0^{t_0} ((t_0-s)^{\alpha-1} - (t-s)^{\alpha-1}) |E_{\alpha,\alpha}(-(t-s)^\alpha A) P f(s)|_{H^{\beta,q}} ds$$

$$+ \int_0^{t_0-\varepsilon} (t_0-s)^{\alpha-1} |(E_{\alpha,\alpha}(-(t-s)^\alpha A) - E_{\alpha,\alpha}(-(t_0-s)^\alpha A)) P f(s)|_{H^{\beta,q}} ds$$

$$+ \int_{t_0-\varepsilon}^{t_0} (t_0-s)^{\alpha-1} |(E_{\alpha,\alpha}(-(t-s)^\alpha A) - E_{\alpha,\alpha}(-(t_0-s)^\alpha A)) P f(s)|_{H^{\beta,q}} ds$$

$$\leq C_1 \int_{t_0}^t (t-s)^{\alpha(1-\beta)-1} |P f(s)|_q ds$$

$$+ C_1 \int_0^{t_0} ((t_0-s)^{\alpha-1} - (t-s)^{\alpha-1})(t-s)^{-\alpha\beta} |P f(s)|_q ds$$

$$+ C_1 \int_0^{t_0-\varepsilon} (t_0-s)^{\alpha-1} |(E_{\alpha,\alpha}(-(t-s)^\alpha A) - E_{\alpha,\alpha}(-(t_0-s)^\alpha A)) P f(s)|_{H^{\beta,q}} ds$$

$$+ 2C_1 \int_{t_0-\varepsilon}^{t_0} (t_0-s)^{\alpha(1-\beta)-1} |P f(s)|_q ds$$

$$\leq C_1 M(t) \int_{t_0}^t (t-s)^{\alpha(1-\beta)-1} s^{-\alpha(1-\beta)} ds$$

$$+ C_1 M(t) \int_0^{t_0} ((t_0-s)^{\alpha-1} - (t-s)^{\alpha-1})(t-s)^{-\alpha\beta} s^{-\alpha(1-\beta)} ds$$

$$+ C_1 M(t) \int_0^{t_0-\varepsilon} (t_0-s)^{\alpha-1} |(E_{\alpha,\alpha}(-(t-s)^\alpha A)$$

$$- E_{\alpha,\alpha}(-(t_0-s)^\alpha A)) s^{-\alpha(1-\beta)}|_{H^{\beta,q}} ds$$

$$+ 2C_1 M(t) \int_{t_0-\varepsilon}^{t_0} (t_0-s)^{\alpha(1-\beta)-1} s^{-\alpha(1-\beta)} ds.$$

By the properties of the Beta function, the first two integrals and the last integral tend to 0 as $t \to t_0$ as well as $\varepsilon \to 0$. In view of Lemma 2.1, the third integral also goes to 0 as $t \to t_0$, which implies

$$|\Phi(t) - \Phi(t_0)|_{H^{\beta,q}} \to 0, \quad \text{as } t \to t_0.$$

The continuity of $\Phi(t)$ evaluated in $H^{\gamma,q}$ follows by the similar argument as above.

On the other hand, we have

$$|\Phi(t)|_{H^{\beta,q}} \leq \left| \int_0^t (t-s)^{\alpha-1} E_{\alpha,\alpha}(-(t-s)^\alpha A) P f(s) ds \right|_{H^{\beta,q}}$$

$$\leq C_1 \int_0^t (t-s)^{\alpha(1-\beta)-1} |P f(s)|_q ds \qquad (2.7)$$

$$\leq C_1 M(t) \int_0^t (t-s)^{\alpha(1-\beta)-1} s^{-\alpha(1-\beta)} ds$$

$$= C_1 M(t) B(\alpha(1-\beta), 1-\alpha(1-\beta))$$

and

$$|\Phi(t)|_{H^{\gamma,q}} \le \left| \int_0^t (t-s)^{\alpha-1} E_{\alpha,\alpha}(-(t-s)^\alpha A)Pf(s)ds \right|_{H^{\gamma,q}}$$

$$\le C_1 \int_0^t (t-s)^{\alpha(1-\gamma)-1}|Pf(s)|_q ds$$

$$\le C_1 M(t) \int_0^t (t-s)^{\alpha(1-\gamma)-1} s^{-\alpha(1-\beta)} ds$$

$$= t^{-\alpha(\gamma-\beta)} C_1 M(t) B(\alpha(1-\gamma), 1-\alpha(1-\beta)).$$

More precisely,

$$t^{\alpha(\gamma-\beta)}|\Phi(t)|_{H^{\gamma,q}} \le C_1 M(t) B(\alpha(1-\gamma), 1-\alpha(1-\beta)) \to 0, \text{ as } t \to 0,$$

since $M(t) \to 0$ as $t \to 0$ due to assumption (f). This ensures that $\Phi(t) \in X_\infty$ and $\|\Phi\|_{X_\infty} \le B_1 M_\infty$.

For $a \in H^{\beta,q}$. By Lemma 2.1, it is easy to see that

$$E_\alpha(-t^\alpha A)a \in C([0,\infty); H^{\beta,q}) \text{ and } E_\alpha(-t^\alpha A)a \in C((0,\infty); H^{\gamma,q}).$$

This, together with Lemma 2.6, implies that for all $t \in (0,T]$,

$$E_\alpha(-t^\alpha A)a \in X_\infty,$$

$$t^{\alpha(\gamma-\beta)} E_\alpha(-t^\alpha A)a \in C([0,\infty); H^{\gamma,q}),$$

$$\|E_\alpha(-t^\alpha A)a\|_{X_\infty} \le C_1 |a|_{H^{\beta,q}}.$$

According to (2.5), the inequality

$$\|E_\alpha(-t^\alpha A)a + \Phi\|_{X_\infty} \le \|E_\alpha(-t^\alpha A)a\|_{X_\infty} + \|\Phi\|_{X_\infty} < \frac{1}{4L}$$

holds. From Lemma 2.4, there exists a $u : [0,\infty) \to H^{\beta,q}$ satisfies (2.4).

Step 4. To show that $u(t) \to a$ in $H^{\beta,q}$ as $t \to 0$.

We need to verify

$$\lim_{t \to 0} \int_0^t (t-s)^{\alpha-1} E_{\alpha,\alpha}(-(t-s)^\alpha A)Pf(s)ds = 0,$$

$$\lim_{t \to 0} \int_0^t (t-s)^{\alpha-1} E_{\alpha,\alpha}(-(t-s)^\alpha A)F(u(s), u(s))ds = 0$$

in $H^{\beta,q}$. In fact, it is obvious that $\lim_{t \to 0} \Phi(t) = 0$ $(\lim_{t \to 0} M(t) = 0)$ owing to (2.7). In addition,

$$\left| \int_0^t (t-s)^{\alpha-1} E_{\alpha,\alpha}(-(t-s)^\alpha A)F(u(s), u(s))ds \right|_{H^{\beta,q}}$$

$$\le C_1 \int_0^t (t-s)^{\alpha(1-\beta)-1}|F(u(s), u(s))|_q ds$$

$$\le MC_1 \int_0^t (t-s)^{\alpha(1-\beta)-1}|u(s)|_{H^{\gamma,q}}^2 ds$$

$$\leq MC_1 \int_0^t (t-s)^{\alpha(1-\beta)-1} s^{-2\alpha(\gamma-\beta)} ds \sup_{s\in[0,t]} \{s^{2\alpha(\gamma-\beta)}|u(s)|^2_{H^{\gamma,q}}\}$$

$$= MC_1 B(\alpha(1-\beta), 1-2\alpha(\gamma-\beta)) \sup_{s\in[0,t]} \{s^{2\alpha(\gamma-\beta)}|u(s)|^2_{H^{\gamma,q}}\} \to 0, \quad \text{as } t \to 0.$$

$\square$

In the following, we study the local mild solution of problem (2.2) in $H^{\beta,q}$.

Theorem 2.2. *Let* $1 < q < \infty$, $0 < \beta < 1$ *and* (f) *hold. Suppose*

$$\frac{n}{2q} - \frac{1}{2} < \beta. \tag{2.8}$$

Then there is a $\gamma > \max\{\beta, \frac{1}{2}\}$ *such that for every* $a \in H^{\beta,q}$ *there exist* $T_* > 0$ *and a unique function* $u : [0, T_*] \to H^{\beta,q}$ *satisfying:*

(a) $u : [0, T_*] \to H^{\beta,q}$ *is continuous and* $u(0) = a$;
(b) $u : (0, T_*] \to H^{\gamma,q}$ *is continuous and* $\lim_{t\to 0} t^{\alpha(\gamma-\beta)}|u(t)|_{H^{\gamma,q}} = 0$;
(c) u *satisfies* (2.4) *for* $t \in [0, T_*]$.

Proof. Let $\gamma = \frac{(1+\beta)}{2}$. Fix $a \in H^{\beta,q}$. Let $X = X[T]$ be the space of all curves $u : (0, T] \to H^{\beta,q}$ such that:

(i) $u : [0, T] \to H^{\beta,q}$ *is continuous*;
(ii) $u : (0, T] \to H^{\gamma,q}$ *is continuous and* $\lim_{t\to 0} t^{\alpha(\gamma-\beta)}|u(t)|_{H^{\gamma,q}} = 0$

with its natural norm

$$\|u\|_X = \sup_{t\in[0,T]} \{t^{\alpha(\gamma-\beta)}|u(t)|_{H^{\gamma,q}}\}.$$

Similar to the proof of Theorem 2.1, it is easy to claim that $\mathcal{G} : X \times X \to X$ is continuous linear map and $\Phi(t) \in X$.

By Lemma 2.1, it is easy to see that for all $t \in (0, T]$,

$$E_\alpha(-t^\alpha A)a \in C([0, T]; H^{\beta,q}),$$
$$E_\alpha(-t^\alpha A)a \in C((0, T]; H^{\gamma,q}).$$

From Lemma 2.6, it follows that

$$E_\alpha(-t^\alpha A)a \in X,$$
$$t^{\alpha(\gamma-\beta)} E_\alpha(-t^\alpha A)a \in C([0, T]; H^{\gamma,q}).$$

Hence, let $T_* > 0$ be sufficiently small such that

$$\|E_\alpha(-t^\alpha A)a + \Phi(t)\|_{X[T_*]} \leq \|E_\alpha(-t^\alpha A)a\|_{X[T_*]} + \|\Phi(t)\|_{X[T_*]} < \frac{1}{4L}.$$

From Lemma 2.4, there exists a $u : [0, T_*] \to H^{\beta,q}$ satisfies (2.4). $\square$

2.2.3 Local Existence

This subsection is devoted to consideration of local mild solution to problem (2.2) in J_q by means of the iteration method. Let $\gamma = \frac{(1+\beta)}{2}$.

Theorem 2.3. *Let $1 < q < \infty$, $0 < \beta < 1$ and (f) hold. Suppose that*

$$a \in H^{\beta,q} \text{ with } \frac{n}{2q} - \frac{1}{2} < \beta.$$

Then problem (2.2) has a unique mild solution u in J_q for $a \in H^{\beta,q}$. Moreover, u is continuous on $[0,T]$, $A^\gamma u$ is continuous in $(0,T]$ and $t^{\alpha(\gamma-\beta)}A^\gamma u(t)$ is bounded as $t \to 0$.

Proof. Step 1. Set

$$K(t) := \sup_{s \in (0,t]} s^{\alpha(\gamma-\beta)}|A^\gamma u(s)|_q$$

and

$$\Psi(t) := \mathcal{G}(u,u)(t) = \int_0^t (t-s)^{\alpha-1}E_{\alpha,\alpha}(-(t-s)^\alpha A)F(u(s),u(s))ds.$$

As an immediate consequences of Step 2 in Theorem 2.1, then $\Psi(t)$ is continuous in $[0,T]$, $A^\gamma\Psi(t)$ exists and is continuous in $(0,T]$ with

$$|A^\gamma\Psi(t)|_q \leq MC_1 B(\alpha(1-\gamma), 1-2\alpha(\gamma-\beta))K^2(t)t^{-\alpha(\gamma-\beta)}. \tag{2.9}$$

We also consider the integral $\Phi(t)$. Since (f) holds, the inequality

$$|Pf(s)|_q \leq M(t)s^{-\alpha(1-\beta)}$$

is satisfied with a continuous function $M(t)$. From Step 3 in Theorem 2.1, we derive that $A^\gamma\Phi(t)$ is continuous in $(0,T]$ with

$$|A^\gamma\Phi(t)|_q \leq C_1 M(t)B(\alpha(1-\gamma), 1-\alpha(1-\beta))t^{-\alpha(\gamma-\beta)}. \tag{2.10}$$

For $|Pf(t)|_q = o(t^{-\alpha(1-\beta)})$ as $t \to 0$, we have $M(t) = 0$. Here (2.10) means $|A^\gamma\Phi(t)|_q = o(t^{-\alpha(\gamma-\beta)})$ as $t \to 0$.

We prove that Φ is continuous in J_q. In fact, take $0 \leq t_0 < t < T$, we have

$$|\Phi(t) - \Phi(t_0)|_q$$

$$\leq C_1 \int_{t_0}^t (t-s)^{\alpha-1}|Pf(s)|_q ds + C_1 \int_0^{t_0} ((t_0-s)^{\alpha-1} - (t-s)^{\alpha-1})|Pf(s)|_q ds$$

$$+ C_1 \int_0^{t_0-\varepsilon} (t_0-s)^{\alpha-1}\|E_{\alpha,\alpha}(-(t-s)^\alpha A) - E_{\alpha,\alpha}(-(t_0-s)^\alpha A)\| |Pf(s)|_q ds$$

$$+ 2C_1 \int_{t_0-\varepsilon}^{t_0} (t_0-s)^{\alpha-1}|Pf(s)|_q ds$$

$$\leq C_1 M(t) \int_{t_0}^t (t-s)^{\alpha-1}s^{-\alpha(1-\beta)}ds$$

$$+ C_1 M(t) \int_0^{t_0} \left((t_0 - s)^{\alpha-1} - (t - s)^{\alpha-1} \right) s^{-\alpha(1-\beta)} ds$$

$$+ C_1 M(t) \int_0^{t_0-\varepsilon} (t_0 - s)^{\alpha-1} s^{-\alpha(1-\beta)} ds$$

$$\times \sup_{s \in [0, t-\varepsilon]} \| E_{\alpha,\alpha}(-(t-s)^{\alpha} A) - E_{\alpha,\alpha}(-(t_0-s)^{\alpha} A) \|$$

$$+ 2C_1 M(t) \int_{t_0-\varepsilon}^{t_0} (t_0 - s)^{\alpha-1} s^{-\alpha(1-\beta)} ds \to 0, \quad \text{as } t \to t_0, \ \varepsilon \to 0$$

by previous discussion.

Further, we consider the function $E_\alpha(-t^\alpha A)a$. It is obvious by Lemma 2.6 that

$$|A^\gamma E_\alpha(-t^\alpha A)a|_q \le C_1 t^{-\alpha(\gamma-\beta)} |A^\beta a|_q = C_1 t^{-\alpha(\gamma-\beta)} |a|_{H^{\beta,q}},$$

$$\lim_{t \to 0} t^{\alpha(\gamma-\beta)} |A^\gamma E_\alpha(-t^\alpha A)a|_q = \lim_{t \to 0} t^{\alpha(\gamma-\beta)} |E_\alpha(-t^\alpha A)a|_{H^{\gamma,q}} - 0.$$

Step 2. Now we construct the solution by the successive approximation:

$$u_0(t) = E_\alpha(-t^\alpha A)a + \Phi(t),$$

$$u_{n+1}(t) = u_0(t) + \mathcal{G}(u_n, u_n)(t), \ n = 0, 1, 2\ldots \ . \tag{2.11}$$

Making use of above results, we know that

$$K_n(t) := \sup_{s \in (0,t]} s^{\alpha(\gamma-\beta)} |A^\gamma u_n(s)|_q$$

are continuous and increasing functions on $[0, T]$ with $K_n(0) = 0$. Furthermore, in virtue of (2.9) and (2.11), $K_n(t)$ fulfills the following inequality

$$K_{n+1}(t) \le K_0(t) + MC_1 B(\alpha(1-\gamma), 1 - 2\alpha(\gamma-\beta)) K_n^2(t). \tag{2.12}$$

For $K_0(0) = 0$, we choose a $T > 0$ such that

$$4MC_1 B(\alpha(1-\gamma), 1 - 2\alpha(\gamma-\beta)) K_0(T) < 1. \tag{2.13}$$

Then a fundamental consideration of (2.12) ensures that the sequence $\{K_n(T)\}$ is bounded, i.e.,

$$K_n(T) \le \rho(T), \ n = 0, 1, 2, \ldots,$$

where

$$\rho(t) = \frac{1 - \sqrt{1 - 4MC_1 B(\alpha(1-\gamma), 1 - 2\alpha(\gamma-\beta)) K_0(t)}}{2MC_1 B(\alpha(1-\gamma), 1 - 2\alpha(\gamma-\beta))}.$$

Analogously, for any $t \in (0, T]$, $K_n(t) \le \rho(t)$ holds. In the same way we note that $\rho(t) \le 2K_0(t)$.

Let us consider the equality

$$w_{n+1}(t) = \int_0^t (t - s)^{\alpha-1} E_{\alpha,\alpha}(-(t-s)^{\alpha} A)$$

$$\times [F(u_{n+1}(s), u_{n+1}(s)) - F(u_n(s), u_n(s))] ds,$$

where $w_n = u_{n+1} - u_n$, $n = 0, 1, ...$, and $t \in (0, T]$. Writing

$$W_n(t) := \sup_{s \in (0,t]} s^{\alpha(\gamma - \beta)} |A^\gamma w_n(s)|_q.$$

On account of (2.6), we have

$$|F(u_{n+1}(s), u_{n+1}(s)) - F(u_n(s), u_n(s))|_q \leq M(K_{n+1}(s) + K_n(s)) W_n(s) s^{-2\alpha(\gamma - \beta)},$$

which follows from Step 2 in Theorem 2.1 that

$$t^{\alpha(\gamma - \beta)} |A^\gamma w_{n+1}(t)|_q \leq 2MC_1 B(\alpha(1 - \gamma), 1 - \alpha(1 - \beta)) \rho(t) W_n(t).$$

This inequality gives

$$\begin{aligned}
W_{n+1}(T) &\leq 2MC_1 B(\alpha(1 - \gamma), 1 - 2\alpha(\gamma - \beta)) \rho(T) W_n(T) \\
&\leq 4MC_1 B(\alpha(1 - \gamma), 1 - 2\alpha(\gamma - \beta)) K_0(T) W_n(T).
\end{aligned} \tag{2.14}$$

According to (2.13) and (2.14), it is easy to see that

$$\lim_{n \to \infty} \frac{W_{n+1}(T)}{W_n(T)} < 4MC_1 K_0(T) B(\alpha(1 - \gamma), 1 - 2\alpha(\gamma - \beta)) < 1,$$

thus the series $\sum_{n=0}^{\infty} W_n(T)$ converges. It shows that the series $\sum_{n=0}^{\infty} t^{\alpha(\gamma - \beta)} A^\gamma w_n(t)$ converges uniformly for $t \in (0, T]$, therefore, the sequence $\{t^{\alpha(\gamma - \beta)} A^\gamma u_n(t)\}$ converges uniformly in $(0, T]$. This implies that

$$\lim_{n \to \infty} u_n(t) = u(t) \in D(A^\gamma)$$

and

$$\lim_{n \to \infty} t^{\alpha(\gamma - \beta)} A^\gamma u_n(t) = t^{\alpha(\gamma - \beta)} A^\gamma u(t) \quad \text{uniformly},$$

since $A^{-\gamma}$ is bounded and A^γ is closed. Accordingly, the function $K(t) = \sup_{s \in (0,t]} s^{\alpha(\gamma - \beta)} |A^\gamma u(s)|_q$ also satisfies

$$K(t) \leq \rho(t) \leq 2K_0(t), \quad t \in (0, T] \tag{2.15}$$

and

$$\begin{aligned}
\varsigma_n &:= \sup_{s \in (0,T]} s^{2\alpha(\gamma - \beta)} |F(u_n(s), u_n(s)) - F(u(s), u(s))|_q \\
&\leq M(K_n(T) + K(T)) \sup_{s \in (0,T]} s^{\alpha(\gamma - \beta)} |A^\gamma(u_n(s) - u(s))|_q \\
&\to 0, \quad \text{as } n \to \infty.
\end{aligned}$$

Finally, it remains to verify that u is a mild solution of problem (2.2) in $[0, T]$. Since

$$\begin{aligned}
|\mathcal{G}(u_n, u_n)(t) - \mathcal{G}(u, u)(t)|_q &\leq C_1 \int_0^t (t - s)^{\alpha - 1} \varsigma_n s^{-2\alpha(\gamma - \beta)} ds \\
&= C_1 B(\alpha, 1 - 2\alpha(\gamma - \beta)) t^{\alpha\beta} \varsigma_n \\
&\to 0, \quad \text{as } n \to \infty,
\end{aligned}$$

we have $\mathcal{G}(u_n, u_n)(t) \to \mathcal{G}(u, u)(t)$. Take the limit on both sides of (2.11), we derive

$$u(t) = u_0(t) + \mathcal{G}(u, u)(t). \tag{2.16}$$

Let $u(0) = a$, we find that (2.16) holds for $t \in [0, T]$ and $u \in C([0, T]; J_q)$. Moreover, the uniform convergence of $t^{\alpha(\gamma-\beta)} A^\gamma u_n(t)$ to $t^{\alpha(\gamma-\beta)} A^\gamma u(t)$ derives the continuity of $A^\gamma u(t)$ on $(0, T]$. From (2.15) and $K_0(0) = 0$, we get that $|A^\gamma u(t)|_q = o(t^{-\alpha(\gamma-\beta)})$ obviously.

Step 3. We prove that the mild solution is unique.

Suppose that u and v are mild solutions of problem (2.2). Let $w = u - v$, we consider the equality

$$w(t) = \int_0^t (t-s)^{\alpha-1} E_{\alpha,\alpha}(-(t-s)^\alpha A)[F(u(s), u(s)) - F(v(s), v(s))]ds.$$

Introducing the function

$$\widetilde{K}(t) := \max\{ \sup_{s \in (0,t]} s^{\alpha(\gamma-\beta)}|A^\gamma u(s)|_q, \sup_{s \in (0,t]} s^{\alpha(\gamma-\beta)}|A^\gamma v(s)|_q\}.$$

By (2.6) and Lemma 2.6, we get

$$|A^\gamma w(t)|_q \le 2MC_1 \widetilde{K}(t) \int_0^t (t-s)^{\alpha(1-\gamma)-1} s^{-\alpha(\gamma-\beta))} |A^\gamma w(s)|_q ds.$$

The Gronwall's inequality shows that $A^\gamma w(t) = 0$ for $t \in (0, T]$. This implies that $w(t) = u(t) - v(t) \equiv 0$ for $t \in [0, T]$. Therefore the mild solution is unique. $\qquad\square$

2.2.4 *Regularity*

In this subsection, we consider the regularity of a solution u which satisfies problem (2.2). Throughout this part we assume that:

(f$_1$) $Pf(t)$ is Hölder continuous with an exponent $\vartheta \in (0, \alpha(1-\gamma)]$, that is, there exists a positive constant $L > 0$ such that

$$|Pf(t) - Pf(s)|_q \le L|t-s|^\vartheta, \quad \text{for all } 0 < t, \, s \le T.$$

Definition 2.3. A function $u : [0, T] \to J_q$ is called a classical solution of problem (2.2), if $u \in C([0, T]; J_q)$ with ${}^C_0 D^\alpha_t u(t) \in C((0, T]; J_q)$, which takes values in $D(A)$ and satisfies (2.2) for all $t \in (0, T]$.

Lemma 2.7. *Let* (f$_1$) *be satisfied. If*

$$\Phi_1(t) := \int_0^t (t-s)^{\alpha-1} E_{\alpha,\alpha}(-(t-s)^\alpha A)(Pf(s) - Pf(t))ds, \quad \text{for } t \in (0, T],$$

then $\Phi_1(t) \in D(A)$ *and* $A\Phi_1(t) \in C^\vartheta([0, T]; J_q)$.

Proof. For fixed $t \in (0, T]$,

$$t^{\alpha-1} A E_{\alpha,\alpha}(-t^\alpha A)$$

$$= \frac{1}{2\pi i} \int_{\Gamma_\theta} t^{\alpha-1} A E_{\alpha,\alpha}(-\mu t^\alpha)(\mu I + A)^{-1} d\mu$$

$$= \frac{1}{2\pi i} \int_{\Gamma_\theta} t^{\alpha-1} E_{\alpha,\alpha}(-\mu t^\alpha) - \frac{1}{2\pi i} \int_{\Gamma_\theta} t^{\alpha-1} \mu E_{\alpha,\alpha}(-\mu t^\alpha)(\mu I + A)^{-1} d\mu$$

$$= \frac{1}{2\pi i} \int_{\Gamma_\theta} -t^{\alpha-1} E_{\alpha,\alpha}(\xi) \frac{1}{t^\alpha} - \frac{1}{2\pi i} \int_{\Gamma_\theta} t^{\alpha-1} E_{\alpha,\alpha}(\xi)(-\frac{\xi}{t^\alpha} I + A)^{-1} \frac{1}{t^\alpha} d\xi.$$

In view of $\|\mu I + A\|^{-1} \leq \frac{C}{|\mu|}$, we derive that there exists a positive constant C_α such that

$$\|t^{\alpha-1} A E_{\alpha,\alpha}(-t^\alpha A)\| \leq C_\alpha t^{-1}. \tag{2.17}$$

From (2.17) and (f_1), we have

$$(t-s)^{\alpha-1} |A E_{\alpha,\alpha}(-(t-s)^\alpha A)(Pf(s) - Pf(t))|_q$$
$$\leq C_\alpha (t-s)^{-1} |Pf(s) - Pf(t)|_q \tag{2.18}$$
$$\leq C_\alpha L (t-s)^{\vartheta-1} \in L^1(0, T; J_q),$$

then

$$|A\Phi_1(t)|_q \leq \int_0^t (t-s)^{\alpha-1} |A E_{\alpha,\alpha}(-(t-s)^\alpha A)(Pf(s) - Pf(t))|_q ds$$

$$\leq C_\alpha L \int_0^t (t-s)^{\vartheta-1} ds = \frac{C_\alpha L}{\vartheta} t^\vartheta < \infty.$$

By the closeness of A, we obtain $\Phi_1(t) \in D(A)$.

We need to show that $A\Phi_1(t)$ is Hölder continuous. Since

$$\frac{d}{dt}\left(t^{\alpha-1} E_{\alpha,\alpha}(-\mu t^\alpha)\right) = t^{\alpha-2} E_{\alpha,\alpha-1}(-\mu t^\alpha),$$

then

$$\frac{d}{dt}\left(t^{\alpha-1} A E_{\alpha,\alpha}(-t^\alpha A)\right)$$

$$= \frac{1}{2\pi i} \int_{\Gamma_\theta} t^{\alpha-2} E_{\alpha,\alpha-1}(-\mu t^\alpha) A(\mu I + A)^{-1} d\mu$$

$$= \frac{1}{2\pi i} \int_{\Gamma_\theta} t^{\alpha-2} E_{\alpha,\alpha-1}(-\mu t^\alpha) d\mu - \frac{1}{2\pi i} \int_{\Gamma_\theta} t^{\alpha-2} \mu E_{\alpha,\alpha-1}(-\mu t^\alpha)(\mu I + A)^{-1} d\mu$$

$$= \frac{1}{2\pi i} \int_{\Gamma'_\theta} -t^{\alpha-2} E_{\alpha,\alpha-1}(\xi) \frac{1}{t^\alpha} d\xi - \frac{1}{2\pi i} \int_{\Gamma'_\theta} t^{\alpha-2} E_{\alpha,\alpha-1}(\xi) \frac{\xi}{t^\alpha}(-\frac{\xi}{t^\alpha} I + A)^{-1} \frac{1}{t^\alpha} d\xi.$$

In view of $\|(\mu I + A)^{-1}\| \leq \frac{C}{|\mu|}$, we derive

$$\left\|\frac{d}{dt}\left(t^{\alpha-1} A E_{\alpha,\alpha}(-t^\alpha A)\right)\right\| \leq C_\alpha t^{-2}, \ 0 < t \leq T.$$

By the mean value theorem, for every $0 < s < t \le T$, we have

$$\|t^{\alpha-1}AE_{\alpha,\alpha}(-t^{\alpha}A) - s^{\alpha-1}AE_{\alpha,\alpha}(-s^{\alpha}A)\|$$

$$= \left\| \int_s^t \frac{d}{d\tau}\left(\tau^{\alpha-1}AE_{\alpha,\alpha}(-\tau^{\alpha}A)\right)d\tau \right\|$$

$$\le \int_s^t \left\| \frac{d}{d\tau}\left(\tau^{\alpha-1}AE_{\alpha,\alpha}(-\tau^{\alpha}A)\right) \right\| d\tau \tag{2.19}$$

$$\le C_{\alpha}\int_s^t \tau^{-2}d\tau = C_{\alpha}\left(s^{-1} - t^{-1}\right).$$

Let $h > 0$ be such that $0 < t < t + h \le T$, then

$$A\Phi_1(t+h) - A\Phi_1(t)$$

$$= \int_0^t \left((t+h-s)^{\alpha-1}AE_{\alpha,\alpha}(-(t+h-s)^{\alpha}A)\right.$$

$$\left. - (t-s)^{\alpha-1}AE_{\alpha,\alpha}(-(t-s)^{\alpha}A)\right)\left(Pf(s) - Pf(t)\right)ds$$

$$+ \int_0^t (t+h-s)^{\alpha-1}AE_{\alpha,\alpha}(-(t+h-s)^{\alpha}A)\left(Pf(t) - Pf(t+h)\right)ds \tag{2.20}$$

$$+ \int_t^{t+h} (t+h-s)^{\alpha-1}AE_{\alpha,\alpha}(-(t+h-s)^{\alpha}A)\left(Pf(s) - Pf(t+h)\right)ds$$

$$=: h_1(t) + h_2(t) + h_3(t).$$

We estimate each of the three terms separately. For $h_1(t)$, from (2.19) and (f_1), we have

$$|h_1(t)|_q \le \int_0^t \|(t+h-s)^{\alpha-1}AE_{\alpha,\alpha}(-(t+h-s)^{\alpha}A)$$

$$- (t-s)^{\alpha-1}AE_{\alpha,\alpha}(-(t-s)^{\alpha}A)\| \|Pf(s) - Pf(t)\|_q ds$$

$$\le C_{\alpha}Lh\int_0^t (t+h-s)^{-1}(t-s)^{\vartheta-1}ds$$

$$= C_{\alpha}Lh\int_0^t (s+h)^{-1}s^{\vartheta-1}ds \tag{2.21}$$

$$\le C_{\alpha}L\int_0^h \frac{h}{s+h}s^{\vartheta-1}ds + C_{\alpha}Lh\int_h^{\infty} \frac{s}{s+h}s^{\vartheta-2}ds$$

$$\le \frac{1}{\vartheta(1-\vartheta)}C_{\alpha}Lh^{\vartheta}.$$

For $h_2(t)$, we use Lemma 2.6 and (f_1),

$$|h_2(t)|_q \le \int_0^t (t+h-s)^{\alpha-1}|AE_{\alpha,\alpha}(-(t+h-s)^{\alpha}A)\left(Pf(t) - Pf(t+h)\right)|_q ds$$

$$\le C_1\int_0^t (t+h-s)^{-1}|Pf(t) - Pf(t+h)|_q ds$$

$$\le C_1Lh^{\vartheta}\int_0^t (t+h-s)^{-1}ds$$

$$= C_1L(\ln(t+h) - \ln h)h^{\vartheta}. \tag{2.22}$$

Furthermore, for $h_3(t)$, by Lemma 2.6 and (f_1), we have

$$|h_3(t)|_q \leq \int_t^{t+h} (t+h-s)^{\alpha-1} |AE_{\alpha,\alpha}(-(t+h-s)^{\alpha}A)(Pf(s)-Pf(t+h))|_q ds$$

$$\leq C_1 \int_t^{t+h} (t+h-s)^{-1} |Pf(s)-Pf(t+h)|_q ds$$

$$\leq C_1 L \int_t^{t+h} (t+h-s)^{\vartheta-1} ds = C_1 L \frac{h^{\vartheta}}{\vartheta}.$$

$$(2.23)$$

Combining (2.21), (2.22) with (2.23), we deduce that $A\Phi_1(t)$ is Hölder continuous.

$\square$

Theorem 2.4. *Let the assumptions of Theorem 2.3 be satisfied. If (f_1) holds, then for every $a \in D(A)$, the mild solution of (2.2) is a classical one.*

Proof. For $a \in D(A)$. Then Lemma 2.2(ii) ensures that $u(t) = E_\alpha(-t^\alpha A)a$ $(t > 0)$ is a classical solution to the following problem

$$\begin{cases} {}^C_0 D_t^\alpha u = -Au, & t > 0, \\ u(0) = a. \end{cases}$$

Step 1. We verify that

$$\Phi(t) = \int_0^t (t-s)^{\alpha-1} E_{\alpha,\alpha}(-(t-s)^\alpha A) Pf(s) ds$$

is a classical solution to the problem

$$\begin{cases} {}^C_0 D_t^\alpha u = -Au + Pf(t), & t > 0, \\ u(0) = 0. \end{cases}$$

It follows from Theorem 2.3 that $\Phi \in C([0,T]; J_q)$. We rewrite $\Phi(t) = \Phi_1(t) + \Phi_2(t)$, where

$$\Phi_1(t) = \int_0^t (t-s)^{\alpha-1} E_{\alpha,\alpha}(-(t-s)^\alpha A)(Pf(s) - Pf(t)) ds,$$

$$\Phi_2(t) = \int_0^t (t-s)^{\alpha-1} E_{\alpha,\alpha}(-(t-s)^\alpha A) Pf(t) ds.$$

According to Lemma 2.7, we know that $\Phi_1(t) \in D(A)$. To prove the same conclusion for $\Phi_2(t)$. By Lemma 2.2(iii), we notice that

$$A\Phi_2(t) = Pf(t) - E_\alpha(-t^\alpha A) Pf(t).$$

Since (f_1) holds, it follows that

$$|A\Phi_2(t)|_q \leq (1 + C_1)|Pf(t)|_q,$$

thus

$$\Phi_2(t) \in D(A) \text{ for } t \in (0,T] \text{ and } A\Phi_2(t) \in C^\vartheta((0,T]; J_q). \tag{2.24}$$

Next, we prove ${}_0^C D_t^\alpha \Phi \in C((0,T]; J_q)$. In view of Lemma 2.2(iv) and $\Phi(0) = 0$, we have

$$
{}_0^C D_t^\alpha \Phi(t) = \frac{d}{dt} \left({}_0 D_t^{\alpha-1} \Phi(t) \right) = \frac{d}{dt} (E_\alpha(-t^\alpha A) * Pf).
$$

It remains to prove that $E_\alpha(-t^\alpha A) * Pf$ is continuously differentiable in J_q. Let $0 < h \le T - t$, one derives the following:

$$
\frac{1}{h} \left(E_\alpha(-(t+h)^\alpha A) * Pf - E_\alpha(-t^\alpha A) * Pf \right)
$$

$$
= \int_0^t \frac{1}{h} \left(E_\alpha(-(t+h-s)^\alpha A) Pf(s) - E_\alpha(-(t-s)^\alpha A) Pf(s) \right) ds
$$

$$
+ \frac{1}{h} \int_t^{t+h} E_\alpha(-(t+h-s)^\alpha A) Pf(s) ds.
$$

Notice that for $0 < \theta < 1$,

$$
\left| \int_0^t \frac{1}{h} E_\alpha(-(t+h-s)^\alpha A) Pf(s) - E_\alpha(-(t-s)^\alpha A) Pf(s) ds \right|_q
$$

$$
= \left| \int_0^t (t-s+\theta h)^{\alpha-1} A E_{\alpha,\alpha}(-(t+\theta h - s)^\alpha A) Pf(s) ds \right|_q
$$

$$
\le \left| \int_0^{t+\theta h} (t-s+\theta h)^{\alpha-1} A E_{\alpha,\alpha}(-(t+\theta h - s)^\alpha A) Pf(s) ds \right|_q.
$$

For any $t \in (0,T]$, due to $\Phi(t) \in D(A)$, one know that

$$
\left| \int_0^t (t-s)^{\alpha-1} A E_{\alpha,\alpha}(-(t-s)^\alpha A) Pf(s) ds \right|_q < \infty,
$$

then using the dominated convergence theorem, we find

$$
\lim_{h \to 0} \int_0^t \frac{1}{h} \left(E_\alpha(-(t+h-s)^\alpha A) Pf(s) - E_\alpha(-(t-s)^\alpha A) Pf(s) \right) ds
$$

$$
= - \int_0^t (t-s)^{\alpha-1} A E_{\alpha,\alpha}(-(t-s)^\alpha A) Pf(s) ds
$$

$$
= - A\Phi(t).
$$

On the other hand,

$$
\frac{1}{h} \int_t^{t+h} E_\alpha(-(t+h-s)^\alpha A) Pf(s) ds
$$

$$
= \frac{1}{h} \int_0^h E_\alpha(-s^\alpha A) Pf(t+h-s) ds
$$

$$
= \frac{1}{h} \int_0^h E_\alpha(-s^\alpha A) \left(Pf(t+h-s) - Pf(t-s) \right) ds
$$

$$
+ \frac{1}{h} \int_0^h E_\alpha(-s^\alpha A) \left(Pf(t-s) - Pf(t) \right) ds + \frac{1}{h} \int_0^h E_\alpha(-s^\alpha A) Pf(t) ds.
$$

From Lemma 2.1 and (f_1), we have

$$\left|\frac{1}{h}\int_0^h E_\alpha(-s^\alpha A)\big(Pf(t+h-s)-Pf(t-s)\big)ds\right|_q \le C_1 Lh^\vartheta,$$

$$\left|\frac{1}{h}\int_0^h E_\alpha(-s^\alpha A)\big(Pf(t-s)-Pf(t)\big)ds\right|_q \le C_1 L\frac{h^\vartheta}{\vartheta+1}.$$

Also Lemma 2.2(i) gives that $\lim_{h\to 0}\frac{1}{h}\int_0^h E_\alpha(s^\alpha A)Pf(t)ds = Pf(t)$. Hence

$$\lim_{h\to 0}\frac{1}{h}\int_t^{t+h} E_\alpha((t+h-s)^\alpha A)Pf(s)ds = Pf(t).$$

We deduce that $E_\alpha(t^\alpha A)*Pf$ is differentiable at t_+ and $\frac{d}{dt}\big(E_\alpha(-t^\alpha A)*Pf\big)_+ = -A\Phi(t)+Pf(t)$. Similarly, $E_\alpha(t^\alpha A)*Pf$ is differentiable at t_- and $\frac{d}{dt}\big(E_\alpha(-t^\alpha A)*Pf\big)_- = -A\Phi(t)+Pf(t)$.

We show that $A\Phi = A\Phi_1+A\Phi_2 \in C((0,T];J_q)$. In fact, it is clear that $A\Phi_2(t) = Pf(t)-E_\alpha(t^\alpha A)Pf(t)$ due to Lemma 2.2(iii), which is continuous in view of Lemma 2.1. Furthermore, according to Lemma 2.7, we know that $A\Phi_1(t)$ is also continuous. Consequently, $^C_0 D_t^\alpha \Phi \in C((0,T];J_q)$.

Step 2. Let u be the mild solution of (2.2). To prove that $F(u,u) \in C^\vartheta((0,T];J_q)$, in view of (2.6), we have to verify that $A^\gamma u$ is Hölder continuous in J_q. Take $h>0$ such that $0<t<t+h$.

Denote $\varphi(t) := E_\alpha(-t^\alpha A)a$, by Lemmas 2.2(iii) and 2.6, then

$$|A^\gamma \varphi(t+h)-A^\gamma\varphi(t)|_q = \left|\int_t^{t+h} -s^{\alpha-1}A^{\gamma+1}E_{\alpha,\alpha}(-s^\alpha A)ads\right|_q$$

$$\le \int_t^{t+h} s^{\alpha-1}|A^\gamma E_{\alpha,\alpha}(-s^\alpha A)Aa|_q ds$$

$$\le C_1\int_t^{t+h} s^{\alpha(1-\gamma)-1}ds|Aa|_q$$

$$= \frac{C_1|a|_{D(A)}}{\alpha(1-\gamma)}\big((t+h)^{\alpha(1-\gamma)}-t^{\alpha(1-\gamma)}\big)$$

$$\le \frac{C_1|a|_{D(A)}}{\alpha(1-\gamma)}h^{\alpha(1-\gamma)}.$$

Thus, $A^\gamma\varphi \in C^\vartheta((0,T];J_q)$.

For every small $\varepsilon>0$, take h such that $\varepsilon \le t<t+h\le T$, since

$$|A^\gamma\Phi(t+h)-A^\gamma\Phi(t)|_q$$

$$\le \left|\int_t^{t+h}(t+h-s)^{\alpha-1}A^\gamma E_{\alpha,\alpha}(-(t+h-s)^\alpha A)Pf(s)ds\right|_q$$

$$+ \left|\int_0^t A^\gamma\big((t+h-s)^{\alpha-1}E_{\alpha,\alpha}(-(t+h-s)^\alpha A)\right.$$

$$\left. -(t-s)^{\alpha-1}E_{\alpha,\alpha}(-(t-s)^\alpha A)\big)Pf(s)ds\right|_q$$

$$=: \phi_1(t)+\phi_2(t).$$

Applying Lemma 2.6 and (f), we get

$$\phi_1(t) \le C_1 \int_t^{t+h} (t+h-s)^{\alpha(1-\gamma)-1} |Pf(s)|_q ds$$

$$\le C_1 M(t) \int_t^{t+h} (t+h-s)^{\alpha(1-\gamma)-1} s^{-\alpha(1-\beta)} ds$$

$$\le M(t) \frac{C_1}{\alpha(1-\gamma)} h^{\alpha(1-\gamma)} t^{-\alpha(1-\beta)}$$

$$\le M(t) \frac{C_1}{\alpha(1-\gamma)} h^{\alpha(1-\gamma)} \varepsilon^{-\alpha(1-\beta)}.$$

To estimate ϕ_2, we give the equations

$$\frac{d}{dt}\left(t^{\alpha-1} A^\gamma E_{\alpha,\alpha}(-t^\alpha A)\right) = \frac{1}{2\pi i} \int_\Gamma \mu^\gamma t^{\alpha-2} E_{\alpha,\alpha-1}(-\mu t^\alpha)(\mu I + A)^{-1} d\mu$$

$$= \frac{1}{2\pi i} \int_{\Gamma'} -\left(-\frac{\xi}{t^\alpha}\right)^\gamma t^{\alpha-2} E_{\alpha,\alpha-1}(\xi)\left(-\frac{\xi}{t^\alpha}I + A\right)^{-1} \frac{1}{t^\alpha} d\xi,$$

this yields that $\left\|\frac{d}{dt}\left(t^{\alpha-1} A^\gamma E_{\alpha,\alpha}(-t^\alpha A)\right)\right\| \le C_\alpha t^{\alpha(1-\gamma)-2}$. The mean value theorem shows

$$\|t^{\alpha-1} A^\gamma E_{\alpha,\alpha}(-t^\alpha A) - s^{\alpha-1} A^\gamma E_{\alpha,\alpha}(-s^\alpha A)\|$$

$$\le \int_s^t \left\|\frac{d}{d\tau}\left(\tau^{\alpha-1} A^\gamma E_{\alpha,\alpha}(-\tau^\alpha A)\right)\right\| d\tau$$

$$\le C_\alpha \int_s^t \tau^{\alpha(1-\gamma)-2} d\tau = \frac{C_\alpha}{1-\alpha(1-\gamma)}\left(s^{\alpha(1-\gamma)-1} - t^{\alpha(1-\gamma)-1}\right), \qquad (2.25)$$

thus

$$\phi_2(t) \le \int_0^t |A^\gamma\big((t+h-s)^{\alpha-1} E_{\alpha,\alpha}(-(t+h-s)^\alpha A)$$

$$- (t-s)^{\alpha-1} E_{\alpha,\alpha}(-(t-s)^\alpha A)\big) Pf(s)|_q ds$$

$$\le \frac{C_\alpha}{1-\alpha(1-\gamma)} \int_0^t \left((t-s)^{\alpha(1-\gamma)-1} - (t+h-s)^{\alpha(1-\gamma)-1}\right) |Pf(s)|_q ds$$

$$\le \frac{C_\alpha}{1-\alpha(1-\gamma)} M(t)\left(\int_0^t (t-s)^{\alpha(1-\gamma)-1} s^{-\alpha(1-\beta)} ds\right.$$

$$\left. - \int_0^{t+h} (t-s+h)^{\alpha(1-\gamma)-1} s^{-\alpha(1-\beta)} ds\right)$$

$$+ \frac{C_\alpha}{1-\alpha(1-\gamma)} M(t) \int_t^{t+h} (t-s+h)^{\alpha(1-\gamma)-1} s^{-\alpha(1-\beta)} ds$$

$$\le \frac{C_\alpha}{1-\alpha(1-\gamma)} M(t)\left(t^{\alpha(\beta-\gamma)} - (t+h)^{\alpha(\beta-\gamma)}\right) B(\alpha(1-\gamma), 1-\alpha(1-\beta))$$

$$+ \frac{C_\alpha}{\alpha(1-\gamma)(1-\alpha(1-\gamma))} M(t) h^{\alpha(1-\gamma)} t^{-\alpha(1-\beta)}$$

$$\leq \frac{C_\alpha}{1 - \alpha(1 - \gamma)} M(t) h^{\alpha(\gamma - \beta)} [\varepsilon(\varepsilon + h)]^{\alpha(\beta - \gamma)} B(\alpha(1 - \gamma), 1 - \alpha(1 - \beta))$$
$$+ \frac{C_\alpha}{\alpha(1 - \gamma)(1 - \alpha(1 - \gamma))} M(t) h^{\alpha(1 - \gamma)} \varepsilon^{-\alpha(1 - \beta)},$$

which ensures that $A^\gamma \Phi \in C^\vartheta([\varepsilon, T]; J_q)$. Therefore $A^\gamma \Phi \in C^\vartheta((0, T]; J_q)$ due to arbitrary ε.

Recall

$$\Psi(t) = \int_0^t (t - s)^{\alpha - 1} E_{\alpha, \alpha}(-(t - s)^\alpha A) F(u(s), u(s)) ds.$$

Since

$$|F(u(s), u(s))|_q \leq M K^2(t) s^{-2\alpha(\gamma - \beta)},$$

where $K(t) := \sup_{s \in [0, t]} s^{\alpha(\gamma - \beta)} |u(s)|_{H^{\gamma, q}}$ is continuous and bounded in $(0, T]$. A similar argument enable us to give the Hölder continuity of $A^\gamma \Psi$ in $C^\vartheta((0, T]; J_q)$. Therefore, we have $A^\gamma u(t) = A^\gamma \varphi(t) + A^\gamma \Phi(t) + A^\gamma \Psi(t) \in C^\vartheta((0, T]; J_q)$.

Since $F(u, u) \in C^\vartheta((0, T]; J_q)$ is proved, according to Step 2, this yields that ${}^C_0 D^\alpha_t \Psi \in C((0, T]; J_q)$, $A\Psi \in C((0, T]; J_q)$ and ${}^C_0 D^\alpha_t \Psi = -A\Psi + F(u, u)$. In this way we obtain that ${}^C_0 D^\alpha_t u \in C((0, T]; J_q)$, $Au \in C((0, T]; J_q)$ and ${}^C_0 D^\alpha_t u = -Au + F(u, u) + Pf$, we conclude that u is a classical solution. $\qquad\square$

Theorem 2.5. *Assume that (f_1) holds and $a \in H^{\beta, q}$. If u is a classical solution of the equation (2.2), then for every $\varepsilon > 0$, $Au \in C^{\min\{\alpha\beta, \alpha(1 - \beta), \vartheta\}}([\varepsilon, T]; J_q)$ and ${}^C_0 D^\alpha_t u \in C^{\min\{\alpha\beta, \alpha(1 - \beta), \vartheta\}}([\varepsilon, T]; J_q)$.*

Proof. If u is a classical solution of (2.2), then $u(t) = \varphi(t) + \Phi(t) + \Psi(t)$. It remains to show that $A\varphi \in C^{\min\{\alpha\beta, \alpha(1 - \beta)\}}([\varepsilon, T]; J_q)$ for every $\varepsilon > 0$, it suffices to prove that $A\varphi \in C^{\min\{\alpha\beta, \alpha(1 - \beta)\}}([\varepsilon, T]; J_q)$. In fact, take h such that $\varepsilon \leq t < t + h \leq T$, by Lemma 2.2(iv) and (2.25), we have

$$|A\varphi(t + h) - A\varphi(t)|_q$$
$$\leq \frac{1}{\Gamma(1 - \alpha)} \left| \int_t^{t+h} s^{-\alpha}(t + h - s)^{\alpha - 1} A^{1 - \beta} E_{\alpha, \alpha}(-(t + h - s)^\alpha A) A^\beta a ds \right|_q$$
$$+ \frac{1}{\Gamma(1 - \alpha)} \left| \int_0^t s^{-\alpha}[(t + h - s)^{\alpha - 1} A^{1 - \beta} E_{\alpha, \alpha}(-(t + h - s)^\alpha A) \right.$$
$$\left. - (t - s)^{\alpha - 1} A^{1 - \beta} E_{\alpha, \alpha}(-(t - s)^\alpha A)] A^\beta a ds \right|_q$$
$$\leq \frac{C_1 |a|_{H^{\beta, q}}}{\Gamma(1 - \alpha)} \int_t^{t+h} s^{-\alpha}(t + h - s)^{\alpha\beta - 1} ds$$
$$+ \frac{C_\alpha |a|_{H^{\beta, q}}}{(1 - \alpha\beta)\Gamma(1 - \alpha)} \int_0^t s^{-\alpha}[(t - s)^{\alpha\beta - 1} - (t + h - s)^{\alpha\beta - 1}] ds.$$

Using the similar arguments as we estimated the two functions $\phi_1(t)$ and $\phi_2(t)$, one can show that

$$\int_t^{t+h} s^{-\alpha}(t+h-s)^{\alpha\beta-1}ds \leq \frac{1}{\alpha\beta}h^{\alpha\beta}\varepsilon^{-\alpha},$$

$$\int_0^t s^{-\alpha}[(t-s)^{\alpha\beta-1}-(t+h-s)^{\alpha\beta-1}]ds \leq h^{\alpha(1-\beta)}[\varepsilon(\varepsilon+h)]^{-\alpha(1-\beta)}B(1-\alpha,\alpha\beta)$$

$$+\frac{1}{\alpha\beta}h^{\alpha\beta}\varepsilon^{-\alpha}.$$

Similar to Lemma 2.7, we write $\Phi(t)$ as

$$\Phi(t) = \Phi_1(t) + \Phi_2(t) = \int_0^t (t-s)^{\alpha-1}E_{\alpha,\alpha}(-(t-s)^\alpha A)\big(Pf(s)-Pf(t)\big)ds$$

$$+\int_0^t (t-s)^{\alpha-1}E_{\alpha,\alpha}(-(t-s)^\alpha A)Pf(t)ds,$$

for $t \in (0,T]$. It follows from Lemma 2.7 and (2.24) that $A\Phi_1(t) \in C^\vartheta([0,T];J_q)$ and $A\Phi_2(t) \in C^\vartheta([\varepsilon,T];J_q)$, respectively.

Since $F(u,u) \in C^\vartheta([\varepsilon,T];J_q)$, the result related to the function $\Psi(t)$ is proved by similar argument, which means that $A\Psi \in C^\vartheta([\varepsilon,T];J_q)$. Therefore $Au \in C^\vartheta([\varepsilon,T];J_q)$ and $_0^CD_t^\alpha u = Au + F(u,u) + Pf \in C^\vartheta([\varepsilon,T];J_q)$. The proof is completed.   □

2.3   Existence and Hölder Continuity

In this section, we discuss the following time-fractional Navier-Stokes equations:

$$\begin{cases} \partial_t^\alpha u(t,x) - \Delta u(t,x) + u(t,x)\cdot\nabla u(t,x) \\ \quad = -\nabla p(t,x) + f(t,x), & (t,x)\in(0,T]\times\Omega, \\ \nabla\cdot u(t,x) = 0, & (t,x)\in(0,T]\times\Omega, \\ u(t,x) = 0, & (t,x)\in(0,T]\times\partial\Omega, \\ u(0,x) = a(x), & x\in\Omega, \end{cases} \quad (2.26)$$

in a bounded set $\Omega \subset \mathbb{R}^N$ ($N \geq 2$) with smooth boundary $\partial\Omega$, where ∂_t^α is the Caputo fractional derivative of order $\alpha \in (0,1)$, $u(t,x)$ represents the velocity field at (t,x), $p(t,x)$ is the pressure, $f(t,x)$ is the external force and $a(x)$ is the initial velocity. Applying the classic approach, the problem of solutions to (2.26) is equivalent to that of the following equations

$$\begin{cases} _0^CD_t^\alpha u = -Au + F(u,u) + Pf, & t\in(0,T], \\ u(0) = a, \end{cases} \quad (2.27)$$

where P is the Helmholtz projector, $F(u,v) = -P(u\cdot\nabla)v$.

The purpose of this section is to investigates the problem of solutions to (2.26). The existence, uniqueness and Hölder continuity of the local mild solutions are established. The section is organized as follows. In Subsection 2.3.1 we introduce some notations, definitions, and preliminary results. Subsection 2.3.2 is concerned with the existence and uniqueness of local mild solutions of problem (2.27) by using the iteration method, then proceed to study Hölder continuity of mild solutions.

2.3.1 Notations and Lemmas

In this subsection, we present notations, definitions, and preliminary facts which are used throughout this section. Let

$H =$ the closure in $(L^r(\Omega))^N$ of $\{u \in (C_0^\infty(\Omega))^N : \nabla \cdot u = 0\}$, $(1 < r < \infty)$.

We denote by P the orthogonal projection $P : (L^r(\Omega))^N \to H$. The Stokes operator A is defined by

$$A : D(A) \subset H \to H, \quad A = -P\Delta, \quad D(A) = H \cap \{v \in (W_r^2(\Omega))^N : v|_{\partial\Omega} = 0\}.$$

Here, $(W_r^2(\Omega))^N$ is a Sobolev space. As such $-A$ generates the bounded analytic semigroup $\{e^{-tA}\}$, see Constantin and Foicas [85]. This allows us to define the fractional powers $A^\beta (\beta \in \mathbb{R})$ in the usual way.

Lemma 2.8. *[138]*

(i) $\|A^\beta e^{-tA}\| \le t^{-\beta}$ *for* $0 < \beta \le e = 2.718..., \ t > 0$.

(ii) $|(e^{-hA} - I)v| \le \frac{1}{\beta} h^\beta |A^\beta v|$ *for* $0 < \beta < 1$, $h > 0$ *and* $v \in D(A^\beta)$.

(iii) *Let* $0 \le \delta < \frac{1}{2} + \frac{N}{2}(1 - \frac{1}{r})$. *Then* $|A^{-\delta} P(u \cdot \nabla)v| \le M|A^\rho u||A^\kappa v|$ *with some constant* $M = M(\delta, r, \rho, \kappa)$, *provided that* $\delta + \rho + \kappa \ge \frac{N}{2r} + \frac{1}{2}, \rho > 0, \kappa > 0, \kappa + \delta > \frac{1}{2}$.

Inspired by Zhou [386, 387], we introduce the following definition of mild solutions to problem (2.27).

Definition 2.4. Let $0 < T \le \infty$. A function $u : [0, T] \to H$ is called a mild solution of problem (2.27), if $u \in C([0, T]; H)$ and

$$u(t) = E_\alpha(-t^\alpha A)a + \int_0^t (t - s)^{\alpha-1}$$
$$\times E_{\alpha,\alpha}(-(t - s)^\alpha A)\big(F(u(s), u(s)) + Pf(s)\big)ds, \quad t \in [0, T], \tag{2.28}$$

where

$$E_\alpha(-t^\alpha A) = \int_0^\infty \mathcal{M}_\alpha(s)e^{-st^\alpha A}ds, \quad E_{\alpha,\alpha}(-t^\alpha A) = \int_0^\infty \alpha s \mathcal{M}_\alpha(s)e^{-st^\alpha A}ds,$$

$$\mathcal{M}_\alpha(\theta) = \sum_{n=0}^\infty \frac{(-\theta)^n}{n!\Gamma(1 - \alpha(1 + n))}.$$

We state some important technical lemmas that contributes to the main results.

Lemma 2.9. *[352] Let* $0 < \alpha < 1$. *Then for* $t > 0$, $E_\alpha(-t^\alpha A)$ *and* $E_{\alpha,\alpha}(-t^\alpha A)$ *are continuous in the uniform operator topology. Moreover, for every* $r > 0$, *the continuity is uniform on* $[r, \infty)$.

Lemma 2.10. *[402] Let $0 < \alpha < 1$ and $\gamma \in [0,1)$. Then there exists a constant $C_\gamma = C(\alpha, \gamma) > 0$ such that*

$$|A^\gamma E_\alpha(-t^\alpha A)v| \le C_\gamma t^{-\alpha\gamma}|v| \text{ and } |A^\gamma E_{\alpha,\alpha}(-t^\alpha A)v| \le C_\gamma t^{-\alpha\gamma}|v|,$$

for all $v \in H$ and $t \in (0,T]$. Furthermore, $\lim_{t\to 0^+} t^{\alpha\gamma}|A^\gamma E_\alpha(-t^\alpha A)v| = 0$.

For convenience, let

$$\Phi(t) = \int_0^t (t-s)^{\alpha-1} E_{\alpha,\alpha}(-(t-s)^\alpha A) P f(s) ds,$$

$$\mathcal{G}(u,v)(t) = \int_0^t (t-s)^{\alpha-1} E_{\alpha,\alpha}(-(t-s)^\alpha A) F(u(s), v(s)) ds.$$

2.3.2   Existence and Uniqueness

This subsection is to show the existence and uniqueness of mild solutions to problem (2.27), this is achieved by iteration methods. Set

$$B_\gamma = B(\alpha(1 - \gamma - \delta), 1 - \alpha(1 - \beta - \delta)).$$

Theorem 2.6. *Fix $\beta < 1$ and choose $\delta > 0$ such that*

$$\frac{N}{2r} - \frac{1}{2} \le \beta, \quad -\beta < \delta < 1 - |\beta|.$$

Suppose that $|A^{-\delta} P f(t)|$ is continuous for $t \in (0,T]$ and $|A^{-\delta} P f(t)| = o(t^{\alpha(\beta+\delta-1)})$ as $t \to 0$. Then for $a \in D(A^\beta)$ there exist $T^* \in (0,T)$ and a unique function $u(t)$ satisfying:

(a) $u \in C([0, T^*]; D(A^\beta))$ and $u(0) = a$;
(b) $u \in C((0, T^*]; D(A^\gamma))$ and $\lim_{t\to 0^+} t^{\alpha(\gamma-\beta)}|A^\gamma u(t)| = 0$ for all γ with $\gamma \in (\beta, 1 - \delta)$;
(c) $u(t) = E_\alpha(-t^\alpha A)a + \int_0^t (t-s)^{\alpha-1} E_{\alpha,\alpha}(-(t-s)^\alpha A)(F(u(s), u(s)) + P f(s)) ds$.

Proof. We construct the solution by means of the successive approximation:

$$u_0(t) = E_\alpha(-t^\alpha A)a + \Phi(t),$$
$$u_{m+1}(t) = u_0(t) + \mathcal{G}(u_m, u_m)(t), \quad m = 0, 1, \dots . \tag{2.29}$$

Step 1. We estimate $|A^\gamma u_m(t)|$ by making use of mathematical induction. For $u_0(t)$, in view of Lemma 2.10, we obtain

$$|A^\gamma u_0(t)| \le |A^\gamma E_\alpha(-t^\alpha A)a| + \int_0^t (t-s)^{\alpha-1}|A^\gamma E_{\alpha,\alpha}(-(t-s)^\alpha A) P f(s)| ds$$

$$\le |A^\gamma E_\alpha(-t^\alpha A)a| + C_{\gamma+\delta}\int_0^t (t-s)^{\alpha(1-\gamma-\delta)-1}|A^{-\delta} P f(s)| ds$$

$$\le |A^\gamma E_\alpha(-t^\alpha A)a| + C_{\gamma+\delta}L\int_0^t (t-s)^{\alpha(1-\gamma-\delta)-1} s^{-\alpha(1-\beta-\delta)} ds$$

$$= K_{\gamma,0} t^{-\alpha(\gamma-\beta)}, \quad \text{for } \beta \le \gamma < 1 - \delta,$$

where

$$L = \sup_{s \in (0,T]} \{s^{\alpha(1-\beta-\delta)}|A^{-\delta}Pf(s)|\},$$

$$K_{\gamma,0} = \sup_{t \in (0,T]} t^{\alpha(\gamma-\beta)}|A^\gamma E_\alpha(-t^\alpha A)a| + B_\gamma C_{\gamma+\delta} L.$$

Suppose that for some $m \geq 0$,

$$|A^\gamma u_m(t)| \leq K_{\gamma,m} t^{-\alpha(\gamma-\beta)}. \tag{2.30}$$

We estimate $|A^\gamma u_{m+1}(t)|$. On the one hand, the inequality $\delta < 1 - |\beta|$ implies the existence of ρ and κ such that

$$\rho + \kappa + \delta = 1 + \beta, \quad \beta < \rho, \kappa < 1 - \delta, \quad \rho > 0, \quad \kappa > 0, \quad \rho + \kappa > \frac{1}{2}. \tag{2.31}$$

Since $\frac{N}{2r} - \frac{1}{2} \leq \beta, N \geq 2$ and $\delta < 1 - |\beta|$, it is easy to know that δ, ρ, κ satisfy the conditions of Lemma 2.8(iii), which together with (2.30) follows that

$$|A^{-\delta}F(u_m(t), u_m(t))| \leq M|A^\rho u_m(t)||A^\kappa u_m(t)| \leq MK_{\rho,m}K_{\kappa,m} t^{-\alpha(1-\beta-\delta)}. \tag{2.32}$$

Therefore, (2.29) and Lemma 2.10 yield that

$$|A^\gamma u_{m+1}(t)|$$
$$\leq |A^\gamma u_0| + \int_0^t (t-s)^{\alpha-1}\|A^{\gamma+\delta}E_{\alpha,\alpha}(-(t-s)^\alpha A)\|\,|A^{-\delta}F(u_m(s), u_m(s))|ds$$
$$\leq K_{\gamma,0} t^{-\alpha(\gamma-\beta)} + C_{\gamma+\delta}MK_{\rho,m}K_{\kappa,m}\int_0^t (t-s)^{\alpha(1-\gamma-\delta)-1} s^{-\alpha(1-\beta-\delta)}ds$$
$$= K_{\gamma,m+1} t^{-\alpha(\gamma-\beta)},$$

where

$$K_{\gamma,m+1} = K_{\gamma,0} + B_\gamma C_{\gamma+\delta} MK_{\rho,m}K_{\kappa,m}. \tag{2.33}$$

Thus $u_m(t)$ satisfies (2.30) with $K_{\gamma,m}$ defined recursively as above.

Step 2. To prove that $u_m \in C((0,T]; D(A^\gamma))$ and $u_m \in C([0,T]; D(A^\beta))$ by using mathematical induction. We need to prove $u_0 \in C((0,T]; D(A^\gamma))$.

For $\varepsilon > 0$ enough small, choose $h > 0$ such that $t + h > t > 0$ (the case $t - h < t$ follows analogously), we have

$$|A^\gamma u_0(t+h) - A^\gamma u_0(t)|$$
$$\leq |(A^{\gamma-\beta}E_\alpha(-(t+h)^\alpha A) - A^{\gamma-\beta}E_\alpha(-t^\alpha A))A^\beta a|$$
$$+ \int_t^{t+h} (t+h-s)^{\alpha-1}|A^{\gamma+\delta}E_{\alpha,\alpha}(-(t+h-s)^\alpha A)A^{-\delta}Pf(s)|ds$$
$$+ \int_0^t ((t-s)^{\alpha-1} - (t+h-s)^{\alpha-1})|A^{\gamma+\delta}E_{\alpha,\alpha}(-(t+h-s)^\alpha A)A^{-\delta}Pf(s)|ds$$

$$+ \int_0^{t-\varepsilon} (t-s)^{\alpha-1} \big| A^{\gamma+\delta} \big(E_{\alpha,\alpha}(-(t+h-s)^\alpha A)$$
$$- E_{\alpha,\alpha}(-(t-s)^\alpha A)\big) A^{-\delta} Pf(s)\big| ds$$
$$+ \int_{t-\varepsilon}^{t} (t-s)^{\alpha-1} \big| A^{\gamma+\delta} \big(E_{\alpha,\alpha}(-(t+h-s)^\alpha A)$$
$$- E_{\alpha,\alpha}(-(t-s)^\alpha A)\big) A^{-\delta} Pf(s)\big| ds$$
$$=: I_{11}(t) + I_{12}(t) + I_{13}(t) + I_{14}(t) + I_{15}(t).$$

We estimate each of the five terms separately. For $I_{11}(t)$, it is clear that $\lim_{h\to 0} I_{11}(t) = 0$ due to Lemma 2.9. For $I_{12}(t)$, in view of Lemma 2.10, one finds

$$I_{12}(t) \le C_{\gamma+\delta} \int_t^{t+h} (t+h-s)^{\alpha(1-\gamma-\delta)-1} |A^{-\delta} Pf(s)| ds$$
$$\le C_{\gamma+\delta} L \int_t^{t+h} (t+h-s)^{\alpha(1-\gamma-\delta)-1} s^{-\alpha(1-\beta-\delta)} ds$$
$$= C_{\gamma+\delta} L (t+h)^{-\alpha(\gamma-\beta)} \int_{t/(t+h)}^{1} (1-s)^{\alpha(1-\gamma-\delta)-1} s^{-\alpha(1-\beta-\delta)} ds.$$

On account of the properties of the Beta function, there exists $\epsilon > 0$ small enough such that for $0 < h < \epsilon$,

$$\int_{t/(t+h)}^{1} (1-s)^{\alpha(1-\gamma-\delta)-1} s^{-\alpha(1-\beta-\delta)} ds \to 0,$$

which implies that $I_{12}(t)$ tends to 0 as $h \to 0$. For $I_{13}(t)$ and $I_{14}(t)$, since

$$I_{13}(t) \le C_{\gamma+\delta} L \int_0^t \big((t-s)^{\alpha-1} - (t+h-s)^{\alpha-1} \big)(t+h-s)^{-\alpha(\gamma+\delta)} s^{-\alpha(1-\beta-\delta)} ds$$
$$\le 2 C_{\gamma+\delta} L \int_0^t (t-s)^{\alpha(1-\gamma-\delta)-1} s^{-\alpha(1-\beta-\delta)} ds = 2 C_{\gamma+\delta} L B_\gamma t^{-\alpha(\gamma-\beta)}$$

and

$$I_{14}(t) \le \int_0^{t-\varepsilon} (t-s)^{\alpha-1} |A^{\gamma+\delta} E_{\alpha,\alpha}(-(t+h-s)^\alpha A) A^{-\delta} Pf(s)| ds$$
$$+ \int_0^{t-\varepsilon} (t-s)^{\alpha-1} |A^{\gamma+\delta} E_{\alpha,\alpha}(-(t-s)^\alpha A) A^{-\delta} Pf(s)| ds$$
$$\le C_{\gamma+\delta} L \int_0^{t-\varepsilon} (t-s)^{\alpha-1} \big((t+h-s)^{-\alpha(\gamma+\delta)} + (t-s)^{-\alpha(\gamma+\delta)} \big) s^{-\alpha(1-\beta-\delta)} ds$$
$$\le 2 C_{\gamma+\delta} L \int_0^{t-\varepsilon} (t-s)^{\alpha(1-\gamma-\delta)-1} s^{-\alpha(1-\beta-\delta)} ds,$$

then we know from the Lebesgue's dominated convergence theorem and Lemma 2.9 that $\lim_{h\to 0} I_{13}(t) = 0$ and

$$\lim_{h\to 0} I_{14}(t) = \int_0^{t-\varepsilon} (t-s)^{\alpha-1} \lim_{h\to 0} \big| A^{\gamma+\delta} \big(E_{\alpha,\alpha}(-(t+h-s)^\alpha A)$$

$$- E_{\alpha,\alpha}(-(t-s)^\alpha A))A^{-\delta}Pf(s)|ds$$
$$=0.$$

For $I_{15}(t)$, we calculate immediately,

$$I_{15}(t) \leq C_{\gamma+\delta} \int_{t-\varepsilon}^{t} (t-s)^{\alpha-1}\left((t+h-s)^{-\alpha(\gamma+\delta)} + (t-s)^{-\alpha(\gamma+\delta)}\right)|A^{-\delta}Pf(s)|ds$$

$$\leq 2C_{\gamma+\delta}L \int_{t-\varepsilon}^{t} (t-s)^{\alpha(1-\gamma-\delta)-1}s^{-\alpha(1-\beta-\delta)}ds$$

$$\to 0, \quad \text{as } \varepsilon \to 0.$$

Therefore, we show

$$|A^\gamma u_0(t+h) - A^\gamma u_0(t)| \to 0, \quad \text{as } h \to 0.$$

From (2.29) and (2.32), we derive $u_m \in C((0,T];D(A^\gamma))$ recursively by the similar discussion as above. So we omit the details. Similarly, we can show $u_m \in C([0,T);D(A^\beta))$.

Step 3. Set

$$K_m = \max\{K_{\rho,m}, K_{\kappa,m}\}, \quad C_1 = \max\{C_{\rho+\delta}, C_{\kappa+\delta}\}, \quad B_1 = \max\{B_\rho, B_\kappa\}.$$

Then (2.33) yields that $K_{m+1} = K_0 + B_1 C_1 M K_m^2$. An elementary calculation shows that if

$$K_0 < \frac{1}{4B_1 C_1 M}, \tag{2.34}$$

then for each $m \geq 1$,

$$K_m < K := \frac{1 - \sqrt{1 - 4B_1 C_1 M}}{2B_1 C_1 M} < \frac{1}{2B_1 C_1 M},$$
$$K_{\gamma,m+1} \leq K_{\gamma,0} + B_\gamma C_1 M K^2 =: K_\gamma', \tag{2.35}$$
$$|A^\gamma u_{m+1}(t)| \leq K_\gamma' t^{-\alpha(\gamma-\beta)}.$$

Let us consider the equality

$$w_{m+1}(t) = \int_0^t (t-s)^{\alpha-1}$$
$$\times E_{\alpha,\alpha}(-(t-s)^\alpha A)[F(u_{m+1}(s), u_{m+1}(s)) - F(u_m(s), u_m(s))]ds,$$

where $w_m = u_{m+1} - u_m$, $m = 0, 1, ...$, and $t \in (0,T]$. By Lemma 2.8(iii), we have

$$\left|A^{-\delta}\big(F(u_{m+1}(s), u_{m+1}(s)) - F(u_m(s), u_m(s))\big)\right|$$
$$\leq M\big(|A^\rho w_m||A^\kappa u_{m+1}| + |A^\rho u_m||A^\kappa w_m|\big),$$

which follows from with an induction on m that

$$|A^\gamma w_{m+1}(t)| \leq 2KB_\gamma C_1(2KMB_1 C_1)^{m-1}t^{-\alpha(\gamma-\beta)},$$

for each γ such that $\gamma < 1 - \delta$. Since $2B_1 C_1 MK < 1$ by (2.35), this means that there exists $u \in C((0,T];D(A^\gamma)) \cap C([0,T);D(A^\beta))$ such that

$$\left|A^\beta\big(u_m(t) - u(t)\big)\right| \to 0 \quad \text{uniformly on } [0,T],$$

$$|A^\gamma(u_m(t) - u(t))| \to 0 \quad \text{uniformly on } [\varepsilon, T] \ (0 < \varepsilon < T) \text{ as } m \to \infty.$$

Moreover, $|A^\gamma u(t)| \le K'_\gamma t^{-\alpha(\gamma-\beta)}$. This in turn gives

$$|A^{-\delta}(F(u_m(s), u_m(s)) - F(u(s), u(s)))|$$

$$\le M(|A^\rho(u_m(t) - u(t))||A^\kappa u_m| + |A^\rho u_m||A^\kappa(u_m(t) - u(t))|) \to 0, \quad \text{as } m \to \infty,$$

and $|A^{-\delta}F(u_m(t), u_m(t))| \le C_2 t^{\alpha(\beta+\delta-1)}$ with a constant $C_2 > 0$ independent of m.

Take the limit on both sides of (2.29), using the dominated convergence theorem, we derive

$$u(t) = u_0(t) + \mathcal{G}(u, u)(t). \tag{2.36}$$

Since $a \in D(A^\beta)$, Lemma 2.10 implies that $t^{\alpha(\gamma-\beta)}|A^\gamma E_\alpha(-t^\alpha A)a| \to 0$ as $t \to 0^+$ for $\gamma > \beta$. If $T > 0$ is chosen sufficiently small then $K_{\gamma,0}$ ($\beta < \gamma < 1 - \delta$) becomes small and K_0 satisfies (2.34). This shows the existence of $T^* > 0$ with the desired property (b).

Step 4. Assume that u and v are mild solutions of problem (2.27). By the above estimates, we suppose without loss of generality that there exists χ, $|\beta| < \chi < 1-\delta$, such that

$$|A^\gamma u(t)| = o(t^{-\alpha(\gamma-\beta)}), \quad |A^\gamma v(t)| = o(t^{-\alpha(\gamma-\beta)}),$$

for all γ with $\beta < \gamma < \chi$. Let $w = u - v$, we consider the equality

$$w(t) = \int_0^t (t - s)^{\alpha-1} E_{\alpha,\alpha}(-(t - s)^\alpha A)[F(u(s), u(s)) - F(v(s), v(s))]ds.$$

Set $\delta' = 1 - \chi$, then $|\beta| < \chi$ implies $\delta' < 1 - |\beta|$. Note that $\beta + \delta' > \beta + \delta > 0$. As Step 1, we can choose ρ and κ, $|\beta| < \rho, \kappa < \chi$ such that

$$|A^\gamma w(t)|$$

$$\le MC_{\gamma+\delta'} \int_0^t (t - s)^{\alpha(1-\gamma-\delta')-1}(|A^\rho w(s)||A^\kappa u(s)| + |A^\rho v(s)||A^\kappa w(s)|)ds \tag{2.37}$$

for $\beta < \gamma < \chi$.

For $\gamma = \rho$ and $\gamma = \kappa$, let $K(t_0)$ (to $t_0 \in (0, T^*]$) be a constant such that

$$|A^\gamma u(t)| \le K(t_0)t^{-\alpha(\gamma-\beta)} \quad \text{and} \quad |A^\gamma v(t)| \le K(t_0)t^{-\alpha(\gamma-\beta)},$$

where $K(t_0) \to 0$ as $t_0 \to 0$. In view of (2.37), we obtain

$$|A^\rho w(t)| \le 2K^2(t_0)MC_{\rho+\delta'} \int_0^t (t - s)^{\alpha(1-\rho-\delta')-1}s^{-\alpha(\rho+\kappa-2\beta)}ds$$

$$+ MC_{\rho+\delta'} \int_0^t (t - s)^{\alpha(1-\rho-\delta')-1}|A^\rho w(s)||A^\kappa u(s)|ds$$

$$\le 2K^2(t_0)MB_2 C_{\rho+\delta'} t^{-\alpha(\rho-\beta)}$$

$$+ K(t_0)MC_{\rho+\delta'} \int_0^t (t - s)^{\alpha(1-\rho-\delta')-1}s^{-\alpha(\kappa-\beta)}|A^\rho w(s)|ds,$$

where $B_2 = B(\alpha(1 - \rho - \delta'), 1 - \alpha(\rho + \kappa - 2\beta))$. By induction, we see that

$$|A^\rho w(t)| \le 2K(t_0)(2K(t_0)MB_2 C_{\rho+\delta'})^m t^{-\alpha(\rho-\beta)}$$

for each $m \ge 1$. Similarly, $|A^\kappa w(t)| \le 2K(t_0)(2K(t_0)MB_2 C_{\rho+\delta'})^m t^{-\alpha(\kappa-\beta)}$ for each $m \ge 1$. Choose $t_0 > 0$ sufficiently small such that $2K(t_0)MB_2 C_{\rho+\delta'} < 1$. Hence $w(t) = 0$ on $[0, t_0]$. Repeating this argument on $[t_0, T^*]$, one finds that $w(t) = 0$ on $[0, T^*]$. Therefore, the mild solution is unique. $\qquad\square$

2.3.3 Hölder Continuous

Theorem 2.7. *Assume that u is the solution given by Theorem 2.6. Then for $\beta < \gamma < 1 - \delta$ and arbitrary $\varepsilon \in (0, T^*)$, $A^\gamma u$ is uniform Hölder continuous on $[\varepsilon, T^*]$.*

Proof. Denote $\varphi(t) := E_\alpha(-t^\alpha A)a$. For arbitrary $s \leq t$ and $0 < \vartheta < \alpha$, by Lemma 2.8, we have

$$
\begin{aligned}
\left| A^\gamma [e^{-t^\alpha \theta A} - e^{-s^\alpha \theta A}]a \right| &\leq \left\| [e^{(s^\alpha \theta - t^\alpha \theta)A} - I]A^{-\frac{\vartheta}{\alpha}} \right\| \cdot \left| A^{\gamma + \frac{\vartheta}{\alpha} - \beta} e^{-s^\alpha \theta A} A^\beta a \right| \\
&\leq \int_0^{t^\alpha \theta - s^\alpha \theta} \| A^{1 - \frac{\vartheta}{\alpha}} e^{-\tau A} \| d\tau \, | A^{\gamma + \frac{\vartheta}{\alpha} - \beta} e^{-s^\alpha \theta A} A^\beta a | \\
&\leq \int_0^{t^\alpha \theta - s^\alpha \theta} \tau^{\frac{\vartheta}{\alpha} - 1} d\tau (s^\alpha \theta)^{\beta - \frac{\vartheta}{\alpha} - \gamma} | A^\beta a | \\
&\leq \frac{\alpha}{\vartheta} (t - s)^\vartheta s^{\alpha(\beta - \gamma) - \vartheta} \theta^{\beta - \gamma} | A^\beta a |.
\end{aligned}
$$

Choose h satisfying $\varepsilon \leq t < t + h \leq T$, by the definition of φ, we have

$$
\begin{aligned}
| A^\gamma \varphi(t + h) - A\varphi(t) | &\leq \int_0^\infty \mathcal{M}_\alpha(\theta) | A^\gamma [e^{-(t+h)^\alpha \theta A} - e^{-t^\alpha \theta A}]a | d\theta \\
&\leq |A^\beta a| h^\vartheta t^{\alpha(\beta - \gamma) - \vartheta} \int_0^\infty \theta^{\beta - \gamma} \mathcal{M}_\alpha(\theta) d\theta \\
&\leq |A^\beta a| h^\vartheta \varepsilon^{\alpha(\beta - \gamma) - \vartheta} \frac{\Gamma(\beta + 1 - \gamma)}{\Gamma(1 + \alpha(\beta - \gamma))}.
\end{aligned}
$$

It suffices to prove the Hölder continuity of $A^\gamma v$, where

$$
v(t) = \int_0^t (t - s)^{\alpha - 1} E_{\alpha, \alpha}(-(t - s)^\alpha A)\big(F(u(s), u(s)) + Pf(s)\big) ds.
$$

Let t_1, t_2 satisfy $0 < t_1 < t_2 \leq T^*$. From Lemma 2.10, we find

$$
\begin{aligned}
&\| t_2^{\alpha - 1} A^{\gamma + \delta} E_{\alpha, \alpha}(-t_2^\alpha A) - t_1^{\alpha - 1} A^{\gamma + \delta} E_{\alpha, \alpha}(-t_1^\alpha A) \| \\
&\leq \| (t_2^{\alpha - 1} - t_1^{\alpha - 1}) A^{\gamma + \delta} E_{\alpha, \alpha}(-t_2^\alpha A) \| \\
&\quad + t_1^{\alpha - 1} \| A^{\gamma + \delta}(E_{\alpha, \alpha}(-t_2^\alpha A) - E_{\alpha, \alpha}(-t_1^\alpha A)) \| \\
&\leq C_{\gamma + \delta}(t_1^{\alpha - 1} - t_2^{\alpha - 1}) t_2^{-\alpha(\gamma + \delta)} \\
&\quad + t_1^{\alpha - 1} \int_0^\infty \alpha \theta \mathcal{M}_\alpha(\theta) \left\| \int_{t_1}^{t_2} \alpha \theta \tau^{\alpha - 1} A^{1 + \gamma + \delta} e^{-\theta \tau^\alpha A} d\tau \right\| d\theta \\
&\leq C_{\gamma + \delta}(t_1^{\alpha - 1} - t_2^{\alpha - 1}) t_2^{-\alpha(\gamma + \delta)} \\
&\quad + C_3 t_1^{\alpha - 1}(t_1^{-\alpha(\gamma + \delta)} - t_2^{-\alpha(\gamma + \delta)}) \int_0^\infty \alpha^2 \theta^{1 - \gamma - \delta} \mathcal{M}_\alpha(\theta) d\theta \\
&\leq C_\alpha(t_1^{\alpha(1 - \gamma - \delta) - 1} - t_2^{\alpha(1 - \gamma - \delta) - 1}),
\end{aligned}
$$

where C_3 is a positive constant depending on α, γ, δ. Take h such that $\varepsilon \leq t < t + h \leq T^*$, then

$$
| A^\gamma v(t + h) - A^\gamma v(t) |
$$

$$\leq \left| \int_t^{t+h} (t+h-s)^{\alpha-1} A^\gamma E_{\alpha,\alpha}(-(t+h-s)^\alpha A)\big(F(u(s),u(s))+Pf(s)\big)ds \right|$$

$$+ \left| \int_0^t A^\gamma\big((t+h-s)^{\alpha-1} E_{\alpha,\alpha}(-(t+h-s)^\alpha A) \right.$$

$$\left. - (t-s)^{\alpha-1} E_{\alpha,\alpha}(-(t-s)^\alpha A)\big)\big(F(u(s),u(s))+Pf(s)\big)ds \right|$$

$$\leq C_{\gamma+\delta} \int_t^{t+h} (t+h-s)^{\alpha(1-\gamma-\delta)-1}\big(|A^{-\delta}F(u(s),u(s))|+|A^{-\delta}Pf(s)|\big)ds$$

$$+ \int_0^t \big\| A^{\gamma+\delta}\big((t+h-s)^{\alpha-1} E_{\alpha,\alpha}(-(t+h-s)^\alpha A)$$

$$- (t-s)^{\alpha-1} E_{\alpha,\alpha}(-(t-s)^\alpha A)\big)\big\|\big(|A^{-\delta}F(u(s),u(s))|+|A^{-\delta}Pf(s)|\big)ds$$

$$\leq C_\varepsilon C_{\gamma+\delta} \int_t^{t+h} (t+h-s)^{\alpha(1-\gamma-\delta)-1}ds$$

$$+ C_{\gamma+\delta} \int_0^t \big((t-s)^{\alpha(1-\gamma-\delta)-1} - (t+h-s)^{\alpha(1-\gamma-\delta)-1}\big)s^{-\alpha(1-\beta-\delta)}ds,$$

where $C_\varepsilon = \sup_{\varepsilon \leq s \leq T^*} \{|A^{-\delta}F(u(s),u(s))|+|A^{-\delta}Pf(s)|\}$. By immediate calculation, the first integral of the last inequality is less than $C_\varepsilon C_3 h^{\alpha(1-\gamma-\delta)}$. Furthermore, since

$$1 - \left(1+\frac{h}{t}\right)^{\alpha(1-\gamma-\delta)-1} = \frac{(1+\frac{h}{t})^{1-\alpha(1-\gamma-\delta)}-1}{(1+\frac{h}{t})^{1-\alpha(1-\gamma-\delta)}} \leq \left(\frac{h}{t}\right)^{1-\alpha(1-\gamma-\delta)},$$

then

$$\int_0^t \big((t-s)^{\alpha(1-\gamma-\delta)-1} - (t+h-s)^{\alpha(1-\gamma-\delta)-1}\big)s^{-\alpha(1-\beta-\delta)}ds$$

$$= t^{\alpha(\beta-\gamma)} \int_0^1 \big(s^{\alpha(1-\gamma-\delta)-1} - (\frac{h}{t}+s)^{\alpha(1-\gamma-\delta)-1}\big)(1-s)^{-\alpha(1-\beta-\delta)}ds$$

$$\leq \varepsilon^{\alpha(\beta-\gamma)}\big(1-(1+\frac{h}{t})^{\alpha(1-\gamma-\delta)-1}\big) \int_0^1 s^{\alpha(1-\gamma-\delta)-1}(1-s)^{-\alpha(1-\beta-\delta)}ds$$

$$\leq B_\gamma \varepsilon^{\alpha(1+\beta-2\gamma-\delta)-1} h^{1-\alpha(1-\gamma-\delta)}.$$

Therefore, $A^\gamma u$ is uniform Hölder continuous on every interval $[\varepsilon, T^*]$.　　□

2.4　Approximations of Solutions

In this section, we mainly discuss the approximation of solutions to the following time-fractional Navier-Stokes equations:

$$\begin{cases} \partial_t^\alpha u(t,x) - \nu\Delta u(t,x) + u(t,x)\cdot\nabla u(t,x) \\ \quad = -\nabla p(t,x) + f(t,x), & (t,x) \in (0,T]\times\Omega, \\ \nabla\cdot u(t,x) = 0, & (t,x) \in (0,T]\times\Omega, \\ u(t,x) = 0, & (t,x) \in (0,T]\times\partial\Omega, \\ u(0,x) = a(x), & x \in \Omega, \end{cases} \quad (2.38)$$

in an open set $\Omega \subset \mathbb{R}^N$ ($N \geq 2$) with smooth boundary $\partial\Omega$, where ∂_t^α is the Caputo fractional derivative of order $\alpha \in (0,1)$, $u(t,x)$ represents the velocity field at (t,x), $p(t,x)$ is the pressure, ν the viscosity, $f(t,x)$ is the external force and $a(x)$ is the initial velocity.

Firstly, we get rid of the pressure term by applying Helmholtz projector P to equations (2.38), which converts equation (2.38) to

$$\begin{cases} \partial_t^\alpha u - \nu P \Delta u + P(u \cdot \nabla)u = Pf, \quad t > 0, \\ \nabla \cdot u = 0, \\ u|_{\partial\Omega} = 0, \\ u(0,x) = a. \end{cases}$$

The operator $-\nu P\Delta$ with Dirichlet's boundary conditions is, basically, the Stokes operator A in the divergence-free function space under consideration. Then the problem of solutions to (2.38) is equivalent to that of the following equations

$$\begin{cases} {}_0^C D_t^\alpha u = -Au + F(u,u) + Pf, \quad t \in (0,T], \\ u(0) = a, \end{cases} \tag{2.39}$$

where $F(u,v) = -P(u \cdot \nabla)v$. Our approach used to prove these results is that numerical and analytic approximations.

The section is organized as follows. In Subsection 2.4.1, we introduce some notations, definitions, and preliminary results. In Subsection 2.4.2, we use fixed point method to obtain the existence and uniqueness of mild solutions to each approximate equation, as well as the convergence of the approximate solutions. Subsection 2.4.3 is devoted to some convergence results for the Faedo-Galerkin approximations of problem (2.39).

2.4.1 *Preliminaries*

Let

$$\mathcal{V} = \{u \in (C_0^\infty(\Omega))^N : \nabla \cdot u = 0\}, \quad H = \text{the closure of } \mathcal{V} \text{ in } (L^2(\Omega))^N,$$
$$V = \text{the closure of } \mathcal{V} \text{ in } (H_0^1(\Omega))^N.$$

We denote by P the orthogonal projection $P : (L^2(\Omega))^N \to H$. Here, $(H_0^1(\Omega))^N$ and $H^2(\Omega)$ are Sobolev spaces.

The Stokes operator A is defined by

$$A : D(A) \subset H \to H, \quad A = -P\Delta, \quad D(A) = H^2(\Omega) \cap V.$$

It is known that the Stokes operator A is symmetric, self-adjoint and A^{-1} is compact operator in H. Therefore, A^{-1} is self-adjoint, injective and compact, and then, there exists a sequence of positive number $\mu_j > 0$, $\mu_{j+1} \leq \mu_j$ and an orthogonal basis of H, $\{\varphi_j\}$ such that $A^{-1}\varphi_j = \mu_j\varphi_j$. We denote $\lambda_j = \mu_j^{-1}$. Since A^{-1} has range in $D(A)$, then

$$A\varphi_j = \lambda_j\varphi_j, \text{ for } \varphi_j \in D(A); \quad 0 < \lambda_1 \leq ...\lambda_j \leq \lambda_{j+1} \leq ...; \quad \lim_{j \to \infty} \lambda_j = \infty.$$

As such $-A$ generates the bounded analytic semigroup $\{e^{-tA}\}$, see Constantin and Foicas [85].

Let X be a Banach space, $\alpha \in (0, 1]$ and $v : [0, T] \to X$. We define the Riemann-Liouville integrals of v:

$$_0D_t^{-\alpha}v(t) = (g_\alpha * v)(t), \quad v \in L^1(0, T; X), \quad t \in [0, T],$$

where $g_\alpha = \frac{t^{\alpha-1}}{\Gamma(\alpha)}, t > 0$ and $*$ denotes the convolution. Further, the Caputo fractional derivative operator of order α for the function v is defined by

$$_0^C D_t^\alpha v(t) = \frac{d}{dt} \left[_0D_t^{\alpha-1}\left(v(t) - v(0)\right)\right].$$

In general, for $u : [0, T] \times \mathbb{R}^N \to \mathbb{R}^N$,

$$\partial_t^\alpha u(t, x) = \partial_t \left[_0D_t^{\alpha-1}\left(u(t, x) - u(0, x)\right)\right].$$

Definition 2.5. Let $0 < T \leq \infty$. A function $u : [0, T] \to H$ is called a mild solution of problem (2.39), if $u \in C([0, T]; H)$ and

$$u(t) = E_\alpha(-t^\alpha A)a + \int_0^t (t-s)^{\alpha-1} E_{\alpha,\alpha}(-(t-s)^\alpha A)Pf(s)ds$$

$$+ \int_0^t (t-s)^{\alpha-1} E_{\alpha,\alpha}(-(t-s)^\alpha A)F(u(s), u(s))ds, \quad t \in [0, T], \tag{2.40}$$

where

$$E_\alpha(-t^\alpha A) = \int_0^\infty \mathcal{M}_\alpha(s)e^{-st^\alpha A}ds, \quad E_{\alpha,\alpha}(-t^\alpha A) = \int_0^\infty \alpha s \mathcal{M}_\alpha(s)e^{-st^\alpha A}ds,$$

where $\mathcal{M}_\alpha(\theta)$ is the Wright-type function defined in Definition 1.8.

We state some important technical lemmas that contribute to the main results.

Lemma 2.11. *[352] Let $0 < \alpha < 1$. Then for $t > 0$, $E_\alpha(-t^\alpha A)$ and $E_{\alpha,\alpha}(-t^\alpha A)$ are continuous in the uniform operator topology. Moreover, for every $r > 0$, the continuity is uniform on $[r, \infty)$.*

Lemma 2.12. *[138] Let $0 \leq \delta < \frac{1}{2} + \frac{N}{4}$. Then*

$$|A^{-\delta}P(u \cdot \nabla)v| \leq M|A^\rho u||A^\kappa v|$$

with some constant $M = M(\delta, \rho, \kappa)$, provided that $\delta + \rho + \kappa \geq \frac{N}{4} + \frac{1}{2}, \rho > 0, \kappa > 0, \kappa + \delta > \frac{1}{2}$.

Lemma 2.13. *[403] Let $1 < q < \infty$ and $\gamma \geq 0$. Then there exists a constant $C_\gamma = C(\alpha, \gamma) > 0$ such that*

$$|A^\gamma E_\alpha(-t^\alpha A)v| \leq C_\gamma t^{-\alpha\gamma}|v| \quad \text{and} \quad |A^\gamma E_{\alpha,\alpha}(-t^\alpha A)v| \leq C_\gamma t^{-\alpha\gamma}|v|,$$

for all $v \in H$ and $t \in (0, T]$. Furthermore, $\lim_{t\to 0+} t^{\alpha\gamma}|A^\gamma E_\alpha(-t^\alpha A)v| = 0$.

Throughout this section, we fix $\beta < 1$ and choose $\delta > 0$ such that

$$\frac{N}{4} - \frac{1}{2} < \beta, \quad -\beta < \delta < 1 - |\beta|, \qquad (2.41)$$

and suppose that

(H) $|A^{-\delta}Pf(t)|$ is continuous for $t \in (0, T^*]$ and $|A^{-\delta}Pf(t)| = o(t^{\alpha(\beta+\delta-1)})$ as $t \to 0$.

For convenience, set

$$L = \sup_{t \in (0, T^*]} \{t^{\alpha(1-\beta-\delta)}|A^{-\delta}Pf(t)|\}, \quad B_\beta = B(\alpha(1-\beta-\delta), 1 - \alpha(1-\beta-\delta))$$

and for all γ with $\gamma \in (\beta, 1-\delta)$, we let

$$B_\gamma = B(\alpha(1-\gamma-\delta), 1 - \alpha(1-\beta-\delta)).$$

Choose

$$\mu \le \frac{1}{4M(B_\beta C_{\beta+\delta} + B_\gamma C_{\gamma+\delta})}.$$

From Lemma 2.13 and the assumption of f, we know that there exist $T^* \in (0, T]$ and $\epsilon > 0$ such that

$$t^{\alpha(\gamma-\beta)}|A^\gamma E_\alpha(-t^\alpha A)a| \le \frac{\mu}{4}, \quad t^{\alpha(1-\beta-\delta)}|A^{-\delta}Pf(t)| \le \frac{\mu}{4B_\gamma C_{\gamma+\delta}}$$

for all $t \in (0, T^*]$ satisfying $|t| < \epsilon$.

Let

$$X_{T^*} := \left\{ u \in C([0, T^*]; D(A^\beta)) : t^{\alpha(\gamma-\beta)}u \in C((0, T^*]; D(A^\gamma)) \right.$$

$$\left. \text{and} \quad \sup_{t \in (0, T^*]} \{t^{\alpha(\gamma-\beta)}|u(t)|_{D(A^\gamma)}\} \le \mu \right\}$$

with the norm

$$\|u\|_{X_{T^*}} = \sup_{t \in [0, T^*]} \{|u(t)|_{D(A^\beta)}\} + \sup_{t \in (0, T^*]} \{t^{\alpha(\gamma-\beta)}|u(t)|_{D(A^\gamma)}\}.$$

2.4.2 Existence and Convergence of Approximate Solutions

Denote by $H_n \subset H$ the finite dimensional subspace spanned by $\{\varphi_1, \varphi_2, \ldots, \varphi_n\}$ for $n = 1, 2, \ldots$ and $Q_n : H \to H_n$ the corresponding projection operators. Then $\|Q_n\| \le 1$. Set

$$F_n(\cdot, \cdot) : H \times H \to H_n, \quad F_n(u(t), u(t)) = F(Q_n u(t), Q_n u(t)).$$

Define a map $\mathcal{F}_n : X_{T^*} \to X_{T^*}$:

$$(\mathcal{F}_n u)(t) = E_\alpha(-t^\alpha A)a$$

$$+ \int_0^t (t-s)^{\alpha-1} E_{\alpha,\alpha}(-(t-s)^\alpha A)\left(F_n(u(s), u(s)) + Pf(s)\right)ds, \quad t \in [0, T^*].$$

Theorem 2.8. *Let (2.41) and the condition (H) be satisfied. Then for $a \in D(A^\beta)$ there exists a unique function $u_n \in X_{T^*}$ satisfying $\mathcal{F}_n u_n = u_n$ and $\lim_{t \to 0+} t^{\alpha(\gamma-\beta)} |A^\gamma u_n(t)| = 0$ for all γ with $\gamma \in (\beta, 1-\delta)$.*

Proof. We first prove that $\mathcal{F}_n u \in C((0, T^*]; D(A^\gamma))$ and $\mathcal{F}_n u \in C([0, T^*]; D(A^\beta))$ for any $u \in X_{T^*}$. For $\varepsilon > 0$ enough small, choosing $h > 0$ such that $t + h > t > 0$ (the case $t - h < t$ follows analogously), we have

$$|A^\gamma \mathcal{F}_n u(t + h) - A^\gamma \mathcal{F}_n u(t)|$$

$$\leq \left|\left(A^{\gamma-\beta} E_\alpha(-(t+h)^\alpha A) - A^{\gamma-\beta} E_\alpha(-t^\alpha A)\right) A^\beta a\right|$$

$$+ \int_t^{t+h} (t + h - s)^{\alpha-1} \left|A^{\gamma+\delta} E_{\alpha,\alpha}(-(t + h - s)^\alpha A) A^{-\delta}\left(F_n(u(s), u(s)) + Pf(s)\right)\right| ds$$

$$+ \int_0^t \left((t-s)^{\alpha-1} - (t+h-s)^{\alpha-1}\right)$$

$$\times \left|A^{\gamma+\delta} E_{\alpha,\alpha}(-(t+h-s)^\alpha A) A^{-\delta}\left(F_n(u(s), u(s)) + Pf(s)\right)\right| ds$$

$$+ \int_0^{t-\varepsilon} (t-s)^{\alpha-1} \left|A^{\gamma+\delta}\left(E_{\alpha,\alpha}(-(t + h - s)^\alpha A)\right.\right.$$

$$\left.\left. - E_{\alpha,\alpha}(-(t-s)^\alpha A)\right) A^{-\delta}\left(F_n(u(s), u(s)) + Pf(s)\right)\right| ds$$

$$+ \int_{t-\varepsilon}^t (t-s)^{\alpha-1} \left|A^{\gamma+\delta}\left(E_{\alpha,\alpha}(-(t + h - s)^\alpha A)\right.\right.$$

$$\left.\left. - E_{\alpha,\alpha}(-(t-s)^\alpha A)\right) A^{-\delta}\left(F_n(u(s), u(s)) + Pf(s)\right)\right| ds$$

$$=: I_{11}(t) + I_{12}(t) + I_{13}(t) + I_{14}(t) + I_{15}(t).$$

We estimate each of the five terms separately. For $I_{11}(t)$, it is clear that $\lim_{h\to 0} I_{11}(t) = 0$ due to Lemma 2.11. On the other hand, the inequality $\delta < 1 - |\beta|$ implies the existence of ρ and κ such that

$$\rho + \kappa + \delta = 1 + \beta, \quad \beta < \rho, \kappa < 1 - \delta, \quad \rho > 0, \ \kappa > 0, \ \rho + \kappa > \frac{1}{2}. \tag{2.42}$$

Since $\frac{N}{4} - \frac{1}{2} < \beta, N \geq 2$ and $\delta < 1 - |\beta|$, it is easy to know that δ, ρ, κ satisfy the conditions of Lemma 2.12, which together with $u \in X_T^*$ follows that

$$|A^{-\delta} F_n(u(t), u(t)))| = |A^{-\delta} F(Q_n u(t), Q_n u(t))|$$

$$\leq M |A^\rho Q_n u(t)| |A^\kappa Q_n u(t)|$$

$$\leq M \|Q_n\|^2 \mu^2 t^{-\alpha(1-\beta-\delta)} \tag{2.43}$$

$$\leq M \mu^2 t^{-\alpha(1-\beta-\delta)}.$$

For $I_{12}(t)$, in view of Lemma 2.13, one finds

$$I_{12}(t) \leq C_{\gamma+\delta} \int_t^{t+h} (t + h - s)^{\alpha(1-\gamma-\delta)-1}\left(|A^{-\delta} Pf(s)| + |A^{-\delta} F_n(u(s), u(s))|\right) ds$$

$$\leq C_{\gamma+\delta}(L + M\mu^2) \int_t^{t+h} (t + h - s)^{\alpha(1-\gamma-\delta)-1} s^{-\alpha(1-\beta-\delta)} ds$$

$$= C_{\gamma+\delta}(L + M\mu^2)(t + h)^{-\alpha(\gamma-\beta)} \int_{t/(t+h)}^1 (1 - s)^{\alpha(1-\gamma-\delta)-1} s^{-\alpha(1-\beta-\delta)} ds.$$

On account of the properties of the Beta function, there exists $\epsilon > 0$ small enough such that for $0 < h < \epsilon$,

$$\int_{t/(t+h)}^{1} (1-s)^{\alpha(1-\gamma-\delta)-1} s^{-\alpha(1-\beta-\delta)} ds \to 0,$$

which implies that $I_{12}(t)$ tends to 0 as $h \to 0$. For $I_{13}(t)$ and $I_{14}(t)$, since

$$I_{13}(t) \le C_{\gamma+\delta}(L + M\mu^2)$$

$$\times \int_0^t \left((t-s)^{\alpha-1} - (t+h-s)^{\alpha-1}\right)(t+h-s)^{-\alpha(\gamma+\delta)} s^{-\alpha(1-\beta-\delta)} ds$$

$$\le 2C_{\gamma+\delta}(L + M\mu^2) \int_0^t (t-s)^{\alpha(1-\gamma-\delta)-1} s^{-\alpha(1-\beta-\delta)} ds$$

$$= 2C_{\gamma+\delta}(L + M\mu^2) B_\gamma t^{-\alpha(\gamma-\beta)}$$

and

$$I_{14}(t) \le \int_0^{t-\varepsilon} (t-s)^{\alpha-1} |A^{\gamma+\delta} E_{\alpha,\alpha}(-(t+h-s)^\alpha A) A^{-\delta}(Pf(s) + F_n(u(s), u(s)))| ds$$

$$+ \int_0^{t-\varepsilon} (t-s)^{\alpha-1} |A^{\gamma+\delta} E_{\alpha,\alpha}(-(t-s)^\alpha A) A^{-\delta}(Pf(s) + F_n(u(s), u(s)))| ds$$

$$\le C_{\gamma+\delta}(L + M\mu^2)$$

$$\times \int_0^{t-\varepsilon} (t-s)^{\alpha-1} \left((t+h-s)^{-\alpha(\gamma+\delta)} + (t-s)^{-\alpha(\gamma+\delta)}\right) s^{-\alpha(1-\beta-\delta)} ds$$

$$\le 2C_{\gamma+\delta}(L + M\mu^2) \int_0^{t-\varepsilon} (t-s)^{\alpha(1-\gamma-\delta)-1} s^{-\alpha(1-\beta-\delta)} ds,$$

then we know from the Lebesgue's dominated convergence theorem and Lemma 2.11 that $\lim_{h\to 0} I_{13}(t) = 0$ and

$$\lim_{h\to 0} I_{14}(t) = \int_0^{t-\varepsilon} (t-s)^{\alpha-1} \lim_{h\to 0} |A^{\gamma+\delta}(E_{\alpha,\alpha}(-(t+h-s)^\alpha A)$$

$$- E_{\alpha,\alpha}(-(t-s)^\alpha A)) A^{-\delta}(Pf(s) + F_n(u(s), u(s)))| ds$$

$$= 0.$$

For $I_{15}(t)$, we calculate immediately

$$I_{15}(t) \le C_{\gamma+\delta} \int_{t-\varepsilon}^t (t-s)^{\alpha-1} \left((t+h-s)^{-\alpha(\gamma+\delta)} + (t-s)^{-\alpha(\gamma+\delta)}\right)$$

$$\times |A^{-\delta}(Pf(s) + F_n(u(s), u(s)))| ds$$

$$\le 2C_{\gamma+\delta}(L + M\mu^2) \int_{t-\varepsilon}^t (t-s)^{\alpha(1-\gamma-\delta)-1} s^{-\alpha(1-\beta-\delta)} ds$$

$$\to 0, \quad \text{as } \varepsilon \to 0.$$

Thenceforth, we show

$$|A^\gamma \mathcal{F}_n u(t+h) - A^\gamma \mathcal{F}_n u(t)| \to 0, \quad \text{as } h \to 0.$$

Similarly, we can show $\mathcal{F}_n u \in C([0, T^*]; D(A^\beta))$.

To prove that $\mathcal{F}_n u \in X_{T^*}$, it remains to verify that

$$\sup_{t \in (0, T^*]} \{t^{\alpha(\gamma-\beta)} |\mathcal{F}_n u(t)|_{D(A^\gamma)}\} \leq \mu, \quad \text{for any } u \in X_{T^*}.$$

Indeed, we have

$$t^{\alpha(\gamma-\beta)} |(\mathcal{F}_n u)(t)|_{D(A^\gamma)}$$

$$\leq t^{\alpha(\gamma-\beta)} |A^\gamma E_\alpha(-t^\alpha A)a|$$

$$+ t^{\alpha(\gamma-\beta)} \int_0^t (t-s)^{\alpha-1} \|A^{\gamma+\delta} E_{\alpha,\alpha}(-(t-s)^\alpha A)\| \|A^{-\delta} Pf(s)| ds$$

$$+ t^{\alpha(\gamma-\beta)} \int_0^t (t-s)^{\alpha-1} \|A^{\gamma+\delta} E_{\alpha,\alpha}(-(t-s)^\alpha A)\| \|A^{-\delta} F_n(u(s), u(s))| ds$$

$$\leq \frac{\mu}{4} + C_{\gamma+\delta} \sup_{s \in (0,t]} \{s^{\alpha(1-\beta-\delta)} |A^{-\delta} Pf(s)|\} t^{\alpha(\gamma-\beta)} \int_0^t (t-s)^{\alpha(1-\gamma-\delta)-1} s^{-\alpha(1-\beta-\delta)} ds$$

$$+ M C_{\gamma+\delta} \|Q_n\|^2 t^{\alpha(\gamma-\beta)} \int_0^t (t-s)^{\alpha(1-\gamma-\delta)-1} |A^\rho u(s)| |A^\kappa u(s)| ds$$

$$\leq \frac{\mu}{4} + B_\gamma C_{\gamma+\delta} \sup_{s \in (0,t]} \{s^{\alpha(1-\beta-\delta)} |A^{-\delta} Pf(s)|\} + M B_\gamma C_{\gamma+\delta} \mu^2$$

$$< \mu.$$

Therefore we guarantee that $\mathcal{F}_n(X_{T^*}) \subseteq X_{T^*}$.

Let $t \in (0, T^*]$ and $u, v \in X_{T^*}$. In view of (2.42) and (2.43), we obtain

$$|(\mathcal{F}_n u)(t) - (\mathcal{F}_n v)(t)|_{D(A^\gamma)}$$

$$\leq \int_0^t (t-s)^{\alpha-1} \|A^{\gamma+\delta} E_{\alpha,\alpha}(-(t-s)^\alpha A)\| \|A^{-\delta} F_n(u(s), u(s)) - A^{-\delta} F_n(v(s), v(s))| ds$$

$$\leq M C_{\gamma+\delta} \|Q_n\|^2 \int_0^t (t-s)^{\alpha(1-\gamma-\delta)-1}$$

$$\times \left(|A^\rho(u(s) - v(s))| |A^\kappa u(s)| + |A^\rho v(s)| |A^\kappa(u(s) - v(s))| \right) ds$$

$$\leq M C_{\gamma+\delta} \int_0^t (t-s)^{\alpha(1-\gamma-\delta)-1} s^{-\alpha(\rho+\kappa-2\beta)} \left(K_\kappa(t) \|u - v\|_{X_{T^*}} + K_\rho(t) \|u - v\|_{X_{T^*}} \right) ds$$

$$\leq 2\mu M B_\gamma C_{\gamma+\delta} t^{-\alpha(\gamma-\beta)} \|u - v\|_{X_{T^*}},$$

where $K_\kappa(t) = \sup_{0 < s \leq t} |A^\kappa u(s)|$, $K_\rho(t) = \sup_{0 < s \leq t} |A^\rho u(s)|$.

Similarly,

$$|(\mathcal{F}_n u)(t) - (\mathcal{F}_n v)(t)|_{D(A^\beta)} \leq 2\mu M B_\beta C_{\beta+\delta} \|u - v\|_{X_{T^*}}.$$

Thus $\|\mathcal{F}_n u - \mathcal{F}_n v\|_{X_{T^*}} \leq 2\mu M (B_\beta C_{\beta+\delta} + B_\gamma C_{\gamma+\delta}) \|u - v\|_{X_{T^*}} \leq \frac{1}{2} \|u - v\|_{X_{T^*}}$. So $\mathcal{F}_n$ is a contraction map, which proves that the map $\mathcal{F}_n$ has a unique fixed point $u_n \in X_{T^*}$ satisfying

$$u_n(t) = E_\alpha(-t^\alpha A)a + \int_0^t (t-s)^{\alpha-1} E_{\alpha,\alpha}(-(t-s)^\alpha A) F_n(u_n(s), u_n(s)) ds$$

$$+ \int_0^t (t-s)^{\alpha-1} E_{\alpha,\alpha}(-(t-s)^\alpha A) Pf(s) ds.$$

$$(2.44)$$

Next we prove that $\lim_{t\to 0^+} t^{\alpha(\gamma-\beta)}|A^\gamma u_n(t)| = 0$. It suffices to verify that

$$\lim_{t\to 0^+} \sup_{s\in(0,t]} \{s^{\alpha(\gamma-\beta)}|A^\gamma u_n(s)|\} = 0, \quad \text{for } \gamma = \rho, \kappa.$$

Indeed, from Lemma 2.13 and the assumption of f, we show that for $\tilde{t} \in (0,t]$,

$$\tilde{t}^{\alpha(\gamma-\beta)}|u_n(t)|_{D(A^\gamma)}$$

$$\leq \tilde{t}^{\alpha(\gamma-\beta)}|A^\gamma E_\alpha(-\tilde{t}^\alpha A)a|$$

$$+ \tilde{t}^{\alpha(\gamma-\beta)} \int_0^{\tilde{t}} (\tilde{t}-s)^{\alpha-1}\|A^{\gamma+\delta}E_{\alpha,\alpha}(-(\tilde{t}-s)^\alpha A)\| |A^{-\delta}Pf(s)|ds$$

$$+ \tilde{t}^{\alpha(\gamma-\beta)} \int_0^{\tilde{t}} (\tilde{t}-s)^{\alpha-1}\|A^{\gamma+\delta}E_{\alpha,\alpha}(-(\tilde{t}-s)^\alpha A)\| |A^{-\delta}F_n(u_n(s), u_n(s))|ds$$

$$\leq \sup_{s\in(0,t]} \{s^{\alpha(\gamma-\beta)}|A^\gamma E_\alpha(-s^\alpha A)a|\} + C_{\gamma+\delta}B_\gamma \sup_{s\in(0,t]} \{s^{\alpha(1-\beta-\delta)}|A^{-\delta}Pf(s)|\}$$

$$+ MC_{\gamma+\delta}\|Q_n\|^2 \tilde{t}^{\alpha(\gamma-\beta)} \int_0^{\tilde{t}} (\tilde{t}-s)^{\alpha(1-\gamma-\delta)-1}|A^\rho u(s)||A^\kappa u(s)|ds$$

$$\leq \sup_{s\in(0,t]} \{s^{\alpha(\gamma-\beta)}|A^\gamma E_\alpha(-s^\alpha A)a|\} + B_\gamma C_{\gamma+\delta} \sup_{s\in(0,t]} \{s^{\alpha(1-\beta-\delta)}|A^{-\delta}Pf(s)|\}$$

$$+ MB_\gamma C_{\gamma+\delta} \sup_{s\in(0,t]} \{s^{\alpha(\rho-\beta)}|A^\rho u(s)|\} \sup_{s\in(0,t]} \{s^{\alpha(\kappa-\beta)}|A^\kappa u(s)|\}.$$

$$(2.45)$$

For $\gamma = \rho$, we obtain that

$$\sup_{s\in(0,t]} \{s^{\alpha(\rho-\beta)}|A^\rho u_n(s)|\} \leq \sup_{s\in(0,t]} \{s^{\alpha(\gamma-\beta)}|A^\gamma E_\alpha(-s^\alpha A)a|\}$$

$$+ B_\gamma C_{\gamma+\delta} \sup_{s\in(0,t]} \{s^{\alpha(1-\beta-\delta)}|A^{-\delta}Pf(s)|\}$$

$$+ \mu M B_\gamma C_{\gamma+\delta} \sup_{s\in(0,t]} \{s^{\alpha(\rho-\beta)}|A^\rho u(s)|\},$$

from which one can see that

$$\sup_{s\in(0,t]} \{s^{\alpha(\rho-\beta)}|A^\rho u_n(s)|\} \leq \frac{4}{3} \sup_{s\in(0,t]} \{s^{\alpha(\gamma-\beta)}|A^\gamma E_\alpha(-s^\alpha A)a|\}$$

$$+ \frac{4}{3}B_\gamma C_{\gamma+\delta} \sup_{s\in(0,t]} \{s^{\alpha(1-\beta-\delta)}|A^{-\delta}Pf(s)|\}$$

$$\to 0, \quad \text{as } t \to 0^+.$$

Clearly, $\sup_{s\in(0,t]}\{s^{\alpha(\kappa-\beta)}|A^\kappa u_n(s)|\}$ has the same conclusion. Finally, (2.45) gives

$$\lim_{t\to 0^+} t^{\alpha(\gamma-\beta)}|A^\gamma u_n(t)| = 0.$$

$\square$

Corollary 2.1. *Let $a \in D(A^\beta)$ and $\beta \leq \eta < 1 - \delta$. Then for any $t_0 \in (0, T^*]$, there exists a positive constant U_0 independent of n such that*

$$u_n(t) \in D(A^\eta) \quad \text{and} \quad |A^\eta u_n(t)| \leq U_0, \quad \text{for all } t \in [t_0, T^*].$$

Proof. By Theorem 2.8, it is clear that $u_n(t) \in D(A^\eta)$. We estimate

$$|A^\eta u_n(t)|$$

$$\leq |A^{\eta-\beta} E_\alpha(-t^\alpha A) A^\beta a|$$

$$+ \int_0^t (t-s)^{\alpha-1} \|A^{\eta+\delta} E_{\alpha,\alpha}(-(t-s)^\alpha A)\| |A^{-\delta}(Pf(s) + F_n(u_n(s), u_n(s)))| ds$$

$$\leq C_{\eta-\beta} t_0^{-\alpha(\eta-\beta)} |A^\beta a| + C_{\eta+\delta} L \int_0^t (t-s)^{\alpha(1-\eta-\delta)-1} s^{\alpha(\beta+\delta-1)} ds$$

$$+ M C_{\eta+\delta} \int_0^t (t-s)^{\alpha(1-\eta-\delta)-1} |A^\rho u_n(s)| |A^\kappa u_n(s)| ds$$

$$\leq C_{\eta-\beta} t_0^{-\alpha(\eta-\beta)} |A^\beta a| + B_\eta C_{\eta+\delta}(L + M\mu^2) t_0^{\alpha(\beta-\eta)}$$

$$\leq U_0.$$

$\square$

Theorem 2.9. *Under the assumptions of Theorem 2.8, we have for any $t_0 \in (0, T^*]$,*

$$\lim_{n \geq m, m \to \infty} \sup_{t \in [t_0, T^*]} t^{\alpha(\gamma-\beta)} |u_n(t) - u_m(t)|_{D(A^\gamma)} = 0,$$

$$\lim_{n \geq m, m \to \infty} \sup_{t \in [t_0, T^*]} |u_n(t) - u_m(t)|_{D(A^\beta)} = 0.$$

Proof. Let $n \geq m \geq n_0$ with n_0 large enough. For arbitrary $t_0' \in (0, t_0)$, we have from Theorem 2.8 that

$$|u_n(t) - u_m(t)|_{D(A^\gamma)} \leq \int_0^{t_0'} (t-s)^{\alpha-1} \|A^{\gamma+\delta} E_{\alpha,\alpha}(-(t-s)^\alpha A)\|$$

$$\times |A^{-\delta} F_n(u_n(s), u_n(s)) - A^{-\delta} F_m(u_m(s), u_m(s))| ds$$

$$+ \int_{t_0'}^t (t-s)^{\alpha-1} \|A^{\gamma+\delta} E_{\alpha,\alpha}(-(t-s)^\alpha A)\|$$

$$\times |A^{-\delta} F_n(u_n(s), u_n(s)) - A^{-\delta} F_m(u_m(s), u_m(s))| ds$$

$$=: I_1(t) + I_2(t).$$

Now we estimate the above inequality as follows: for $s \in [t_0', t]$,

$$|A^{-\delta} F_n(u_n(s), u_n(s)) - A^{-\delta} F_m(u_m(s), u_m(s))|$$

$$\leq |A^{-\delta} F_n(u_n(s), u_n(s)) - A^{-\delta} F_n(u_m(s), u_m(s))|$$

$$+ |A^{-\delta} F_n(u_m(s), u_m(s)) - A^{-\delta} F_m(u_m(s), u_m(s))|$$

$$\leq M(|A^\rho(u_n(s) - u_m(s))| |A^\kappa u_n(s)| + |A^\rho u_m(s)| |A^\kappa(u_n(s) - u_m(s))|)$$

$$+ M(|A^\rho(Q_n - Q_m)u_m(s)| |A^\kappa Q_n u_m(s)| + |A^\rho Q_m u_m(s)| |A^\kappa(Q_n - Q_m)u_m(s)|)$$

$$\leq M\mu s^{-\alpha(\rho+\kappa-2\beta)}(w_\rho(s) + w_\kappa(s)) + M\mu(s^{-\alpha(\kappa-\beta)} |A^\rho(Q_n - Q_m)u_m(s)|$$

$$+ s^{-\alpha(\rho-\beta)} |A^\kappa(Q_n - Q_m)u_m(s)|),$$

$$(2.46)$$

where we denote $w_\gamma(s) = s^{\alpha(\gamma-\beta)}|A^\gamma(u_n(s) - u_m(s))|$. Meanwhile, it follows that for $\rho, \kappa < \eta$,

$$|A^\rho(Q_n - Q_m)u_m(s)| \leq |A^{\rho-\eta}(Q_n - Q_m)A^\eta u_m(s)| \leq \lambda_m^{(\rho-\eta)}|A^\eta u_m(s)| \leq \lambda_m^{(\rho-\eta)}U_0,$$

$$|A^\kappa(Q_n - Q_m)u_m(s)| \leq |A^{\kappa-\eta}(Q_n - Q_m)A^\eta u_m(s)| \leq \lambda_m^{(\kappa-\eta)}|A^\eta u_m(s)| \leq \lambda_m^{(\kappa-\eta)}U_0.$$

Consequently, the inequality (2.46) turns into

$$|A^{-\delta}F_n(u_n(s), u_n(s)) - A^{-\delta}F_m(u_m(s), u_m(s))|$$
$$\leq M\mu s^{-\alpha(\rho+\kappa-2\beta)}\left(w_\rho(s) + w_\kappa(s)\right)$$
$$+ M\mu s^{-\alpha(\rho-\beta)}\lambda_m^{(\kappa-\eta)}U_0 + M\mu s^{-\alpha(\kappa-\beta)}\lambda_m^{(\rho-\eta)}U_0.$$

For $I_1(t)$, we have

$$I_1(t) \leq C_{\gamma+\delta}\int_0^{t_0'}(t - s)^{\alpha(1-\gamma-\delta)-1}\left(|A^{-\delta}F_n(u_n(s), u_n(s))|\right.$$
$$\left. + |A^{-\delta}F_m(u_m(s), u_m(s))|\right)ds$$
$$\leq 2MC_{\gamma+\delta}\mu^2\int_0^{t_0'}(t - s)^{\alpha(1-\gamma-\delta)-1}s^{-\alpha(\rho+\kappa-2\beta)}ds$$
$$\leq 2MC_{\gamma+\delta}\mu^2 t^{-\alpha(\gamma-\beta)}\int_0^{t_0'/t}(1 - s)^{\alpha(1-\gamma-\delta)-1}s^{-\alpha(\rho+\kappa-2\beta)}ds.$$

For $I_2(t)$, we have

$$I_2(t) \leq MC_{\gamma+\delta}\mu\int_{t_0'}^t(t - s)^{\alpha(1-\gamma-\delta)-1}s^{-\alpha(\rho+\kappa-2\beta)}(w_\rho(s) + w_\kappa(s))ds$$
$$+ MC_{\gamma+\delta}\mu\lambda_m^{(\kappa-\eta)}U_0\int_{t_0'}^t(t - s)^{\alpha(1-\gamma-\delta)-1}s^{-\alpha(\rho-\beta)}ds$$
$$+ MC_{\gamma+\delta}\mu\lambda_m^{(\rho-\eta)}U_0\int_{t_0'}^t(t - s)^{\alpha(1-\gamma-\delta)-1}s^{-\alpha(\kappa-\beta)}ds$$
$$\leq MC_{\gamma+\delta}\mu\int_{t_0'}^t(t - s)^{\alpha(1-\gamma-\delta)-1}s^{-\alpha(\rho+\kappa-2\beta)}(w_\rho(s) + w_\kappa(s))ds$$
$$+ MC_{\gamma+\delta}\mu B_{\gamma,\rho}\lambda_m^{(\kappa-\eta)}U_0 t^{\alpha(\kappa-\gamma)} + MC_{\gamma+\delta}\mu B_{\gamma,\kappa}\lambda_m^{(\rho-\eta)}U_0 t^{\alpha(\rho-\gamma)},$$

where

$$B_{\gamma,\rho} = B(\alpha(1 - \gamma - \delta), 1 - \alpha(\rho - \beta)), \quad B_{\gamma,\kappa} = B(\alpha(1 - \gamma - \delta), 1 - \alpha(\kappa - \beta)).$$

This ensures

$$t^{\alpha(\gamma-\beta)}|u_n(t) - u_m(t)|_{D(A^\gamma)}$$
$$\leq 2MC_{\gamma+\delta}\mu^2\int_0^{t_0'/t_0}(1 - s)^{\alpha(1-\gamma-\delta)-1}s^{-\alpha(\rho+\kappa-2\beta)}ds$$
$$+ MC_{\gamma+\delta}\mu B_{\gamma,\rho}\lambda_m^{(\kappa-\eta)}U_0 T^{\alpha(\kappa-\beta)} + MC_{\gamma+\delta}\mu B_{\gamma,\kappa}\lambda_m^{(\rho-\eta)}U_0 T^{\alpha(\rho-\beta)}$$

$$+ MC_{\gamma+\delta}\mu t^{\alpha(\gamma-\beta)}\int_{t_0'}^{t}(t-s)^{\alpha(1-\gamma-\delta)-1}s^{-\alpha(\rho+\kappa-2\beta)}(w_\rho(s)+w_\kappa(s))ds$$

$$=:a_\gamma + b_\gamma t^{\alpha(\gamma-\beta)}\int_{t_0'}^{t}(t-s)^{\alpha(1-\gamma-\delta)-1}s^{-\alpha(\rho+\kappa-2\beta)}(w_\rho(s)+w_\kappa(s))ds.$$

For $\gamma = \rho, \kappa$, we obtain

$$w_\rho(t) + w_\kappa(t)$$

$$\leq (a_\rho + a_\kappa) + b_\rho t^{\alpha(\rho-\beta)}\int_{t_0'}^{t}(t-s)^{\alpha(1-\rho-\delta)-1}s^{-\alpha(\rho+\kappa-2\beta)}(w_\rho(s)+w_\kappa(s))ds$$

$$+ b_\kappa t^{\alpha(\kappa-\beta)}\int_{t_0'}^{t}(t-s)^{\alpha(1-\kappa-\delta)-1}s^{-\alpha(\rho+\kappa-2\beta)}(w_\rho(s)+w_\kappa(s))ds.$$

By the generalized Gronwall's inequality, we get

$$w_\rho(t) + w_\kappa(t) \leq (a_\rho + a_\kappa) + (a_\rho + a_\kappa)\exp(b_\rho B_\rho + b_\kappa B_\kappa)$$

$$\times \left(\int_{t_0'}^{t} b_\rho t^{\alpha(\rho-\beta)}(t-s)^{\alpha(1-\rho-\delta)-1}s^{-\alpha(\rho+\kappa-2\beta)}ds \right.$$

$$\left. + b_\kappa t^{\alpha(\kappa-\beta)}\int_{t_0'}^{t}(t-s)^{\alpha(1-\kappa-\delta)-1}s^{-\alpha(\rho+\kappa-2\beta)}ds \right)$$

$$\leq (a_\rho + a_\kappa) + (a_\rho + a_\kappa)(b_\rho B_\rho + b_\kappa B_\kappa)\exp(b_\rho B_\rho + b_\kappa B_\kappa).$$

Thus

$$t^{\alpha(\gamma-\beta)}|u_n(t) - u_m(t)|_{D(A^\gamma)}$$
$$\leq a_\gamma + b_\gamma B_\gamma[(a_\rho + a_\kappa) + (a_\rho + a_\kappa)(b_\rho B_\rho + b_\kappa B_\kappa)\exp(b_\rho B_\rho + b_\kappa B_\kappa)]. \tag{2.47}$$

We also get

$$|u_n(t) - u_m(t)|_{D(A^\beta)} \leq a_\beta + b_\beta B_\beta[(a_\rho + a_\kappa)$$
$$+ (a_\rho + a_\kappa)(b_\rho B_\rho + b_\kappa B_\kappa)\exp(b_\rho B_\rho + b_\kappa B_\kappa)] \tag{2.48}$$

in the same way as we derived $t^{\alpha(\gamma-\beta)}|u_n(t) - u_m(t)|_{D(A^\gamma)}$.

Take t_0' sufficiently small, together with $\lambda_m^{(\kappa-\eta)} \to 0$ and $\lambda_m^{(\rho-\eta)} \to 0$ as $m \to \infty$, we can derive that a_γ tends to 0 for $\beta \leq \gamma < 1 - \delta$. It yields that the right-hand side of (2.47) and (2.48) tend to 0 as $m \to \infty$. The conclusion follows. $\square$

Theorem 2.10. *Let the assumptions of Theorem 2.8 be satisfied. Then there exists a unique function $u \in X_{T_0}$ such that $u_n \to u$ as $n \to \infty$ in X_{T^*} and u satisfies (2.40) on $[0, T^*]$.*

Proof. Since $A^\beta u_n(t) \to A^\beta u(t)$ and $t^{\alpha(\gamma-\beta)}A^\gamma u_n(t) \to t^{\alpha(\gamma-\beta)}A^\gamma u(t)$ as $n \to \infty$ for $0 < t \leq T^*$ due to Theorem 2.9, then $u_n \in X_{T^*}$ follows that $u \in X_{T^*}$. Further, for any $0 < t_0 \leq T^*$, we have

$$\lim_{n\to\infty} \sup_{t\in[t_0,T^*]} \{t^{\alpha(\gamma-\beta)}|u_n(t) - u(t)|_{D(A^\gamma)}\} = 0,$$

$$\lim_{n\to\infty} \sup_{t\in[t_0,T^*]} \{|u_n(t) - u(t)|_{D(A^\beta)}\} = 0.$$

Also, for $s \in [t_0, T^*]$,

$$|A^{-\delta}F_n(u_n(s), u_n(s)) - A^{-\delta}F(u(s), u(s))|$$

$$\leq M\mu s^{-\alpha(\rho+\kappa-2\beta)} \Big(\sup_{s\in[t_0,T^*]} s^{\alpha(\rho-\beta)}|A^\rho(u_n(s) - u(s))|$$

$$+ \sup_{s\in[t_0,T^*]} \{s^{\alpha(\kappa-\beta)}|A^\kappa(u_n(s) - u(s))|\}\Big)$$

$$+ M\mu s^{-\alpha(\rho+\kappa-2\beta)}\big(|(Q_n - I)u|_{D(A^\rho)} + |(Q_n - I)u|_{D(A^\kappa)}\big)$$

$$\to 0, \quad \text{as } n \to \infty.$$

Now, for $0 < t_0 < t$, we have

$$u_n(t) = E_\alpha(-t^\alpha A)a + \int_0^{t_0} (t - s)^{\alpha-1}E_{\alpha,\alpha}(-(t - s)^\alpha A)\big(F_n(u_n(s), u_n(s)) + Pf(s)\big)ds$$

$$+ \int_{t_0}^t (t - s)^{\alpha-1}E_{\alpha,\alpha}(-(t - s)^\alpha A)\big(F_n(u_n(s), u_n(s)) + Pf(s)\big)ds.$$

Moreover,

$$\Big|u_n(t) - E_\alpha(-t^\alpha A)a$$

$$- \int_{t_0}^t (t - s)^{\alpha-1}E_{\alpha,\alpha}(-(t - s)^\alpha A)\big(F_n(u_n(s), u_n(s)) + Pf(s)\big)ds\Big|_{D(A^\gamma)}$$

$$= \Big|\int_0^{t_0} (t - s)^{\alpha-1}E_{\alpha,\alpha}(-(t - s)^\alpha A)\big(F_n(u_n(s), u_n(s)) + Pf(s)\big)ds\Big|_{D(A^\gamma)}$$

$$\leq C_{\gamma+\delta}(L + M\mu^2)\int_0^{t_0} (t - s)^{\alpha(1-\gamma-\delta)-1}s^{\alpha(\beta+\delta-1)}ds$$

$$\leq B_\gamma C_{\gamma+\delta}(L + M\mu^2)t^{-\alpha(\gamma-\beta)}.$$

Taking the limit in the above inequality, the Lebesgue's dominated convergence theorem shows

$$t^{\alpha(\gamma-\beta)}\Big|u(t) - E_\alpha(-t^\alpha A)a$$

$$- \int_{t_0}^t (t - s)^{\alpha-1}E_{\alpha,\alpha}(-(t - s)^\alpha A)\big(F(u(s), u(s)) + Pf(s)\big)ds\Big|_{D(A^\gamma)}$$

$$\leq C_{\gamma+\delta}(L + M\mu^2)t^{\alpha(\gamma-\beta)}\int_0^{t_0} (t - s)^{\alpha(1-\gamma-\delta)-1}s^{\alpha(\beta+\delta-1)}ds$$

$$\leq C_{\gamma+\delta}(L + M\mu^2)\int_0^{t_0/t} (1 - s)^{\alpha(1-\gamma-\delta)-1}s^{\alpha(\beta+\delta-1)}ds.$$

Similarly,

$$\Big|u(t) - E_\alpha(-t^\alpha A)a$$

$$- \int_{t_0}^t (t - s)^{\alpha-1}E_{\alpha,\alpha}(-(t - s)^\alpha A)\big(F(u(s), u(s)) + Pf(s)\big)ds\Big|_{D(A^\beta)}$$

$$\leq C_{\beta+\delta}(L+M\mu^2)\int_0^{t_0}(t-s)^{\alpha(1-\beta-\delta)-1}s^{\alpha(\beta+\delta-1)}ds$$

$$\leq C_{\beta+\delta}(L+M\mu^2)\int_0^{t_0/t}(1-s)^{\alpha(1-\beta-\delta)-1}s^{\alpha(\beta+\delta-1)}ds.$$

Since $t_0 \in (0,T^*]$ is arbitrary, we get that u satisfies the integral equation (2.40). □

2.4.3 *Faedo-Galerkin Approximation*

This subsection is concerned with some convergence results of the Faedo-Galerkin approximations of mild solutions.

From Theorem 2.10 we know that the solution $u(t)$ of (2.40) exists on $0 \leq t \leq T^*$, then it has the representation

$$u(t) = \sum_{i=1}^{\infty} \xi_i \varphi_i,$$

where $\xi_i = (u(t), \varphi_i)$.

With Q_n as above, the Faedo-Galerkin approximation $u_n(t)$ by considering the projection of (2.40) on H_n, that is,

$$Q_n u_n(t) = E_\alpha(-t^\alpha A)Q_n a + \int_0^t (t-s)^{\alpha-1} E_{\alpha,\alpha}(-(t-s)^\alpha A)Q_n Pf(s)ds \tag{2.49}$$
$$+ \int_0^t (t-s)^{\alpha-1} E_{\alpha,\alpha}(-(t-s)^\alpha A)Q_n F_n(u_n(s), u_n(s))ds,$$

where F_n is the same as defined earlier.

If we set

$$v_n(t) := Q_n u_n(t) = \sum_{i=1}^{n} \xi_i^n \varphi_i, \quad \xi_i^n = (v_n(t), \varphi_i),$$

then (2.49) is equivalent to the following fractional differential equations

$$\begin{cases} {}_0^C D_t^\alpha \xi_i^n(t) + \lambda_j \xi_i^n(t) + F_i^n(\xi_i^1, \ldots, \xi_i^n) = \eta_i, \ i = 1, \ldots, n, \\ \xi_i^n(0) = a_i^0, \end{cases} \tag{2.50}$$

where

$$F_i^n(\xi_i^1, \ldots, \xi_i^n) = \left(F\left(\sum_{i=1}^{n} \xi_i^n \varphi_i, \sum_{i=1}^{n} \xi_i^n \varphi_i \right), \varphi_i \right)$$

and $\eta_i = (f, \varphi_i)$, $a_i^0 = (a, \varphi_i)$.

The following convergence results are stated as consequences of Theorems 2.8 and 2.9.

Theorem 2.11. *Let the assumptions of Theorem 2.8 hold. Then there exist a unique function $\{v_n\} \subset X_{T^*}$ satisfying (2.44) and $u \in X_{T^*}$ such that $v_n \to u$ as $n \to \infty$ in X_{T^*} and u satisfies (2.40) on $[0, T^*]$.*

Equation (2.50) determines the $\xi_i^n(t)$ and in turn the $v_n(t)$. The question of convergence means: in what sense does $v_n(t) \to u(t)$ as $n \to +\infty$?

Theorem 2.12. *Under the assumptions of Theorem 2.8, we have for any $t_0 \in (0, T^*]$,*

$$\lim_{n \to \infty} \sup_{t \in [t_0, T^*]} \left\{ t^{2\alpha(\gamma - \beta)} \sum_{i=1}^n \lambda_i^{2\gamma} (\xi_i^n(t) - \xi_i(t))^2 \right\} = 0,$$

$$\lim_{n \to \infty} \sup_{t \in [t_0, T^*]} \left\{ \sum_{i=1}^n \lambda_i^{2\beta} (\xi_i^n(t) - \xi_i(t))^2 \right\} = 0.$$

Proof. Consider

$$A^\gamma(v_n(t) - u(t)) = A^\gamma \sum_{i=1}^\infty (\xi_i^n(t) - \xi_i(t))\varphi_i = \sum_{i=1}^\infty \lambda_i^\gamma (\xi_i^n(t) - \xi_i(t))\varphi_i,$$

then

$$|A^\gamma(v_n(t) - u(t))|^2 \geq \sum_{i=1}^n \lambda_i^{2\gamma} (\xi_i^n(t) - \xi_i(t))^2.$$

Therefore

$$\sup_{t \in [t_0, T^*]} \left\{ t^{2\alpha(\gamma - \beta)} \sum_{i=1}^n \lambda_i^{2\gamma} (\xi_i^n(t) - \xi_i(t))^2 \right\} \leq \sup_{t \in [t_0, T^*]} \left\{ t^{2\alpha(\gamma - \beta)} |A^\gamma(v_n(t) - u(t))|^2 \right\}$$

$$\to 0, \quad \text{as } n \to \infty.$$

Similarly,

$$\sup_{t \in [t_0, T^*]} \left\{ \sum_{i=1}^n \lambda_i^{2\beta} (\xi_i^n(t) - \xi_i(t))^2 \right\} \leq \sup_{t \in [t_0, T^*]} \left\{ |A^\beta(v_n(t) - u(t))|^2 \right\} \to 0, \quad \text{as } n \to \infty.$$

$\square$

Theorem 2.13. *Under the assumptions of Theorem 2.8, we have for any $t_0 \in (0, T^*]$,*

$$\lim_{n \geq m, m \to \infty} \sup_{t \in [t_0, T^*]} t^{\alpha(\gamma - \beta)} |v_n(t) - v_m(t)|_{D(A^\gamma)} = 0,$$

$$\lim_{n \geq m, m \to \infty} \sup_{t \in [t_0, T^*]} |v_n(t) - v_m(t)|_{D(A^\beta)} = 0.$$

Proof. For $\eta > \gamma$, we have

$$t^{\alpha(\gamma - \beta)} |v_n(t) - v_m(t)|_{D(A^\gamma)}$$

$$= t^{\alpha(\gamma - \beta)} |Q_n u_n(t) - Q_m u_m(t)|_{D(A^\gamma)}$$

$$\leq t^{\alpha(\gamma - \beta)} |Q_n(u_n(t) - u_m(t))|_{D(A^\gamma)} + t^{\alpha(\gamma - \beta)} |(Q_n - Q_m)u_m(t)|_{D(A^\gamma)}$$

$$\leq t^{\alpha(\gamma - \beta)} |u_n(t) - u_m(t)|_{D(A^\gamma)} + t^{\alpha(\gamma - \beta)} \lambda_m^{(\gamma - \eta)} |A^\eta u_m(s)|.$$

Similarly, for the same η,

$$|v_n(t) - v_m(t)|_{D(A^\beta)} \leq |u_n(t) - u_m(t)|_{D(A^\beta)} + \lambda_m^{(\beta - \eta)} |A^\eta u_m(s)|.$$

From Theorem 2.9, the conclusion follows. The proof is completed. $\square$

2.5   Weak Solution and Optimal Control

In this section we study weak solutions of the following time-fractional Navier-Stokes equations in an open set $\Omega \subset \mathbb{R}^n$ $(n \le 4)$ with smooth boundary $\partial\Omega$:

$$\begin{cases} \partial_t^\alpha u - \nu\Delta u + (u \cdot \nabla)u = -\nabla p + f, & (t,x) \in [0,T] \times \Omega, \\ \nabla \cdot u = 0, & (t,x) \in [0,T] \times \Omega, \\ u(t,x) = 0, & (t,x) \in [0,T] \times \partial\Omega, \\ u(0,x) = a, & x \in \Omega, \end{cases} \tag{2.51}$$

where ∂_t^α is the Caputo fractional derivative of order $\alpha \in (0,1)$, $u = (u_1(t,x), u_2(t,x), ..., u_n(t,x))$ represents the velocity field at a point $x \in \Omega$ and time $t > 0$, $\nu > 0$ is the viscosity, $p = p(t,x)$ is the pressure, $f = (f_1(t,x), f_2(t,x), ..., f_n(t,x))$ is the external force and $a = (a_1(x), a_2(x), ..., a_n(x))$ is the initial velocity.

Then we proceed to consider the systems with control in $\Omega \subset \mathbb{R}^2$:

$$\begin{cases} \partial_t^\alpha u - \nu\Delta u + (u \cdot \nabla)u = -\nabla p + C_0 w + f, & (t,x) \in [0,T] \times \Omega, \\ \nabla \cdot u = 0, & (t,x) \in [0,T] \times \Omega, \\ u(t,x) = 0, & (t,x) \in [0,T] \times \partial\Omega, \\ u(0,x) = a, & x \in \Omega, \end{cases} \tag{2.52}$$

where U is a real Hilbert space, $w : [0,T] \to U$ and the operator $C_0 : U \to (L^2(\Omega))^2$ is linear and continuous.

Since many good properties satisfied in integer-order differential equations aren't generalized directly to fractional-order case, we found it more challenging in dealing with weak solutions of equation (2.51). As we all know, the main difficulty to study weak solutions is how to give an appropriate definition of weak solutions, introduce a suitable "work space" and establish the estimates of inequality by using the Galerkin approximations.

We begin in Subsection 2.5.1 with some notations and definitions. Subsection 2.5.2 is devoted to the proof of the existence for weak solutions to equation (2.51), then proceed to study the uniqueness of the weak solution for such equations in $\mathbb{R}^2$. Finally, Subsection 2.5.3 is concerned with an existence result of the optimal control of systems (2.52).

2.5.1   *Notations and Definitions*

In this subsection, we introduce notations, definitions, and preliminary facts which are used throughout this section.

Assume that X is a Banach space. Let $\alpha \in (0,1]$ and $v : [0,T] \to X$. We recall the left and right Riemann-Liouville integrals of v:

$$_0D_t^{-\alpha}v(t) = \int_0^t g_\alpha(t-s)v(s)ds, \quad _tD_T^{-\alpha}v(t) = \int_t^T g_\alpha(s-t)v(s)ds, \quad t > 0,$$

provided the integrals are point-wise defined on $[0, \infty)$, where g_α denotes the Riemann-Liouville kernel $g_\alpha(t) = \frac{t^{\alpha-1}}{\Gamma(\alpha)}$, $t > 0$. Further, $_0^C D_t^\alpha$ and $_t D_T^\alpha$ stand the left Caputo and right Riemann-Liouville fractional derivative operators of order α, respectively; they are defined by

$$_0^C D_t^\alpha v(t) = \int_0^t g_{1-\alpha}(t-s)\frac{d}{ds}v(s)ds, \quad t > 0,$$

$$_t D_T^\alpha v(t) = -\frac{d}{dt}\left(\int_t^T g_{1-\alpha}(s-t)v(s)ds\right), \quad t > 0.$$

More generally, for $u : [0, \infty) \times \mathbb{R}^n \to \mathbb{R}^n$, the left Caputo fractional derivative with respect to time of the function u can be written as

$$\partial_t^\alpha u(t, x) = \int_0^t g_{1-\alpha}(t-s)\partial_s u(s, x)ds, \quad t > 0.$$

Let $v : \mathbb{R} \to X$. We define the Liouville-Weyl fractional integral and the Caputo fractional derivative on the real axis:

$$_{-\infty} D_t^{-\alpha} v(t) = \int_{-\infty}^t g_\alpha(t-s)v(s)ds, \qquad _{-\infty}^C D_t^\alpha v(t) = \int_{-\infty}^t g_{1-\alpha}(t-s)\frac{d}{ds}v(s)ds,$$

respectively.

We also need for our purposes the fractional integration by parts in the formula (see [7]):

$$\int_0^T (\partial_t^\alpha u(t), \psi(t))dt = \int_0^T (u(t), {}_t D_T^\alpha \psi(t))dt + (u(t), {}_t D_T^{\alpha-1}\psi(t))|_0^T,$$

since $\lim_{t \to T} {}_t D_T^{\alpha-1}\psi(t) = 0$ for $\psi \in C_0^\infty([0, T]; X)$, then

$$\int_0^T (\partial_t^\alpha u(t), \psi(t))dt = \int_0^T (u(t), {}_t D_T^\alpha \psi(t))dt - (u(0), {}_0 D_T^{\alpha-1}\psi(t)). \qquad (2.53)$$

We can refer to Kilbas et al. [197] and Zhou [386, 387] for more details.

Next we declare a compactness theorem in Hilbert spaces.

Let X_0, X, X_1 be Hilbert spaces with

$$X_0 \hookrightarrow X \hookrightarrow X_1 \text{ being continuous and } X_0 \hookrightarrow X \text{ is compact.} \qquad (2.54)$$

Assume that $v : \mathbb{R} \to X_1$, $\hat{v}$ denotes its Fourier transform

$$\hat{v}(\tau) = \int_{-\infty}^{+\infty} e^{-2i\pi t\tau} v(t)dt.$$

We have

$$\widehat{_{-\infty}^C D_t^\alpha v}(\tau) = (2i\pi\tau)^\alpha \hat{v}(\tau).$$

We introduce a space for given $0 < \gamma \leq 1$,

$$\mathcal{W}^\gamma(\mathbb{R}, X_0, X_1) = \left\{v \in L^2(\mathbb{R}, X_0) : {}_{-\infty}^C D_t^\gamma v \in L^2(\mathbb{R}, X_1)\right\}.$$

Clearly, it is a Hilbert space for the norm

$$\|v\|_{\mathcal{W}^\gamma} = \left\{ \|v\|^2_{L^2(\mathbb{R}, X_0)} + \||\tau|^\gamma \hat{v}\|^2_{L^2(\mathbb{R}, X_1)} \right\}^{\frac{1}{2}}.$$

For any set $J \subset \mathbb{R}$, we associate with the subspace $\mathcal{W}^\gamma_J \subset \mathcal{W}^\gamma$ defined as:

$$\mathcal{W}^\gamma_J(\mathbb{R}, X_0, X_1) = \{ v \in \mathcal{W}^\gamma(\mathbb{R}, X_0, X_1) : \text{ support } v \subset J \}.$$

By similar discussion as the proof of Theorem 2.2 in Temam [330], we state the following compactness result.

Theorem 2.14. *[330] Let X_0, X, X_1 be Hilbert spaces and (2.54) hold. Then $\mathcal{W}^\gamma_J(\mathbb{R}, X_0, X_1) \hookrightarrow L^2(\mathbb{R}, X)$ is compact for any bounded set J and $\gamma > 0$.*

In order to get the desired results, we introduce some necessary spaces.

$$\mathcal{V} = \{ u \in (C_0^\infty(\Omega))^n : \nabla \cdot u = 0 \}.$$

Let us denote by H and V the following sets:

$$H = \text{the closure of } \mathcal{V} \text{ in } (L^2(\Omega))^n, \quad V = \text{the closure of } \mathcal{V} \text{ in } (H_0^1(\Omega))^n.$$

Here $(\cdot, \cdot)$ is duality in H, $\langle \cdot, \cdot \rangle$ is a duality pairing of V' and V. Also, we have

$$\langle f, u \rangle = (f, u), \ \forall \ f \in H, \ \forall \ u \in V. \tag{2.55}$$

$\mathcal{L}(U, H)$ stands for the set of all bounded linear operators from U to H.

Denote by P the orthogonal projection $P : (L^2(\Omega))^n \to H$. Here, the space $H_0^1(\Omega)$ is also a Hilbert space with the associated scalar product

$$((u, v)) = \sum_{i=1}^n (D_i u, D_i v).$$

The Stokes operator A is defined by

$$A : D(A) \subset H \to H, \quad A = -P\Delta, \quad D(A) = H^2(\Omega) \cap V.$$

Clearly,

$$\langle Au, v \rangle = ((u, v)), \ \forall \ u, v \in D(A) \tag{2.56}$$

and $A\varphi_j = \lambda_j \varphi_j$, for $\varphi_j \in D(A)$; $0 < \lambda_1 \leq ...\lambda_j \leq \lambda_{j+1} \leq ...$; $\lim_{j\to\infty} \lambda_j = \infty$.

Setting

$$b(u, v, w) = \sum_{i,j=1}^n \int_\Omega u_i (D_i v_j) w_j dx,$$

it is easy to know that b is trilinear continuous. If $u \in V$, then

$$b(u, v, v) = 0, \quad \text{for any } v \in H_0^1(\Omega).$$

For more details, we can refer the reader to Constantin and Foicas [85] and Temam [330].

We introduce an important space

$$W^\alpha([0, T], V, V') := \{ u \in L^2(0, T; V) : \partial_t^\alpha u \in L^2(0, T; V') \}.$$

Lemma 2.14. $\partial_t^\alpha(u(t), v) = \langle \partial_t^\alpha u(t), v \rangle$ *in* $\mathcal{L}(C_0^\infty([0, T]; \Omega))$ *for* $u \in W^\alpha([0, T], V, V')$, $v \in V$.

Proof. Let $\psi \in C_0^\infty([0, T]; \Omega)$. Notice that $u(t)$, $v \in V \subset H$, we know that $\langle \cdot, \cdot \rangle$ is consistent with $(\cdot, \cdot)$ by (2.55). It follows

$$\langle v, u(t) \rangle = (v, u(t)) = (u(t), v).$$

From (2.53) we obtain

$$\int_0^T \langle \partial_t^\alpha u(t), v \rangle \psi(t) dt = \int_0^T \langle u(t), v \rangle_t D_T^\alpha \psi(t) dt - \langle u(0), v \rangle_0 D_T^{\alpha-1} \psi(t)$$

$$= \int_0^T (u(t), v)_t D_T^\alpha \psi(t) dt - (u(0), v)_0 D_T^{\alpha-1} \psi(t)$$

$$= \int_0^T \partial_t^\alpha (u(t), v) \psi(t) dt.$$

Hence, $\partial_t^\alpha (u(t), v) = \langle \partial_t^\alpha u(t), v \rangle$ in $\mathcal{L}(C_0^\infty([0, T]; \Omega))$. $\qquad \square$

We assume that u is a classical solution of (2.51) associated with p, it means that $u \in C^2([0, T] \times \Omega)$, $p \in C^1([0, T] \times \Omega)$. Obviously $u \in L^2(0, T; V)$ and if $v \in V$, it is easy to check

$$\partial_t^\alpha (u, v) + \nu((u, v)) + b(u, u, v) = \langle f, v \rangle. \tag{2.57}$$

This presents the following weak formulation of equation (2.51). Let $\alpha_1 \in (0, \alpha)$ and

$$C_w([0, T]; H) := \{u \in L^\infty(0, T; H) : u \text{ is weakly continuous}\}.$$

Definition 2.6. Let $a \in H$, $f \in L^{\frac{2}{\alpha_1}}(0, T; V')$. A function $u \in L^2(0, T; V) \cap C_w([0, T]; H)$ is said a weak solution of equation (2.51) if

$$\begin{cases} \partial_t^\alpha (u, v) + \nu((u, v)) + b(u, u, v) = \langle f, v \rangle, & \forall v \in V, \\ u(0) = a. \end{cases} \tag{2.58}$$

Lemma 2.15. *[330] Let $u \in L^2(0, T; V)$. Then the function $Bu(t)$ defined by*

$$(B(u(t), u(t)), v) = b(u(t), u(t), v), \text{ for any } v \in V, \text{ a.e. in } t \in [0, T] \tag{2.59}$$

belongs to V'.

Now if u satisfies the problem (2.58), then based on (2.55), (2.56) and Lemma 2.15, we rewrite the first equality of (2.58) as

$$\partial_t^\alpha \langle u, v \rangle = \langle f - \nu Au - B(u, u), v \rangle, \ \forall v \in V.$$

Since $Au \in L^2(0, T; V')$, then $f(t) - \nu Au(t) - Bu(t) \in V'$. Thus Lemma 2.14 shows that

$$\partial_t^\alpha u = f - \nu Au - B(u, u).$$

An equivalent form of problem (2.58) is

$$\begin{cases} {}^C_0D^\alpha_t u + \nu Au + B(u,u) = f, \\ u(0) = a. \end{cases} \tag{2.60}$$

Lemma 2.16. *Suppose that X is a real Hilbert space and $v : [0,T] \to X$ has a derivative, then there holds*

$$(v(t), {}^C_0D^\alpha_t v(t)) \geq \frac{1}{2}{}^C_0D^\alpha_t |v(t)|^2.$$

Note the above lemma is a generalization of the inequality proved in [10].

Lemma 2.17. *Suppose that a nonnegative function $v(t)$ satisfies*

$${}^C_0D^\alpha_t v(t) + c_1 v(t) \leq c_2(t)$$

for almost all $t \in [0,T]$, where $c_1 > 0$ and the function $c_2(t)$ is nonnegative and integrable for $t \in [0,T]$. Then

$$v(t) \leq v(0) + \frac{1}{\Gamma(\alpha)}\int_0^t (t-s)^{\alpha-1}c_2(s)ds.$$

Proof. Let ${}^C_0D^\alpha_t v(t) + c_1 v(t) = g(t)$, then by [197] we have the following expression:

$$v(t) = v(0)E_\alpha(-c_1 t^\alpha) + \int_0^t (t-s)^{\alpha-1}E_{\alpha,\alpha}(-c_1(t-s)^\alpha)g(s)ds,$$

where

$$E_\alpha(t) = \sum_{k=0}^\infty \frac{t^k}{\Gamma(\alpha k + 1)} \quad \text{and} \quad E_{\alpha,\alpha}(t) = \sum_{k=0}^\infty \frac{t^k}{\Gamma(\alpha k + \alpha)}$$

are Mittag-Leffler functions. Since $0 < E_\alpha(-c_1 t^\alpha) \leq 1$, $0 < E_{\alpha,\alpha}(-c_1 t^\alpha) \leq \frac{1}{\Gamma(\alpha)}$ and $g(t) \leq c_2(t)$, then

$$v(t) \leq v(0)E_\alpha(-c_1 t^\alpha) + \frac{1}{\Gamma(\alpha)}\int_0^t (t-s)^{\alpha-1}g(s)ds$$

$$\leq v(0) + \frac{1}{\Gamma(\alpha)}\int_0^t (t-s)^{\alpha-1}c_2(s)ds.$$

$\square$

2.5.2 *Existence and Uniqueness*

Theorem 2.15. *Let $a \in H$, $f \in L^{\frac{2}{\alpha_1}}(0,T;V')$. Then there exists at least one function $u \in L^2(0,T;V) \cap C_w([0,T];H)$ which satisfies (2.60).*

Proof. Step 1. Let $H_m \subset H$ denote the finite dimensional subspace spanned by $\{\varphi_1, \varphi_2, ..., \varphi_m\}$ and let $P_m : H \to H_m$ be the corresponding projection operators for $m = 1, 2, ...$. Applying P_m to (2.60) yields the equation

$$_0^C D_t^\alpha (P_m u) + \nu A(P_m u) + P_m B(u, u) = P_m f.$$

The Galerkin system of order m is the system

$$\begin{cases} _0^C D_t^\alpha u_m + \nu A u_m + P_m B(u_m, u_m) = f_m, \\ u_m(0) = a_m, \end{cases} \tag{2.61}$$

where $f_m = P_m f$ and $a_m = P_m a$. The function $u_m(t)$ belongs to $P_m H$. Taking the scalar product of (2.61) with φ_j we obtain

$$\begin{cases} (_0^C D_t^\alpha u_m, \varphi_j) + \nu(A u_m, \varphi_j) + (P_m B(u_m, u_m), \varphi_j) = \langle f_m, \varphi_j \rangle, \\ (u_m(0), \varphi_j) = (a_m, \varphi_j). \end{cases} \tag{2.62}$$

More precisely, let $\xi_j = \xi_j(t)$ denote the jth component of $u_m(t)$: $\xi_j(t) = (u_m(t), \varphi_j)$. Also, let $\eta_j(t) = (f_m(t), \varphi_j)$ be component of $f_m(t)$. Then (2.62) is equivalent to a nonlinear differential system for the functions

$$\begin{cases} _0^C D_t^\alpha \xi_j + \nu \lambda_j \xi_j + \sum_{k,l=1}^m b(\varphi_k, \varphi_l, \varphi_j) \xi_k \xi_l = \eta_j, \quad j = 1, ..., m, \\ \xi_j(0) = \xi_j^0, \end{cases} \tag{2.63}$$

where $\xi_j^0 = (a_m, \varphi_j)$.

The nonlinear differential system (2.63) has a maximal solution on the interval $[0, t_m]$. If $t_m < T$, then $|\xi(t)|$ must go to $+\infty$ as $t \to t_m$, the estimates that we show indicate that this does not happen. Indeed, taking the scalar product of (2.63) with ξ in $\mathbb{R}$, we obtain

$$(_0^C D_t^\alpha \xi(t), \xi(t)) + \nu \sum_{j=1}^m \lambda_j \xi_j^2(t) = \langle \eta(t), \xi(t) \rangle,$$

because of

$$\sum_{k,l,j} b(\varphi_k, \varphi_l, \varphi_j) \xi_k \xi_l \xi_j = 0.$$

From Lemma 2.16 and the Young's inequality, we know

$$\frac{1}{2} {}_0^C D_t^\alpha |\xi(t)|^2 + \nu \lambda_1 |\xi(t)|^2 \leq \langle \eta(t), \xi(t) \rangle \tag{2.64}$$

$$\leq \frac{1}{2\nu\lambda_1} |\eta(t)|^2 + \frac{\nu\lambda_1}{2} |\xi(t)|^2.$$

Using Lemma 2.17 and the fact that $\lambda_1 > 0$, we have the estimation

$$|\xi(t)|^2 \leq |\xi(0)|^2 + \frac{1}{\nu\lambda_1\Gamma(\alpha)} \int_0^t (t-s)^{\alpha-1} |\eta(s)|^2 ds$$

$$\leq |\xi(0)|^2 + \frac{1}{\nu\lambda_1\Gamma(\alpha)} \int_0^t |\eta(s)|^{2/\alpha_1} ds + \frac{1}{\nu\lambda_1\Gamma(\alpha)} \int_0^t (t-s)^{\frac{\alpha-1}{1-\alpha_1}} ds$$

$$\leq |\xi(0)|^2 + \frac{1}{\nu\lambda_1\Gamma(\alpha)} \int_0^T |\eta(s)|^{2/\alpha_1} ds + \frac{T^{1+b}}{(1+b)\nu\lambda_1\Gamma(\alpha)}$$

$$=: M,$$

where $\alpha_1 \in (0, \alpha)$, $b = \frac{\alpha-1}{1-\alpha_1}$. Therefore $t_m = T$.

Step 2. We also estimate the right-hand side of (2.64) differently. Since

$$|\langle \eta(t), \xi(t) \rangle| \leq \left(\sum_{j=1}^m \lambda_j \xi_j^2 \right)^{\frac{1}{2}} \left(\sum_{j=1}^m \lambda_j^{-1} \eta_j^2 \right)^{\frac{1}{2}},$$

and we return to the notation

$$u_m(t) = \sum_{j=1}^m \xi_j \varphi_j, \quad f_m(t) = \sum_{j=1}^m \eta_j \varphi_j,$$

then from (2.64) we get

$$\frac{1}{2}{}_0^C D_t^\alpha |u_m(t)|^2 + \nu\|u_m(t)\|_V^2 \leq \langle f_m(t), u_m(t) \rangle$$

$$\leq |A^{-\frac{1}{2}} f_m(t)| \|u_m(t)\|_V \qquad (2.65)$$

$$\leq \frac{1}{2\nu} |A^{-\frac{1}{2}} f_m(t)|^2 + \frac{\nu}{2} \|u_m(t)\|_V^2.$$

We use the notation $|h|_{V'} = |A^{-\frac{1}{2}} h|$, meaning that we identify the dual V' of V with $D(A^{-\frac{1}{2}})$. Integrating (2.65) (with order α) and applying the Young's inequality, we get

$$|u_m(t)|^2 + \frac{\nu}{\Gamma(\alpha)} \int_0^t (t-s)^{\alpha-1} \|u_m(s)\|_V^2 ds$$

$$\leq |u_m(0)|^2 + \frac{1}{\nu\Gamma(\alpha)} \int_0^t (t-s)^{\alpha-1} |f_m(s)|_{V'}^2 ds$$

$$\leq |u_m(0)|^2 + \frac{1}{\nu\Gamma(\alpha)} \int_0^t |f_m(s)|_{V'}^{2/\alpha_1} ds + \frac{1}{\nu\Gamma(\alpha)} \int_0^t (t-s)^{\frac{\alpha-1}{1-\alpha_1}} ds$$

$$\leq |u_m(0)|^2 + \frac{1}{\nu\Gamma(\alpha)} \int_0^T |f_m(s)|_{V'}^{2/\alpha_1} ds + \frac{T^{1+b}}{(1+b)\nu\Gamma(\alpha)}.$$

It follows that for a.e. $t \in [0, T]$,

$$|u_m(t)|^2 \leq |u_m(0)|^2 + \frac{1}{\nu\Gamma(\alpha)} \int_0^T |f(s)|_{V'}^{2/\alpha_1} ds + \frac{T^{1+b}}{(1+b)\nu\Gamma(\alpha)},$$

$$\frac{\nu}{\Gamma(\alpha)} \int_0^t (t-s)^{\alpha-1} \|u_m(s)\|_V^2 ds \qquad (2.66)$$

$$\leq |u_m(0)|^2 + \frac{1}{\nu\Gamma(\alpha)} \int_0^T |f(s)|_{V'}^{2/\alpha_1} ds + \frac{T^{1+b}}{(1+b)\nu\Gamma(\alpha)}.$$

This ensures that

$$\text{the sequence } \{u_m\} \text{ is a bounded set of } L^\infty(0,T;H). \tag{2.67}$$

Moreover,

$$\frac{\nu}{\Gamma(\alpha)}T^{\alpha-1}\int_0^t \|u_m(s)\|_V^2 ds \leq \frac{\nu}{\Gamma(\alpha)}\int_0^t (t-s)^{\alpha-1}\|u_m(s)\|_V^2 ds$$

$$\leq |u_m(0)|^2 + \frac{1}{\nu\Gamma(\alpha)}\int_0^T |f(s)|_{V'}^{2/\alpha_1} ds + \frac{T^{1+b}}{(1+b)\nu\Gamma(\alpha)}.$$

Hence

$$\text{the sequence } \{u_m\} \text{ is a bounded set of } L^2(0,T;V). \tag{2.68}$$

Step 3. Let $\tilde{u}_m : \mathbb{R} \to V$ denote the function defined as

$$\tilde{u}_m(t) = \begin{cases} u_m(t), & t \in [0,T], \\ 0, & t \in \mathbb{R} \setminus [0,T], \end{cases}$$

and $\hat{u}_m$ denote the Fourier transform of $\tilde{u}_m$. To show that

$$\text{the sequence } \{\tilde{u}_m\} \text{ remains to be a bounded set of } \mathcal{W}^\gamma(\mathbb{R},V,H). \tag{2.69}$$

Applying Theorem 2.14 along with (2.68), we need to verify that

$$\int_{-\infty}^{+\infty} |\tau|^{2\gamma}|\hat{u}_m(\tau)|^2 d\tau \leq \text{const.}, \quad \text{for some } \gamma > 0. \tag{2.70}$$

In order to prove (2.70), we observe that (2.62) can be written as

$$({}_0^C D_t^\alpha \tilde{u}_m, \varphi_j) = \langle \tilde{g}_m, \varphi_j \rangle + (u_m(0), \varphi_j)_{-\infty} D_t^{\alpha-1}\delta_0 - (u_m(T), \varphi_j)_{-\infty} D_t^{\alpha-1}\delta_T, \tag{2.71}$$

where

$$\tilde{g}_m(t) = \begin{cases} g_m(t), & t \in [0,T], \\ 0, & t \in \mathbb{R} \setminus [0,T], \end{cases}$$

here $g_m = f_m - \nu Au_m - Bu_m$, δ_0 and δ_T are the Dirac distributions at 0 and T.

Indeed, it is classical that since $\tilde{u}_m$ has two discontinuities at 0 and T, the Caputo derivative of $\tilde{u}_m$ is given by

$$_{-\infty}^C D_t^\alpha \tilde{u}_m = {}_{-\infty} D_t^{\alpha-1}\left(\frac{d}{dt}\tilde{u}_m\right)$$

$$= {}_{-\infty} D_t^{\alpha-1}\left(\frac{d}{dt}u_m + u_m(0)\delta_0 - u_m(T)\delta_T\right)$$

$$= {}_0^C D_t^\alpha u_m + {}_{-\infty} D_t^{\alpha-1}\left(u_m(0)\delta_0 - u_m(T)\delta_T\right).$$

By the Fourier transform, (2.71) yields

$$(2i\pi\tau)^\alpha(\hat{u}_m, \varphi_j) = \langle \hat{g}_m, \varphi_j \rangle + (u_m(0), \varphi_j)(2i\pi\tau)^{\alpha-1} \\ - (u_m(T), \varphi_j)(2i\pi\tau)^{\alpha-1}e^{-2i\pi T\tau}, \tag{2.72}$$

here $\hat{u}_m$ and $\hat{g}_m$ denote the Fourier transforms of $\tilde{u}_m$ and $\tilde{g}_m$, respectively.

We multiply (2.72) by $\hat{\xi}_j(\tau)$ (i.e., Fourier transform of ξ_j) and plus the resulting equations, it follows

$$(2i\pi\tau)^\alpha |\hat{u}_m(\tau)|^2 = \langle \hat{g}_m, \hat{u}_m(\tau) \rangle + (u_m(0), \hat{u}_m(\tau))(2i\pi\tau)^{\alpha-1}$$
$$- (u_m(T), \hat{u}_m(\tau))(2i\pi\tau)^{\alpha-1}e^{-2i\pi T\tau}. \tag{2.73}$$

In view of the inequality $\|Bu\|_{V'} \le c\|u\|_V^2$, for all $u \in V$,

$$\int_0^T \|g_m(t)\|_{V'}dt \le \int_0^T \left(\|f_m(t)\|_{V'} + \nu\|u_m(t)\|_V + c\|u_m(t)\|_V^2\right)dt,$$

this remains bounded based on (2.68). Therefore

$$\sup_{\tau \in \mathbb{R}} \|\hat{g}_m(\tau)\|_{V'} \le \text{const.}, \ \forall\, m.$$

On account of (2.67),

$$|u_m(0)| \le \text{const.}, \ |u_m(T)| \le \text{const.},$$

and we conclude from (2.73) that

$$|\tau|^\alpha |\hat{u}_m(\tau)|^2 \le c_2 \|\hat{u}_m(\tau)\|_V + c_3|\tau|^{\alpha-1}|\hat{u}_m(\tau)|$$

or

$$|\tau|^\alpha |\hat{u}_m(\tau)|^2 \le c_4 \max\{1, |\tau|^{\alpha-1}\}\|\hat{u}_m(\tau)\|_V.$$

For γ fixed, $\gamma < \alpha/4$, we see that

$$|\tau|^{2\gamma} \le c_5(\gamma)\frac{1+|\tau|^\alpha}{1+|\tau|^{\alpha-2\gamma}}.$$

Accordingly,

$$\int_{-\infty}^{+\infty} |\tau|^{2\gamma}|\hat{u}_m(\tau)|^2 d\tau \le c_5(\gamma)\int_{-\infty}^{+\infty} \frac{1+|\tau|^\alpha}{1+|\tau|^{\alpha-2\gamma}}|\hat{u}_m(\tau)|^2 d\tau$$
$$\le c_6(\gamma)\int_{-\infty}^{+\infty} \|\hat{u}_m(\tau)\|_V^2 d\tau + c_7(\gamma)\int_{-\infty}^{+\infty} \frac{|\tau|^{\alpha-1}\|\hat{u}_m(\tau)\|_V}{1+|\tau|^{\alpha-2\gamma}}d\tau.$$

Applying the Parseval's equality, the first integral is bounded as $m \to +\infty$, thus (2.70) is showed if one proves that

$$\int_{-\infty}^{+\infty} \frac{|\tau|^{\alpha-1}\|\hat{u}_m(\tau)\|_V}{1+|\tau|^{\alpha-2\gamma}}d\tau \le \text{const.} \tag{2.74}$$

Making use of the Schwarz's inequality we estimate above integral by

$$\int_{-\infty}^{+\infty} \frac{|\tau|^{\alpha-1}\|\hat{u}_m(\tau)\|_V}{1+|\tau|^{\alpha-2\gamma}}d\tau$$
$$\le \left(\int_{-\infty}^{+\infty} \frac{|\tau|^{\alpha-1}d\tau}{(1+|\tau|^{\alpha-2\gamma})^2}\right)^{\frac{1}{2}}\left(\int_{-\infty}^{+\infty} |\tau|^{\alpha-1}\|\hat{u}_m(\tau)\|_V^2 d\tau\right)^{\frac{1}{2}}.$$

Since

$$\int_{-\infty}^{+\infty} \frac{|\tau|^{\alpha-1}d\tau}{(1+|\tau|^{\alpha-2\gamma})^2} = 2\left(\int_0^1 \frac{|\tau|^{\alpha-1}d\tau}{(1+|\tau|^{\alpha-2\gamma})^2} + \int_1^{+\infty} \frac{|\tau|^{\alpha-1}d\tau}{(1+|\tau|^{\alpha-2\gamma})^2}\right),$$

it is clear that the first integral is finite and the second integral is also finite due to $\gamma < \alpha/4$. On the other hand, it follows from the Parseval's equality that

$$\int_{-\infty}^{+\infty} |\tau|^{\alpha-1} \|\hat{u}_m(\tau)\|_V^2 d\tau = \int_{-\infty}^{+\infty} \left(\|_{-\infty} D_t^{\frac{\alpha}{2}-\frac{1}{2}} \tilde{u}_m(t)\|_V\right)^2 dt$$

$$= \int_0^T \|_0 D_t^{\frac{\alpha}{2}-\frac{1}{2}} u_m(t)\|_V^2 dt$$

$$\leq \left(\frac{T^{\frac{1}{2}-\frac{\alpha}{2}}}{\Gamma(\frac{3}{2}-\frac{\alpha}{2})}\right)^2 \int_0^T \|u_m(t)\|_V^2 dt,$$

which implies that (2.74) holds by (2.68). In a word, (2.70) and (2.69) hold.

Step 4. (2.67) and (2.68) enable us to say the existence of an element in $L^2(0,T;V) \cap L^\infty(0,T;H)$ and a subsequence $\{u_{m'}\}$ such that

$$u_{m'} \rightharpoonup u \quad \text{weakly in } L^2(0,T;V) \text{ and weak} - \text{star in } L^\infty(0,T;H), \text{ as } m \to \infty.$$

Owing to (2.69) and Theorem 2.14, we get

$$u_{m'} \to u \quad \text{strongly in } L^2(0,T;H).$$

Let $v \in V$. Take the scalar product of (2.61) with v and integrate

$$(u_{m'}(t),v) - (u_{m'}(t_0),v)$$

$$= \frac{1}{\Gamma(\alpha)} \int_0^{t_0} \left((t_0-s)^{\alpha-1} - (t-s)^{\alpha-1}\right)$$

$$\times [\nu((u_{m'}(s),v)) + b(u_{m'}(s),u_{m'}(s),P_{m'}v) - \langle f_{m'}(s),v\rangle] ds$$

$$- \frac{1}{\Gamma(\alpha)} \int_{t_0}^t (t-s)^{\alpha-1} [\nu((u_{m'}(s),v)) + b(u_{m'}(s),u_{m'}(s),P_{m'}v) - \langle f_{m'}(s),v\rangle] ds.$$

$$(2.75)$$

Since $u_{m'} \rightharpoonup u$ weakly in $L^2(0,T;V)$, by extracting a subsequence, relabeled $u_{m'}$, we assume that $u_{m'}(t_0) \to u(t_0)$ in V for all $t_0 \in [0,T]\backslash K$, for some K satisfying $\mathrm{mes}(K) = 0$. Thus $\lim_{m\to\infty} u_{m'}(t_0) = u(t_0)$ in H for $t_0 \notin K$. Now by (2.66) and using the Lebesgue's dominated convergence theorem, we have

$$\lim_{m\to\infty} \int_{t_0}^t (t-s)^{\alpha-1}((u_{m'}(s),v)) ds = \int_{t_0}^t (t-s)^{\alpha-1}((u(s),v)) ds,$$

$$\lim_{m\to\infty} \int_0^{t_0} \left((t_0-s)^{\alpha-1} - (t-s)^{\alpha-1}\right)((u_{m'}(s),v)) ds$$

$$= \int_0^{t_0} \left((t_0-s)^{\alpha-1} - (t-s)^{\alpha-1}\right)((u(s),v)) ds.$$

A simple argument shows that

$$\lim_{m\to\infty} \int_{t_0}^t (t-s)^{\alpha-1} b(u_{m'}(s),u_{m'}(s),P_{m'}v) ds = \int_{t_0}^t (t-s)^{\alpha-1} b(u(s),u(s),v) ds,$$

$$\lim_{m\to\infty} \int_0^{t_0} \left((t_0-s)^{\alpha-1} - (t-s)^{\alpha-1}\right) b(u_{m'}(s),u_{m'}(s),P_{m'}v) ds$$

$$= \int_0^{t_0} \left((t_0-s)^{\alpha-1} - (t-s)^{\alpha-1}\right) b(u(s),u(s),v) ds.$$

Passing to the limit of (2.75) with the sequence m', one finds that the equality for $t \geq t_0$ and $t, t_0 \notin K$,

$$(u(t), v) - (u(t_0), v)$$

$$= \frac{1}{\Gamma(\alpha)} \int_0^{t_0} \left((t_0 - s)^{\alpha-1} - (t - s)^{\alpha-1} \right) [\nu((u(s), v)) + b(u(s), u(s), v) - \langle f(s), v \rangle] ds$$

$$- \frac{1}{\Gamma(\alpha)} \int_{t_0}^t (t - s)^{\alpha-1} [\nu((u(s), v)) + b(u(s), u(s), v) - \langle f(s), v \rangle] ds.$$

$$(2.76)$$

It is clear that $(u(t), v) - (u(t_0), v) \to 0$ as $t - t_0 \to 0$ after a fundamental calculation. This follows the weak continuity of $u(t)$ in H since V is dense in H. Applying the Caputo fractional derivative of order α on the sides of (2.76) for $t_0 = 0$, we obtain (2.58). $\qquad\square$

We address the uniqueness of weak solutions in the case that $n = 2$.

Theorem 2.16. *The solution u of the problem (2.60) given by Theorem 2.15 is unique.*

Proof. Let us assume that u_1 and u_2 are two solutions of (2.60) and let $u = u_1 - u_2$. We obtain for u the equation

$$\begin{cases} {}^C_0 D_t^\alpha u + \nu A u + B(u_1, u) + B(u, u_2) = 0, \\ u(0) = 0. \end{cases} \qquad (2.77)$$

Taking the scalar product of (2.77) with u we obtain

$$\langle {}^C_0 D_t^\alpha u, u \rangle + \nu \|u\|_V^2 + b(u, u_2, u) = 0.$$

We proceed similarly as the proof of Theorem 2.15 to show that

$${}^C_0 D_t^\alpha |u|^2 + 2\nu \|u\|_V^2 \leq 2b(u, u, u_2). \qquad (2.78)$$

By the Hölder's inequality,

$$|b(u, u, u_2)| \leq c_0 |u| \|u\|_V \|u_2\|_V.$$

Then we bound the right-hand side of (2.78) by

$$|b(u, u, u_2)| \leq c_0 |u| \|u\|_V \|u_2\|_V \leq \nu \|u\|_V^2 + c_1 |u|^2 \|u_2\|_V^2,$$

we get

$${}^C_0 D_t^\alpha |u(t)|^2 \leq 2c_1 |u(t)|^2 \|u_2(t)\|_V^2$$

or

$$|u(t)|^2 \leq |u(0)|^2 + \frac{c_1}{\Gamma(\alpha)} \int_0^t (t - s)^{\alpha-1} |u(s)|^2 \|u_2(s)\|_V^2 ds.$$

By the Gronwall's inequality, (2.66) and (2.67),

$$|u(t)|^2 \leq |u(0)|^2 \exp\left(\frac{c_1}{\Gamma(\alpha)} \int_0^t (t - s)^{\alpha-1} \|u_2(s)\|_V^2 ds \right)$$

$$\leq |u(0)|^2 \exp\left[\frac{c_1}{\nu} \left(|u_2(0)|^2 + \frac{1}{\nu\Gamma(\alpha)} \int_0^T |f(s)|_{V'}^{2/\alpha_1} ds + \frac{T^{1+b}}{(1+b)\nu\Gamma(\alpha)} \right) \right].$$

Since $|u(0)| = 0$, it follows that $|u| \equiv 0$. The proof is completed. $\qquad\square$

2.5.3 Optimal Control

In this subsection we study the optimal control problem:

$$\text{minimize} \quad \frac{1}{2}\int_0^T \int_\Omega (u(t,x) - z(t,x))^2 dxdt + \int_0^T h(w(t))dt \qquad (2.79)$$

over $w \in L^{\frac{2}{\alpha_1}}(0,T;U)$ and $u \in L^2([0,T] \times \Omega; \mathbb{R}^2)$ subject to systems (2.52).

Let $C \in \mathcal{L}(U,H)$ be given by $C = PC_0$. By the same way as we derived (2.58), then we give the concept of weak solutions of systems (2.52).

Definition 2.7. Let $a \in H$, $f \in L^{\frac{2}{\alpha_1}}(0,T;V')$ and $w \in L^{\frac{2}{\alpha_1}}(0,T;U)$. A function $u \in L^2(0,T;V) \cap C_w([0,T];H)$ is said a weak solution of systems (2.52) if

$$\begin{cases} \partial_t^\alpha(u,v) + \nu((u,v)) + b(u,u,v) = \langle C_0 w + f, v\rangle, & \forall\, v \in V, \\ u(0) = a. \end{cases}$$

Similarly we give the equivalent form as follows:

$$\begin{cases} {}_0^C D_t^\alpha u + \nu Au + B(u,u) = Cw + f, \\ u(0) = a. \end{cases} \qquad (2.80)$$

If $z \in L^2(0,T;H)$, rewriting the problem (2.79) in the abstract form, we have

$$\text{(P)} \quad \text{minimize } J(u,w) = \frac{1}{2}\int_0^T (u(t) - z(t))^2 dt + \int_0^T h(w(t))dt$$

over $(u,w) \in (L^2(0,T;V) \cap C_w([0,T];H)) \times L^{\frac{2}{\alpha_1}}(0,T;U)$ subject to (2.80).

We assume that:

(i) $z \in L^2(0,T;H)$;

(ii) the function $h: U \to \mathbb{R}$ is convex, lower semi-continuous and satisfies

$$|h(w)| \geq b_1 |w|_U^{\frac{2}{\alpha_1}} + b_2,$$

for some $b_1 > 0, b_2 \in \mathbb{R}$.

Theorem 2.17. *If hypotheses (i) and (ii) hold, then the problem (P) has at least one solution* $(u^*, w^*) \in (L^2(0,T;V) \cap C_w([0,T];H)) \times L^{\frac{2}{\alpha_1}}(0,T;U)$.

Proof. According to Theorems 2.15 and 2.16, by replacing f with $Cw + f$ we show that equation (2.80) has a weak solution corresponding to w. Suppose that (u_m, w_m) are a minimizing sequence of problem (P), that is,

$$\begin{aligned} & J(u_m, w_m) \to \inf J(u,w), \\ & {}_0^C D_t^\alpha u_m + \nu Au_m + B(u_m, u_m) = Cw_m + f, \ t \in [0,T], \qquad (2.81) \\ & u_m(0) = a. \end{aligned}$$

Hypothesis (ii) ensures that $\{w_m\}$ is bounded in $L^{\frac{2}{\alpha_1}}(0,T;U)$, then there exists a subsequence, relabeled as $\{w_m\}$ and $w^* \in L^{\frac{2}{\alpha_1}}(0,T;U)$ such that

$$w_m \rightharpoonup w^* \quad \text{weakly in } L^{\frac{2}{\alpha_1}}(0,T;U).$$

Similar to (2.65), we get

$$\frac{1}{2}{}_0^C D_t^\alpha |u_m(t)|^2 + \nu\|u_m(t)\|_V^2 \leq \langle Cw_m + f(t), u_m(t)\rangle.$$

This yields

$$|u_m(t)|^2 + \frac{\nu}{\Gamma(\alpha)}\int_0^t (t-s)^{\alpha-1}\|u_m(s)\|_V^2 ds$$

$$\leq |u_m(0)|^2 + \frac{1}{\nu\Gamma(\alpha)}\int_0^T |Cw_m(s) + f(s)|_{V'}^{2/\alpha_1} ds + \frac{T^{1+b}}{(1+b)\nu\Gamma(\alpha)}.$$

Hence,

$$\frac{\nu}{\Gamma(\alpha)}T^{\alpha-1}\int_0^t \|u_m(s)\|_V^2 ds$$

$$\leq \frac{\nu}{\Gamma(\alpha)}\int_0^t (t-s)^{\alpha-1}\|u_m(s)\|_V^2 ds$$

$$\leq |u_m(0)|^2 + \frac{1}{\nu\Gamma(\alpha)}\int_0^T |Cw_m(s) + f(s)|_{V'}^{2/\alpha_1} ds + \frac{T^{1+b}}{(1+b)\nu\Gamma(\alpha)}.$$

Therefore

the sequence $\{u_m\}$ is bounded in $L^\infty(0,T;H)$,

the sequence $\{u_m\}$ is bounded in $L^2(0,T;V)$,

thus there exists an element $u^* \in L^2(0,T;V) \cap L^\infty(0,T;H)$ such that

$$u_m \rightharpoonup u^* \quad \text{weakly in } L^2(0,T;V) \text{ and weak} - \text{star in } L^\infty(0,T;H), \text{ as } m \to \infty.$$

Meanwhile, we derive from similar argument as given in the Step 3 of Theorem 2.15 that

$$u_m \to u^* \quad \text{strongly in } L^2(0,T;H).$$

We need to give some uniform bounds on Au_m and $B(u_m, u_m)$. Since $\{u_m\} \subseteq L^2(0,T;V)$ is bounded, then $\{Au_m\} \subseteq L^2(0,T;V')$ is bounded. Actually,

$$\int_0^t |Au_m(s)|_{V'}^2 ds = \int_0^t \|u_m(s)\|_V^2 ds.$$

Therefore, choosing a subsequence, if necessary, we have

$$Au_m \rightharpoonup Au^* \quad \text{weakly in } L^2(0,T;V').$$

It remains to investigate the term $P_m B(u_m, u_m)$. Now, consider

$$|(A^{-\frac{1}{2}}B(u_m, u_m), v)| \leq \tilde{c}|u_m|\|u_m\|_V\|v\|_V, \ \tilde{c} > 0,$$

then

$$|B(u_m, u_m)|_{V'} = |A^{-\frac{1}{2}}B(u_m, u_m)| \leq \tilde{c}|u_m|\|u_m\|_V.$$

Since $|u_m(s)|$ is uniformly bounded and $\|u_m(s)\|_V^2$ is uniformly integrable, we obtain that

$$\int_0^t |B(u_m(s), u_m(s))|_{V'}^2\, ds \leq \tilde{c} \int_0^t |u_m(s)|^2 \|u_m(s)\|_V^2\, ds$$

$$\leq \tilde{c} \sup |u_m|^2 \int_0^t \|u_m(s)\|_V^2\, ds < \infty.$$

This implies that there exists η^* such that

$$B(u_m, u_m) \rightharpoonup \eta^* \quad \text{weakly in } L^2(0, T; V').$$

Now by virtue of (2.59) we have

$$|(B(u_m(t), u_m(t)) - B(u^*(t), u^*(t)), v)|$$
$$\leq |b(u_m(t) - u^*(t), u_m(t), v)| + |b(u^*(t), u_m(t) - u^*(t), v)|$$
$$\leq C|u_m(t) - u^*(t)|^{\frac{1}{2}} \|u_m(t) - u^*(t)\|_V^{\frac{1}{2}} |u_m(t)|^{\frac{1}{2}} \|u_m(t)\|_V^{\frac{1}{2}} \|v\|_V$$
$$+ C|u^*(t)|^{\frac{1}{2}} \|u^*(t)\|_V^{\frac{1}{2}} |u_m(t) - u^*(t)|^{\frac{1}{2}} \|u_m(t) - u^*(t)\|_V^{\frac{1}{2}} \|v\|_V, \ \forall\, v \in V,$$

we infer that

$$B(u_m, u_m) \rightharpoonup B(u^*, u^*) \quad \text{weakly in } L^2(0, T; V').$$

Thus $\eta^*(t) = B(u^*(t), u^*(t))$ a.e. $t \in [0, T]$.

Let $v \in V$. Take the scalar product of (2.81) with v and integrate, it follows

$$(u_m(t), v) - (u_m(t_0), v)$$
$$= \frac{1}{\Gamma(\alpha)} \int_0^{t_0} ((t_0 - s)^{\alpha-1} - (t - s)^{\alpha-1}) [\nu((u_m(s), v))$$
$$+ b(u_m(s), u_m(s), P_m v) - \langle Cw_m(s) + f(s), v \rangle] ds$$
$$- \frac{1}{\Gamma(\alpha)} \int_{t_0}^t (t - s)^{\alpha-1}$$
$$\times [\nu((u_m(s), v)) + b(u_m(s), u_m(s), P_m v) - \langle Cw_m(s) + f(s), v \rangle] ds.$$

By similar arguments with Step 4 in Theorem 2.15 and taking the limits as $m \to \infty$, we know that

$$(u^*(t), v) - (u^*(t_0), v)$$
$$= \frac{1}{\Gamma(\alpha)} \int_0^{t_0} ((t_0 - s)^{\alpha-1} - (t - s)^{\alpha-1}) [\nu((u^*(s), v))$$
$$+ b(u^*(s), u^*(s), v) - \langle Cw^*(s) + f(s), v \rangle] ds$$
$$- \frac{1}{\Gamma(\alpha)} \int_{t_0}^t (t - s)^{\alpha-1} [\nu((u^*(s), v)) + b(u^*(s), u^*(s), v) - \langle Cw^*(s) + f(s), v \rangle] ds.$$

This similarly yields the weak continuity of u^*. By applying the Caputo fractional derivative of order α on the sides of the above equality for $t_0 = 0$, we obtain

$$\begin{cases} \langle {}_0^C D_t^\alpha u^*, v \rangle + \nu((u^*, v)) + b(u^*, u^*, v) = \langle Cw^* + f, v \rangle, \quad t \in [0, T], \text{ for all } v \in V, \\ u^*(0) = a. \end{cases}$$

Clearly, an equivalent form of the last equality is

$$\begin{cases} {}^{C}_{0}D^{\alpha}_{t}u^* + \nu Au^* + B(u^*, u^*) = Cw^* + f, \\ u^*(0) = a. \end{cases}$$

Hence, we see that (u^*, w^*) satisfies system (2.81).

From weak lower semi-continuity of the function $(u, w) \mapsto J(u, w)$ we deduce $J(u^*, w^*) = \inf J(u, w)$. Consequently, (u^*, w^*) is an optimal pair for problem (P). This completes the proof. $\qquad \square$

2.6 Energy Methods

In this section we consider the following Navier-Stokes equations with time-fractional derivatives in $\mathbb{R}^3$:

$$\begin{cases} \partial^{\alpha}_t u + (u \cdot \nabla)u = -\nabla p + \nu \Delta u, \quad t > 0, \\ \nabla \cdot u = 0, \\ u(0, x) = u_0, \end{cases} \tag{2.82}$$

where ∂^{α}_t is the Caputo fractional derivative of order $\alpha \in (0, 1)$, $u = (u_1(t, x), u_2(t, x), u_3(t, x))$ represents the velocity field at a point $x \in \mathbb{R}^3$ and time $t > 0$, $p = p(t, x)$ is the pressure, ν the viscosity and $u_0 = u_0(x)$ is the initial velocity. It is worth-mentioning that the purpose of this section is to develop a new global result when we relax the smallness condition on the initial data.

The section is organized as follows. In Subsection 2.6.1 we recall some notations, definitions, and preliminary facts. In Subsection 2.6.2 we construct a regularized equation, and show the local existence of its solutions. Subsection 2.6.3 deals with the global existence and continuation of solutions for a kind of fractional differential equations in a Banach space. Via energy methods, we also obtain the convergence of approximate solutions and the global existence and uniqueness of mild solutions of problem (2.83) in different spaces.

2.6.1 *Notations and Lemmas*

Here we recall some notations, definitions, and preliminary facts.

Denote by $\| \cdot \|$ the L^2 norm on $\mathbb{R}^3$, where

$$\|u\| = \left(\int_{\mathbb{R}^3} |u(x)|^2 dx \right)^{\frac{1}{2}}.$$

By $H^m(\mathbb{R}^3)$, $m \in \mathbb{N}_0$, we denote the Sobolev space consisting of functions $u \in L^2(\mathbb{R}^3)$ such that $D^{\beta}u \in L^2(\mathbb{R}^3)$, $0 \leq |\beta| \leq m$ with the norm $\| \cdot \|_m$ defined by

$$\|u\|_m = \left(\sum_{0 \leq |\beta| \leq m} \|D^{\beta}u\|^2 \right)^{\frac{1}{2}}.$$

For $s \in \mathbb{R}$, the Sobolev space $H^s(\mathbb{R}^3)$ is the completion of $\mathcal{S}(\mathbb{R}^3)$ with respect to the norm

$$\|u\|_s = \left(\int_{\mathbb{R}^3} (1 + |\xi|^2)^s |\widehat{u}(\xi)|^2 d\xi \right)^{\frac{1}{2}}.$$

Here $\widehat{u}$ represents the Fourier transform of u and $\mathcal{S}(\mathbb{R}^3)$ denotes the Schwarz space of rapidly decreasing smooth functions. It is clear that the two norms are equivalent for $s = m$.

Let ϱ be the standard mollifier satisfying

$$\varrho(|x|) \in C_0^\infty(\mathbb{R}^3), \quad 0 \le \varrho \le 1, \quad \int_{\mathbb{R}^3} \varrho(x)dx = 1.$$

We define the mollification $\mathcal{J}_\varepsilon u$ of the function $u \in L^q(\mathbb{R}^3)$ $(1 \le q \le \infty)$ as

$$(\mathcal{J}_\varepsilon u)(x) = \varepsilon^{-3} \int_{\mathbb{R}^3} \varrho\left(\frac{x-y}{\varepsilon}\right) u(y)dy.$$

We introduce the space

$$V^s := \{u \in H^s(\mathbb{R}^3) : \nabla \cdot u = 0\}$$

and call the operator $P : H^s(\mathbb{R}^3) \to V^s$ the Leray projector. Clearly, P commutes with $\mathcal{J}_\varepsilon$. For more details, we refer the reader to Majda and Bertozzi [256].

Let us denote an important space: for $m \in \mathbb{Z}$,

$$\mathcal{X}^m = \left\{ u \in \mathcal{D}'(\mathbb{R}^3) : \int_{\mathbb{R}^3} |\xi|^m |\widehat{u}|d\xi < \infty \right\}.$$

Here $\mathcal{D}'(\mathbb{R}^3)$ stands for the space of distributions with the norm $\|\cdot\|_{\mathcal{X}^m}$ given by

$$\|u\|_{\mathcal{X}^m} = \int_{\mathbb{R}^3} |\xi|^m |\widehat{u}|d\xi.$$

It is not difficult to show that if $u \in H^1$, then $u \in \mathcal{X}^{-1}$.

For $u : [0, \infty) \times \mathbb{R}^3 \to \mathbb{R}^3$, Caputo time-fractional derivative of the function u can generally be written as

$$\partial_t^\alpha u(t, x) = \frac{\partial}{\partial t}\left[{}_0D_t^{\alpha-1}\big(u(t, x) - u(0, x)\big)\right], \quad t > 0.$$

Before proceeding further, we present two important results which play a key role in proving the main results.

Lemma 2.18. *[188] Let $T > 0$. Then, for any $v \in L^2((0, T] \times \mathbb{R}^3; \mathbb{R}^3)$, there holds*

$$\int_{\mathbb{R}^3} v(t, x)\partial_t^\alpha v(t, x)dx \ge \|v\|\|\partial_t^\alpha v\|.$$

The following result is a generalization of Lemma 6.1 in Kemppainen et al. [188].

Lemma 2.19. *Let the function $v : (0, T] \times \mathbb{R}^3 \to \mathbb{C}^3$ and $\partial_t^\alpha v$ exist. Then*

$$v(t)\partial_t^\alpha \overline{v(t)} + \overline{v(t)}\partial_t^\alpha v(t) \ge 2|v||\partial_t^\alpha|v|,$$

where $\overline{v(t)}$ denotes the dual of $v(t)$.

Proof. Let $v(t) = a(t) + b(t)i$. Then

$$v(t)\partial_t^\alpha \overline{v(t)} + \overline{v(t)}\partial_t^\alpha v(t)$$

$$= 2\big(a(t)\partial_t^\alpha a(t) + b(t)\partial_t^\alpha b(t)\big)$$

$$= 2\Big[\partial_t^\alpha a^2(t) + \partial_t^\alpha b^2(t) + t^{-\alpha}\big(a^2(t) + b^2(t)\big)$$

$$+ \int_0^t |a(t) - a(t-s)|^2 [\alpha s^{-\alpha-1}]ds + \int_0^t |b(t) - b(t-s)|^2 [\alpha s^{-\alpha-1}]ds\Big]$$

$$\geq 2\Big(\partial_t^\alpha |v(t)|^2 + t^{-\alpha}|v(t)|^2 + \int_0^t \big(|v(t)| - |v(t-s)|\big)^2 [\alpha s^{-\alpha-1}]ds\Big)$$

$$= 2|v(t)|\partial_t^\alpha |v(t)|.$$

$\qquad\qquad\qquad\qquad\qquad\qquad\qquad\qquad\qquad\qquad\qquad\qquad\qquad\qquad$ □

2.6.2   *Local Existence*

This subsection is concerned with the construction of an approximate (regularized) equation for the time-fractional Navier-Stokes equations and obtain the existence and some properties of its solution. To do this, we use the mollifier $\mathcal{J}_\varepsilon$ to regularize equation (2.82):

$$\begin{cases} \partial_t^\alpha u^\varepsilon + \mathcal{J}_\varepsilon[(\mathcal{J}_\varepsilon u^\varepsilon) \cdot \nabla(\mathcal{J}_\varepsilon u^\varepsilon)] = -\nabla p^\varepsilon + \nu \mathcal{J}_\varepsilon \Delta(\mathcal{J}_\varepsilon u^\varepsilon), \\ \nabla \cdot u^\varepsilon = 0, \\ u^\varepsilon(0) = \mathcal{J}_\varepsilon u_0. \end{cases} \qquad (2.83)$$

By applying Leray projector P to equation (2.83), we get rid of the pressure term in equation (2.83), which reduces to an ordinary differential equation (ODE) in V^s:

$$\begin{cases} \partial_t^\alpha u^\varepsilon + P\mathcal{J}_\varepsilon[(\mathcal{J}_\varepsilon u^\varepsilon) \cdot \nabla(\mathcal{J}_\varepsilon u^\varepsilon)] = \nu \mathcal{J}_\varepsilon^2 \Delta u^\varepsilon, \\ u^\varepsilon(0) = \mathcal{J}_\varepsilon u_0. \end{cases} \qquad (2.84)$$

We set

$$F_\varepsilon(u^\varepsilon) = \nu \mathcal{J}_\varepsilon^2 \Delta u^\varepsilon - P\mathcal{J}_\varepsilon[(\mathcal{J}_\varepsilon u^\varepsilon) \cdot \nabla(\mathcal{J}_\varepsilon u^\varepsilon)].$$

From the argument of Proposition 3.6 in Majda and Bertozzi [256], we know that

$$\|F_\varepsilon(u^1) - F_\varepsilon(u^2)\|_m \leq L(\|u^j\|, \varepsilon, M)\|u^1 - u^2\|_m, \quad \text{for } u^j \in B_M, \ j = 1, 2, \quad (2.85)$$

where $m \in \mathbb{N}_0$ and

$$B_M = \{u \in V^m : \|u\|_m < M\}.$$

It means that F_ε is locally Lipschitz continuous on any open set.

Let

$$C^\alpha([0,T]; V^m) = \{u \in C([0,T]; V^m) : \partial_t^\alpha u \in C([0,T]; V^m)\}.$$

Theorem 2.18. *Assume the initial data $u_0 \in V^0$. Then, for any $\varepsilon > 0$,*

(i) *there exists $T_\varepsilon = T(\|\mathcal{J}_\varepsilon u_0\|_m, \varepsilon)$ such that equation (2.84) has a unique solution $u^\varepsilon \in C^\alpha([0, T_\varepsilon); V^m)$;*

(ii) *$u^\varepsilon \in C^\alpha([0, T]; V^0)$ on any interval $[0, T]$ with*

$$\sup_{t \in [0,T]} \|u^\varepsilon\| \leq \|u_0\|. \tag{2.86}$$

Proof. (i) If $u_0 \in V^0$, then $\mathcal{J}_\varepsilon u_0 \in V^m$, $m \in \mathbb{N}_0$. For given $r > 0$, define

$$B(r, T_\varepsilon) = \left\{ u^\varepsilon \in C([0, T_\varepsilon); V^m) : \sup_{t \in [0, T_\varepsilon)} \|u^\varepsilon(t) - \mathcal{J}_\varepsilon u_0\|_m \leq r \right\}.$$

Notice that

$$M = \sup_{(t, u^\varepsilon) \in [0, T_\varepsilon) \times B(1, T_\varepsilon)} \|F_\varepsilon(u^\varepsilon)\|_m < +\infty. \tag{2.87}$$

Indeed,

$$\begin{aligned} \|F_\varepsilon(u^\varepsilon)\|_m &\leq \|F_\varepsilon(u^\varepsilon) - F_\varepsilon(\mathcal{J}_\varepsilon u_0)\|_m + \|F_\varepsilon(\mathcal{J}_\varepsilon u_0)\|_m \\ &\leq L\|u^\varepsilon - \mathcal{J}_\varepsilon u_0\|_m + \|F_\varepsilon(\mathcal{J}_\varepsilon u_0)\|_m \\ &\leq Lr + \|F_\varepsilon(\mathcal{J}_\varepsilon u_0)\|_m < +\infty. \end{aligned}$$

Consider the operator $\mathcal{T}$:

$$\mathcal{T}u^\varepsilon(t) = \mathcal{J}_\varepsilon u_0 + \frac{1}{\Gamma(\alpha)} \int_0^t (t - s)^{\alpha - 1} F_\varepsilon(u^\varepsilon(s)) ds.$$

Obviously $\mathcal{T}(B(r, T_\varepsilon)) \subset B(r, T_\varepsilon)$. Moreover, for $u^1, u^2 \in B(r, T_\varepsilon)$,

$$\|\mathcal{T}u^1(t) - \mathcal{T}u^2(t)\|_m \leq \frac{Lt^\alpha}{\Gamma(\alpha + 1)} \sup_{s \in [0, t]} \|u^1(s) - u^2(s)\|_m, \quad \text{for } t \in [0, T_\varepsilon).$$

Fixing $T_\varepsilon < \min \left\{ \left(\frac{\Gamma(\alpha + 1)r}{M} \right)^{\frac{1}{\alpha}}, \left(\frac{\Gamma(\alpha + 1)}{L} \right)^{\frac{1}{\alpha}} \right\}$, it is easy to show that $\mathcal{T}$ is a strict contraction mapping on $B_r(r, T_\varepsilon)$. In consequence, we deduce that $\mathcal{T}$ has a fixed point.

On account of (2.85) and $u^\varepsilon \in C([0, T_\varepsilon); V^m)$, it is clear that $\partial_t^\alpha u^\varepsilon \in C([0, T_\varepsilon); V^m)$.

(ii) Take the L^2 inner product of equation (2.84) with u^ε and apply Lemma 2.18 to obtain

$$\|u^\varepsilon\|\partial_t^\alpha\|u^\varepsilon\| \leq \nu \int_{\mathbb{R}^3} u^\varepsilon \mathcal{J}_\varepsilon^2 \Delta u^\varepsilon dx - \int_{\mathbb{R}^3} u^\varepsilon P \mathcal{J}_\varepsilon[(\mathcal{J}_\varepsilon u^\varepsilon) \cdot \nabla(\mathcal{J}_\varepsilon u^\varepsilon)] dx.$$

Integrating by parts and using the fact that $\nabla \cdot u^\varepsilon = 0$, we get

$$\|u^\varepsilon\|\partial_t^\alpha\|u^\varepsilon\| + \nu\|\mathcal{J}_\varepsilon \nabla u^\varepsilon\|^2 \leq 0.$$

Since $\nu \geq 0$, we have

$$\partial_t^\alpha\|u^\varepsilon\| \leq 0.$$

Then the definition of $\partial_t^\alpha u$ implies that

$$\sup_{t \in [0,T]} \|u^\varepsilon\| \leq \|\mathcal{J}_\varepsilon u_0\| \leq \|u_0\|.$$

$\square$

In the next lemma, we derive a key estimate which plays an important role in proving the main results.

Lemma 2.20. *The regularized solution u^ε of equation (2.84) satisfies the inequality:*

$$\partial_t^\alpha \|u^\varepsilon\|_m \le c_m |\mathcal{J}_\varepsilon \nabla u^\varepsilon|_{L^\infty} \|u^\varepsilon\|_m.$$

Proof. We take the derivative D^β of equation (2.84) and L^2 inner product with $D^\beta u^\varepsilon$ to obtain

$$\langle D^\beta \partial_t^\alpha u^\varepsilon, D^\beta u^\varepsilon \rangle$$
$$= \nu \langle D^\beta \mathcal{J}_\varepsilon^2 \Delta u^\varepsilon, D^\beta u^\varepsilon \rangle - \langle D^\beta P \mathcal{J}_\varepsilon [(\mathcal{J}_\varepsilon u^\varepsilon) \cdot \nabla(\mathcal{J}_\varepsilon u^\varepsilon)], D^\beta u^\varepsilon \rangle$$
$$= - \nu \|\mathcal{J}_\varepsilon D^\beta \nabla u^\varepsilon\|^2 - \langle P \mathcal{J}_\varepsilon [(\mathcal{J}_\varepsilon u^\varepsilon) \cdot \nabla(D^\beta \mathcal{J}_\varepsilon u^\varepsilon)], D^\beta u^\varepsilon \rangle$$
$$- \langle D^\beta P \mathcal{J}_\varepsilon [(\mathcal{J}_\varepsilon u^\varepsilon) \cdot \nabla(\mathcal{J}_\varepsilon u^\varepsilon)] - P \mathcal{J}_\varepsilon [(\mathcal{J}_\varepsilon u^\varepsilon) \cdot \nabla(D^\beta \mathcal{J}_\varepsilon u^\varepsilon)], D^\beta u^\varepsilon \rangle.$$

From the discussion of Majda and Bertozzi [256] and Lemma 2.18, we deduce that

$$\|u^\varepsilon\|_m \partial_t^\alpha \|u^\varepsilon\|_m + \nu \|\mathcal{J}_\varepsilon \nabla u^\varepsilon\|_m^2 \le c_m |\mathcal{J}_\varepsilon \nabla u^\varepsilon|_{L^\infty} \|u^\varepsilon\|_m^2,$$

which takes the following form for $\nu > 0$,

$$\partial_t^\alpha \|u^\varepsilon\|_m \le c_m |\mathcal{J}_\varepsilon \nabla u^\varepsilon|_{L^\infty} \|u^\varepsilon\|_m.$$

$\square$

2.6.3   Global Existence

Here we discuss the global existence of solutions by means of mathematical analysis and previous estimate when the initial data being less than the viscosity.

Firstly, we consider the following autonomous equation

$$_0^C D_t^\alpha z(t) = F(z), \quad z(0) = z_0. \tag{2.88}$$

Lemma 2.21. *Let X be a Banach space and $U \subset X$ be an open set. Suppose that the mapping $F : U \to X$ satisfies the locally Lipschitz condition, that is, for any $z \in U$ there is a constant $L > 0$ and an open neighborhood $U_z \subset U$ of z such that*

$$|F(z_1) - F(z_2)|_X \le L|z_1 - z_2|_X, \quad \text{for all } z_1, z_2 \in U_z.$$

Then for any $z_0 \in U$, there exists a time T such that equation (2.88) has a unique (local) solution $z \in C([0, T); U)$.

Proof. The method for showing the local existence of solutions for equation (2.88) is similar to that of Theorem 2.18 (i), so we omit it. $\square$

Now we establish the global existence of solutions for a kind of fractional differential equations in Banach space, which furnishes as the key instrument in the proof of the global existence of solutions for the time-fractional Navier-Stokes equations.

Lemma 2.22. *Under the conditions of Lemma 2.21, the unique solution $z = z(t)$ ($t \in [0, T)$) of the fractional differential equations (2.88) either exists globally in time, or $T < \infty$ and $z(t)$ preserves the open set U as $t \to T^-$.*

Proof. Suppose that the maximum existing interval for the solution $z(t)$ is $[0, T)$, that is,

$$T = \sup\{T_z : z(t) \text{ is defined on } [0, T_z] \text{ and } z(t) \text{ is a solution of (2.88) on } [0, T_z]\}.$$

Then $T = \infty$ or $T < \infty$. If $T = \infty$, the conclusion holds. If $T < \infty$, we show that $z(t)$ preserves the open set U as $t \to T^-$. Observe that there exists a sequence $\{t_k\}$ and a positive constant $K > 0$ such that

$$t_n \leq t_{n+1} \text{ for } n \in \mathbb{N}^+, \quad \lim_{n \to \infty} t_n = T, \quad |z(t_n)| \leq K. \tag{2.89}$$

According to the equicontinuity of $z(t)$ and Lemma 2.21, $\{z(t_n)\}$ has a convergent subsequence. Without loss of generality, let $\lim_{n \to \infty} z(t_n) = z^*$. This together with (2.89) implies that for sufficiently small $\tau > 0$, there exists n_0 such that $T - \tau < t_{n_0} < T$ and for $n \geq n_0$, we have $|z(t_n) - z^*| \leq \frac{\varepsilon}{2}$.

We show that $\lim_{t \to T^-} z(t) = z^*$. On the contrary, for $n \geq n_0$, there exists $\eta_n \in (t_n, T)$ such that $|z(\eta_n) - z^*| \geq \varepsilon$ and $|z(t) - z^*| < \varepsilon$, for $t \in (t_n, \eta_n)$. By the continuity of F on $[0, T)$, we denote $\overline{M} = \sup_{s \in [0, \eta_n]} |F(z(s))|$. Thus,

$$\varepsilon \leq |z(\eta_n) - z^*| \leq |z(t_n) - z^*| + |z(\eta_n) - z(t_n)|$$

$$\leq \frac{\varepsilon}{2} + \frac{1}{\Gamma(\alpha)} \int_0^{t_n} \left((t_n - s)^{\alpha-1} - (\eta_n - s)^{\alpha-1}\right)|F(z(s))| ds$$

$$+ \frac{1}{\Gamma(\alpha)} \int_{t_n}^{\eta_n} (\eta_n - s)^{\alpha-1} |F(z(s))| ds$$

$$\leq \frac{\varepsilon}{2} + \frac{1}{\Gamma(1+\alpha)} \sup_{s \in [0, t_n]} |F(z(s))| \left((\eta_n - t_n)^\alpha + t_n^\alpha - \eta_n^\alpha\right)$$

$$+ \frac{1}{\Gamma(1+\alpha)} \sup_{s \in [t_n, \eta_n]} |F(z(s))| (\eta_n - t_n)^\alpha$$

$$\leq \varepsilon,$$

which is a contradiction in view of

$$\frac{1}{\Gamma(1+\alpha)} \sup_{s \in [0, \eta_n]} |F(z(s))| \left(2(\eta_n - t_n)^\alpha + t_n^\alpha - \eta_n^\alpha\right)$$

$$\leq \frac{\overline{M}}{\Gamma(1+\alpha)} \left(2(\eta_n - t_n)^\alpha + t_n^\alpha - \eta_n^\alpha\right)$$

$$\leq \frac{\varepsilon}{2}, \quad \text{for sufficiently large } n \geq n_0.$$

Thus $\lim_{t \to T^-} z(t)$ exists.

Next we show that $z^* \in \partial U$. On the contrary, we assume that $z^* \notin \partial U$. Since $z(t) \in U$ $(t \in [0, T))$, therefore, $z^* \in U$. Let $z(t)$ be equal to z^* for $t = T$ and itself for $0 \leq t < T$.

Denote

$$\bar{\varphi}(t) = z_0 + \frac{1}{\Gamma(\alpha)} \int_0^T (t - s)^{\alpha-1} F(z(s)) ds, \quad t \in [T, T_1],$$

with $\bar{\varphi} \in C([T, T_1]; U)$, and define the operator $\mathcal{S}$ as follow:

$$\mathcal{S}y(t) = \bar{\varphi}(t) + \frac{1}{\Gamma(\alpha)} \int_T^t (t-s)^{\alpha-1} F(y(s)) ds, \quad t \in [T, T_1],$$

where $y \in C([T, T_1]; U)$.

Let

$$B_r = \left\{ (t, y) : t \in [T, T_1], \ |y(t)| \leq \sup_{t \in [T, T_1]} |\bar{\varphi}(t)| + r \right\}.$$

On account of the continuity of F on B_r, let $M = \sup_{y \in B_r} |F(y(s))|$. Consider

$$B_r(\bar{\varphi}, T+h) = \left\{ y \in C([T, T+h]; U) : \sup_{t \in [T, T+h]} |y(t) - \bar{\varphi}(t)| \leq r, \ y(T) = \bar{\varphi}(T) \right\},$$

where $h \in \left(0, \min \left\{ T_1 - T, \ \left(\frac{\Gamma(1+\alpha) r}{M} \right)^{\frac{1}{\alpha}} \right\} \right)$.

As argued in Lemma 2.21, we can derive that $\mathcal{S}$ is a strict contraction mapping on $B_r(\bar{\varphi}, T+h)$. Therefore, $\mathcal{S}$ has a fixed point $\tilde{z}(t) \in B_r(\bar{\varphi}, T+h)$, that is,

$$\tilde{z}(t) = \bar{\varphi}(t) + \frac{1}{\Gamma(\alpha)} \int_T^t (t-s)^{\alpha-1} F(\tilde{z}(s)) ds$$

$$= z_0 + \frac{1}{\Gamma(\alpha)} \int_0^t (t-s)^{\alpha-1} F(\bar{z}(s)) ds, \quad t \in [T, T+h],$$

with

$$\bar{z}(t) = \begin{cases} z(t), & t \in [0, T), \\ \tilde{z}(t), & t \in [T, T+h]. \end{cases}$$

Evidently $\bar{z} \in C([0, T+h]; U)$. Thus we find that $\bar{z}(t)$ is the solution of (2.88) on the interval $[0, T+h]$, which contradicts the assumption that $[0, T)$ is the maximum existing interval. $\qquad \square$

Let us introduce

$$L_\alpha^\infty(\mathbb{R}_+, \mathfrak{X}^1) = \left\{ u : \mathbb{R}_+ \to \mathfrak{X}^1 : \int_0^t (t-s)^{\alpha-1} \|u(s)\|_{\mathfrak{X}^1} ds \in L^\infty(\mathbb{R}_+, \mathfrak{X}^1) \right\}.$$

Theorem 2.19. *For $u_0 \in \mathfrak{X}^{-1}$ satisfying*

$$\|u_0\|_{\mathfrak{X}^{-1}} < \nu, \tag{2.90}$$

the problem (2.82) has a unique solution u existing globally in time. Moreover, $u \in C([0, \infty); \mathfrak{X}^{-1}) \cap L_\alpha^\infty(\mathbb{R}_+, \mathfrak{X}^1)$ and the following estimate holds:

$$\sup_{0 \leq t < \infty} \left(\|u(t)\|_{\mathfrak{X}^{-1}} + (\nu - \|u_0\|_{\mathfrak{X}^{-1}}) \int_0^t (t-s)^{\alpha-1} \|\nabla u(s)\|_{L^\infty} ds \right) \leq \|u_0\|_{\mathfrak{X}^{-1}}.$$

Proof. For $u_0 \in \mathfrak{X}^{-1}$, we have $\|\mathcal{J}_\varepsilon u_0\|_{\mathfrak{X}^{-1}} \leq \|u_0\|_{\mathfrak{X}^{-1}}$ by the properties of ϱ. By the arguments similar to the ones employed in Theorem 2.18, there exists a unique locally continuous solution $u^\varepsilon(t,x)$ on the interval $[0, T_\varepsilon)$. The associated pressure p^ε is given by

$$p^\varepsilon = \sum_{i,j=1} R_i R_j u^i u^j = (R \otimes R)(u^\varepsilon \otimes u^\varepsilon),$$

where R is the Riesz operator.

Firstly we have the estimate

$$\|u^\varepsilon(t)\|_{\mathfrak{X}^{-1}} = \int_{|\xi| \leq 1} |\xi|^{-1} |\widehat{u^\varepsilon}(t)| d\xi + \int_{|\xi| \geq 1} |\xi|^{-2} |\widehat{\sqrt{-\Delta} u^\varepsilon}(t)| d\xi \leq C \|u^\varepsilon(t)\|_1. \quad (2.91)$$

This implies that $u^\varepsilon \in L^\infty(0, T_\varepsilon; \mathfrak{X}^{-1})$.

Next, by applying the Fourier transform to (2.83), we obtain

$$\begin{cases} \partial_t^\alpha \widehat{u^\varepsilon} - i \int \widehat{u^\varepsilon}(\eta) \otimes \widehat{u^\varepsilon}(\xi - \eta) d\eta \cdot \xi - i\xi \widehat{p^\varepsilon} + \nu |\xi|^2 \widehat{u^\varepsilon} = 0, \\ \xi \cdot \widehat{u^\varepsilon} = 0. \end{cases} \quad (2.92)$$

From (2.92), one infers that

$$\overline{\widehat{u^\varepsilon}} \cdot \partial_t^\alpha \widehat{u^\varepsilon} - i \int \overline{\widehat{u^\varepsilon}(\xi)} \cdot \widehat{u^\varepsilon}(\eta) \otimes \widehat{u^\varepsilon}(\xi - \eta) d\eta \cdot \xi + \nu |\xi|^2 |\widehat{u^\varepsilon}|^2 = 0$$

and

$$\widehat{u^\varepsilon} \cdot \partial_t^\alpha \overline{\widehat{u^\varepsilon}} + i \int \widehat{u^\varepsilon}(\xi) \cdot \overline{\widehat{u^\varepsilon}(\eta) \otimes \widehat{u^\varepsilon}(\xi - \eta)} d\eta \cdot \xi + \nu |\xi|^2 |\widehat{u^\varepsilon}|^2 = 0.$$

Thus it follows that

$$\widehat{u^\varepsilon} \cdot \partial_t^\alpha \overline{\widehat{u^\varepsilon}} + \overline{\widehat{u^\varepsilon}} \cdot \partial_t^\alpha \widehat{u^\varepsilon} + 2\nu |\xi|^2 |\widehat{u^\varepsilon}|^2$$

$$= i \left(\int \left(\overline{\widehat{u^\varepsilon}(\xi)} \cdot \widehat{u^\varepsilon}(\eta) \otimes \widehat{u^\varepsilon}(\xi - \eta) - \widehat{u^\varepsilon}(\xi) \cdot \overline{\widehat{u^\varepsilon}(\eta) \otimes \widehat{u^\varepsilon}(\xi - \eta)} \right) d\eta \cdot \xi \right).$$

Using Lemma 2.19, multiplying by $|\xi|^{-1} |\widehat{u^\varepsilon}|^{-1}$ and integrating with respect to ξ, we get

$$\partial_t^\alpha \int |\xi|^{-1} |\widehat{u^\varepsilon}| d\xi + \nu \int |\xi| |\widehat{u^\varepsilon}| d\xi$$

$$\leq \frac{i}{2} \iint \left[\left(\widehat{u^\varepsilon}(\eta) \cdot |\widehat{u^\varepsilon}|^{-1} \overline{\widehat{u^\varepsilon}(\xi)} \right) \widehat{u^\varepsilon}(\xi - \eta) \right.$$

$$\left. - \left(\overline{\widehat{u^\varepsilon}(\eta)} \cdot |\widehat{u^\varepsilon}|^{-1} \widehat{u^\varepsilon}(\xi) \right) \overline{\widehat{u^\varepsilon}(\xi - \eta)} \right] \cdot |\xi|^{-1} \xi d\eta d\xi$$

$$\leq \iint |\widehat{u^\varepsilon}(\eta)| |\widehat{u^\varepsilon}(\xi - \eta)| d\eta d\xi$$

$$\leq \frac{1}{2} \iint \left(|\eta|^{-1} |\xi - \eta| + |\eta| |\xi - \eta|^{-1} \right) |\widehat{u^\varepsilon}(\eta)| |\widehat{u^\varepsilon}(\xi - \eta)| d\eta d\xi$$

$$\leq \int |\xi|^{-1} |\widehat{u^\varepsilon}(\xi)| d\xi \int |\xi| |\widehat{u^\varepsilon}(\xi)| d\xi. \quad (2.93)$$

From (2.90) and (2.91), we see that there exists a small enough δ with $\delta \in (0, T_\varepsilon)$ such that $\|u^\varepsilon(t)\|_{\mathcal{X}^{-1}} < \nu$ for $t \in [0, \delta]$, since $u^\varepsilon(t)$ is local smooth solution on $[0, T_\varepsilon)$, that is, $u^\varepsilon \in C([0, T_\varepsilon); V^m)$. Therefore, we have

$$\partial_t^\alpha \|u^\varepsilon(t)\|_{\mathcal{X}^{-1}} \leq 0 \quad \text{on } [0, \delta],$$

which, according to the definition of ∂_t^α, yields

$$\|u^\varepsilon(t)\|_{\mathcal{X}^{-1}} \leq \|\mathcal{J}^\varepsilon u_0\|_{\mathcal{X}^{-1}} \leq \|u_0\|_{\mathcal{X}^{-1}} < \nu \quad \text{on } [0, \delta].$$

Similarly, the above procedure can be applied to small intervals $[\delta, 2\delta]$, $[2\delta, 3\delta]$, ... This also ensures that $\|u^\varepsilon(t)\|_{\mathcal{X}^{-1}} \leq \|u_0\|_{\mathcal{X}^{-1}} < \nu$ for each $t \in [0, T_\varepsilon)$. Integrating (2.93) from 0 to t, we get

$$\|u^\varepsilon(t)\|_{\mathcal{X}^{-1}} + (\nu - \|u_0\|_{\mathcal{X}^{-1}}) \int_0^t (t - s)^{\alpha-1} \|u^\varepsilon(s)\|_{\mathcal{X}^1} ds \leq \|u_0\|_{\mathcal{X}^{-1}}, \quad \text{for } t \in [0, T_\varepsilon).$$

This, together with

$$\|\nabla u^\varepsilon(t)\|_{L^\infty} = \left| \int \exp(ix \cdot \xi) \widehat{\nabla u^\varepsilon}(\xi) d\xi \right| \leq \int |\widehat{\nabla u^\varepsilon}(\xi)| d\xi = \int |\xi| |\widehat{u^\varepsilon}(\xi)| d\xi,$$

shows that

$$\int_0^t (t - s)^{\alpha-1} \|\nabla u^\varepsilon(s)\|_{L^\infty} ds \leq \int_0^t (t - s)^{\alpha-1} \|u^\varepsilon(s)\|_{\mathcal{X}^1} ds \leq \frac{\|u_0\|_{\mathcal{X}^{-1}}}{\nu - \|u_0\|_{\mathcal{X}^{-1}}}.$$

On the other hand, in view of Lemma 2.20, we obtain

$$\|u^\varepsilon(t)\|_m \leq \|u^\varepsilon(0)\|_m + c_m \int_0^t (t - s)^{\alpha-1} |\mathcal{J}_\varepsilon \nabla u^\varepsilon(s)|_{L^\infty} \|u^\varepsilon(s)\|_m ds$$

$$\leq \|\mathcal{J}_\varepsilon u_0\|_m + c_m \int_0^t (t - s)^{\alpha-1} |\nabla u^\varepsilon(s)|_{L^\infty} \|u^\varepsilon(s)\|_m ds.$$

Using the Gronwall's inequality again, we get

$$\|u^\varepsilon(t)\|_m \leq \|\mathcal{J}_\varepsilon u_0\|_m \exp\left(c_m \int_0^t (t - s)^{\alpha-1} |\nabla u^\varepsilon(s)|_{L^\infty} ds \right)$$

$$\leq \|\mathcal{J}_\varepsilon u_0\|_m \exp\left(\frac{c_m \|u_0\|_{\mathcal{X}^{-1}}}{\nu - \|u_0\|_{\mathcal{X}^{-1}}} \right), \tag{2.94}$$

for all $t \in [0, T_\varepsilon)$. It follows from Lemma 2.22 that $T_\varepsilon = \infty$. Moreover, we have the following uniform estimate on u^ε:

$$\sup_{0 \leq t < \infty} \left\{ \|u^\varepsilon(t)\|_{\mathcal{X}^{-1}} + (\nu - \|u_0\|_{\mathcal{X}^{-1}}) \int_0^t (t - s)^{\alpha-1} \|u^\varepsilon(s)\|_{\mathcal{X}^1} ds \right\} \leq \|u_0\|_{\mathcal{X}^{-1}}. \tag{2.95}$$

Estimate (2.95) implies that there exists a subsequence of $\{u^\varepsilon\}$, relabeled as $\{u^\varepsilon\}$, and

$$u \in L^\infty(\mathbb{R}_+, \mathcal{X}^{-1}) \cap L_\alpha^\infty(\mathbb{R}_+, \mathcal{X}^1)$$

such that

$$u^\varepsilon \rightharpoonup u \text{ weakly}^* \text{ in } L^\infty_\alpha(\mathbb{R}_+, \mathfrak{X}^1), \quad u^\varepsilon \rightharpoonup u \text{ weakly}^* \text{ in } L^\infty(\mathbb{R}_+, \mathfrak{X}^{-1}), \quad \text{as } \varepsilon \to 0.$$
(2.96)

For the relation between $u^\varepsilon(0)$ and u_0, we notice that

$$\|u^\varepsilon(0) - u_0\|_{\mathfrak{X}^{-1}} \to 0, \quad \text{as } \varepsilon \to 0.$$
(2.97)

In order to prove the strong convergence of u^ε, we proceed with the same argument as in (2.93) to have the estimate:

$$\partial_t^\alpha \int |\xi|^{-1}|\widehat{u^{\varepsilon_1}} - \widehat{u^{\varepsilon_2}}|d\xi + \nu \int |\xi||\widehat{u^{\varepsilon_1}} - \widehat{u^{\varepsilon_2}}|d\xi$$

$$\leq \iint \left(|\widehat{u^{\varepsilon_1}}(\eta)| + |\widehat{u^{\varepsilon_2}}(\eta)|\right)|\widehat{u^{\varepsilon_1}}(\xi - \eta) - \widehat{u^{\varepsilon_2}}(\xi - \eta)|d\eta d\xi$$

$$\leq \frac{1}{2} \iint \left(|\eta|^{-1}|\xi - \eta| + |\eta||\xi - \eta|^{-1}\right)\left(|\widehat{u^{\varepsilon_1}}(\eta)| \right.$$

$$\left. + |\widehat{u^{\varepsilon_2}}(\eta)|\right)|\widehat{u^{\varepsilon_1}}(\xi - \eta) - \widehat{u^{\varepsilon_2}}(\xi - \eta)|d\eta d\xi$$

$$\leq \frac{1}{2}\left(\|u^{\varepsilon_1}\|_{\mathfrak{X}^{-1}} + \|u^{\varepsilon_2}\|_{\mathfrak{X}^{-1}}\right)\|u^{\varepsilon_1} - u^{\varepsilon_2}\|_{\mathfrak{X}^1} + \frac{1}{2}\left(\|u^{\varepsilon_1}\|_{\mathfrak{X}^1} + \|u^{\varepsilon_2}\|_{\mathfrak{X}^1}\right)\|u^{\varepsilon_1} - u^{\varepsilon_2}\|_{\mathfrak{X}^{-1}}.$$

Combining the above inequality and (2.95), we obtain

$$\partial_t^\alpha \int |\xi|^{-1}|\widehat{u^{\varepsilon_1}} - \widehat{u^{\varepsilon_2}}|d\xi + \left(\nu - \|u_0\|_{\mathfrak{X}^{-1}}\right)\int |\xi||\widehat{u^{\varepsilon_1}} - \widehat{u^{\varepsilon_2}}|d\xi$$

$$\leq \frac{1}{2}\left(\|u^{\varepsilon_1}\|_{\mathfrak{X}^1} + \|u^{\varepsilon_2}\|_{\mathfrak{X}^1}\right)\|u^{\varepsilon_1} - u^{\varepsilon_2}\|_{\mathfrak{X}^{-1}},$$

that is,

$$\partial_t^\alpha\|u^{\varepsilon_1} - u^{\varepsilon_2}\|_{\mathfrak{X}^{-1}} + \left(\nu - \|u_0\|_{\mathfrak{X}^{-1}}\right)\|u^{\varepsilon_1} - u^{\varepsilon_2}\|_{\mathfrak{X}^1}$$

$$\leq \frac{1}{2}\left(\|u^{\varepsilon_1}\|_{\mathfrak{X}^1} + \|u^{\varepsilon_2}\|_{\mathfrak{X}^1}\right)\|u^{\varepsilon_1} - u^{\varepsilon_2}\|_{\mathfrak{X}^{-1}}.$$

Using the Gronwall's inequality again and noting that $\nu - \|u_0\|_{\mathfrak{X}^{-1}} > 0$, we get

$$\|u^{\varepsilon_1}(t) - u^{\varepsilon_2}(t)\|_{\mathfrak{X}^{-1}}$$

$$\leq \|u^{\varepsilon_1}(0) - u^{\varepsilon_2}(0)\|_{\mathfrak{X}^{-1}} \exp\left(\frac{1}{2}\int_0^t (t-s)^{\alpha-1}\left(\|u^{\varepsilon_1}\|_{\mathfrak{X}^1} + \|u^{\varepsilon_2}\|_{\mathfrak{X}^1}\right)d\xi\right)$$

$$\leq \|u^{\varepsilon_1}(0) - u^{\varepsilon_2}(0)\|_{\mathfrak{X}^{-1}} \exp\left(\frac{\|u_0\|_{\mathfrak{X}^{-1}}}{\nu - \|u_0\|_{\mathfrak{X}^{-1}}}\right)$$

and

$$\left(\nu - \|u_0\|_{\mathfrak{X}^{-1}}\right)\int_0^t (t-s)^{\alpha-1}\|u^{\varepsilon_1} - u^{\varepsilon_2}\|_{\mathfrak{X}^1}ds$$

$$\leq \frac{1}{2}\int_0^t (t-s)^{\alpha-1}\left(\|u^{\varepsilon_1}\|_{\mathfrak{X}^1} + \|u^{\varepsilon_2}\|_{\mathfrak{X}^1}\right)\|u^{\varepsilon_1} - u^{\varepsilon_2}\|_{\mathfrak{X}^{-1}}ds$$

$$\leq \frac{1}{2}\|u^{\varepsilon_1}(0) - u^{\varepsilon_2}(0)\|_{\mathfrak{X}^{-1}} \exp\left(\frac{\|u_0\|_{\mathfrak{X}^{-1}}}{\nu - \|u_0\|_{\mathfrak{X}^{-1}}}\right)\int_0^t (t-s)^{\alpha-1}\left(\|u^{\varepsilon_1}\|_{\mathfrak{X}^1} + \|u^{\varepsilon_2}\|_{\mathfrak{X}^1}\right)ds$$

$$\leq \|u^{\varepsilon_1}(0) - u^{\varepsilon_2}(0)\|_{\mathfrak{X}^{-1}} \frac{\|u_0\|_{\mathfrak{X}^{-1}}}{\nu - \|u_0\|_{\mathfrak{X}^{-1}}} \exp\left(\frac{\|u_0\|_{\mathfrak{X}^{-1}}}{\nu - \|u_0\|_{\mathfrak{X}^{-1}}}\right).$$

Thus we conclude that $\{u^{\varepsilon}\}$ is a Cauchy sequence in $L^{\infty}(\mathbb{R}_+, \mathfrak{X}^{-1}) \cap L^{\infty}_{\alpha}(\mathbb{R}_+, \mathfrak{X}^1)$ by (2.95), and that the convergence in (2.96) is a strong one. In practice, the above estimate also ensures that the solution in the space $L^{\infty}(\mathbb{R}_+, \mathfrak{X}^{-1}) \cap L^{\infty}_{\alpha}(\mathbb{R}_+, \mathfrak{X}^1)$ is unique under the condition (2.90).

To establish continuity of $u(t, x)$, we return to equation (2.92). It is obvious that

$$\int_0^t (t-s)^{\alpha-1} \|\Delta u^{\varepsilon}(s)\|_{\mathfrak{X}^{-1}} ds = \int_0^t (t-s)^{\alpha-1} \int |\xi|^{-1} |\widehat{\Delta u^{\varepsilon}}(\xi)| d\xi ds$$

$$= \int_0^t (t-s)^{\alpha-1} \int |\xi| |\widehat{u^{\varepsilon}}(\xi)| d\xi ds$$

$$= \int_0^t (t-s)^{\alpha-1} \|u^{\varepsilon}\|_{\mathfrak{X}^1} ds.$$

In addition, like the estimate (2.93), we can obtain

$$\int_0^t (t-s)^{\alpha-1} \|\nabla \cdot (u^{\varepsilon}(s) \otimes u^{\varepsilon}(s))\|_{\mathfrak{X}^{-1}} ds$$

$$\leq \int_0^t (t-s)^{\alpha-1} \iint |\widehat{u^{\varepsilon}}(\eta)| |\widehat{u^{\varepsilon}}(\xi - \eta)| d\xi d\eta ds$$

$$\leq \int_0^t (t-s)^{\alpha-1} \|u^{\varepsilon}\|_{\mathfrak{X}^{-1}} \|u^{\varepsilon}\|_{\mathfrak{X}^1} ds$$

$$\leq \sup_{0 \leq t < \infty} \|u^{\varepsilon}(t)\|_{\mathfrak{X}^{-1}} \int_0^t (t-s)^{\alpha-1} \|u^{\varepsilon}\|_{\mathfrak{X}^1} ds.$$

The pressures term can be handled in a similar manner, that is, for $\widehat{R_j u^{\varepsilon}}(\xi) = -i\frac{\xi_j}{|\xi|} \cdot \widehat{u^{\varepsilon}}(\xi)$, we have

$$\int_0^t (t-s)^{\alpha-1} \|\nabla p^{\varepsilon}\|_{\mathfrak{X}^{-1}} ds \leq \int_0^t (t-s)^{\alpha-1} \iint |\widehat{R u^{\varepsilon}}(\eta)| |\widehat{R u^{\varepsilon}}(\xi - \eta)| d\xi d\eta ds$$

$$\leq \int_0^t (t-s)^{\alpha-1} \iint |\widehat{u^{\varepsilon}}(\eta)| |\widehat{u^{\varepsilon}}(\xi - \eta)| d\xi d\eta ds$$

$$\leq \sup_{0 \leq t < \infty} \|u^{\varepsilon}(t)\|_{\mathfrak{X}^{-1}} \int_0^t (t-s)^{\alpha-1} \|u^{\varepsilon}\|_{\mathfrak{X}^1} ds.$$

Observe that

$$\|u^{\varepsilon}(t)\|_{\mathfrak{X}^{-1}} - \|u^{\varepsilon}(t_0)\|_{\mathfrak{X}^{-1}}$$

$$\leq \int_0^{t_0} \left((t_0 - s)^{\alpha-1} - (t-s)^{\alpha-1}\right)$$

$$\times \left(\|\Delta u^{\varepsilon}(s)\|_{\mathfrak{X}^{-1}} + \|\nabla \cdot (u^{\varepsilon}(s) \otimes u^{\varepsilon}(s))\|_{\mathfrak{X}^{-1}} + \|\nabla p^{\varepsilon}\|_{\mathfrak{X}^{-1}}\right) ds$$

$$+ \int_{t_0}^t (t-s)^{\alpha-1} \left(\|\Delta u^{\varepsilon}(s)\|_{\mathfrak{X}^{-1}} + \|\nabla \cdot (u^{\varepsilon}(s) \otimes u^{\varepsilon}(s))\|_{\mathfrak{X}^{-1}} + \|\nabla p^{\varepsilon}\|_{\mathfrak{X}^{-1}}\right) ds$$

$$=: I_1(t) + I_2(t),$$

for $0 < t_0 \le t$. Now we estimate each term in the above expression. For $I_1(t)$, notice that

$$\int_0^{t_0} \left((t-s)^{\alpha-1} - (t_0-s)^{\alpha-1}\right)$$

$$\times \left(\|\Delta u^\varepsilon(s)\|_{\mathfrak{X}^{-1}} + \|\nabla \cdot (u^\varepsilon(s) \otimes u^\varepsilon(s))\|_{\mathfrak{X}^{-1}} + \|\nabla p^\varepsilon\|_{\mathfrak{X}^{-1}}\right) ds$$

$$\le 2 \int_0^{t_0} (t_0-s)^{\alpha-1} \left(\|\Delta u^\varepsilon(s)\|_{\mathfrak{X}^{-1}} + \|\nabla \cdot (u^\varepsilon(s) \otimes u^\varepsilon(s))\|_{\mathfrak{X}^{-1}} + \|\nabla p^\varepsilon\|_{\mathfrak{X}^{-1}}\right) ds$$

$$\le 2 \int_0^{t_0} (t_0-s)^{\alpha-1} \|u^\varepsilon(s)\|_{\mathfrak{X}^1} ds + 4 \sup_{0 \le s < t_0} \|u^\varepsilon(s)\|_{\mathfrak{X}^{-1}} \int_0^{t_0} (t_0-s)^{\alpha-1} \|u^\varepsilon(s)\|_{\mathfrak{X}^1} ds$$

$$< \infty,$$

due to the estimate (2.95). Applying the Lebesgue's dominated convergence theorem, we get

$$\lim_{t-t_0 \to 0} I_1(t) = \int_0^{t_0} \lim_{t-t_0 \to 0} \left((t_0-s)^{\alpha-1} - (t-s)^{\alpha-1}\right)$$

$$\times \left(\|\Delta u^\varepsilon(s)\|_{\mathfrak{X}^{-1}} + \|\nabla \cdot (u^\varepsilon(s) \otimes u^\varepsilon(s))\|_{\mathfrak{X}^{-1}} + \|\nabla p^\varepsilon\|_{\mathfrak{X}^{-1}}\right) ds$$

$$= 0.$$

For $I_2(t)$, by an elementary calculation and the fact that $u^\varepsilon \in L_\alpha^\infty(\mathbb{R}_+, \mathfrak{X}^1)$, we obtain the following inequality

$$I_2(t) \le \left(1 + 2 \sup_{t_0 \le s < t} \|u^\varepsilon(s))\|_{\mathfrak{X}^{-1}}\right) \int_{t_0}^t (t-s)^{\alpha-1} \|u^\varepsilon(s)\|_{\mathfrak{X}^1} ds$$

$$\to 0, \quad \text{as } t - t_0 \to 0.$$

From the foregoing arguments, it follows that $u \in C([0,\infty); \mathfrak{X}^{-1}) \cap L_\alpha^\infty(\mathbb{R}_+, \mathfrak{X}^1)$. This completes the proof. $\square$

The following results deal with the global existence of approximate solutions and mild solutions, respectively.

Theorem 2.20. *Let $0 < T < \infty$. If $u_0 \in V^m$, $m \in \mathbb{N}_0$, then there exists a unique solution $u^\varepsilon \in C^\alpha([0,T]; V^m)$ of the problem (2.84) for any $\varepsilon > 0$.*

Proof. We show an a priori bound on $\|u^\varepsilon\|_m$. To do this, we note that the relation (2.85) with $u^2(t,x) \equiv 0$ and (2.86) yields

$$\partial_t^\alpha \|u^\varepsilon\|_m \le L(\|u^\varepsilon\|, \varepsilon, M) \|u^\varepsilon\|_m \le L(\|u_0\|, \varepsilon, M) \|u^\varepsilon\|_m.$$

Thanks to the Gronwall's inequality, we have

$$\|u^\varepsilon(t)\|_m \le \|u_0\|_m \exp\left(L(\|u_0\|, \varepsilon, M) \frac{T^\alpha}{\alpha}\right),$$

which shows a priori bound on $\|u^\varepsilon\|_m$. By Theorem 2.18, the regularized equation (2.84) has a unique solution $u^\varepsilon \in C^\alpha([0,T]; V^m)$. $\square$

Theorem 2.21. *Assume that $u_0 \in V^m$, $m \geq 3$. Then the following results hold.*

(i) *If $\|u_0\|_1 < \frac{\nu}{C}$ and $0 < T < \infty$, there exists a unique solution*

$$u \in C([0,T]; C^2(\mathbb{R}^3)) \cap C^\alpha([0,T]; C(\mathbb{R}^3))$$

of the time-fractional Navier-Stokes equations (2.84). Moreover, u is the limit of a subsequence of $\{u^\varepsilon\}$, where u^ε are the approximate solutions given by Theorem 2.20.

(ii) *The following estimates hold:*

$$\sup_{t \in [0,T]} \|u^\varepsilon(t)\|_m \leq \|u_0\|_m \exp\left(\frac{c_m C \|u_0\|_1}{\nu - C\|u_0\|_1}\right),$$

$$\sup_{t \in [0,T]} \|u(t)\|_m \leq \|u_0\|_m \exp\left(\frac{c_m C \|u_0\|_1}{\nu - C\|u_0\|_1}\right).$$

(iii) *u^ε and u are uniformly bounded in $L^\infty(0,T; H^m(\mathbb{R}^3)) \cap C_w([0,T]; H^m(\mathbb{R}^3))$.*

Proof. In the first step, we show that $\{u^\varepsilon(t)\}$ is a uniformly bounded set of $H^m(\mathbb{R}^3)$ independent of ε. By Lemma 2.20, we have

$$\partial_t^\alpha \|u^\varepsilon\|_m \leq c_m |\mathcal{J}_\varepsilon \nabla u^\varepsilon|_{L^\infty} \|u^\varepsilon\|_m.$$

Equation (2.94) and $\|u_0\|_{\mathfrak{x}-1} \leq C\|u_0\|_1$ ensure that

$$\|u^\varepsilon(t)\|_m \leq \|u_0\|_m \exp\left(\frac{c_m \|u_0\|_{\mathfrak{x}-1}}{\nu - \|u_0\|_{\mathfrak{x}-1}}\right) \leq \|u_0\|_m \exp\left(\frac{c_m C\|u_0\|_1}{\nu - C\|u_0\|_1}\right). \qquad (2.98)$$

In consequence, $\{u^\varepsilon\}$ is a uniformly bounded set of $C([0,T]; H^m(\mathbb{R}^3))$.

Next we prove that $\{u^\varepsilon\}$ constitutes a Cauchy sequence in $C([0,T]; L^2(\mathbb{R}^3))$. Using the equation (2.84), we have

$$\|u^\varepsilon - u^{\varepsilon'}\| \partial_t^\alpha \|u^\varepsilon - u^{\varepsilon'}\|$$

$$\leq \nu \left\langle \mathcal{J}_\varepsilon^2 \Delta u^\varepsilon - \mathcal{J}_{\varepsilon'}^2 \Delta u^{\varepsilon'}, u^\varepsilon - u^{\varepsilon'} \right\rangle$$

$$- \left\langle P\mathcal{J}_\varepsilon[(\mathcal{J}_\varepsilon u^\varepsilon) \cdot \nabla(\mathcal{J}_\varepsilon u^\varepsilon)] - P\mathcal{J}_{\varepsilon'}[(\mathcal{J}_{\varepsilon'} u^{\varepsilon'}) \cdot \nabla(\mathcal{J}_{\varepsilon'} u^{\varepsilon'})], u^\varepsilon - u^{\varepsilon'} \right\rangle.$$

Using the argument of [256, Lemma 3.7], we obtain

$$\left\langle \mathcal{J}_\varepsilon^2 \Delta u^\varepsilon - \mathcal{J}_{\varepsilon'}^2 \Delta u^{\varepsilon'}, u^\varepsilon - u^{\varepsilon'} \right\rangle$$

$$\leq \left\langle (\mathcal{J}_\varepsilon^2 - \mathcal{J}_{\varepsilon'}^2) \Delta u^\varepsilon, u^\varepsilon - u^{\varepsilon'} \right\rangle - \|\mathcal{J}_{\varepsilon'} \nabla(u^\varepsilon - u^{\varepsilon'})\|^2$$

$$\leq C_1 \max\{\varepsilon, \varepsilon'\} \|u^\varepsilon\|_3 \|u^\varepsilon - u^{\varepsilon'}\|$$

and

$$\left| \left\langle P\mathcal{J}_\varepsilon[(\mathcal{J}_\varepsilon u^\varepsilon) \cdot \nabla(\mathcal{J}_\varepsilon u^\varepsilon)] - P\mathcal{J}_{\varepsilon'}[(\mathcal{J}_{\varepsilon'} u^{\varepsilon'}) \cdot \nabla(\mathcal{J}_{\varepsilon'} u^{\varepsilon'})], u^\varepsilon - u^{\varepsilon'} \right\rangle \right|$$

$$= \left| \left\langle (\mathcal{J}_\varepsilon - \mathcal{J}_{\varepsilon'})[(\mathcal{J}_\varepsilon u^\varepsilon) \cdot \nabla(\mathcal{J}_\varepsilon u^\varepsilon)], u^\varepsilon - u^{\varepsilon'} \right\rangle \right|$$

$$+ \left| \left\langle \mathcal{J}_{\varepsilon'}[(\mathcal{J}_\varepsilon - \mathcal{J}_{\varepsilon'}) u^\varepsilon \cdot \nabla(\mathcal{J}_{\varepsilon'} u^{\varepsilon'})], u^\varepsilon - u^{\varepsilon'} \right\rangle \right|$$

$$+ \left| \left\langle \mathcal{J}_{\varepsilon'}[\mathcal{J}_{\varepsilon'}(u^{\varepsilon} - u^{\varepsilon'}) \cdot \nabla(\mathcal{J}_{\varepsilon'}u^{\varepsilon'})], u^{\varepsilon} - u^{\varepsilon'} \right\rangle \right|$$

$$+ \left| \left\langle \mathcal{J}_{\varepsilon'}\{\mathcal{J}_{\varepsilon'}u^{\varepsilon'} \cdot \nabla[(\mathcal{J}_{\varepsilon} - \mathcal{J}_{\varepsilon'})u^{\varepsilon'}]\}, u^{\varepsilon} - u^{\varepsilon'} \right\rangle \right|$$

$$+ \left| \left\langle \mathcal{J}_{\varepsilon'}\{\mathcal{J}_{\varepsilon'}u^{\varepsilon'} \cdot \nabla[\mathcal{J}_{\varepsilon'}(u^{\varepsilon} - u^{\varepsilon'})]\}, u^{\varepsilon} - u^{\varepsilon'} \right\rangle \right|$$

$$\leq C_1 \max\{\varepsilon, \varepsilon'\} \|u^{\varepsilon}\|_m^2 \|u^{\varepsilon} - u^{\varepsilon'}\| + C_1 \max\{\varepsilon, \varepsilon'\} \|u^{\varepsilon}\|_m \|u^{\varepsilon'}\|_m \|u^{\varepsilon} - u^{\varepsilon'}\|$$

$$+ C_1 \|u^{\varepsilon} - u^{\varepsilon'}\|^2 \|u^{\varepsilon}\|_m + C_1 \max\{\varepsilon, \varepsilon'\} \|u^{\varepsilon}\|_m^2 \|u^{\varepsilon} - u^{\varepsilon'}\|,$$

which imply that

$$\partial_t^{\alpha} \|u^{\varepsilon} - u^{\varepsilon'}\| \leq C_1(M)\left(\max\{\varepsilon, \varepsilon'\} + \|u^{\varepsilon} - u^{\varepsilon'}\|\right),$$

where $M = \sup_{\varepsilon}\{\|u^{\varepsilon}\|_m\}$ from (2.98). Integrating the above inequality and then using the Gronwall's inequality yields

$$\|u^{\varepsilon}(t) - u^{\varepsilon'}(t)\| \leq \left(\|u^{\varepsilon}(0) - u^{\varepsilon'}(0)\| + C_1(M)\max\{\varepsilon, \varepsilon'\}\frac{T^{\alpha}}{\alpha}\right)$$

$$\times \exp\left(C_1(M)\int_0^t (t-s)^{\alpha-1}ds\right)$$

$$\leq \left(\|u^{\varepsilon}(0) - u^{\varepsilon'}(0)\| + C_1(M)\max\{\varepsilon, \varepsilon'\}\frac{T^{\alpha}}{\alpha}\right)\exp\left(C_1(M)\frac{T^{\alpha}}{\alpha}\right).$$

Thus we have

$$\sup_{t\in[0,T]} \|u^{\varepsilon}(t) - u^{\varepsilon'}(t)\| \leq \left(\|u^{\varepsilon}(0) - u^{\varepsilon'}(0)\| + C_1(M)\max\{\varepsilon, \varepsilon'\}\frac{T^{\alpha}}{\alpha}\right)$$

$$\times \exp\left(C_1(M)\frac{T^{\alpha}}{\alpha}\right),$$

which, in view of (2.97), implies that $\{u^{\varepsilon}\}$ is a Cauchy sequence in $C([0,T]; L^2(\mathbb{R}^3))$. Hence there exists $u \in C([0,T]; L^2(\mathbb{R}^3))$ such that the sequence $\{u^{\varepsilon}\}$ converges strongly to u in $C([0,T]; L^2(\mathbb{R}^3))$, that is,

$$\sup_{t\in[0,T]} \|u^{\varepsilon}(t) - u(t)\| \leq C_2\varepsilon. \tag{2.99}$$

Finally we verify that $u \in C([0,T]; C^2(\mathbb{R}^3)) \cap C^{\alpha}([0,T]; C(\mathbb{R}^3))$. By the inequalities (2.98) and (2.99), we get

$$\sup_{t\in[0,T]} \{\|u^{\varepsilon}(t) - u^{\varepsilon'}(t)\|_{m'}\} \leq \sup_{t\in[0,T]} \left\{\|u^{\varepsilon}(t) - u(t)\|^{1-\frac{m'}{m}} \|u^{\varepsilon}(t) - u^{\varepsilon'}(t)\|_m^{\frac{m'}{m}}\right\}$$

$$\leq C(\|u_0\|_m)\varepsilon^{1-\frac{m'}{m}}, \quad \text{for } 0 < m' < m, \tag{2.100}$$

which implies strong convergence in $C([0,T]; H^{m'}(\mathbb{R}^3))$ for all $m' < m$. With $\frac{7}{2} < m' < m$, we also have strong convergence in $C([0,T]; C^2(\mathbb{R}^3))$. Moreover, from the equation

$$\partial_t^{\alpha} u^{\varepsilon} = \nu \mathcal{J}_{\varepsilon}^2 \Delta u^{\varepsilon} - P\mathcal{J}_{\varepsilon}[(\mathcal{J}_{\varepsilon}u^{\varepsilon}) \cdot \nabla(\mathcal{J}_{\varepsilon}u^{\varepsilon})],$$

we deduce that $\partial_t^\alpha u^\varepsilon$ converges to $\nu\Delta u - P(u\cdot\nabla u)$ in $C([0,T];C(\mathbb{R}^3))$. Note that $\int_0^T \partial_t^\alpha u(t)\phi(t)dt = \int_0^T \left(u(t) - u(0)\right){}_t^C D_T^\alpha \phi(t)dt$ and $u^\varepsilon \to u$ as $\varepsilon \to 0$, it follows that for $\phi \in C_0^\infty([0,T];C(\mathbb{R}^3))$,

$$\int_0^T \partial_t^\alpha u^\varepsilon(t)\phi(t)dt = \int_0^T \left(u^\varepsilon(t) - u^\varepsilon(0)\right){}_t^C D_T^\alpha\phi(t)dt$$

$$\to \int_0^T \partial_t^\alpha u(t)\phi(t)dt, \quad \text{as } \varepsilon \to 0.$$

Hence the distribution limit of $\partial_t^\alpha u^\varepsilon$ must be $\partial_t^\alpha u$.

(iii) Recalling that for any bounded sequence $\{v_\varepsilon\}$ in $H^m(\mathbb{R}^3)$, there exists a subsequence converging weakly to a limit in $H^m(\mathbb{R}^3)$, that is, $v_\varepsilon \rightharpoonup u$, as $\varepsilon \to 0$.

From the fact

$$\sup_{t\in[0,T]} \|u^\varepsilon(t)\|_m \le M. \tag{2.101}$$

Accordingly, u^ε is uniformly bounded in $L^2(0,T;H^m(\mathbb{R}^3))$ and hence there exists a subsequence, relabeled as $\{u^\varepsilon\}$, and $u \in L^2(0,T;H^m(\mathbb{R}^3))$ such that

$$u^\varepsilon \rightharpoonup u \ \text{ in } L^2(0,T;H^m(\mathbb{R}^3)).$$

Moreover, for any given $t \in [0,T]$, the sequence $\{u^\varepsilon(t,\cdot)\}$ is a uniformly bounded set of $H^m(\mathbb{R}^3)$. Hence there is also a subsequence of $\{u^\varepsilon(t,\cdot)\}$, relabeled as $\{u^\varepsilon(t,\cdot)\}$ such that

$$u^\varepsilon(t,\cdot) \rightharpoonup u(t,\cdot) \ \text{ in } H^m(\mathbb{R}^3).$$

Hence we deduce that $\|u(t)\|_m$ is bounded for each t, which yields

$$u \in L^\infty(0,T;H^m(\mathbb{R}^3)).$$

For $\phi \in H^{-m}(\mathbb{R}^3)$, Denote by $[\phi,u]$ the dual pairing of $H^{-m}(\mathbb{R}^3)$ and $H^m(\mathbb{R}^3)$. From (2.100), we have

$$u^\varepsilon \to u \ \text{ in } u \in C([0,T];H^{m'}(\mathbb{R}^3)),$$

which implies that for any $\phi \in H^{-m'}(\mathbb{R}^3)$,

$$[\phi,u^\varepsilon(t,\cdot)] \to [\phi,u(t,\cdot)] \ \text{ uniformly for } t \in [0,T].$$

Applying (2.101) and the conclusion that for $m' < m$, $H^{-m'}(\mathbb{R}^3)$ is dense in $H^{-m}(\mathbb{R}^3)$, we obtain that for any $\phi \in H^{-m}(\mathbb{R}^3)$,

$$[\phi,u^\varepsilon(t,\cdot)] \to [\phi,u(t,\cdot)] \ \text{ uniformly for } t \in [0,T].$$

Thus $u \in C_w([0,T];H^m(\mathbb{R}^3))$. This completes the proof. $\qquad\square$

2.7 Cauchy Problem in Sobolev Space

Consider the following the fractional Navier-Stokes equations in $\mathbb{R}^d$ $(d \geq 2)$:

$$
\begin{cases}
\partial_t^\alpha u + (u \cdot \nabla)u = \Delta u - \nabla \pi, & t > 0, \\
\nabla \cdot u = 0, \\
u(0, x) = u_0,
\end{cases}
\tag{2.102}
$$

where ∂_t^α is the Caputo fractional derivative of order $\alpha \in (0, 1)$, $u = (u_1(t, x), u_2(t, x), ..., u_d(t, x))$ represents the velocity field at a point $x \in \mathbb{R}^d$ and time $t > 0$, while $\pi(t, x)$ denotes the pressure.

The section is organized as follows. In Subsection 2.7.1 we recall some notations, definitions, and preliminary facts, then we give the concept of mild solutions for problem (2.102) and some properties of solution operators via the tools from harmonic analytic. Subsections 2.7.2 and 2.7.3 are dedicated to studying the local existence of solutions for problem (2.102) with the help of an assistant space, and two main results are presented.

2.7.1 Definitions and Lemmas

In this subsection, we review some definitions, and preliminary facts which are used throughout this section. Also, we obtain an integral formula which is formally equivalent to (2.102).

We denote by $L^q(\mathbb{R}^d)$ the space consisting of q-integral functions with the norm $\| \cdot \|_q$. $\mathcal{F}$ and $\mathcal{F}^{-1}$ represent the Fourier transform and its inverse respectively. We define $\dot{\Lambda}^s := \mathcal{F}^{-1}|\xi|^s \mathcal{F}$ and the homogeneous Sobolev space $\dot{H}_q^s := \dot{\Lambda}^{-s} L^q(\mathbb{R}^d)$ equipped with the norm $\|f\|_{\dot{H}_q^s} := \|\dot{\Lambda}^s f\|_q$.

Initially, we recall the well-known embedding mapping for homogeneous Sobolev spaces.

Lemma 2.23. *Assume that $s_1 > s_2$, $1 < q_1, q_2 < \infty$ with $s_1 - \frac{d}{q_1} = s_2 - \frac{d}{q_2}$. Then*

$$
\dot{H}_{q_1}^{s_1} \hookrightarrow \dot{H}_{q_2}^{s_2}.
$$

Let f be complex-valued measurable functions in a bounded domain Ω. Then

$$
\mu_f(a) = \text{mes}\{x \in \Omega : |f(x)| > a\}, \quad a > 0
$$

is the distribution function of f and

$$
f^*(t) = \inf\{a : |\mu_f(a)| \leq t\}, \quad t \geq 0.
$$

Let $0 < p < \infty$ and $0 < q \leq \infty$. The Lorentz space $L^{p,q}(\Omega)$ collects all measurable functions f in Ω satisfying

$$
\|f\|_{L^{p,q}} = \left(\int_0^{|\Omega|} \left(t^{\frac{1}{p}} f^*(t) \right)^q \frac{dt}{t} \right)^{\frac{1}{q}} < \infty,
$$

when $0 < q < \infty$ and

$$\|f\|_{L^{p,q}} = \sup_{t\in(0,|\Omega|]} t^{\frac{1}{p}} f^*(t) < \infty$$

when $q = \infty$. For an introduction to Lorentz spaces, see e.g., Bergh and Lofstrom [45].

Throughout this section, the notation $U \lesssim V$ means that there exists a constant M such that $U \leq MV$. We will need the following useful result of convolution in Lorentz spaces.

Lemma 2.24. *[221] Assume that $1 < p, p_1, p_2 < \infty$, $1 \leq q, q_1, q_2 \leq \infty$ satisfy*

$$\frac{1}{p_2} + 1 = \frac{1}{p} + \frac{1}{p_1} \quad and \quad \frac{1}{q_2} = \frac{1}{q} + \frac{1}{q_1}.$$

Then for any $f \in L^{p,q}$ and $g \in L^{p_1,q_1}$, we have $\|f * g\|_{L^{p_2,q_2}} \lesssim \|f\|_{L^{p,q}} \|g\|_{L^{p_1,q_1}}$. Here $*$ denotes the convolution.

Denote R_j by the Riesz transforms defined as

$$R_j = \frac{\partial_j}{\sqrt{-\Delta}}, \quad \text{i.e.,} \quad \mathcal{F}(R_j g)(\xi) = \frac{i\xi_j}{|\xi|} \mathcal{F}(g)(\xi).$$

Then the operator $\mathbb{P}$ is defined by

$$(\mathbb{P}f)_j = f_j + \sum_{1 \leq k \leq d} R_j R_k f_k,$$

which is called the Helmholtz-Leray projection onto the divergence-free fields.

Let $v : [0, \infty) \times \mathbb{R}^d \to \mathbb{R}^d$. The fractional integral of order $\alpha \in (0, 1]$ for the function v is defined as

$$_0D_t^{-\alpha}v(t,x) = \frac{1}{\Gamma(\alpha)} \int_0^t (t-\tau)^{\alpha-1} v(\tau, x)d\tau, \quad t > 0.$$

Further, $\partial_t^\alpha v$ represents the Caputo fractional derivative of order α for the function v; it is defined by

$$\partial_t^\alpha v(t,x) = \frac{1}{\Gamma(1-\alpha)} \int_0^t (t-\tau)^{-\alpha} \frac{\partial}{\partial \tau} v(\tau, x)d\tau, \quad t > 0.$$

For more details, we refer the reader to Kilbas et al. [197].

Next we study the mild formulation of solutions to equation (2.102).

Let $E_\alpha(t), E_{\alpha,\alpha}(t)$ be the Mittag-Leffler functions as in Definition 1.7 and $\mathcal{M}_\alpha(\theta)$ be the Wright-type function as in Definition 1.8.

Lemma 2.25. *If u satisfies equation (2.102), then we have*

$$u(t,x) = E_{\alpha,1}(t^\alpha \Delta)u_0(x) - \int_0^t (t-\tau)^{\alpha-1} E_{\alpha,\alpha}((t-\tau)^\alpha \Delta)\mathbb{P}\nabla \cdot (u(\tau,x) \otimes u(\tau,x))d\tau,$$

$$(2.103)$$

where $E_{\alpha,1}(t^\alpha \Delta)$ and $E_{\alpha,\alpha}(t^\alpha \Delta)$ are expressed as:

$$E_{\alpha,1}(t^\alpha \Delta)v(x) = \left((4\pi t^\alpha)^{-\frac{d}{2}} \int_0^\infty \theta^{-\frac{d}{2}} \mathcal{M}_\alpha(\theta) \exp(-\frac{|\cdot|^2}{4\theta t^\alpha})d\theta * v \right)(x),$$

$$E_{\alpha,\alpha}(t^\alpha \Delta)v(x) = \left((4\pi t^\alpha)^{-\frac{d}{2}} \int_0^\infty \alpha\theta^{1-\frac{d}{2}} \mathcal{M}_\alpha(\theta) \exp(-\frac{|\cdot|^2}{4\theta t^\alpha})d\theta * v \right)(x).$$

Proof. We apply Helmholtz-Leray projector $\mathbb{P}$ to equations (2.102), which removes the pressure term. Then by using the space Fourier transform, the equations formally reduce to the form

$$\begin{cases} \partial_t^\alpha \mathcal{F}(u)(t,\xi) = -|\xi|^2 \mathcal{F}(u)(t,\xi) - \mathcal{F}[\mathbb{P}\nabla \cdot (u \otimes u)](t,\xi), & t > 0, \\ \mathcal{F}(u)(0,\xi) = \mathcal{F}(u_0). \end{cases}$$

Making use of [197], we have the following expression

$$\begin{aligned} \mathcal{F}(u)(t,\xi) =& E_{\alpha,1}(-t^\alpha |\xi|^2) \mathcal{F}(u_0) \\ &- \int_0^t (t-\tau)^{\alpha-1} E_{\alpha,\alpha}(-(t-\tau)^\alpha |\xi|^2) \mathcal{F}[\mathbb{P}\nabla \cdot (u \otimes u)](t,\xi) d\tau. \end{aligned}$$

Afterwards, using the Fourier inverse transform allows us to derive that

$$\begin{aligned} u(t,x) =& \mathcal{F}^{-1}[E_{\alpha,1}(-t^\alpha |\xi|^2) \mathcal{F}(u_0)] \\ &- \int_0^t (t-\tau)^{\alpha-1} \mathcal{F}^{-1}\big[E_{\alpha,\alpha}(-(t-\tau)^\alpha |\xi|^2) \mathcal{F}[\mathbb{P}\nabla \cdot (u \otimes u)](\tau,\xi)\big] d\tau. \end{aligned}$$

According to the relationship between convolutions and Fourier transforms, we obtain

$$\mathcal{F}^{-1}[E_{\alpha,1}(-t^\alpha |\xi|^2) \mathcal{F}(u_0)] = \{\mathcal{F}^{-1}[E_{\alpha,1}(-t^\alpha |\xi|^2)]\} * u_0(x),$$

$$\begin{aligned} &\mathcal{F}^{-1}\big[E_{\alpha,\alpha}(-t^\alpha |\xi|^2) \mathcal{F}[\mathbb{P}\nabla \cdot (u \otimes u)](t,\xi)\big] \\ &= \{\mathcal{F}^{-1}[E_{\alpha,\alpha}(-t^\alpha |\xi|^2)]\} * [\mathbb{P}\nabla \cdot (u \otimes u)](t,x), \end{aligned}$$

it follows from the equality $E_{\alpha,1}(-z) = \int_0^\infty \mathcal{M}_\alpha(\theta) \exp(-z\theta) d\theta$ and the Fubini's theorem that

$$\begin{aligned} \mathcal{F}^{-1}[E_{\alpha,1}(-t^\alpha |\xi|^2)] =& \left(\frac{1}{2\pi}\right)^d \int_{\mathbb{R}^d} \exp(ix \cdot \xi) \int_0^\infty \mathcal{M}_\alpha(\theta) \exp(-\theta t^\alpha |\xi|^2) d\theta d\xi \\ =& \left(\frac{1}{2\pi}\right)^d \int_0^\infty \mathcal{M}_\alpha(\theta) d\theta \int_{\mathbb{R}^d} \exp(ix \cdot \xi - \theta t^\alpha |\xi|^2) d\xi \\ =& (4\pi t^\alpha)^{-\frac{d}{2}} \int_0^\infty \theta^{-\frac{d}{2}} \mathcal{M}_\alpha(\theta) \exp\left(-\frac{|x|^2}{4\theta t^\alpha}\right) d\theta. \end{aligned}$$

Notice that $E_{\alpha,\alpha}(-z) = \int_0^\infty \alpha \theta \mathcal{M}_\alpha(\theta) \exp(-z\theta) d\theta$, a similar argument indicates that

$$\mathcal{F}^{-1}[E_{\alpha,\alpha}(-t^\alpha |\xi|^2)] = (4\pi t^\alpha)^{-\frac{d}{2}} \int_0^\infty \alpha \theta^{1-\frac{d}{2}} \mathcal{M}_\alpha(\theta) \exp\left(-\frac{|x|^2}{4\theta t^\alpha}\right) d\theta.$$

Consequently,

$$u(t,x) = E_{\alpha,1}(t^\alpha \Delta) u_0(x) - \int_0^t (t-\tau)^{\alpha-1} E_{\alpha,\alpha}((t-\tau)^\alpha \Delta) \mathbb{P}\nabla \cdot (u(\tau,x) \otimes u(\tau,x)) d\tau.$$

$\square$

Motivated by above discussion, we can define the concept of mild solutions to the objective equations by means of the above representation formula.

Definition 2.8. For $T > 0$, by a mild solution of the equation (2.102) on $[0, T]$ corresponding to a divergence-free initial value u_0, we mean that u satisfies integral equation (2.103).

Before going further, we present the representation and basic estimates for the composition of the operator $\dot{\Lambda}^s$ and the solution operator $E_{\alpha,1}(t^\alpha \Delta)$, which will be used to show that the initial value part belongs to a particular space given in the later. The main tool we use comes from harmonic analysis.

Lemma 2.26. *Let $s > -1$. Then the kernel function of $\dot{\Lambda}^s E_{\alpha,1}(t^\alpha \Delta)$ is the function*

$$K_{\alpha,s}(t,x) = t^{-\frac{\alpha(s+d)}{2}} \int_0^\infty \theta^{-\frac{s+d}{2}} M_\alpha(\theta) K\left(\frac{x}{\sqrt{\theta t^\alpha}}\right) d\theta.$$

Here the function K denotes the kernel function of $\dot{\Lambda}^s \exp(\Delta)$ and the following estimate holds

$$|K(x)| \lesssim \frac{1}{1 + |x|^{s+d}}.$$

Proof. By the definition of $\dot{\Lambda}^s$, we have

$$\dot{\Lambda}^s E_{\alpha,1}(t^\alpha \Delta)v = \mathcal{F}^{-1}\{|\xi|^s \mathcal{F}[E_{\alpha,1}(t^\alpha \Delta)v]\}$$
$$= \mathcal{F}^{-1}[|\xi|^s E_{\alpha,1}(-t^\alpha |\xi|^2)\mathcal{F}(v)] = \{\mathcal{F}^{-1}[|\xi|^s E_{\alpha,1}(-t^\alpha |\xi|^2)]\} * v.$$

Then we proceed to deduce $\mathcal{F}^{-1}[|\xi|^s E_{\alpha,1}(-t^\alpha |\xi|^2)]$ as follows:

$$\mathcal{F}^{-1}[|\xi|^s E_{\alpha,1}(-t^\alpha |\xi|^2)]$$
$$= \left(\frac{1}{2\pi}\right)^d \int_{\mathbb{R}^d} \exp(ix \cdot \xi)|\xi|^s \int_0^\infty M_\alpha(\theta) \exp(-\theta t^\alpha |\xi|^2) d\theta d\xi$$
$$= \left(\frac{1}{2\pi}\right)^d \int_0^\infty M_\alpha(\theta) d\theta \int_{\mathbb{R}^d} |\xi|^s \exp(ix \cdot \xi - \theta t^\alpha |\xi|^2) d\xi$$
$$= t^{-\frac{\alpha(s+d)}{2}} \left(\frac{1}{2\pi}\right)^d \int_0^\infty \theta^{-\frac{s+d}{2}} M_\alpha(\theta) d\theta \int_{\mathbb{R}^d} \exp\left(i\frac{x}{\sqrt{\theta t^\alpha}} \cdot \xi\right)|\xi|^s \exp(-|\xi|^2) d\xi$$
$$= t^{-\frac{\alpha(s+d)}{2}} \int_0^\infty \theta^{-\frac{s+d}{2}} M_\alpha(\theta) K\left(\frac{x}{\sqrt{\theta t^\alpha}}\right) d\theta,$$

where the change $\xi \mapsto \frac{\xi}{\sqrt{\theta t^\alpha}}$ is used. $\square$

Next we also state the similar properties for the composition of the operator $\dot{\Lambda}^s$ and the solution operator $E_{\alpha,\alpha}(t^\alpha \Delta)\mathbb{P}\nabla\cdot$, it will play an important role in obtaining the bound for nonlinear part of the integral equations.

Lemma 2.27. *Let $s > -1$. Then the kernel function of $\dot{\Lambda}^s E_{\alpha,\alpha}(t^\alpha \Delta)\mathbb{P}\nabla\cdot$ is the function*

$$P_{\alpha,s}(t,x) = t^{-\frac{\alpha(s+d+1)}{2}} \int_0^\infty \alpha\theta^{\frac{1-s-d}{2}} M_\alpha(\theta) P\left(\frac{x}{\sqrt{\theta t^\alpha}}\right) d\theta.$$

Here the function P denotes the kernel function of $\dot\Lambda^s \exp(\Delta)\mathbb{P}\nabla\cdot$ and the following estimate holds

$$|P(x)| \lesssim \frac{C}{1+|x|^{s+d+1}}.$$

Moreover, $P_{\alpha,s}(t,x)$ satisfies the inequality

$$|P_{\alpha,s}(t,x)| \lesssim \frac{1}{t^{\alpha\lambda_2}|x|^{\lambda_1}}, \quad \text{for } \lambda_1 > 0,\ 0 < \lambda_2 < 2 \text{ and } \lambda_1 + 2\lambda_2 = d + s + 1.$$

Proof. Arguing as in the proof of Lemma 2.26, we infer

$$\begin{aligned}
\dot\Lambda^s E_{\alpha,\alpha}(t^\alpha\Delta)\mathbb{P}\nabla\cdot v &= \mathcal{F}^{-1}\{|\xi|^s \mathcal{F}[E_{\alpha,\alpha}(t^\alpha\Delta)\mathbb{P}\nabla\cdot v]\} \\
&= \mathcal{F}^{-1}[|\xi|^s E_{\alpha,\alpha}(-t^\alpha|\xi|^2)\mathcal{F}(F(x)v)] \\
&= \{\mathcal{F}^{-1}[|\xi|^s E_{\alpha,\alpha}(-t^\alpha|\xi|^2)]\mathcal{F}[F](\xi)\} * v,
\end{aligned}$$

where $F(x)$ is the tensor kernel associated with the operator $\mathbb{P}\nabla\cdot$, that is

$$\mathcal{F}[F_{l,k,j}](\xi) = i\left(\delta_{j,k} - \frac{\xi_j\xi_k}{|\xi|^2}\right)\xi_l, \quad j,k,l = 1,...,d.$$

Proceeding as before, one has

$$\begin{aligned}
&\mathcal{F}^{-1}[|\xi|^s E_{\alpha,\alpha}(-t^\alpha|\xi|^2)\mathcal{F}[F](\xi)] \\
&= \left(\frac{1}{2\pi}\right)^d \int_{\mathbb{R}^d} \exp(ix\cdot\xi)\mathcal{F}[F](\xi)|\xi|^s \int_0^\infty \alpha\theta M_\alpha(\theta)\exp(-\theta t^\alpha|\xi|^2)d\theta d\xi \\
&= \left(\frac{1}{2\pi}\right)^d \int_0^\infty \alpha\theta M_\alpha(\theta)d\theta \int_{\mathbb{R}^d} \mathcal{F}[F](\xi)|\xi|^s \exp(ix\cdot\xi - \theta t^\alpha|\xi|^2)d\xi \\
&= t^{-\frac{\alpha(s+d+1)}{2}} \left(\frac{1}{2\pi}\right)^d \int_0^\infty \alpha\theta^{\frac{1-s-d}{2}} M_\alpha(\theta)d\theta \\
&\quad \times \int_{\mathbb{R}^d} \exp\left(i\frac{x}{\sqrt{\theta t^\alpha}}\cdot\xi\right)\mathcal{F}[F](\xi)|\xi|^s \exp(-|\xi|^2)d\xi \\
&= t^{-\frac{\alpha(s+d+1)}{2}} \int_0^\infty \alpha\theta^{\frac{1-s-d}{2}} M_\alpha(\theta)P\left(\frac{x}{\sqrt{\theta t^\alpha}}\right)d\theta.
\end{aligned}$$

On the other hand, by the Young's inequality, we deduce

$$\begin{aligned}
|P_{\alpha,s}(t,x)| &\leq t^{-\frac{\alpha(s+d+1)}{2}} \int_0^\infty \alpha\theta^{\frac{1-s-d}{2}} M_\alpha(\theta)\frac{1}{1+|\frac{x}{\sqrt{\theta t^\alpha}}|^{d+s+1}}d\theta \\
&= \int_0^\infty \alpha\theta M_\alpha(\theta)\frac{1}{(\theta t^\alpha)^{\frac{(s+d+1)}{2}} + |x|^{d+s+1}}d\theta \\
&\leq \frac{1}{t^{\alpha\lambda_2}|x|^{\lambda_1}} \int_0^\infty \alpha\theta^{1-\lambda_2} M_\alpha(\theta)d\theta \\
&\lesssim \frac{1}{t^{\alpha\lambda_2}|x|^{\lambda_1}},
\end{aligned}$$

for $\lambda_1 > 0$, $0 < \lambda_2 < 2$ and $\lambda_1 + 2\lambda_2 = d + s + 1$. $\square$

2.7.2   Local Existence I

The purpose of this subsection is to study the existence of solutions for equation (2.102). In order to prove the version of Khai and Duong [193] in the fractional framework, a fixed point technique is applied. In this way, we define the bilinear operator $\mathcal{B}$ as

$$\mathcal{B}(u,v)(t,x) = \int_0^t (t-\tau)^{\alpha-1} E_{\alpha,\alpha}((t-\tau)^\alpha \Delta) \mathbb{P}\nabla \cdot (u(\tau,x) \otimes v(\tau,x)) d\tau.$$

Before starting our theorems, let $\tilde{q} \geq q \geq d$, $\beta = d\left(\frac{1}{q} - \frac{1}{\tilde{q}}\right)$ with $0 \leq \beta < 1$. We introduce a suitable spaces $\mathcal{H}_{q,T}^{\tilde{q}}$ consisting of the functions $u(t,x)$ such that

$$\|u\|_{\mathcal{H}_{q,T}^{\tilde{q}}} := \left\| \sup_{0<t<T} t^{\frac{\alpha\beta}{2}} |u(t,x)| \right\|_{\tilde{q}} < \infty \quad \text{and} \quad \lim_{t\to 0} \left\| \sup_{0<\tau<t} \tau^{\frac{\alpha\beta}{2}} |u(\tau,x)| \right\|_{\tilde{q}} = 0.$$

Let $p > 1$ and $s \geq \frac{d}{p} - 1$. We also introduce the space $\mathcal{G}_{p,T}^s$ consisting of the functions $u(t,x)$ such that

$$\|u\|_{\mathcal{G}_{p,T}^s} := \left\| \sup_{0<t<T} |\dot{\Lambda}^s u(t,x)| \right\|_p < \infty \quad \text{and} \quad \lim_{t\to 0} \left\| \sup_{0<\tau<t} |\dot{\Lambda}^s u(\tau,x)| \right\|_p = 0.$$

We continue by studying the solution operator on the space $\mathcal{H}_{q,T}^{\tilde{q}}$, which is given in the following lemma.

Lemma 2.28. *Let $p > 1$ and $\frac{d}{p} - 1 \leq s < \frac{d}{p}$. Assume that $u_0 \in \dot{H}_p^s(\mathbb{R}^d)$. Then for all $\tilde{q}$ satisfying $\tilde{q} > \max\{p,q\}$ with $\frac{1}{\tilde{q}} = \frac{1}{p} - \frac{s}{d}$, we have*

$$E_{\alpha,1}(t^\alpha \Delta) u_0 \in \mathcal{H}_{q,T}^{\tilde{q}}.$$

Proof. First, we consider the case $p \leq q$. In this case $s \geq 0$, using Lemma 2.23 for $s_1 = s$, $s_2 = 0$, $q_1 = p$, $q_2 = q$, we have $\dot{H}_p^s \hookrightarrow \dot{H}_q^0 = L^q$. Thus $u_0 \in L^q$.

On the one hand, note that

$$\beta = d\left(\frac{1}{q} - \frac{1}{\tilde{q}}\right) = \frac{d}{p} - s - \frac{d}{\tilde{q}} \leq 1 - \frac{d}{\tilde{q}} < 1.$$

We have the following estimates

$$t^{\frac{\alpha\beta}{2}} |E_{\alpha,1}(t^\alpha \Delta) u_0| = t^{\frac{\alpha\beta}{2}} (4\pi t^\alpha)^{-\frac{d}{2}} \left| \int_0^\infty \theta^{-\frac{d}{2}} \mathcal{M}_\alpha(\theta) \exp\left(-\frac{|\cdot|^2}{4\theta t^\alpha}\right) d\theta * u_0 \right|$$

$$\lesssim t^{-\frac{\alpha(d-\beta)}{2}} \int_0^\infty \theta^{-\frac{d}{2}} \mathcal{M}_\alpha(\theta) \exp\left(-\frac{|\cdot|^2}{4\theta t^\alpha}\right) d\theta * |u_0|$$

$$= \int_0^\infty \theta^{-\frac{\beta}{2}} \mathcal{M}_\alpha(\theta) \frac{1}{|\cdot|^{d-\beta}} \left(\frac{|\cdot|}{\sqrt{\theta t^\alpha}}\right)^{d-\beta} \exp\left(-\frac{|\cdot|^2}{4\theta t^\alpha}\right) d\theta * |u_0|$$

$$\leq \int_0^\infty \theta^{-\frac{\beta}{2}} \mathcal{M}_\alpha(\theta) d\theta \sup_{x\in\mathbb{R}^d} \left\{ \left(\frac{|x|}{\sqrt{\theta t^\alpha}}\right)^{d-\beta} \right.$$

$$\left. \times \exp\left(-\frac{|x|^2}{4\theta t^\alpha}\right) \right\} \frac{1}{|\cdot|^{d-\beta}} * |u_0|$$

$$\lesssim \frac{1}{|\cdot|^{d-\beta}} * |u_0|.$$

$$(2.104)$$

Notice that $L^{q,q} \hookrightarrow L^{q,\tilde{q}}$ for $\tilde{q} > q$ and $\frac{1}{|\cdot|^s} \in L^{\frac{d}{s},\infty}$ for $0 < s \le d$. We obtain from $L^{\tilde{q},\tilde{q}} = L^{\tilde{q}}$ and Lemma 2.24 that

$$\left\| \sup_{0<t<T} t^{\frac{\alpha\beta}{2}} |E_{\alpha,1}(t^\alpha \Delta)u_0| \right\|_{\tilde{q}} \lesssim \left\| \frac{1}{|\cdot|^{d-\beta}} * |u_0| \right\|_{\tilde{q}} \lesssim \left\| \frac{1}{|x|^{d-\beta}} \right\|_{L^{\frac{d}{d-\beta},\infty}} \|u_0\|_{L^{q,\tilde{q}}} \lesssim \|u_0\|_q,$$

which shows

$$\left\| \sup_{0<t<T} t^{\frac{\alpha\beta}{2}} |E_{\alpha,1}(t^\alpha \Delta)u_0| \right\|_{\tilde{q}} < \infty.$$

On the other hand, we consider the function $\mathcal{Y}_n$:

$$\mathcal{Y}_n(x) = \begin{cases} 0, & \text{if } x \in \{x : |x| < n\} \cap \{x : |u_0(x)| < n\}, \\ 1, & \text{otherwise.} \end{cases}$$

Since

$$t^{\frac{\alpha\beta}{2}} |E_{\alpha,1}(t^\alpha \Delta)u_0| \lesssim t^{-\frac{\alpha(d-\beta)}{2}} \int_0^\infty \theta^{-\frac{d}{2}} \mathcal{M}_\alpha(\theta) \exp\left(-\frac{|\cdot|^2}{4\theta t^\alpha}\right) d\theta * |\mathcal{Y}_n u_0|$$

$$+ t^{-\frac{\alpha(d-\beta)}{2}} \int_0^\infty \theta^{-\frac{d}{2}} \mathcal{M}_\alpha(\theta) \exp\left(-\frac{|\cdot|^2}{4\theta t^\alpha}\right) d\theta * |(1 - \mathcal{Y}_n)u_0|.$$

Similar to the proof of (2.104), we derive

$$t^{-\frac{\alpha(d-\beta)}{2}} \int_0^\infty \theta^{-\frac{d}{2}} \mathcal{M}_\alpha(\theta) \exp\left(-\frac{|\cdot|^2}{4\theta t^\alpha}\right) d\theta * |\mathcal{Y}_n u_0| \lesssim \frac{1}{|\cdot|^{d-\beta}} * |\mathcal{Y}_n u_0|,$$

as well as for fixed $\bar{q}$ such that $q < \bar{q} < \tilde{q}$,

$$t^{-\frac{\alpha(d-\beta)}{2}} \int_0^\infty \theta^{-\frac{d}{2}} \mathcal{M}_\alpha(\theta) \exp\left(-\frac{|\cdot|^2}{4\theta t^\alpha}\right) d\theta * |(1 - \mathcal{Y}_n)u_0|$$

$$= t^{\frac{\alpha(\beta-\gamma)}{2}} \int_0^\infty \theta^{-\frac{d}{2}} \mathcal{M}_\alpha(\theta) \frac{1}{|\cdot|^{d-\gamma}} \left(\frac{|\cdot|}{\sqrt{\theta t^\alpha}}\right)^{d-\gamma} \exp\left(-\frac{|\cdot|^2}{4\theta t^\alpha}\right) d\theta * |(1 - \mathcal{Y}_n)u_0|$$

$$\le t^{\frac{\alpha(\beta-\gamma)}{2}} \int_0^\infty \theta^{-\frac{\gamma}{2}} \mathcal{M}_\alpha(\theta) d\theta \sup_{x\in\mathbb{R}^d} \left\{\left(\frac{|x|}{\sqrt{\theta t^\alpha}}\right)^{d-\gamma}\right.$$

$$\times \left. \exp\left(-\frac{|x|^2}{4\theta t^\alpha}\right)\right\} \frac{1}{|\cdot|^{d-\gamma}} * |(1 - \mathcal{Y}_n)u_0|$$

$$\lesssim n t^{\frac{\alpha d}{2}(\frac{1}{q} - \frac{1}{\tilde{q}})} \frac{1}{|\cdot|^{d-\gamma}} * |1 - \mathcal{Y}_n|,$$

where $\gamma = d\left(\frac{1}{q} - \frac{1}{\tilde{q}}\right)$. The above discussion shows that

$$\left\| \sup_{0<\tau<t} \tau^{\frac{\alpha\beta}{2}} |E_{\alpha,1}(\tau^\alpha \Delta)u_0| \right\|_{\tilde{q}}$$

$$\lesssim \left\| \frac{1}{|\cdot|^{d-\beta}} * |\mathcal{Y}_n u_0| \right\|_{\tilde{q}} + n t^{\frac{\alpha d}{2}(\frac{1}{q} - \frac{1}{\tilde{q}})} \left\| \frac{1}{|\cdot|^{d-\gamma}} * |1 - \mathcal{Y}_n| \right\|_{\tilde{q}}$$

$$\lesssim \left\| \frac{1}{|x|^{d-\beta}} \right\|_{L^{\frac{d}{d-\beta},\infty}} \|\mathcal{Y}_n u_0\|_q + n t^{\frac{\alpha d}{2}(\frac{1}{q} - \frac{1}{\tilde{q}})} \left\| \frac{1}{|x|^{d-\gamma}} \right\|_{L^{\frac{d}{d-\gamma},\infty}} \|1 - \mathcal{Y}_n\|_{\bar{q}}$$

$$\lesssim \|\mathcal{Y}_n u_0\|_q + n t^{\frac{\alpha d}{2}(\frac{1}{q} - \frac{1}{\tilde{q}})} \|1 - \mathcal{Y}_n\|_{\bar{q}}.$$

It is clear that for any $\varepsilon > 0$, we make n sufficiently large such that

$$\|\mathcal{Y}_n u_0\|_q < \frac{\varepsilon}{2}.$$

Then we choose one of such n, there exists $t_0 = t_0(n)$ small enough such that for $t < t_0$,

$$nt^{\frac{\alpha d}{2}(\frac{1}{q}-\frac{1}{\tilde{q}})}\|1 - \mathcal{Y}_n\|_{\tilde{q}} < \frac{\varepsilon}{2}.$$

This ensures

$$\left\|\sup_{0<\tau<t} \tau^{\frac{\alpha\beta}{2}}|E_{\alpha,1}(\tau^\alpha\Delta)u_0|\right\|_{\tilde{q}} \lesssim \|\mathcal{Y}_n u_0\|_q + nt^{\frac{\alpha d}{2}(\frac{1}{q}-\frac{1}{\tilde{q}})}\|1 - \mathcal{Y}_n\|_{\tilde{q}} < \varepsilon, \quad \text{for } t < t_0.$$

Therefore we show that for $\tilde{q} > q$,

$$\lim_{t\to 0}\left\|\sup_{0<\tau<t} \tau^{\frac{\alpha\beta}{2}}|E_{\alpha,1}(\tau^\alpha\Delta)u_0|\right\|_{\tilde{q}} = 0. \tag{2.105}$$

Second, it remains to consider the case $p > q$. At this point $s < 0$. By Lemma 2.26, we have

$$E_{\alpha,1}(t^\alpha\Delta)u_0 = E_{\alpha,1}(t^\alpha\Delta)\dot{\Lambda}^{-s}\dot{\Lambda}^s u_0$$

$$= t^{-\frac{\alpha(d-s)}{2}}\int_0^\infty \theta^{-\frac{d-s}{2}}M_\alpha(\theta)K\left(\frac{\cdot}{\sqrt{\theta t^\alpha}}\right)d\theta * (\dot{\Lambda}^s u_0),$$

then

$$t^{\frac{\alpha\beta}{2}}|E_{\alpha,1}(t^\alpha\Delta)u_0|$$

$$\lesssim t^{\frac{\alpha(\beta-d+s)}{2}}\int_0^\infty \theta^{-\frac{d-s}{2}}M_\alpha(\theta)K\left(\frac{\cdot}{\sqrt{\theta t^\alpha}}\right)d\theta * |\dot{\Lambda}^s u_0|$$

$$= \int_0^\infty \theta^{-\frac{\beta}{2}}M_\alpha(\theta)\frac{1}{|\cdot|^{d-s-\beta}}\left(\frac{|\cdot|}{\sqrt{\theta t^\alpha}}\right)^{d-s-\beta}K\left(\frac{\cdot}{\sqrt{\theta t^\alpha}}\right)d\theta * |\dot{\Lambda}^s u_0|$$

$$\leq \int_0^\infty \theta^{-\frac{\beta}{2}}M_\alpha(\theta)d\theta \sup_{x\in\mathbb{R}^d}\left\{\left(\frac{|x|}{\sqrt{\theta t^\alpha}}\right)^{d-s-\beta}\left|K\left(\frac{x}{\sqrt{\theta t^\alpha}}\right)\right|\right\}\frac{1}{|\cdot|^{d-s-\beta}} * |\dot{\Lambda}^s u_0|$$

$$\lesssim \frac{1}{|\cdot|^{d-s-\beta}} * |\dot{\Lambda}^s u_0|.$$

Notice that $L^{\tilde{q},p} \hookrightarrow L^{\tilde{q},\tilde{q}} = L^{\tilde{q}}$ for $\tilde{q} > p$. We obtain from Lemma 2.24 that

$$\left\|\sup_{0<t<T} t^{\frac{\alpha\beta}{2}}|E_{\alpha,1}(t^\alpha\Delta)u_0|\right\|_{\tilde{q}} \lesssim \left\|\frac{1}{|\cdot|^{d-s-\beta}} * |\dot{\Lambda}^s u_0|\right\|_{\tilde{q}}$$

$$\lesssim \left\|\frac{1}{|\cdot|^{d-s-\beta}} * |\dot{\Lambda}^s u_0|\right\|_{L^{\tilde{q},p}}$$

$$\lesssim \left\|\frac{1}{|x|^{d-s-\beta}}\right\|_{L^{\frac{d}{d-s-\beta},\infty}}\|\dot{\Lambda}^s u_0\|_p \lesssim \|u_0\|_{\dot{H}^s_p},$$

this shows

$$\left\|\sup_{0<t<T} t^{\frac{\alpha\beta}{2}}|E_{\alpha,1}(t^\alpha\Delta)u_0|\right\|_{\tilde{q}} < \infty.$$

On the other hand, we consider the function $\mathcal{Y}_{n,s}$:

$$\mathcal{Y}_{n,s}(x) = \begin{cases} 0, & \text{if } x \in \{x : |x| < n\} \cap \{x : |\dot{\Lambda}^s u_0(x)| < n\}, \\ 1, & \text{otherwise.} \end{cases}$$

For fixed $\bar{q}$ such that $p < \bar{q} < \tilde{q}$, let $\gamma = d\left(\frac{1}{p} - \frac{1}{\bar{q}}\right)$. By the same way as we derived (2.105), for any $\varepsilon > 0$, there exists a sufficiently large n and a sufficiently small $t_0 = t_0(n)$ such that for $t < t_0$,

$$\left\| \sup_{0<\tau<t} \tau^{\frac{\alpha\beta}{2}} |E_{\alpha,1}(\tau^\alpha \Delta) u_0| \right\|_{\bar{q}} \lesssim \left\| \frac{1}{|x|^{d-s-\beta}} \right\|_{L^{\frac{d}{d-s-\beta},\infty}} \|\mathcal{Y}_{n,s} \dot{\Lambda}^s u_0\|_p$$

$$+ nt^{\frac{\alpha d}{2}\left(\frac{1}{p} - \frac{1}{\bar{q}}\right)} \left\| \frac{1}{|x|^{d-s+\gamma-\beta}} \right\|_{L^{\frac{d}{d-s-\beta+\gamma},\infty}} \|1 - \mathcal{Y}_{n,s}\|_{\bar{q}}$$

$$\lesssim \|\mathcal{Y}_{n,s} \dot{\Lambda}^s u_0\|_p + nt^{\frac{\alpha d}{2}\left(\frac{1}{p} - \frac{1}{\bar{q}}\right)} \|1 - \mathcal{Y}_{n,s}\|_{\bar{q}}$$

$$< \varepsilon.$$

This yields that for $\tilde{q} > p$,

$$\lim_{t \to 0} \left\| \sup_{0<\tau<t} \tau^{\frac{\alpha\beta}{2}} |E_{\alpha,1}(\tau^\alpha \Delta) u_0| \right\|_{\bar{q}} = 0.$$

Consequently, we complete the proof. $\qquad\square$

After that we focus main attention on the investigation of estimates corresponding to the bilinear operator $\mathcal{B}(u, v)$ which plays an important role in proving the existence of solutions.

Lemma 2.29. *Let $p > \frac{d}{2}$ and $\frac{d}{p} - 1 \le s < \frac{d}{p}$. Then for all $\tilde{q}$ satisfying $q < \tilde{q} < 2p$ with $\frac{1}{q} = \frac{1}{p} - \frac{s}{d}$ the bilinear operator $\mathcal{B}$ is continuous from $\mathcal{H}^{\tilde{q}}_{q,T} \times \mathcal{H}^{\tilde{q}}_{q,T}$ into $\mathcal{G}^s_{p,T}$. Moreover, there exists a constant L dependent of $q, \tilde{q}, d$ and independent of T such that*

$$\|\mathcal{B}(u, v)\|_{\mathcal{G}^s_{p,T}} \le LT^{\frac{\alpha}{2}\left(1+s-\frac{d}{p}\right)} \|u\|_{\mathcal{H}^{\tilde{q}}_{q,T}} \|v\|_{\mathcal{H}^{\tilde{q}}_{q,T}}.$$

Proof. From Lemma 2.27, we have that for $u, v \in \mathcal{H}^{\tilde{q}}_{q,T}$,

$$|\dot{\Lambda}^s \mathcal{B}(u, v)(t, x)| \le \int_0^t (t-\tau)^{\alpha-1} |\dot{\Lambda}^s E_{\alpha,\alpha}((t-\tau)^\alpha \Delta) \mathbb{P} \nabla \cdot (u(\tau, x) \otimes v(\tau, x))| d\tau$$

$$= \int_0^t (t-\tau)^{\alpha-1} |P_{\alpha,s}(t-\tau, \cdot) * (u(\tau, \cdot) \otimes v(\tau, \cdot))| d\tau$$

$$\lesssim \int_0^t (t-\tau)^{\alpha-\alpha\lambda_2-1} \frac{1}{|\cdot|^{\lambda_1}} * |(u(\tau, \cdot) \otimes v(\tau, \cdot)| d\tau,$$

where $\lambda_1 > 0, 0 < \lambda_2 < 2$ and $\lambda_1 + 2\lambda_2 = d + s + 1$. We choose

$$\lambda_1 = d\left(1 + \frac{1}{p} - \frac{2}{\tilde{q}}\right), \quad \lambda_2 = \frac{1}{2} - \frac{d}{2p} + \frac{s}{2} + \frac{d}{\tilde{q}},$$

then

$$|\dot{\Lambda}^s \mathcal{B}(u,v)(t,x)|$$

$$\lesssim \int_0^t (t-\tau)^{\alpha(1-\lambda_2)-1} \frac{1}{|\cdot|^{\lambda_1}} * |(u(\tau,\cdot) \otimes v(\tau,\cdot)| d\tau$$

$$\leq \frac{1}{|\cdot|^{\lambda_1}} * \int_0^t (t-\tau)^{\alpha(1-\lambda_2)-1} \tau^{-\alpha\beta} \sup_{0<\tau<t} \{\tau^{\frac{\alpha\beta}{2}}|u(\tau,\cdot)|\} \sup_{0<\tau<t} \{\tau^{\frac{\alpha\beta}{2}}|v(\tau,\cdot)|\} d\tau$$

$$= \frac{1}{|\cdot|^{\lambda_1}} * \left(\sup_{0<\tau<t} \{\tau^{\frac{\alpha\beta}{2}}|u(\tau,\cdot)|\} \sup_{0<\tau<t} \{\tau^{\frac{\alpha\beta}{2}}|v(\tau,\cdot)|\} \right) \int_0^t (t-\tau)^{\alpha(1-\lambda_2)-1} \tau^{-\alpha\beta} d\tau$$

$$\lesssim t^{\frac{\alpha}{2}(1+s-\frac{d}{p})} \frac{1}{|\cdot|^{d\left(1+\frac{1}{p}-\frac{2}{q}\right)}} * \left(\sup_{0<\tau<t} \{\tau^{\frac{\alpha\beta}{2}}|u(\tau,\cdot)|\} \sup_{0<\tau<t} \{\tau^{\frac{\alpha\beta}{2}}|v(\tau,\cdot)|\} \right).$$

Note that $L^{p,\frac{\tilde{q}}{2}} \hookrightarrow L^p$ for $p > \frac{\tilde{q}}{2}$. We further obtain from Lemma 2.24 and the Hölder's inequality that

$$\left\| \sup_{0<\tau<t} |\dot{\Lambda}^s \mathcal{B}(u,v)(\tau,x)| \right\|_p$$

$$\lesssim \left\| \sup_{0<\tau<t} |\dot{\Lambda}^s \mathcal{B}(u,v)(\tau,x)| \right\|_{L^{p,\frac{\tilde{q}}{2}}}$$

$$\lesssim t^{\frac{\alpha}{2}(1+s-\frac{d}{p})} \left\| \frac{1}{|\cdot|^{d(1+\frac{1}{p}-\frac{2}{q})}} * \left(\sup_{0<\tau<t} \{\tau^{\frac{\alpha\beta}{2}}|u(\tau,\cdot)|\} \sup_{0<\tau<t} \{\tau^{\frac{\alpha\beta}{2}}|v(\tau,\cdot)|\} \right) \right\|_{L^{p,\frac{\tilde{q}}{2}}}$$

$$\lesssim t^{\frac{\alpha}{2}(1+s-\frac{d}{p})} \left\| \frac{1}{|x|^{d(1+\frac{1}{p}-\frac{2}{q})}} \right\|_{L^{\frac{1}{1+\frac{1}{p}-\frac{2}{q}},\infty}} \left\| \sup_{0<\tau<t} \{\tau^{\frac{\alpha\beta}{2}}|u(\tau,x)|\} \sup_{0<\tau<t} \{\tau^{\frac{\alpha\beta}{2}}|v(\tau,x)|\} \right\|_{\frac{\tilde{q}}{2}}$$

$$\lesssim t^{\frac{\alpha}{2}(1+s-\frac{d}{p})} \left\| \sup_{0<\tau<t} \{\tau^{\frac{\alpha\beta}{2}}|u(\tau,x)|\} \right\|_{\tilde{q}} \left\| \sup_{0<\tau<t} \{\tau^{\frac{\alpha\beta}{2}}|v(\tau,x)|\} \right\|_{\tilde{q}} < \infty.$$

$$(2.106)$$

Finally, it is easy to verify that

$$\lim_{t\to 0} \left\| \sup_{0<\tau<t} |\dot{\Lambda}^s \mathcal{B}(u,v)(\tau,x)| \right\|_p = 0,$$

because of $u,v \in \mathcal{H}_{q,T}^{\tilde{q}}$ and (2.106). The proof is completed. $\square$

Lemma 2.30. *Let q and q_1 satisfy $d \leq q < q_1 < \infty$. Then for all q_2 satisfying*

$$\frac{1}{q_2} \in \left(0, \frac{1}{q}\right] \bigcap \left(\frac{2}{q_1} - \frac{1}{d}, \frac{2}{q_1}\right),$$

the bilinear operator $\mathcal{B}$ is continuous from $\mathcal{H}_{q,T}^{q_1} \times \mathcal{H}_{q,T}^{q_1}$ into $\mathcal{H}_{q,T}^{q_2}$. Moreover, there exists a constant L dependent of q_1, q_2, d and independent of T such that

$$\|\mathcal{B}(u,v)\|_{\mathcal{H}_{q,T}^{q_2}} \leq LT^{\frac{\alpha}{2}(1-\frac{d}{q})} \|u\|_{\mathcal{H}_{q,T}^{q_1}} \|v\|_{\mathcal{H}_{q,T}^{q_1}}.$$

Proof. From Lemma 2.27, for $u,v \in \mathcal{H}_{q,T}^{q_1}$, we have that

$$|\mathcal{B}(u,v)(t,x)| \leq \int_0^t (t-\tau)^{\alpha-1} |E_{\alpha,\alpha}((t-\tau)^\alpha \Delta) \mathbb{P} \nabla \cdot (u(\tau,x) \otimes v(\tau,x))| d\tau$$

$$= \int_0^t (t-\tau)^{\alpha-1} |P_{\alpha,0}(t-\tau,\cdot) * (u(\tau,\cdot) \otimes v(\tau,\cdot))| d\tau$$

$$\lesssim \int_0^t (t-\tau)^{\alpha-\alpha\lambda_2-1} \frac{1}{|\cdot|^{\lambda_1}} * |u(\tau,\cdot) \otimes v(\tau,\cdot)| d\tau,$$

where $\lambda_1 > 0, 0 < \lambda_2 < 2$ and $\lambda_1 + 2\lambda_2 = d+1$. Write

$$\beta_1 = d\Big(\frac{1}{q} - \frac{1}{q_1}\Big), \quad \beta_2 = d\Big(\frac{1}{q} - \frac{1}{q_2}\Big).$$

Choose

$$\lambda_1 = d\Big(1 + \frac{1}{q_2} - \frac{2}{q_1}\Big), \quad \lambda_2 = \frac{1}{2} + \frac{d}{q_1} - \frac{d}{2q_2},$$

then

$$\Big| t^{\frac{\alpha\beta_2}{2}} \mathcal{B}(u,v)(t,x) \Big|$$

$$\lesssim t^{\frac{\alpha\beta_2}{2}} \int_0^t (t-\tau)^{\alpha(1-\lambda_2)-1} \frac{1}{|\cdot|^{\lambda_1}} * |u(\tau,\cdot) \otimes v(\tau,\cdot)| d\tau$$

$$\lesssim t^{\frac{\alpha\beta_2}{2}} \frac{1}{|\cdot|^{\lambda_1}} * \int_0^t (t-\tau)^{\alpha(1-\lambda_2)-1} \tau^{-\alpha\beta_1} \sup_{0<\tau<t} \{\tau^{\frac{\alpha\beta_1}{2}} |u(\tau,\cdot)|\} \sup_{0<\tau<t} \{\tau^{\frac{\alpha\beta_1}{2}} |v(\tau,\cdot)|\} d\tau$$

$$= \frac{1}{|\cdot|^{\lambda_1}} * \Big(\sup_{0<\tau<t} \{\tau^{\frac{\alpha\beta_1}{2}} |u(\tau,\cdot)|\}$$

$$\times \sup_{0<\tau<t} \{\tau^{\frac{\alpha\beta_1}{2}} |v(\tau,\cdot)|\} \Big) t^{\frac{\alpha\beta_2}{2}} \int_0^t (t-\tau)^{\alpha(1-\lambda_2)-1} \tau^{-\alpha\beta_1} d\tau$$

$$\lesssim t^{\frac{\alpha}{2}(1-\frac{d}{q})} \frac{1}{|\cdot|^{d\left(1+\frac{1}{q_2}-\frac{2}{q_1}\right)}} * \Big(\sup_{0<\tau<t} \{\tau^{\frac{\alpha\beta_1}{2}} |u(\tau,\cdot)|\} \sup_{0<\tau<t} \{\tau^{\frac{\alpha\beta_1}{2}} |v(\tau,\cdot)|\} \Big).$$

Note that $L^{q_2,\frac{q_1}{2}} \hookrightarrow L^{q_2}$ for $\frac{q_1}{2} < q_2$. We further obtain from Lemma 2.24 and the Hölder's inequality that

$$\Big\| \sup_{0<\tau<t} t^{\frac{\alpha\beta_2}{2}} |\mathcal{B}(u,v)(t,x)| \Big\|_{q_2}$$

$$\leq \Big\| \sup_{0<\tau<t} t^{\frac{\alpha\beta_2}{2}} |\mathcal{B}(u,v)(\tau,x)| \Big\|_{L^{q_2,\frac{q_1}{2}}}$$

$$\lesssim t^{\frac{\alpha}{2}(1-\frac{d}{q})} \Big\| \frac{1}{|\cdot|^{d\left(1+\frac{1}{q_2}-\frac{2}{q_1}\right)}} * \Big(\sup_{0<\tau<t} \{\tau^{\frac{\alpha\beta_1}{2}} |u(\tau,\cdot)|\} \sup_{0<\tau<t} \{\tau^{\frac{\alpha\beta_1}{2}} |v(\tau,\cdot)|\} \Big) \Big\|_{L^{q_2,\frac{q_1}{2}}}$$

$$\lesssim t^{\frac{\alpha}{2}(1-\frac{d}{q})} \Big\| \frac{1}{|x|^{d\left(1+\frac{1}{q_2}-\frac{2}{q_1}\right)}} \Big\|_{L^{\frac{1}{1+\frac{1}{q_2}-\frac{2}{q_1}},\infty}}$$

$$\times \Big\| \sup_{0<\tau<t} \{\tau^{\frac{\alpha\beta_1}{2}} |u(\tau,x)|\} \sup_{0<\tau<t} \{\tau^{\frac{\alpha\beta_1}{2}} |v(\tau,x)|\} \Big\|_{\frac{q_1}{2}}$$

$$\lesssim t^{\frac{\alpha}{2}(1-\frac{d}{q})} \Big\| \sup_{0<\tau<t} \{\tau^{\frac{\alpha\beta_1}{2}} |u(\tau,x)|\} \Big\|_{q_1} \Big\| \sup_{0<\tau<t} \{\tau^{\frac{\alpha\beta_1}{2}} |v(\tau,x)|\} \Big\|_{q_1} < \infty.$$

In addition, it is easy to verify that

$$\lim_{t\to 0} \Big\| \sup_{0<\tau<t} t^{\frac{\alpha\beta_2}{2}} |\mathcal{B}(u,v)(\tau,x)| \Big\|_{q_2} = 0,$$

because of $u, v \in \mathcal{H}_{q,T}^{q_1}$. The proof is completed. $\qquad\square$

The following results present the existence of mild solutions.

Theorem 2.22. *Let $p > \frac{d}{2}$ and $\frac{d}{p} - 1 \leq s < \frac{d}{p}$. Assume that for all $u_0 \in \dot{H}_p^s(\mathbb{R}^d)$ and $\tilde{q}$ satisfying $\tilde{q} > \max\{p, q\}$ with $\frac{1}{q} = \frac{1}{p} - \frac{s}{d}$, there exists a constant δ such that*

$$\left\| \sup_{0<t<T} t^{\frac{\alpha d}{2}(\frac{1}{p} - \frac{s}{d} - \frac{1}{\tilde{q}})} |E_{\alpha,1}(t^\alpha \Delta) u_0| \right\|_{\tilde{q}} < \delta.$$

Then the equation (2.102) has a unique solution $u \in L^\infty(0, T; \dot{H}_p^s(\mathbb{R}^d))$. Moreover, the following estimate holds:

$$\left\| \sup_{0<t<T} t^{\frac{\alpha d}{2}(\frac{1}{q} - \frac{1}{r})} |u(t, x)| \right\|_r < \infty, \quad \text{for all } r > \max\{p, q\}.$$

Proof. Making use of Lemma 2.30 and choosing $q_1 = q_2 = \tilde{q}$, we deduce that $\mathcal{B}$ is continuous from $\mathcal{H}_{q,T}^{\tilde{q}} \times \mathcal{H}_{q,T}^{\tilde{q}}$ into $\mathcal{H}_{q,T}^{\tilde{q}}$ and we have the inequality

$$\|\mathcal{B}(u, v)\|_{\mathcal{H}_{q,T}^{\tilde{q}}} \leq L_{q,\tilde{q},d} T^{\frac{\alpha}{2}(1 - \frac{d}{\tilde{q}})} \|u\|_{\mathcal{H}_{q,T}^{\tilde{q}}} \|v\|_{\mathcal{H}_{q,T}^{\tilde{q}}} = L_{q,\tilde{q},d} T^{\frac{\alpha}{2}(1 + s - \frac{d}{p})} \|u\|_{\mathcal{H}_{q,T}^{\tilde{q}}} \|v\|_{\mathcal{H}_{q,T}^{\tilde{q}}}.$$

Since for any $u_0 \in \dot{H}_p^s(\mathbb{R}^d)$, we can take one $\delta = \dfrac{1}{4 L_{q,\tilde{q},d} T^{\frac{\alpha}{2}(1+s-\frac{d}{p})}}$ such that

$$\|E_{\alpha,1}(t^\alpha \Delta) u_0\|_{\mathcal{H}_{q,T}^{\tilde{q}}} = \left\| \sup_{0<t<T} t^{\frac{\alpha d}{2}(\frac{1}{q} - \frac{1}{\tilde{q}})} |E_{\alpha,1}(t^\alpha \Delta) u_0| \right\|_{\tilde{q}}$$

$$= \left\| \sup_{0<t<T} t^{\frac{\alpha d}{2}(\frac{1}{p} - \frac{s}{d} - \frac{1}{\tilde{q}})} |E_{\alpha,1}(t^\alpha \Delta) u_0| \right\|_{\tilde{q}} < \delta,$$

then equation (2.102) has a unique solution $u \in \mathcal{H}_{q,T}^{\tilde{q}}$.

Next we show that

$$u \in \bigcap_{r > \max\{p, q\}} \mathcal{H}_{q,T}^r.$$

Using Lemma 2.30 for $q_1 = \tilde{q}$, this gives that for all r_1 satisfying $\frac{1}{r_1} \in (0, \frac{1}{q}] \cap (\frac{2}{\tilde{q}} - \frac{1}{d}, \frac{2}{\tilde{q}})$, $\mathcal{B}(u, u) \in \mathcal{H}_{q,T}^{r_1}$. In addition, Lemma 2.28 ensures that $E_{\alpha,1}(t^\alpha \Delta) u_0 \in \mathcal{H}_{q,T}^{r_1}$ for all r_1 satisfying $r_1 > \max\{p, q\}$. So $u = E_{\alpha,1}(t^\alpha \Delta) u_0 + \mathcal{B}(u, u)$ yields that $u \in \mathcal{H}_{q,T}^{r_1}$. Here r_1 fulfills $\frac{1}{r_1} \in (0, \frac{1}{\max\{p,q\}}] \cap (\frac{2}{\tilde{q}} - \frac{1}{d}, \frac{2}{\tilde{q}})$.

We replace q_1 with such r_1 satisfying $\frac{1}{r_1} \in (0, \frac{1}{\max\{p,q\}}] \cap (\frac{2}{\tilde{q}} - \frac{1}{d}, \frac{2}{\tilde{q}})$, using again Lemmas 2.28 and 2.30, this also ensures that for all new r_2 satisfying

$$\frac{1}{r_2} \in \left(0, \frac{1}{\max\{p, q\}}\right] \cap \left(\frac{2}{r_1} - \frac{1}{d}, \frac{2}{r_1}\right),$$

we obtain that $u \in \mathcal{H}_{q,T}^{r_2}$ when $u \in \mathcal{H}_{q,T}^{r_1}$. By fundamental calculation, one sees that

$$\left(\frac{2}{r_1} - \frac{1}{d}, \frac{2}{r_1}\right) \subset \left(\frac{2^2}{\tilde{q}} - \frac{3}{d}, \frac{2^2}{\tilde{q}}\right) = \left(\frac{1}{d} - 2^2(\frac{1}{d} - \frac{1}{\tilde{q}}), \frac{2^2}{\tilde{q}}\right).$$

Hence r_2 satisfy

$$\frac{1}{r_2} \in \left(0, \frac{1}{\max\{p, q\}}\right] \cap \left(\frac{1}{d} - 2^2(\frac{1}{d} - \frac{1}{\tilde{q}}), \frac{2^2}{\tilde{q}}\right).$$

Repeating above procedures, we derive that $u \in \mathcal{H}_{q,T}^{r_n}$ for all r_n satisfying

$$\frac{1}{r_n} \in \left(0, \frac{1}{\max\{p,q\}}\right] \cap \left(\frac{1}{d} - 2^n\left(\frac{1}{d} - \frac{1}{\tilde{q}}\right), \frac{2^n}{\tilde{q}}\right).$$

This, together with $\frac{1}{d} - \frac{1}{\tilde{q}} > 0$ shows that there exists n large enough such that

$$\left(0, \frac{1}{\max\{p,q\}}\right] \cap \left(\frac{1}{d} - 2^n\left(\frac{1}{d} - \frac{1}{\tilde{q}}\right), \frac{2^n}{\tilde{q}}\right) = \left(0, \frac{1}{\max\{p,q\}}\right].$$

Thus, for all $r > \max\{p,q\}$, we have $u \in \mathcal{H}_{q,T}^r$.

Finally, it remains to verify that $u \in L^\infty(0,T; \dot{H}_p^s(\mathbb{R}^d))$. In fact, from Lemma 2.29, we know that $\mathcal{B}(u,u) \in \mathcal{G}_{p,T}^s \subset L^\infty(0,T; \dot{H}_p^s(\mathbb{R}^d))$ for $u \in \mathcal{H}_{q,T}^r$ with $r > \max\{p,q\}$. Furthermore, for $u_0 \in \dot{H}_p^s(\mathbb{R}^d)$, we have

$$\|E_{\alpha,1}(t^\alpha \Delta)u_0\|_{\dot{H}_p^s} = \|\dot{\Lambda}^s E_{\alpha,1}(t^\alpha \Delta)u_0\|_p = \||\xi|^s E_{\alpha,1}(-t^\alpha|\xi|^2)\mathcal{F}u_0\|_p$$

$$\leq C\||\xi|^s \mathcal{F}u_0\|_p = C\|\dot{\Lambda}^s u_0\|_p = C\|u_0\|_{\dot{H}_p^s}.$$

This gives $E_{\alpha,1}(t^\alpha \Delta)u_0 \in L^\infty(0,T; \dot{H}_p^s(\mathbb{R}^d))$. Consequently,

$$u = E_{\alpha,1}(t^\alpha \Delta)u_0 + \mathcal{B}(u,u) \in L^\infty(0,T; \dot{H}_p^s(\mathbb{R}^d)).$$

This complete the proof. $\qquad\square$

As an immediate consequence of Theorem 2.22, we obtain the existence and uniqueness of the objective equations in the case of the critical index.

Corollary 2.2. *Let $p > \frac{d}{2}$. Assume that for all $u_0 \in \dot{H}_p^{\frac{d}{p}-1}(\mathbb{R}^d)$ and $\tilde{q}$ satisfying $\tilde{q} > \max\{p,d\}$, there exists a constant δ such that*

$$\left\| \sup_{0<t<T} t^{\frac{\alpha}{2}(1-\frac{d}{\tilde{q}})}|E_{\alpha,1}(t^\alpha \Delta)u_0| \right\|_{\tilde{q}} < \delta.$$

Then equation (2.102) has a unique solution $u \in L^\infty(0,T; \dot{H}_p^{\frac{d}{p}-1}(\mathbb{R}^d))$. Moreover, the following estimate holds:

$$\left\| \sup_{0<t<T} t^{\frac{\alpha d}{2}(\frac{1}{d}-\frac{1}{r})}|u(t,x)| \right\|_r < \infty, \quad \text{for all } r > \max\{p,d\}.$$

2.7.3 Local Existence II

In this subsection, let $\tilde{q} \geq q \geq d$, $\beta = d\left(\frac{1}{q} - \frac{1}{\tilde{q}}\right)$ and $0 \leq \beta < 1$. We introduce a suitable space $\mathcal{N}_{q,T}^{\tilde{q}}$ consisting of the functions $u(t,x)$ such that

$$t^{\frac{\alpha\beta}{2}}u \in C([0,T]; L^{\tilde{q}}(\mathbb{R}^d)),$$

$$\|u\|_{\mathcal{N}_{q,T}^{\tilde{q}}} := \sup_{0<t<T} t^{\frac{\alpha\beta}{2}}\|u(t,x)\|_{\tilde{q}} < \infty \text{ and } \lim_{t\to 0} t^{\frac{\alpha\beta}{2}}\|u(t,x)\|_{\tilde{q}} = 0.$$

Let $p > 1$ and $s \geq \frac{d}{p} - 1$. We also introduce the space $\mathcal{K}_{p,T}^s$ consisting of the functions $u(t,x) \in C([0,T]; \dot{H}_p^s(\mathbb{R}^d))$ such that

$$\|u\|_{\mathcal{K}_{p,T}^s} := \sup_{0<t<T} \|u(t,x)\|_{\dot{H}_p^s} < \infty \text{ and } \lim_{t\to 0} \|u(t,x)\|_{\dot{H}_p^s} = 0.$$

Next we study the solution operator on the space $\mathcal{N}_{q,T}^{\tilde{q}}$ which is presented below.

Lemma 2.31. *Under the assumptions of Lemma 2.28, we have*
$$E_{\alpha,1}(t^\alpha\Delta)u_0 \in \mathcal{N}_{q,T}^{\tilde{q}}.$$

Proof. By referring to the proof of Lemma 2.28, we also divide the proof into two steps. Firstly, when $p \leq q$, we have $s \geq 0$. Let $h \in [1,\infty]$ satisfy
$$1 + \frac{1}{\tilde{q}} = \frac{1}{h} + \frac{1}{q}.$$

The Young's inequality yields that

$$
\begin{aligned}
t^{\frac{\alpha\beta}{2}}\|E_{\alpha,1}(t^\alpha\Delta)u_0\|_{\tilde{q}} &= t^{\frac{\alpha\beta}{2}}(4\pi t^\alpha)^{-\frac{d}{2}}\left\|\left|\int_0^\infty \theta^{-\frac{d}{2}}\mathcal{M}_\alpha(\theta)\exp\left(-\frac{|\cdot|^2}{4\theta t^\alpha}\right)d\theta * u_0\right|\right\|_{\tilde{q}}\\
&\lesssim t^{\frac{\alpha(\beta-d)}{2}}\int_0^\infty \theta^{-\frac{d}{2}}\mathcal{M}_\alpha(\theta)\left\|\left|\exp\left(-\frac{|\cdot|^2}{4\theta t^\alpha}\right) * u_0\right|\right\|_{\tilde{q}}d\theta\\
&\lesssim t^{\frac{\alpha(\beta-d)}{2}}\int_0^\infty \theta^{-\frac{d}{2}}\mathcal{M}_\alpha(\theta)\left\|\left|\exp\left(-\frac{|x|^2}{4\theta t^\alpha}\right)\right|\right\|_h\|u_0\|_q d\theta\\
&= \int_0^\infty \theta^{-\frac{\beta}{2}}\mathcal{M}_\alpha(\theta)d\theta\left\|\left|\exp\left(-\frac{|x|^2}{4}\right)\right|\right\|_h\|u_0\|_q\\
&\lesssim \|u_0\|_q \lesssim \|u_0\|_{\dot{H}_p^s}.
\end{aligned}
$$

$$(2.107)$$

On the other hand, for the same $\mathcal{Y}_n$ we have

$$
\begin{aligned}
&t^{\frac{\alpha\beta}{2}}\|E_{\alpha,1}(t^\alpha\Delta)u_0\|_{\tilde{q}}\\
&\lesssim t^{\frac{\alpha(\beta-d)}{2}}\int_0^\infty \theta^{-\frac{d}{2}}\mathcal{M}_\alpha(\theta)\left\|\left|\exp\left(-\frac{|\cdot|^2}{4\theta t^\alpha}\right) * (\mathcal{Y}_n u_0)\right|\right\|_{\tilde{q}}d\theta\\
&+ t^{\frac{\alpha(\beta-d)}{2}}\int_0^\infty \theta^{-\frac{d}{2}}\mathcal{M}_\alpha(\theta)\left\|\left|\exp\left(-\frac{|\cdot|^2}{4\theta t^\alpha}\right) * [(1-\mathcal{Y}_n)u_0]\right|\right\|_{\tilde{q}}d\theta.
\end{aligned}
$$

Similar to the proof of (2.107), we derive

$$
t^{\frac{\alpha(\beta-d)}{2}}\int_0^\infty \theta^{-\frac{d}{2}}\mathcal{M}_\alpha(\theta)\left\|\left|\exp\left(-\frac{|\cdot|^2}{4\theta t^\alpha}\right) * (\mathcal{Y}_n u_0)\right|\right\|_{\tilde{q}}d\theta \lesssim \|\mathcal{Y}_n u_0\|_q,
$$

as well as

$$
\begin{aligned}
&t^{\frac{\alpha(\beta-d)}{2}}\int_0^\infty \theta^{-\frac{d}{2}}\mathcal{M}_\alpha(\theta)\left\|\left|\exp\left(-\frac{|\cdot|^2}{4\theta t^\alpha}\right) * [(1-\mathcal{Y}_n)u_0]\right|\right\|_{\tilde{q}}d\theta\\
&\lesssim t^{\frac{\alpha(\beta-d)}{2}}\int_0^\infty \theta^{-\frac{d}{2}}\mathcal{M}_\alpha(\theta)\left\|\left|\exp\left(-\frac{|x|^2}{4\theta t^\alpha}\right)\right|\right\|_1\|(1-\mathcal{Y}_n)u_0\|_{\tilde{q}}d\theta\\
&= t^{\frac{\alpha\beta}{2}}\int_0^\infty \mathcal{M}_\alpha(\theta)d\theta\left\|\left|\exp\left(-\frac{|x|^2}{4}\right)\right|\right\|_1\|(1-\mathcal{Y}_n)u_0\|_{\tilde{q}}\\
&\lesssim t^{\frac{\alpha\beta}{2}}\|n(1-\mathcal{Y}_n)\|_{\tilde{q}}.
\end{aligned}
$$

For any $\varepsilon > 0$, we make n large enough and $t_0 = t_0(n)$ sufficiently small such that

$$t^{\frac{\alpha\beta}{2}}\|E_{\alpha,1}(t^\alpha\Delta)u_0\|_{\tilde{q}} \lesssim \|\mathcal{Y}_n u_0\|_q + n t^{\frac{\alpha\beta}{2}}\|(1-\mathcal{Y}_n)\|_{\tilde{q}} < \varepsilon, \quad \text{for } t < t_0.$$

This also shows the continuity at $t = 0$ of $t^{\frac{\alpha\beta}{2}}E_{\alpha,1}(t^\alpha\Delta)u_0$.

Furthermore, for $0 < t_1 < t_2 \leq T$,

$$\left\| t_2^{\frac{\alpha\beta}{2}}E_{\alpha,1}(t_2^\alpha\Delta)u_0 - t_1^{\frac{\alpha\beta}{2}}E_{\alpha,1}(t_1^\alpha\Delta)u_0 \right\|_{\tilde{q}}$$

$$\leq \left\| \left(t_2^{\frac{\alpha\beta}{2}} - t_1^{\frac{\alpha\beta}{2}}\right)E_{\alpha,1}(t_2^\alpha\Delta)u_0 \right\|_{\tilde{q}} + t_1^{\frac{\alpha\beta}{2}}\left\| E_{\alpha,1}(t_2^\alpha\Delta)u_0 - E_{\alpha,1}(t_1^\alpha\Delta)u_0 \right\|_{\tilde{q}}$$

$$\lesssim \left(t_2^{\frac{\alpha\beta}{2}} - t_1^{\frac{\alpha\beta}{2}}\right)t_2^{-\frac{\alpha\beta}{2}}\|u_0\|_q$$

$$+ t_1^{\frac{\alpha\beta}{2}}\int_0^\infty \theta^{-\frac{d}{2}}\mathcal{M}_\alpha(\theta)\left\| \left[t_2^{-\frac{\alpha d}{2}}\exp\left(-\frac{|\cdot|^2}{4\theta t_2^\alpha}\right) - t_1^{-\frac{\alpha d}{2}}\exp\left(-\frac{|\cdot|^2}{4\theta t_1^\alpha}\right)\right] * u_0 \right\|_{\tilde{q}}d\theta.$$

We estimate each term. The first term is equal to $\left[1 - \left(\frac{t_1}{t_2}\right)^{\frac{\alpha\beta}{2}}\right]\|u_0\|_q$, which tends to zero as $t_2 \to t_1$. For the second term we proceed to estimate

$$t_1^{\frac{\alpha\beta}{2}}\int_0^\infty \theta^{-\frac{d}{2}}\mathcal{M}_\alpha(\theta)\left\| \left[t_2^{-\frac{\alpha d}{2}}\exp\left(-\frac{|\cdot|^2}{4\theta t_2^\alpha}\right) - t_1^{-\frac{\alpha d}{2}}\exp\left(-\frac{|\cdot|^2}{4\theta t_1^\alpha}\right)\right] * u_0 \right\|_{\tilde{q}}d\theta$$

$$\leq t_1^{\frac{\alpha\beta}{2}}\int_0^\infty \theta^{-\frac{d}{2}}\mathcal{M}_\alpha(\theta)\left\| t_2^{-\frac{\alpha d}{2}}\exp\left(-\frac{|x|^2}{4\theta t_2^\alpha}\right) - t_1^{-\frac{\alpha d}{2}}\exp\left(-\frac{|x|^2}{4\theta t_1^\alpha}\right)\right\|_h \|u_0\|_q d\theta$$

$$=: J(t_1, t_2).$$

Notice that

$$\left\| t_2^{-\frac{\alpha d}{2}}\exp\left(-\frac{|x|^2}{4\theta t_2^\alpha}\right) - t_1^{-\frac{\alpha d}{2}}\exp\left(-\frac{|x|^2}{4\theta t_1^\alpha}\right)\right\|_h$$

$$\leq \left(t_1^{-\frac{\alpha d}{2}} - t_2^{-\frac{\alpha d}{2}}\right)\left\| \exp\left(-\frac{|x|^2}{4\theta t_2^\alpha}\right)\right\|_h + t_1^{-\frac{\alpha d}{2}}\left\| \exp\left(-\frac{|x|^2}{4\theta t_2^\alpha}\right) - \exp\left(-\frac{|x|^2}{4\theta t_1^\alpha}\right)\right\|_h$$

$$\leq \left(t_1^{-\frac{\alpha d}{2}} - t_2^{-\frac{\alpha d}{2}}\right)(\theta t_2^\alpha)^{\frac{d}{2h}}\left\| \exp\left(-\frac{|x|^2}{4}\right)\right\|_h$$

$$+ t_1^{-\frac{\alpha d}{2}}(\theta t_1^\alpha)^{\frac{d}{2h}}\left\| \exp\left(-\frac{|x|^2}{4}\frac{t_1^\alpha}{t_2^\alpha}\right) - \exp\left(-\frac{|x|^2}{4}\right)\right\|_h.$$

(2.108)

Then we obtain

$$J(t_1, t_2) \leq \left[\left(\frac{t_2}{t_1}\right)^{\frac{\alpha}{2}(d-\beta)} - \left(\frac{t_1}{t_2}\right)^{\frac{\alpha\beta}{2}}\right]\int_0^\infty \theta^{-\frac{\beta}{2}}\mathcal{M}_\alpha(\theta)d\theta\left\| \exp\left(-\frac{|x|^2}{4}\right)\right\|_h \|u_0\|_q$$

$$+ \left\| \exp\left(-\frac{|x|^2}{4}\frac{t_1^\alpha}{t_2^\alpha}\right) - \exp\left(-\frac{|x|^2}{4}\right)\right\|_h \|u_0\|_q \int_0^\infty \theta^{-\frac{\beta}{2}}\mathcal{M}_\alpha(\theta)d\theta$$

$$\to 0, \quad \text{as } t_2 \to t_1.$$

Therefore we show that for $\tilde{q} > q$,

$$E_{\alpha,1}(t^\alpha\Delta)u_0 \in \mathcal{N}_{q,T}^{\tilde{q}}.$$

Secondly, when $p > q$, we have $s < 0$. Let $h \in [1, \infty]$ satisfy

$$1 + \frac{1}{\tilde{q}} = \frac{1}{h} + \frac{1}{p}.$$

Then

$$t^{\frac{\alpha\beta}{2}} \|E_{\alpha,1}(t^\alpha \Delta)u_0\|_{\tilde{q}} = t^{\frac{\alpha\beta}{2} - \frac{\alpha(d-s)}{2}} \left\| \int_0^\infty \theta^{-\frac{d-s}{2}} \mathcal{M}_\alpha(\theta) K\left(\frac{\cdot}{\sqrt{\theta t^\alpha}}\right) d\theta * (\dot{\Lambda}^s u_0) \right\|_{\tilde{q}}$$

$$\lesssim t^{\frac{\alpha\beta}{2} - \frac{\alpha(d-s)}{2}} \int_0^\infty \theta^{-\frac{d-s}{2}} \mathcal{M}_\alpha(\theta) \left\| K\left(\frac{x}{\sqrt{\theta t^\alpha}}\right) \right\|_h \|\dot{\Lambda}^s u_0\|_p d\theta$$

$$\lesssim \int_0^\infty \theta^{-\frac{\beta}{2}} \mathcal{M}_\alpha(\theta) d\theta \left\| \frac{1}{(1 + |x|^{d-s})} \right\|_h \|\dot{\Lambda}^s u_0\|_p$$

$$\lesssim \|u_0\|_{\dot{H}_p^s}.$$

Besides, taking the same $\mathcal{Y}_{n,s}$ as above, we obtain for any $\varepsilon > 0$, there exists a sufficiently large n and a sufficiently small $t_0 = t_0(n)$ such that for $t < t_0$,

$$t^{\frac{\alpha\beta}{2}} \left\| E_{\alpha,1}(\tau^\alpha \Delta)u_0 \right\|_{\tilde{q}} \lesssim \left\| \frac{1}{(1 + |x|^{d-s})} \right\|_h \|\mathcal{Y}_{n,s}\dot{\Lambda}^s u_0\|_p$$

$$+ t^{\frac{\alpha d}{2}(\frac{1}{p} - \frac{1}{q})} \left\| \frac{1}{(1 + |x|^{d-s})} \right\|_1 \|n(1 - \mathcal{Y}_{n,s})\|_{\tilde{q}}$$

$$\lesssim \|\mathcal{Y}_{n,s}\dot{\Lambda}^s u_0\|_p + nt^{\frac{\alpha d}{2}(\frac{1}{p} - \frac{1}{q})} \|1 - \mathcal{Y}_{n,s}\|_{\tilde{q}}$$

$$< \varepsilon.$$

Finally, from the above discussion we know that for $0 < t_1 < t_2 \leq T$,

$$\left\| t_2^{\frac{\alpha\beta}{2}} E_{\alpha,1}(t_2^\alpha \Delta)u_0 - t_1^{\frac{\alpha\beta}{2}} E_{\alpha,1}(t_1^\alpha \Delta)u_0 \right\|_{\tilde{q}}$$

$$\leq \left(t_2^{\frac{\alpha\beta}{2}} - t_1^{\frac{\alpha\beta}{2}} \right) t_2^{-\frac{\alpha\beta}{2}} \|\dot{\Lambda}^s u_0\|_p + t_1^{\frac{\alpha\beta}{2}} \int_0^\infty \theta^{-\frac{d-s}{2}} \mathcal{M}_\alpha(\theta)$$

$$\times \left\| t_2^{-\frac{\alpha(d-s)}{2}} K\left(\frac{\cdot}{\sqrt{\theta t_2^\alpha}}\right) - t_1^{-\frac{\alpha(d-s)}{2}} K\left(\frac{\cdot}{\sqrt{\theta t_1^\alpha}}\right) d\theta * (\dot{\Lambda}^s u_0) \right\|_{\tilde{q}} d\theta$$

$$\leq \left[1 - \left(\frac{t_1}{t_2}\right)^{\frac{\alpha\beta}{2}} \right] \|\dot{\Lambda}^s u_0\|_p$$

$$+ \left[\left(\frac{t_2}{t_1}\right)^{\frac{\alpha\beta}{2}} - \left(\frac{t_1}{t_2}\right)^{\frac{\alpha\beta}{2h}} \right] \int_0^\infty \theta^{-\frac{\beta}{2}} \mathcal{M}_\alpha(\theta) d\theta \|K(x)\|_h \|\dot{\Lambda}^s u_0\|_p$$

$$+ \left\| K\left(\left(\frac{t_1}{t_2}\right)^{\frac{\alpha}{2}} x\right) - K(x) \right\|_h \|\dot{\Lambda}^s u_0\|_p \int_0^\infty \theta^{-\frac{\beta}{2}} \mathcal{M}_\alpha(\theta) d\theta$$

$$\to 0, \quad \text{as } t_2 \to t_1.$$

This yields that for $\tilde{q} > p$,

$$E_{\alpha,1}(t^\alpha \Delta)u_0 \in \mathcal{N}_{q,T}^{\tilde{q}}.$$

Consequently, we complete the proof. $\qquad \square$

Similarly, we also state the estimates of the bilinear operator.

Lemma 2.32. *Under the assumptions of Lemma 2.29, the bilinear operator $\mathcal{B}$ is continuous from $\mathcal{N}_{q,T}^{\tilde{q}} \times \mathcal{N}_{q,T}^{\tilde{q}}$ into $\mathcal{K}_{p,T}^s$. Moreover, there exists a constant M such that*

$$\|\mathcal{B}(u,v)\|_{\mathcal{K}_{p,T}^s} \leq MT^{\frac{\alpha}{2}(1+s-\frac{d}{p})}\|u\|_{\mathcal{N}_{q,T}^{\tilde{q}}}\|v\|_{\mathcal{N}_{q,T}^{\tilde{q}}}.$$

Proof. From Lemma 2.27, we have that for $u, v \in \mathcal{N}_{q,T}^{\tilde{q}}$,

$$\|\dot{\Lambda}^s \mathcal{B}(u,v)(t,x)\|_p$$

$$\leq \int_0^t (t-\tau)^{\alpha-1}\|\dot{\Lambda}^s E_{\alpha,\alpha}((t-\tau)^\alpha \Delta)\mathbb{P}\nabla \cdot (u(\tau,x) \otimes v(\tau,x))\|_p d\tau$$

$$= \int_0^t (t-\tau)^{\alpha-1}\|P_{\alpha,s}(t-\tau,\cdot) * (u(\tau,\cdot) \otimes v(\tau,\cdot))\|_p d\tau$$

$$\lesssim \int_0^t (t-\tau)^{-\frac{\alpha(s+d-1)}{2}-1} \int_0^\infty \alpha\theta^{\frac{1-s-d}{2}}\mathcal{M}_\alpha(\theta)$$

$$\times \left\| P\left(\frac{\cdot}{\sqrt{\theta(t-\tau)^\alpha}}\right) * (u(\tau,\cdot) \otimes v(\tau,\cdot)) \right\|_p d\theta d\tau.$$

We choose $k \in [1,\infty]$ satisfying $1 + \frac{1}{p} = \frac{1}{k} + \frac{2}{\tilde{q}}$. Then

$$\left\| P\left(\frac{\cdot}{\sqrt{\theta(t-\tau)^\alpha}}\right) * (u(\tau,\cdot) \otimes v(\tau,\cdot)) \right\|_p$$

$$\leq \left\| P\left(\frac{x}{\sqrt{\theta(t-\tau)^\alpha}}\right) \right\|_k \|u(\tau,x) \otimes v(\tau,x)\|_{\frac{\tilde{q}}{2}} \tag{2.109}$$

$$\leq [\theta(t-\tau)^\alpha]^{\frac{d}{2k}}\|P(x)\|_k\|u(\tau,x)\|_{\tilde{q}}\|v(\tau,x)\|_{\tilde{q}}$$

$$\lesssim [\theta(t-\tau)^\alpha]^{\frac{d}{2k}}\|u(\tau,x)\|_{\tilde{q}}\|v(\tau,x)\|_{\tilde{q}}.$$

This allows us to infer that

$$\|\dot{\Lambda}^s \mathcal{B}(u,v)(t,x)\|_p$$

$$\lesssim \int_0^t (t-\tau)^{-\frac{\alpha(s-1)}{2}+\alpha d(\frac{1}{2p}-\frac{1}{\tilde{q}})-1}$$

$$\times \|u(\tau,x)\|_{\tilde{q}}\|v(\tau,x)\|_{\tilde{q}}d\tau \int_0^\infty \alpha\theta^{\frac{1-s}{2}+d(\frac{1}{2p}-\frac{1}{\tilde{q}})}\mathcal{M}_\alpha(\theta)d\theta$$

$$\lesssim \int_0^t (t-\tau)^{-\frac{\alpha(s-1)}{2}+\alpha d(\frac{1}{2p}-\frac{1}{\tilde{q}})-1}\tau^{-\alpha\beta}$$

$$\times \sup_{0<\tau<t}\{\tau^{\frac{\alpha\beta}{2}}\|u(\tau,x)\|_{\tilde{q}}\} \sup_{0<\tau<t}\{\tau^{\frac{\alpha\beta}{2}}\|v(\tau,x)\|_{\tilde{q}}\}d\tau$$

$$= \sup_{0<\tau<t}\{\tau^{\frac{\alpha\beta}{2}}\|u(\tau,x)\|_{\tilde{q}}\} \sup_{0<\tau<t}\{\tau^{\frac{\alpha\beta}{2}}\|v(\tau,x)\|_{\tilde{q}}\}$$

$$\times \int_0^t (t-\tau)^{-\frac{\alpha(s-1)}{2}+\alpha d(\frac{1}{2p}-\frac{1}{\tilde{q}})-1}\tau^{-\alpha\beta}d\tau$$

$$\lesssim t^{\frac{\alpha}{2}(1+s-\frac{d}{p})} \sup_{0<\tau<t}\{\tau^{\frac{\alpha\beta}{2}}\|u(\tau,x)\|_{\tilde{q}}\} \sup_{0<\tau<t}\{\tau^{\frac{\alpha\beta}{2}}\|v(\tau,x)\|_{\tilde{q}}\}.$$

This immediately leads to the fact that

$$\lim_{t \to 0} \|\mathcal{B}(u,v)(t)\|_{\dot{H}_p^s} = 0.$$

Finally, we show the continuity of $\mathcal{B}(u,v)$. For $0 \le t_1 < t_2 \le T$, we have

$$\|\mathcal{B}(u,v)(t_2,x) - \mathcal{B}(u,v)(t_1,x)\|_{\dot{H}_p^s}$$

$$\le \int_{t_1}^{t_2} (t_2-\tau)^{\alpha-1} \|\dot{\Lambda}^s E_{\alpha,\alpha}((t_2-\tau)^\alpha \Delta) \mathbb{P} \nabla \cdot (u(\tau,x) \otimes v(\tau,x))\|_p d\tau$$

$$+ \int_0^{t_1} \left[(t_1-\tau)^{\alpha-1} - (t_2-\tau)^{\alpha-1} \right]$$

$$\times \|\dot{\Lambda}^s E_{\alpha,\alpha}((t_2-\tau)^\alpha \Delta) \mathbb{P} \nabla \cdot (u(\tau,x) \otimes v(\tau,x))\|_p d\tau$$

$$+ \int_0^{t_1} (t_1-\tau)^{\alpha-1}$$

$$\times \|\dot{\Lambda}^s [E_{\alpha,\alpha}((t_2-\tau)^\alpha \Delta) - E_{\alpha,\alpha}((t_1-\tau)^\alpha \Delta)] \mathbb{P} \nabla \cdot (u(\tau,x) \otimes v(\tau,x))\|_p d\tau$$

$$= \int_{t_1}^{t_2} (t_2-\tau)^{\alpha-1} \|P_{\alpha,s}(t_2-\tau,\cdot) * (u(\tau,\cdot) \otimes v(\tau,\cdot))\|_p d\tau$$

$$+ \int_0^{t_1} \left[(t_1-\tau)^{\alpha-1} - (t_2-\tau)^{\alpha-1} \right] \|P_{\alpha,s}(t_2-\tau,\cdot) * (u(\tau,\cdot) \otimes v(\tau,\cdot))\|_p d\tau$$

$$+ \int_0^{t_1} (t_1-\tau)^{\alpha-1} \|[P_{\alpha,s}(t_2-\tau,\cdot) - P_{\alpha,s}(t_1-\tau,\cdot)] * (u(\tau,\cdot) \otimes v(\tau,\cdot))\|_p d\tau$$

$$=: \Gamma_1(t_1,t_2) + \Gamma_2(t_1,t_2) + \Gamma_3(t_1,t_2).$$

We estimate each term as follows. From the similar argument and the properties of the Beta function, we deduce that

$$\Gamma_1(t_1,t_2) \le \sup_{0<\tau<t_2} \left\{ \tau^{\frac{\alpha\beta}{2}} \|u(\tau,x)\|_{\tilde{q}} \right\}$$

$$\times \sup_{0<\tau<t_2} \left\{ \tau^{\frac{\alpha\beta}{2}} \|v(\tau,x)\|_{\tilde{q}} \right\} \int_{t_1}^{t_2} (t_2-\tau)^{-\frac{\alpha(s-1)}{2}+\alpha d(\frac{1}{2p}-\frac{1}{\tilde{q}})-1} \tau^{-\alpha\beta} d\tau$$

$$= t_2^{\frac{\alpha}{2}(1+s-\frac{d}{p})} \|u\|_{\mathcal{N}_{q,t}^{\tilde{q}}} \|v\|_{\mathcal{N}_{q,T}^{\tilde{q}}} \int_{t_1/t_2}^1 (1-\tau)^{-\frac{\alpha(s-1)}{2}+\alpha d(\frac{1}{2p}-\frac{1}{\tilde{q}})-1} \tau^{-\alpha\beta} d\tau$$

$$\to 0, \quad \text{as } t_2 \to t_1.$$

And $\Gamma_2(t_1,t_2)$ is bound by

$$\|u\|_{\mathcal{N}_{q,t}^{\tilde{q}}} \|v\|_{\mathcal{N}_{q,T}^{\tilde{q}}} \int_0^{t_1} \left[(t_1-\tau)^{\alpha-1} - (t_2-\tau)^{\alpha-1} \right] (t_2-\tau)^{-\alpha(\frac{1}{2}-\frac{d}{2q}+\frac{d}{q})} \tau^{-\alpha\beta} d\tau.$$

Since

$$\int_0^{t_1} \left[(t_1-\tau)^{\alpha-1} - (t_2-\tau)^{\alpha-1} \right] (t_2-\tau)^{-\alpha(\frac{1}{2}-\frac{d}{2q}+\frac{d}{q})} \tau^{-\alpha\beta} d\tau$$

$$\leq \int_0^{t_1} \left[(t_1 - \tau)^{\alpha-1}(t_2 - \tau)^{-\alpha(\frac{1}{2} - \frac{d}{2q} + \frac{d}{q})} + (t_2 - \tau)^{-\alpha(-\frac{d}{2q} + \frac{d}{q} - \frac{1}{2}) - 1} \right] \tau^{-\alpha\beta} d\tau$$

$$\leq \int_0^{t_1} \left[(t_1 - \tau)^{-\alpha(-\frac{d}{2q} + \frac{d}{q} - \frac{1}{2}) - 1} + (t_2 - \tau)^{-\alpha(-\frac{d}{2q} + \frac{d}{q} - \frac{1}{2}) - 1} \right] \tau^{-\alpha\beta} d\tau < \infty,$$

then it follows from the Lebesgue's dominated convergence theorem that $\Gamma_2(t_1, t_2) \to 0$ as $t_2 \to t_1$. Before calculating $\Gamma_3(t_1, t_2)$, we insert the following inequality

$$\left\| t_2^{-\frac{\alpha(s+d+1)}{2}} P\left(\frac{x}{\sqrt{\theta t_2^\alpha}}\right) - t_1^{-\frac{\alpha(s+d+1)}{2}} P\left(\frac{x}{\sqrt{\theta t_1^\alpha}}\right) \right\|_k$$

$$\leq \left(t_1^{-\frac{\alpha(s+d+1)}{2}} - t_2^{-\frac{\alpha(s+d+1)}{2}} \right) \left\| P\left(\frac{x}{\sqrt{\theta t_2^\alpha}}\right) \right\|_k$$

$$+ t_1^{-\frac{\alpha(s+d+1)}{2}} \left\| P\left(\frac{x}{\sqrt{\theta t_2^\alpha}}\right) - P\left(\frac{x}{\sqrt{\theta t_1^\alpha}}\right) \right\|_k$$

$$\leq \left(t_1^{-\frac{\alpha(s+d+1)}{2}} - t_2^{-\frac{\alpha(s+d+1)}{2}} \right) (\theta t_2^\alpha)^{\frac{d}{2k}} \| P(x) \|_k$$

$$+ t_1^{-\frac{\alpha(s+d+1)}{2}} (\theta t_1^\alpha)^{\frac{d}{2k}} \left\| P\left(\left(\frac{t_1}{t_2}\right)^{\frac{\alpha}{2}} x \right) - P(x) \right\|_k.$$

This allows us to obtain the estimate of $\Gamma_3(t_1, t_2)$, which is

$$\Gamma_3(t_1, t_2) \leq \int_0^{t_1} (t_1 - \tau)^{\alpha-1} \| u(\tau, x) \|_{\tilde{q}} \| v(\tau, x) \|_{\tilde{q}}$$

$$\times \int_0^\infty \alpha \theta^{\frac{1-s-d}{2}} M_\alpha(\theta) \left\| (t_2 - \tau)^{-\frac{\alpha(s+d+1)}{2}} P\left(\frac{x}{\sqrt{\theta(t_2 - \tau)^\alpha}}\right) \right.$$

$$\left. - (t_1 - \tau)^{-\frac{\alpha(s+d+1)}{2}} P\left(\frac{x}{\sqrt{\theta(t_1 - \tau)^\alpha}}\right) \right\|_k d\theta d\tau$$

$$\leq \sup_{0 < \tau < t} \left\{ \tau^{\frac{\alpha\beta}{2}} \| u(\tau, x) \|_{\tilde{q}} \right\}$$

$$\times \sup_{0 < \tau < t} \left\{ \tau^{\frac{\alpha\beta}{2}} \| v(\tau, x) \|_{\tilde{q}} \right\} \| P(x) \|_k \int_0^\infty \alpha \theta^{\frac{d}{2}(\frac{1}{q} + \frac{1}{d} - \frac{2}{\tilde{q}})} M_\alpha(\theta) d\theta$$

$$\times \left\{ \int_0^{t_1} \left[(t_1 - \tau)^{-\frac{\alpha(s+d-1)}{2} - 1} (t_2 - \tau)^{\frac{\alpha d}{2k}} \right. \right.$$

$$\left. - (t_1 - \tau)^{\alpha-1} (t_2 - \tau)^{-\frac{\alpha(s+d+1)}{2} + \frac{\alpha d}{2k}} \right] \tau^{-\alpha\beta} d\tau$$

$$\left. + \int_0^{t_1} (t_1 - \tau)^{-\frac{\alpha(s-1)}{2} + \alpha d(\frac{1}{2p} - \frac{1}{\tilde{q}}) - 1} \tau^{-\alpha\beta} \left\| P\left(\left(\frac{t_1 - \tau}{t_2 - \tau}\right)^{\frac{\alpha}{2}} x \right) - P(x) \right\|_k d\tau \right\}.$$

Using the Lebesgue's dominated convergence theorem again, one shows that $\Gamma_3(t_1, t_2) \to 0$ as $t_2 \to t_1$. This proves the continuity. The proof is completed. $\square$

Lemma 2.33. *Let q and $\tilde{q}$ satisfy $d \leq q < \tilde{q}$. Then the bilinear operator $\mathcal{B}$ is continuous from $\mathcal{N}_{q,T}^{\tilde{q}} \times \mathcal{N}_{q,T}^{\tilde{q}}$ into $\mathcal{N}_{q,T}^{\tilde{q}}$. Moreover, there exists a constant M such that*

$$\| \mathcal{B}(u, v) \|_{\mathcal{N}_{q,T}^{\tilde{q}}} \leq M T^{\frac{\alpha}{2}(1 - \frac{d}{q})} \| u \|_{\mathcal{N}_{q,T}^{\tilde{q}}} \| v \|_{\mathcal{N}_{q,T}^{\tilde{q}}}.$$

Proof. Using the inequality (2.109), we obtain

$$t^{\frac{\alpha\beta}{2}}\|\mathcal{B}(u,v)(t,x)\|_{\tilde{q}}$$

$$\leq t^{\frac{\alpha\beta}{2}}\int_0^t (t-\tau)^{\alpha-1}\|P_{\alpha,0}(t-\tau,\cdot)*(u(\tau,\cdot)\otimes v(\tau,\cdot))\|_{\tilde{q}}d\tau$$

$$\lesssim t^{\frac{\alpha\beta}{2}}\int_0^t (t-\tau)^{-\frac{\alpha(d-1)}{2}-1}\int_0^\infty \alpha\theta^{\frac{1-d}{2}}\mathcal{M}_\alpha(\theta)$$

$$\times\left\|P\left(\frac{\cdot}{\sqrt{\theta(t-\tau)^\alpha}}\right)*(u(\tau,\cdot)\otimes v(\tau,\cdot))\right\|_{\tilde{q}}d\theta d\tau$$

$$\lesssim t^{\frac{\alpha\beta}{2}}\int_0^t (t-\tau)^{\frac{\alpha}{2}(1-\frac{d}{q})-1}\|u(\tau,x)\|_{\tilde{q}}\|v(\tau,x)\|_{\tilde{q}}d\tau\int_0^\infty \alpha\theta^{\frac{1}{2}(1-\frac{d}{q})}\mathcal{M}_\alpha(\theta)d\theta$$

$$\lesssim \sup_{0<\tau<t}\{\tau^{\frac{\alpha\beta}{2}}\|u(\tau,x)\|_{\tilde{q}}\}\sup_{0<\tau<t}\{\tau^{\frac{\alpha\beta}{2}}\|v(\tau,x)\|_{\tilde{q}}\}t^{\frac{\alpha\beta}{2}}\int_0^t (t-\tau)^{\frac{\alpha}{2}(1-\frac{d}{q})-1}\tau^{-\alpha\beta}d\tau$$

$$\lesssim t^{\frac{\alpha}{2}(1-\frac{d}{q})}\sup_{0<\tau<t}\{\tau^{\frac{\alpha\beta}{2}}\|u(\tau,x)\|_{\tilde{q}}\}\sup_{0<\tau<t}\{\tau^{\frac{\alpha\beta}{2}}\|v(\tau,x)\|_{\tilde{q}}\}.$$

Hence,

$$\|\mathcal{B}(u,v)\|_{\mathcal{N}_{q,T}^{\tilde{q}}}=\sup_{t\in(0,T)}\{t^{\frac{\alpha\beta}{2}}\|\mathcal{B}(u,v)(t)\|_{\tilde{q}}\}\leq MT^{\frac{\alpha}{2}(1-\frac{d}{q})}\|u\|_{\mathcal{N}_{q,T}^{\tilde{q}}}\|v\|_{\mathcal{N}_{q,T}^{\tilde{q}}},$$

and $\lim_{t\to 0}t^{\frac{\alpha\beta}{2}}\|\mathcal{B}(u,v)(t)\|_{\tilde{q}}=0$.

It remains to show the continuity of $\mathcal{B}(u,v)$. Applying the proof of continuity in Lemma 2.32 for $s=0$ and $p=\tilde{q}$, we can easily obtain $t^{\frac{\alpha\beta}{2}}\mathcal{B}(u,v)\in C([0,T];L^{\tilde{q}}(\mathbb{R}^d))$. This completes the proof. $\qquad\square$

As a consequence of Lemma 2.33, one can obtain the following result, the proof of which is omitted.

Lemma 2.34. *Let* $d\leq q\leq \tilde{q}_2<\infty$ *and* $q<\tilde{q}_1<\infty$. *If one of the following conditions holds:*

(i) $q\leq \tilde{q}_1<2d$, $q\leq \tilde{q}_2<\frac{d\tilde{q}_1}{2d-\tilde{q}_1}$;
(ii) $2d\leq \tilde{q}_1<2q$, $q\leq \tilde{q}_2<\infty$;
(iii) $2q\leq \tilde{q}_1<\infty$, $\frac{\tilde{q}_1}{2}\leq \tilde{q}_2<\infty$,

then the bilinear operator $\mathcal{B}$ *is continuous from* $\mathcal{N}_{q,T}^{\tilde{q}_1}\times\mathcal{N}_{q,T}^{\tilde{q}_1}$ *into* $\mathcal{N}_{q,T}^{\tilde{q}_2}$. *Moreover, there exists a constant* M *such that*

$$\|\mathcal{B}(u,v)\|_{\mathcal{N}_{q,T}^{\tilde{q}_2}}\leq MT^{\frac{\alpha}{2}(1-\frac{d}{q})}\|u\|_{\mathcal{N}_{q,T}^{\tilde{q}_1}}\|v\|_{\mathcal{N}_{q,T}^{\tilde{q}_1}}.$$

Finally, we present the last main result.

Theorem 2.23. *Under the assumptions of Theorem 2.22, we have that the equation* (2.102) *has a unique solution* $u\in C([0,T];\dot{H}_p^s(\mathbb{R}^d))$. *Moreover, the following estimate holds:*

$$t^{\frac{\alpha d}{2}(\frac{1}{q}-\frac{1}{r})}u\in C([0,T];L^r(\mathbb{R}^d)),\quad \text{for all}\ \ r>\max\{p,q\}.$$

Proof. Making use of Lemma 2.33 we deduce that $\mathcal{B}$ is continuous from $\mathcal{N}_{q,T}^{\tilde{q}} \times \mathcal{N}_{q,T}^{\tilde{q}}$ into $\mathcal{N}_{q,T}^{\tilde{q}}$ and the following inequality holds:

$$\|\mathcal{B}(u,v)\|_{\mathcal{N}_{q,T}^{\tilde{q}}} \le M_{q,\tilde{q},d} T^{\frac{\alpha}{2}(1-\frac{d}{\tilde{q}})} \|u\|_{\mathcal{N}_{q,T}^{\tilde{q}}} \|v\|_{\mathcal{N}_{q,T}^{\tilde{q}}} = M_{q,\tilde{q},d} T^{\frac{\alpha}{2}(1+s-\frac{d}{p})} \|u\|_{\mathcal{N}_{q,T}^{\tilde{q}}} \|v\|_{\mathcal{N}_{q,T}^{\tilde{q}}}.$$

Since for any $u_0 \in \dot{H}_p^s(\mathbb{R}^d)$, we can take one $\delta = \dfrac{1}{4M_{q,\tilde{q},d} T^{\frac{\alpha}{2}(1+s-\frac{d}{p})}}$ such that

$$\|E_{\alpha,1}(t^\alpha \Delta)u_0\|_{\mathcal{N}_{q,T}^{\tilde{q}}} = \sup_{0<t<T} t^{\frac{\alpha d}{2}(\frac{1}{q}-\frac{1}{\tilde{q}})} \|E_{\alpha,1}(t^\alpha \Delta)u_0\|_{\tilde{q}}$$

$$= \sup_{0<t<T} t^{\frac{\alpha d}{2}(\frac{1}{p}-\frac{s}{d}-\frac{1}{\tilde{q}})} \|E_{\alpha,1}(t^\alpha \Delta)u_0\|_{\tilde{q}} < \delta,$$

then equation (2.102) has a unique solution $u \in \mathcal{N}_{q,T}^{\tilde{q}}$.

Next we show that

$$u \in \bigcap_{r > \max\{p,q\}} \mathcal{N}_{q,T}^r.$$

We can divide the process into three cases:

(i) $q \le \tilde{q} < 2d$,

(ii) $2d \le \tilde{q} < 2q$,

(iii) $2q \le \tilde{q} < \infty$.

If case (i) holds, then there exist two cases $\tilde{q} > \frac{4d}{3}$ and $\tilde{q} \le \frac{4d}{3}$.

For the case $\tilde{q} > \frac{4d}{3}$, we use case (i) of Lemma 2.34 for $\tilde{q}_1 = \tilde{q}$ and $\tilde{q}_2 = 2d$, it is clear that $\max\{p,q\} < 2d < \frac{d\tilde{q}}{2d-\tilde{q}}$, so one has $\mathcal{B}(u,u) \in \mathcal{N}_{q,T}^{2d}$. This also gives from $E_{\alpha,1}(t^\alpha \Delta)u_0 \in \mathcal{N}_{q,T}^{2d}$ that $u = E_{\alpha,1}(t^\alpha \Delta)u_0 + \mathcal{B}(u,u) \in \mathcal{N}_{q,T}^{2d}$. Using again Lemma 2.31 and the case (ii) of Lemma 2.34 for $\tilde{q}_1 = 2d$ and $\tilde{q}_2 = r$ with $r > \max\{p,q\}$, we have $u \in \mathcal{N}_{q,T}^r$.

For the case $\tilde{q} \le \frac{4d}{3}$, we construct an increase sequence $\{q_j\}_{j=0}^N$ as follows: let $q_0 = \tilde{q}$ and $q_1 = \frac{d\tilde{q}}{2d-\tilde{q}}$. If $q_1 > \frac{4d}{3}$, then we stop it and choose $N = 1$. Otherwise, let $q_2 = \frac{dq_1}{2d-q_1}$. If $q_2 > \frac{4d}{3}$, then we stop it and choose $N = 2$. Repeat above procedures until there exists an $i \ge 0$ such that $q_i \le \frac{4d}{3}$, $q_{i+1} > \frac{4d}{3}$. In this moment $N = i+1$ and

$$q_0 = \tilde{q}, \quad q_{j+1} = \frac{dq_j}{2d-q_j}, \quad j = 0,1,...,N-1, \quad q_{N-1} \le \frac{4d}{3} < q_N \le 2d.$$

It is evident that $q_j < q_{j+1}$. Using case (i) of Lemma 2.34 for $\tilde{q}_1 = q_0$, $\tilde{q}_2 = r_0$ with $\max\{p,q\} < r_0 < q_1$, we have $u \in \mathcal{N}_{q,T}^{r_0}$. Using case (i) of Lemma 2.34 for $\tilde{q}_1 = r_0$, $\tilde{q}_2 = r_1$ with $\max\{p,q\} < r_1 < q_2$ again, one has $u \in \mathcal{N}_{q,T}^{r_1}$. Repeating the procedures, this also gives that $u \in \mathcal{N}_{q,T}^{r_{N-1}}$ for $\max\{p,q\} < r_{N-1} < q_N$. Besides, when $\frac{4d}{3} < q_N \le r$, we also obtain that $u \in \mathcal{N}_{q,T}^r$. Therefore we only need $r > \max\{p,q\}$ to ensure $u \in \mathcal{N}_{q,T}^r$.

If case (ii) holds, it immediately yields that $u \in \mathcal{N}_{q,T}^r$ for $r > \max\{p,q\}$ from case (ii) of Lemma 2.34 for $\tilde{q}_1 = \tilde{q}$, $\tilde{q}_2 = r$ with $q \le r < \infty$.

If case (iii) holds, then there exists an $n_0 \in \mathbb{N}^+$ such that $\frac{\tilde{q}}{2^{n_0}} < \max\{2q, p\} \le \frac{\tilde{q}}{2^{n_0-1}}$. Since $\tilde{q} > \max\{p, q\}$ and $\tilde{q} > 2q$, we obtain $n_0 \ge 1$. Hence we use case (iii) of Lemma 2.34 for $\tilde{q}_1 = \tilde{q}$, $\tilde{q}_2 = r_1$ with $\frac{\tilde{q}}{2} < r_1 < \infty$ to show $u = E_{\alpha,1}(t^\alpha \Delta) u_0 + \mathcal{B}(u, v) \in \mathcal{N}^{r_1}_{q,T}$. Use again case (iii) of Lemma 2.34 for $\tilde{q}_1 = r_1$, $\tilde{q}_2 = r_2$ with $\frac{\tilde{q}}{2^2} < \frac{r_1}{2} < r_2 < \infty$ to show $u \in \mathcal{N}^{r_2}_{q,T}$. Repeating this procedure until $u \in \mathcal{N}^{r_{n_0-1}}_{q,T}$ for $\frac{\tilde{q}}{2^{n_0-1}} < r_{n_0-1} < \infty$. When $\frac{\tilde{q}}{2^{n_0}} < r < \frac{\tilde{q}}{2^{n_0-1}}$ and $r > \max\{p, q\}$ one has $u \in \mathcal{N}^r_{q,T}$. In this moment if $\max\{p, q\} \ge \frac{\tilde{q}}{2^{n_0}}$, then we only need $r > \max\{p, q\}$ to have $u \in \mathcal{N}^r_{q,T}$. If $\max\{p, q\} < \frac{\tilde{q}}{2^{n_0}}$, then from $\frac{\tilde{q}}{2^{n_0}} < \max\{2q, p\}$ we know $2q > \frac{\tilde{q}}{2^{n_0}}$. Hence $u \in \mathcal{N}^{2q}_{q,T}$. We use case (iii) of Lemma 2.34 for $\tilde{q}_1 = 2q$, $\tilde{q}_2 = r$ with $r > q$ to show $u \in \mathcal{N}^r_{q,T}$. By this time $r > \max\{p, q\}$.

Finally, it remains to verify that $u \in C([0, T]; \dot{H}^s_p(\mathbb{R}^d))$. In fact, from Lemma 2.32, we know that $\mathcal{B}(u, u) \in \mathcal{K}^s_{p,T} \subset C([0, T]; \dot{H}^s_p(\mathbb{R}^d))$ for $u \in \mathcal{N}^r_{q,T}$ with $r > \max\{p, q\}$. Furthermore, we show the continuity of $E_{\alpha,1}(t^\alpha \Delta) u_0$. Indeed, the continuity at $t = 0$ of $E_{\alpha,1}(t^\alpha \Delta) u_0$ is given as follows:

$$
\lim_{t \to 0} E_{\alpha,1}(t^\alpha \Delta) u_0 = \lim_{t \to 0} \int_0^\infty M_\alpha(\theta)(4\pi\theta t^\alpha)^{-\frac{d}{2}} \exp(-\frac{|\cdot|^2}{4\theta t^\alpha}) * u_0 d\theta
$$

$$
= \int_0^\infty M_\alpha(\theta) \lim_{t \to 0} (4\pi\theta t^\alpha)^{-\frac{d}{2}} \exp(-\frac{|\cdot|^2}{4\theta t^\alpha}) * u_0 d\theta
$$

$$
= u_0 \int_0^\infty M_\alpha(\theta) d\theta
$$

$$
= u_0.
$$

The continuity on $(0, T]$ of $E_{\alpha,1}(t^\alpha \Delta) u_0$ is presented as below. For $0 < t_1 < t_2 \le T$, using the inequality (2.108) for $h = 1$ we have

$$
\|E_{\alpha,1}(t_2^\alpha \Delta) u_0 - E_{\alpha,1}(t_1^\alpha \Delta) u_0\|_{\dot{H}^s_p}
$$

$$
\lesssim \int_0^\infty \theta^{-\frac{d}{2}} M_\alpha(\theta) \left\| \left(t_2^{-\frac{\alpha d}{2}} \exp(-\frac{|\cdot|^2}{4\theta t_2^\alpha}) - t_1^{-\frac{\alpha d}{2}} \exp(-\frac{|\cdot|^2}{4\theta t_1^\alpha}) \right) * (\dot{\Lambda}^s u_0) \right\|_p d\theta
$$

$$
\lesssim \int_0^\infty \theta^{-\frac{d}{2}} M_\alpha(\theta) \left\| t_2^{-\frac{\alpha d}{2}} \exp(-\frac{|x|^2}{4\theta t_2^\alpha}) - t_1^{-\frac{\alpha d}{2}} \exp(-\frac{|x|^2}{4\theta t_1^\alpha}) \right\|_1 \|\dot{\Lambda}^s u_0\|_p d\theta
$$

$$
\lesssim \int_0^\infty M_\alpha(\theta) d\theta \left[\left(\left(\frac{t_2}{t_1}\right)^{\frac{\alpha d}{2}} - 1 \right) \left\| \exp\left(-\frac{|x|^2}{4}\right) \right\|_1 \right.
$$

$$
\left. + \left\| \exp\left(-\frac{|x|^2}{4} \frac{t_1^\alpha}{t_2^\alpha}\right) - \exp\left(-\frac{|x|^2}{4}\right) \right\|_1 \right]
$$

$$
\to 0, \quad \text{as } t_2 \to t_1.
$$

This gives $E_{\alpha,1}(t^\alpha \Delta) u_0 \in C([0, T]; \dot{H}^s_p(\mathbb{R}^d))$. Consequently,

$$
u = E_{\alpha,1}(t^\alpha \Delta) u_0 + \mathcal{B}(u, u) \in C([0, T]; \dot{H}^s_p(\mathbb{R}^d)).
$$

This completes the proof.　　　　　　　　　　　　　　　　　　　　　　□

2.8   Notes and Remarks

The results of Section 2.2 due to Zhou and Peng [402]. The contents in Section 2.3 are taken from Zhou, Peng and Huang [404]. The material in Section 2.4 is taken from Peng, Debbouche and Zhou [293]. The results in Section 2.5 are adopted from Zhou and Peng [403]. The contents of Section 2.6 due to Zhou and Peng [406]. Section 2.7 is taken from Zhou, Peng, Ahmad et al. [298].

Chapter 3

Fractional Rayleigh-Stokes Equations

The Rayleigh-Stokes problem has gained much attention with the further study of non-Newtonain fluids. In Section 3.1, we are interested in discussing the existence of solutions for nonlinear Rayleigh-Stokes problem for a generalized second grade fluid with the Riemann-Liouville fractional derivative. We firstly show that the solution operator of the problem is compact, and continuous in the uniform operator topology. Furthermore, we give an existence result of mild solutions for the nonlinear problem. In Section 3.2, we are devoted to the global/local well-posedness of mild solutions for a semilinear time-fractional Rayleigh-Stokes problem on $\mathbb{R}^N$. We are concerned with, the approaches rely on the Gagliardo-Nirenberg's inequalities, operator theory, standard fixed point technique and harmonic analysis methods. We also present several results on the continuation, a blow-up alternative with the blow-up rate and the integrability in Lebesgue spaces. In Section 3.3, we consider the fractional Rayleigh-Stokes problem with the nonlinearity term satisfies certain critical conditions. The local existence, uniqueness and continuous dependence upon the initial data of ϵ-regular mild solutions are obtained. Furthermore, a unique continuation result and a blow-up alternative result of ϵ-regular mild solutions are given. In Section 3.4, we are devoted to the study of the existence, uniqueness and regularity of weak solutions in $L^\infty(0, b; L^2(\Omega)) \bigcap L^2(0, b; H_0^1(\Omega))$ of the Rayleigh-Stokes problem. Furthermore, we prove an improved regularity result of weak solutions. In Section 3.5, we prove a long time existence result for fractional Rayleigh-Stokes equations. More precisely, we discuss the existence, uniqueness, continuous dependence on initial value and asymptotic behavior of global solutions in Besov-Morrey spaces. The results are formulated that allows for a larger class in initial value than the previous works and the approach is also suitable for fractional diffusion cases. In Section 3.6, we study a fractional nonlinear Rayleigh-Stokes problem with final value condition. By means of the finite dimensional approximation, we first obtain the compactness of solution operators. Moreover, we handle the problem in weighted continuous function spaces, and then the existence result of solutions is established. Finally, because of the ill-posedness of backward problem in the sense of Hadamard, the quasi-boundary value method is utilized to get the regularized solutions, and the corresponding convergence rate is obtained.

3.1   Initial Value Problem

3.1.1   *Introduction*

In recent years, as the development of chemical, biorheology, petroleum industries and geophysics, the interest in non-Newtonian fluids has grown considerably. There are many examples of non-Newtonian fluids in our daily life, such as blood, drilling mud, shampoo, oil paint, ketchup, clay and so on. Many models are constructed to describe different non-Newtonian fluids by reason of the difficulty of finding a simple model. Recently, researchers pay their attention on how to converge the constitutive equations of non-Newtonian fluids to the mathematical models, and then, they obtained the exact solutions of the classical Rayleigh-Stokes problem by using the integral transformations. Fetecau et al. found the exact solutions of Rayleigh-Stokes problem for heated second grade fluid [122], Oldroyd-B fluid [123] and Maxwell fluid [124]. Nadeem et al. [275] presented the exact solutions of the Rayleigh-Stokes problem for rectangular pipe in Maxwell and second grade fluid. Salah et al. [314] obtained an exact solution for Rayleigh-Stokes problem of Maxwell fluid in a porous medium and rotating frame.

Fractional calculus has proved a powerful tool to describe the viscoelasticity of fluids and anomalous diffusion phenomena, such as the constitutive relationship of the fluid models, basic random walk models and so on. It is also known that it is more difficult to construct a simple mathematical model for describing many different behavior of non-Newtonian fluids in a generalized second grade fluid, that is, in the second grade fluid, the employed constitutive relationship has the following form:

$$\sigma = -pI + \mu\epsilon_1 + \varrho_1\epsilon_2 + \varrho_2\epsilon_1^2,$$

where σ is the Cauchy stress tensor, p is the hydrostatic pressure, I is the identity tensor. $\mu \geq 0$, ϱ_1 and ϱ_2 are normal stress moduli. ϵ_1 and ϵ_2 are the kinematical tensors defined by

$$\epsilon_1 = \nabla V + (\nabla V)^{\mathrm{T}}, \quad \epsilon_2 = \frac{d\epsilon_1}{dt} + \epsilon_1(\nabla V) + (\nabla V)^{\mathrm{T}}\epsilon_1,$$

where V is the velocity, ∇ is the gradient operator and the superscript T denotes a transpose operation. Generally the constitutive relationship of viscoelastic second grade fluids has the form as follows:

$$\epsilon_2 = \partial_t^\alpha \epsilon_1 + \epsilon_1(\nabla V) + (\nabla V)^{\mathrm{T}}\epsilon_1,$$

where ∂_t^α is the Riemann-Liouville fractional derivative of order $\alpha \in (0,1)$ defined by

$$\partial_t^\alpha u(t,x) = \frac{1}{\Gamma(1-\alpha)}\partial_t \int_0^t (t-s)^{-\alpha}u(s,x)ds, \quad t > 0,$$

where $\Gamma(\cdot)$ stands for the Gamma function. The form of the model was selected for its ability to portray accurately the temperature distribution in a generalized second grade fluid subject to a linear flow on a heated flat plate and within a heated edge, see [321].

Such problems play a central role for describing the viscoelasticity of non-Newtonian fluids behavior and characteristic. There are many scholars taking more strong interesting in this problem. Shen et al. [321] gained the exact solution of fractional Rayleigh-Stokes problem for a heated generalized second grade fluid by using double Fourier sine transform and fractional Laplace transform. In [366, 367], Xue et al. studied the Rayleigh-Stokes problem for generalized Maxwell fluid and heated generalized second grade fluid in a porous half-space. Bazhlekove et al. [39] obtained the solutions of fractional Rayleigh-Stokes problem for a generalized second grade fluid by using eigenfunction expansion and Laplace transform. The existence and regularity for fractional Rayleigh-Stokes equation are derived by Nguyen et al. [278, 339]. For more results, we can refer to other papers [125, 196, 315].

Consider fractional Rayleigh-Stokes problem

$$\begin{cases} \partial_t u - (1 + \gamma \partial_t^\alpha)\Delta u = f(t, u), & x \in \Omega, \ t \in (0, b], \\ u(x, t) = 0, & x \in \partial\Omega, \ t \in (0, b], \\ u(x, 0) = u_0(x), & x \in \Omega, \end{cases} \tag{3.1}$$

where ∂_t^α is the Riemann-Liouville fractional derivative of order $\alpha \in (0, 1)$, $\Omega \subset \mathbb{R}^d (d \geq 1)$ is a bounded domain with smooth boundary $\partial\Omega$, Δ is the Laplacian operator, $b > 0$ is a given time, $\gamma > 0$ is a constant, $u_0(x)$ is the initial data for u in $L^2(\Omega)$, $f : [0, b] \times \mathbb{R} \to \mathbb{R}$ is a given function satisfying some assumptions in the later subsection.

Let $J = [0, b], b > 0$ be a finite interval in $\mathbb{R}$, $X := L^2(\Omega)$ denotes the Banach space of all measurable functions on $\Omega \subset \mathbb{R}^d (d \geq 1)$ with the inner product $(\cdot, \cdot)$ and norm $\| \cdot \|_X$. $\mathcal{B}(X)$ stands for the space of all bounded linear operators from X into X. Let $C(J, X)$ be the Banach space of continuous functions from J into X equipped with the norm

$$\|u\| = \sup_{t \in J} \|u(t)\|_X.$$

The left Riemann-Liouville fractional derivative of order $\alpha \in (0, 1)$ is defined by

$$\partial_t^\alpha u(x, t) = \frac{1}{\Gamma(1 - \alpha)} \frac{\partial}{\partial t} \int_0^t (t - s)^{-\alpha} u(x, s) ds, \quad t > 0.$$

Lemma 3.1. *[176] Let B and C be Banach spaces and assume that $T : B \to C$ is linear, if the dimension of the range of T is finite, T is compact.*

Lemma 3.2. *[176] Let B, C be Banach spaces, and $\mathcal{B}(B, C)$ stands for the space of all bounded linear operators from B into C, if $\{T_n\}$ is a sequence of compact operators in $\mathcal{B}(B, C)$ that converge uniformly to T, then T is compact, that is the set of operators in $\mathcal{B}(B, C)$ is closed.*

In this section, we firstly prove that the solution operator of the equation (3.1) is compact and continuous in the uniform operator topology, which are key properties for discussing the existence of mild solutions for nonlinear problem. Further, we obtain an existence result of mild solutions of problem (3.1).

3.1.2   *Existence*

In this subsection, we first consider the spectral problem

$$\begin{cases} -\Delta\varphi_n(x) = \lambda_n\varphi_n(x), & x \in \Omega, \\ \varphi_n(x) = 0, & x \in \partial\Omega, \end{cases} \tag{3.2}$$

where the eigenvalues $0 < \lambda_1 \le \lambda_2 \le ... \le \lambda_n \le ...$, with $\lambda_n \to \infty$ as $n \to \infty$. The corresponding eigenfunctions $\varphi_n \in H_0^1(\Omega)$.

Let $u : \Omega \times J \,\rightarrow\, \mathbb{R}$ be a solution of problem (3.1), and $u(x,t)$ can be described as $u(x,t) = \sum_{j=1}^{\infty} u_j(t)\varphi_j(x)$, where $u_j(t) = (u(x,t), \varphi_j(x))$. Denote $f_j(t, u) = (f(t, u), \varphi_j(x))$, then the problem (3.1) can be rewritten as

$$\begin{cases} D_t u_j(t) + (1 + \gamma D_t^\alpha)\lambda_j u_j(t) = f_j(t, u), & t \in (0, b), \\ u_j(0) = u_{0,j}, \end{cases} \tag{3.3}$$

where $j = 1, 2, ...$ Taking the Laplace transform of (3.3), we have

$$z\mathcal{L}\{u_j(t)\}(z) - u_{0,j} + (1 + \gamma z^\alpha)\lambda_j\mathcal{L}\{u_j(t)\}(z) = \mathcal{L}\{f_j(t, u)\}(z),$$

where $\mathcal{L}\{v\}(z)$ stands for the Laplace transform of function v. It follows that

$$\mathcal{L}\{u_j(t)\}(z) = \frac{u_{0,j} + \mathcal{L}\{f_j(t, u)\}(z)}{z + \lambda_j + \lambda_j\gamma z^\alpha},$$

then by the uniqueness of Laplace transform, we have

$$u_j(t) = P_{\alpha,j}(t)u_{0,j} + P_{\alpha,j}(t) * f_j(t, u),$$

where $*$ denotes the convolution and

$$P_{\alpha,j}(t) = \mathcal{L}^{-1}\left(\frac{1}{z + \lambda_j + \lambda_j\gamma z^\alpha}\right)(t),$$

that is

$$\mathcal{L}\{P_{\alpha,j}(t)\}(z) = \frac{1}{z + \lambda_j + \lambda_j\gamma z^\alpha},$$

thus

$$u(x,t) = \sum_{j=1}^{\infty} P_{\alpha,j}(t)u_{0,j}\varphi_j(x) + \sum_{j=1}^{\infty} \int_0^t P_{\alpha,j}(t - s)f_j(s, u(s))ds\varphi_j(x). \tag{3.4}$$

Let

$$S_\alpha(t)v = \sum_{j=1}^{\infty} P_{\alpha,j}(t)(v, \varphi_j)\varphi_j, \quad \text{for} \ \ v \in X.$$

Then the equation (3.4) can be rewritten as

$$u(t) = S_\alpha(t)u_0 + \int_0^t S_\alpha(t-s)f(s, u(s))ds.$$

Due to the discussion of above, we give the definition of mild solutions of problem (3.1) as follows.

Definition 3.1. By a mild solution of problem (3.1), we mean a function $u \in C(J, X)$ satisfying

$$u(t) = S_\alpha(t)u_0 + \int_0^t S_\alpha(t-s)f(s, u(s))ds, \quad t \in J.$$

Now, we give some properties of operator $S_\alpha(t)$, they are important for us to prove the existence of mild solutions of problem (3.1).

Lemma 3.3. *For any $t > 0$, the operator $S_\alpha(t)$ is compact.*

Proof. Firstly, we estimate $P_{\alpha,j}(t), j = 1, 2, \ldots$ By using the results of [39], we have

$$P_{\alpha,j}(t) = \int_0^\infty e^{-\xi t} K_j(\xi)d\xi,$$

where

$$K_j(\xi) = \frac{\gamma}{\pi} \frac{\lambda_j \xi^\alpha \sin(\alpha\pi)}{(-\xi + \lambda_j \gamma \xi^\alpha \cos(\alpha\pi) + \lambda_j)^2 + (\lambda_j \gamma \xi^\alpha \sin(\alpha\pi))^2},$$

it follows that

$$K_j(\xi) \le \frac{\gamma}{\pi} \frac{\lambda_j \xi^\alpha \sin(\alpha\pi)}{(\lambda_j \gamma \xi^\alpha \sin(\alpha\pi))^2} = \frac{\xi^{-\alpha}}{\pi \lambda_j \gamma \sin(\alpha\pi)},$$

thus

$$P_{\alpha,j}(t) \le \frac{1}{\pi \lambda_j \gamma \sin(\alpha\pi)} \int_0^\infty e^{-\xi t}\xi^{-\alpha}d\xi = \frac{\Gamma(1-\alpha)}{\pi \lambda_j \gamma \sin(\alpha\pi)}t^{\alpha-1}.$$

On the other hand, by the fact that $0 < P_{\alpha,j}(t) \le 1$ (see [39]), we have

$$P_{\alpha,j}(t)(1 + \lambda_j t^{1-\alpha}) \le \frac{\Gamma(1-\alpha)}{\pi \gamma \sin(\alpha\pi)} + 1,$$

and then

$$P_{\alpha,j}(t) \le \left(\frac{\Gamma(1-\alpha)}{\pi \gamma \sin(\alpha\pi)} + 1 \right)(1 + \lambda_j t^{1-\alpha})^{-1} =: \frac{M}{1 + \lambda_j t^{1-\alpha}}. \tag{3.5}$$

Next, we construct a sequence of operators defined on a finite dimensional space and prove the compactness of them.

For any $n \in \mathbb{N}^+$ and $t > 0$, let

$$S_\alpha^n(t)u = \sum_{j=1}^n P_{\alpha,j}(t)(u, \varphi_j)\varphi_j,$$

where $u \in X$ and $\varphi_n \in H_0^1(\Omega)$, from (3.5), we have

$$\|S_\alpha^n(t)u\|_X^2 = \sum_{j=1}^n |P_{\alpha,j}(t)(u,\varphi_j)|^2$$

$$\leq \sum_{j=1}^n \left(\frac{M}{1+\lambda_j t^{1-\alpha}}\right)^2 (u,\varphi_j)^2$$

$$\leq \left(\frac{M}{1+\lambda_1 t^{1-\alpha}}\right)^2 \|u\|_X^2.$$

Therefore $S_\alpha^n(t)$ is a linear bounded operator and the range of $S_\alpha^n(t)$ is finite dimensional. By Lemma 3.1, we conclude that $S_\alpha^n(t)$ is a compact operator.

Finally, we prove that the operator $S_\alpha(t)$ is compact. In view of the properties of eigenvalues and (3.5), we have

$$\|S_\alpha(t)u - S_\alpha^n(t)u\|_X^2 = \sum_{j=n+1}^\infty |P_{\alpha,j}(t)(u,\varphi_j)|^2$$

$$\leq \sum_{j=n+1}^\infty \left(\frac{M}{1+\lambda_j t^{1-\alpha}}\right)^2 (u,\varphi_j)^2$$

$$\leq \left(\frac{M}{1+\lambda_{n+1} t^{1-\alpha}}\right)^2 \sum_{j=n+1}^\infty (u,\varphi_j)^2$$

$$\leq \left(\frac{M}{1+\lambda_{n+1} t^{1-\alpha}}\right)^2 \|u\|_X^2$$

$$\to 0, \quad \text{as } n \to \infty.$$

Consequently, from Lemma 3.2, we conclude that the operator $S_\alpha(t)$ is compact. □

Lemma 3.4. *The operator $S_\alpha(t)$ is continuous in the uniform operator topology for $t > 0$.*

Proof. For any $t_2 > t_1 > 0$, by using the inequalities $1 - e^{-t} \leq t$, $e^{-t} \leq \frac{1}{t}$ for $t > 0$, and noting that $P_{\alpha,j}(0) = 1$, we have

$$\|S_\alpha(t_2)u - S_\alpha(t_1)u\|_X^2 \leq \sum_{j=1}^\infty |P_{\alpha,j}(t_2) - P_{\alpha,j}(t_1)|^2 |u_j|^2$$

$$\leq \sum_{j=1}^\infty \left(\int_0^\infty |e^{-\xi t_2} - e^{-\xi t_1}| K_j(\xi) d\xi\right)^2 |u_j|^2$$

$$= \sum_{j=1}^\infty \left(\int_0^\infty e^{-\xi t_1}\left(1 - e^{-\xi(t_2-t_1)}\right) K_j(\xi) d\xi\right)^2 |u_j|^2$$

$$\leq \sum_{j=1}^\infty \left(\int_0^\infty e^{-\xi t_1} \xi(t_2 - t_1) K_j(\xi) d\xi\right)^2 |u_j|^2$$

$$\leq \sum_{j=1}^{\infty} \left(\frac{t_2 - t_1}{t_1}\right)^2 \left(\int_0^{\infty} K_j(\xi)d\xi\right)^2 |u_j|^2$$

$$= \left(\frac{t_2 - t_1}{t_1}\right)^2 \|u\|_X^2,$$

thus

$$\|S_\alpha(t_2)u - S_\alpha(t_1)u\|_X \leq \frac{t_2 - t_1}{t_1}\|u\|_X \to 0, \quad \text{as } t_2 \to t_1,$$

which implies $S_\alpha(t)$ is strongly continuous for $t > 0$. Therefore, we conclude that the conclusion of Lemma 3.4 holds. □

We introduce the following assumptions:

(H1) $f(t, u)$ is measurable with respect to $t \in J$;
(H2) $f(t, u)$ is continuous with respect to $u \in C(J, X)$;
(H3) there exists a positive real-value function $a(t) \in L^1(J)$ such that for any $u \in X$, $\|f(t, u)\|_X \leq a(t)$ for $t \in J$.

Choose a constant $r > 0$ such that

$$M\|u_0\|_X + M\|a\|_{L^1(J)} \leq r. \tag{3.6}$$

Let $B_r(J) = \{u \in C(J, X) : \|u\| \leq r\}$. Obviously, $B_r(J)$ is a bounded closed and convex subset of $C(J, X)$.

Theorem 3.1. *Assume that (H1)-(H3) hold. Then the problem (3.1) has at least one mild solution in $C(J, X)$.*

Proof. For $u \in C(J, X)$, we define an operator T as follows

$$(Tu)(t) = S_\alpha(t)u_0 + \int_0^t S_\alpha(t - s)f(s, u(s))ds.$$

It is easy to see that the problem (3.1) has a mild solution in $C(J, X)$ if and only if the operator T has a fixed point in $C(J, X)$.

By (3.5), for any $t \geq 0$ and $u \in X$, we have

$$\|S_\alpha(t)u\|_X \leq \frac{M}{1 + \lambda_1 t^{1-\alpha}}\|u\|_X \leq M\|u\|_X. \tag{3.7}$$

Claim I. The set $\{Tu : u \in B_r(J)\}$ is equicontinuous.

In fact, for any $u \in B_r(J)$, $t_1 = 0$ and $0 < t_2 \leq b$, it follows from the fact $P_{\alpha,j}(0) = 1$ that $S_\alpha(0)u = u$, and then, by (3.7), (H3) and Lemma 3.4, we have

$$\|(Tu)(t_2) - (Tu)(0)\|_X = \left\|S_\alpha(t_2)u_0 - u_0 + \int_0^{t_2} S_\alpha(t_2 - s)f(s, u(s))ds\right\|_X$$

$$\leq \|S_\alpha(t_2)u_0 - u_0\|_X + M\int_0^{t_2} a(s)ds$$

$$\to 0, \quad \text{as } t_2 \to 0.$$

For $0 < t_1 < t_2 \leq b$, by (3.7) and (H3), we have

$$\|(Tu)(t_2) - (Tu)(t_1)\|_X$$
$$\leq \|S_\alpha(t_2)u_0 - S_\alpha(t_1)u_0\|_X$$
$$+ \left\| \int_0^{t_2} S_\alpha(t_2 - s)f(s, u(s))ds - \int_0^{t_1} S_\alpha(t_1 - s)f(s, u(s))ds \right\|_X$$
$$\leq \|S_\alpha(t_2)u_0 - S_\alpha(t_1)u_0\|_X + \left\| \int_0^{t_1 - \delta} (S_\alpha(t_2 - s) - S_\alpha(t_1 - s))f(s, u(s))ds \right\|_X$$
$$+ \left\| \int_{t_1 - \delta}^{t_1} (S_\alpha(t_2 - s) - S_\alpha(t_1 - s))f(s, u(s))ds \right\|_X$$
$$+ \left\| \int_{t_1}^{t_2} S_\alpha(t_2 - s)f(s, u(s))ds \right\|_X$$
$$=: I_1 + I_2 + I_3 + I_4.$$

In view of Lemma 3.4, we have that $I_1 \to 0$ as $t_2 \to t_1$, and

$$I_2 \leq \int_0^{t_1 - \delta} \|S_\alpha(t_2 - s) - S_\alpha(t_1 - s)\|_{\mathcal{B}(X)} \|f(s, u(s))\|_X ds$$
$$\leq \sup_{s \in [0, t_1 - \delta]} \|S_\alpha(t_2 - s) - S_\alpha(t_1 - s)\|_{\mathcal{B}(X)} \|a\|_{L^1(J)}$$
$$\to 0, \quad \text{as } t_2 \to t_1.$$

Noting that $\delta \to 0$ as $t_2 \to t_1$, by (3.7) and (H3) we have

$$I_3 \leq \int_{t_1 - \delta}^{t_1} \|S_\alpha(t_2 - s) - S_\alpha(t_1 - s)\|_{\mathcal{B}(X)} \|f(s, u(s))\|_X ds$$
$$\leq 2M \int_{t_1 - \delta}^{t_1} a(s)ds \to 0, \quad \text{as } t_2 \to t_1,$$

and it is easy to see that

$$I_4 \leq \int_{t_1}^{t_2} \|S_\alpha(t_2 - s)\|_{\mathcal{B}(X)} \|f(s, u(s))\|_X ds$$
$$\leq M \int_{t_1}^{t_2} a(s)ds \to 0, \quad \text{as } t_2 \to t_1.$$

Therefore

$$\|(Tu)(t_2) - (Tu)(t_1)\|_X \to 0, \quad \text{as } t_2 \to t_1,$$

and the limit is independent of $u \in B_r(J)$. Thus the set $\{Tu : u \in B_r(J)\}$ is equicontinuous.

Claim II. The nonlinear operator T is continuous in $B_r(J)$ and T maps $B_r(J)$ into itself.

For any $\{u_m\}_{m=1}^\infty \subseteq B_r(J), u \in B_r(J)$, we assume that $\lim_{m \to \infty} \|u_m - u\|_X = 0$. It follows by (3.7) that

$$\|(Tu_m)(t) - (Tu)(t)\|_X = \left\|\int_0^t S_\alpha(t-s)(f(s,u_m(s)) - f(s,u(s)))ds\right\|_X$$

$$\leq M \int_0^t \|f(s,u_m(s)) - f(s,u(s))\|_X ds.$$

By (H1) − (H3), we have

$$\lim_{m \to \infty} f(t,u_m) = f(t,u)$$

and

$$\|f(s,u_m(s)) - f(s,u(s))\|_X \leq 2a(s).$$

Then, by using the Lebesgue's dominated convergence theorem, we have

$$\|(Tu_m)(t) - (Tu)(t)\|_X \leq M \int_0^t \|f(s,u_m(s)) - f(s,u(s))\|_X ds$$

$$\to 0, \quad \text{as } m \to \infty,$$

thus $Tu_m \to Tu$ pointwise on J as $m \to \infty$, by which implies that $Tu_m \to Tu$ uniformly on J as $m \to \infty$ and so T is continuous in $B_r(J)$.

For any $u \in B_r(J)$, from (3.6), (3.7) and (H1) − (H3), we have

$$\|(Tu)(t)\|_X = \left\|S_\alpha(t)u_0 + \int_0^t S_\alpha(t-s)f(s,u(s))ds\right\|_X$$

$$\leq \|S_\alpha(t)u_0\|_X + \left\|\int_0^t S_\alpha(t-s)f(s,u(s))ds\right\|_X$$

$$\leq M\|u_0\|_X + M \int_0^t a(s)ds$$

$$\leq r.$$

Hence, $\|Tu\| \leq r$ for any $u \in B_r(J)$, which implies that T maps $B_r(J)$ into $B_r(J)$.

Claim III. $T(B_r(J))$ is relatively compact in X.

Let $V(t) = \{(Tu)(t) : u \in B_r(J)\}$, from the definition of T, it is easy to see that $V(0)$ is relatively compact in X. Thus, it remains to prove that $V(t)$ is relatively compact for every $t \in (0,b]$. In fact, for fixed $t \in (0,b]$ and $0 < \varepsilon < t$, from (3.7) and (H3), we have

$$\left\|\int_0^{t-\varepsilon} S_\alpha(t-s)f(s,u(s))ds\right\|_X \leq M\|a\|_{L^1(J)},$$

and then, from Lemma 3.3, we know that

$$\left\{S_\alpha(\varepsilon)\int_0^{t-\varepsilon} S_\alpha(t-s)f(s,u(s))ds : u \in B_r(J)\right\}$$

is relatively compact in X. From Lemma 3.4 and the fact that $a(t) \in L^1(J)$, we have

$$\left\|S_\alpha(\varepsilon)\int_0^{t-\varepsilon} S_\alpha(t-s)f(s,u(s))ds - \int_0^t S_\alpha(t-s)f(s,u(s))ds\right\|_X$$

$$\leq \|S_\alpha(\varepsilon) - I\|_{\mathcal{B}(X)}\left\|\int_0^{t-\varepsilon} S_\alpha(t-s)f(s,u(s))ds\right\|_X + \left\|\int_{t-\varepsilon}^t S_\alpha(t-s)f(s,u(s))ds\right\|_X$$

$$\leq M\|S_\alpha(\varepsilon) - I\|_{\mathcal{B}(X)}\|a\|_{L^1(J)} + M \int_{t-\varepsilon}^{t} a(s)ds$$

$$\to 0, \quad \text{as } \varepsilon \to 0,$$

which implies that the set $\left\{ \int_0^t S_\alpha(t-s)f(s,u(s))ds : u \in B_r(J) \right\}$ is relatively compact for all $t \in (0,b]$. Noting that $S_\alpha(t)$ is compact, we obtain that $V(t)$ is relatively compact in X for every $t \in J$. Hence the operator T is compact by the Arzelà-Ascoli theorem. Therefore, the Schauder fixed point theorem shows that the operator T has a fixed point on $B_r(J)$, that is the problem (3.1) has at least one mild solution. □

3.2 Well-posedness on $\mathbb{R}^N$

3.2.1 *Introduction*

In this section, we concern with the following semilinear time-fractional Rayleigh-Stokes problem

$$\begin{cases} \partial_t u - \partial_t^\alpha \Delta u - \Delta u = f(u), & t > 0, \ x \in \mathbb{R}^N, \\ u(0,x) = \varphi(x), & x \in \mathbb{R}^N, \end{cases} \tag{3.8}$$

where Δ is the Laplacian operator, f is a semilinear function and φ is a given initial condition in $L^p(\mathbb{R}^N)$.

Such type of problems plays a central role in describing the behavior and characteristic viscoelasticity of non-Newtonian fluids. Many researchers showed strong interest in this issue and they also obtained some satisfactory results. Here is a short description on the closely related works and comparison to our results. The Rayleigh-Stokes problem for a generalized second grade fluid subject to a flow on a heated flat plate and within a heated edge was introduced by Shen et al. [321], they considered the exact solutions of the velocity and temperature fields. As for a viscous Newtonian fluid, they revealed that the solutions of the Stokes' first problem appear in the limits of these exact solutions. The results in [321] were generalized later by Xue and Nie [367] in a porous half-space with a heated flat plate, they also obtained exact solutions of the velocity field and temperature fields, from which some classical results can be recovered. The methods of above two papers are based on the Fourier sine transform and the fractional Laplace transform. For the smooth and nonsmooth initial data on a bounded domain $\Omega \subset \mathbb{R}^N(N = 1,2,3)$, Bazhlekova et al. [39] considered solutions of the homogeneous problem on $C([0,T];L^2(\Omega)) \cap C((0,T];H_0^1(\Omega) \cap H^2(\Omega))$ for the initial value $\varphi \in L^2(\Omega)$, moreover, some operator theory and spectrum technique are also used to establish the related Sobolev regularity of solutions. In the meantime, Bazhlekova [40] obtained a well-posedness result under the abstract analysis framework of a subordination identity on the solution operator associated with a bounded C_0-semigroup and a two parameters probability density function.

Zhou and Wang [409] also established the existence results on $C([0,T];L^2(\Omega))$of non-linear problem in operator theory on a bounded domain Ω with smooth boundary. As for the numerical solution of Rayleigh-Stokes problem with fractional derivatives, several scholars have considered and developed it, for example, Chen et al. [73], Yang and Jiang [369], Zaky [378], etc.. Additionally, most fluid flow and transport processes use more distribution parameters to establish equations, the inverse problem of parameter identification has been proposed to deal with this matter, see Nguyen et al. [239, 280].

In view of the results in above works, it is natural to consider whether the well-posedness result on $\mathbb{R}^N$ can be extended to the mixed norm $L^p(L^q)$ spaces and whether it is possible to obtain the integrability of solutions. Unfortunately, it turns out that these extensions cannot be established by applying the technique of subordination principle in [40]. Moreover, some useful L^p-L^q inequality estimates about solution operator generated by the Eq. (3.8) are not easy to obtain, in which it is not immediately suitable to build an integral equation when we consider that the time-fractional derivative depends on all the past states, and we cannot apply the approach of getting classical solution operators to derive the relevant estimates, while it is readily to achieve at the classical solution operators of type heat operator, like fractional diffusion equations [225] and fractional Navier-Stokes equations [298]. To overcome the difficulty from the effect of time-fractional derivative, we propose a different technique to estimate the solution operator by means of the Gagliardo-Nirenberg's inequality and generalized Gagliardo-Nirenberg's inequality. For the proof, by using an admissible triplet concept that depends on the time-fractional derivative order $\alpha \in (0,1)$ and the exponents p,q in $L^p(L^q)$ spaces and their dimensions, we shall use the standard fixed point argument to establish main well-posedness results. We also consider a contain special space to make the local existence, that is because the decay exponent depends only on the order of time-fractional derivative derived from the solution operator. We find that the local solution will blow up in $L^r(\mathbb{R}^N)$, and then the rate of the blow up solution may depend on the exponent of nonlinearity and $r \geq 2$ in $L^r(\mathbb{R}^N)$. Also, based on the standard harmonic analysis methods, such as Marcinkiewicz interpolation theorem and doubly weighted Hardy-Littlewood-Sobolev's inequality, some new conclusions such as the integrability of global mild solutions in Lebesgue spaces $L^\mu(0,\infty;L^r(\mathbb{R}^N))$ are investigated.

This section is organized as follows. In Subsection 3.2.2, we give some concepts about fractional calculus in Banach space and we introduce several useful analytic properties of Laplacian operator. By a rigorous analysis of solution operator $S(t)$, we establish two crucial estimates that will be used throughout this section. After introducing a definition of mild solution in Subsection 3.2.3, the first part in this subsection shows the global and local well-posedness of the semilinear problem (3.8). Further, we obtain continuation and blow-up alternative of local mild solution of problem (3.8). In the last subsection, we show several integrability results of the global mild solution in Lebesgue space.

3.2.2   *Preliminaries*

Let $(X, |\cdot|)$ be a Banach space and let $\mathcal{L}(X)$ stand for the space of all linear bounded operators maps Banach space X into Banach space X, equipped with the norm $\|\cdot\|_{\mathcal{L}(X)}$, we remark that $C_b(\mathbb{R}_+, X)$ stands for the space of bounded continuous functions which is defined on $\mathbb{R}_+$ and takes values in X, equipped with the norm $\sup_{t \in \mathbb{R}_+} |\cdot|$ and $C(J, X)$ stands for the space of continuous functions which is defined on an interval $J \subseteq \mathbb{R}_+$ and takes values in X. If A is a linear closed operator, the symbols $\rho(A)$ and $\sigma(A)$ are called the resolvent set and the spectral set of A, respectively, $R(\lambda, A) = (\lambda I - A)^{-1}$ is the resolvent operator of A. We denote by $D(A^\alpha)$, $\alpha \in (0, 1)$, the fractional power spaces associated with the linear closed operator A.

An operator A is called the sectorial operator, if it follows the next concept.

Definition 3.2. Let A be a densely defined linear closed operator on Banach space X, then A is called a sectorial operator if there exist $M > 0$ and $\theta \in (0, \pi/2)$ such that

$$\Sigma'_\theta = \{z \in \mathbb{C} : z \neq 0, \ \theta \leq |\mathrm{arg} z| \leq \pi\} \cup \{0\} \subset \rho(A)$$

and $\|R(z, A)\|_{\mathcal{L}(X)} \leq M/|z|$ for $z \in \Sigma'_\theta$, $z \neq 0$.

Additionally, from [63, Theorem 2.3.2], it is not difficult to check that the Laplacian operator Δ with maximal domain $D(\Delta) = \{u \in X : \Delta u \in X\}$ generates a bounded analytic semigroup of the spectral angle less than or equal to $\pi/2$ on $X := L^p(\mathbb{R}^N)$ with $1 < p < +\infty$. Moreover, the spectrum is given by $\sigma(-\Delta) = [0, +\infty)$.

For $\delta > 0$ and $\theta \in (0, \pi/2)$ we introduce the contour $\Gamma_{\delta, \theta}$ defined by

$$\Gamma_{\delta, \theta} = \{re^{-i\theta} : r \geq \delta\} \cup \{\delta e^{i\psi} : |\psi| \leq \theta\} \cup \{re^{i\theta} : r \geq \delta\},$$

where the circular arc is oriented counterclockwise, and the two rays are oriented with an increasing imaginary part. In the sequel, let $A = -\Delta$, then A is a densely defined linear closed operator on Banach space $L^p(\mathbb{R}^N)$ with $1 < p < +\infty$, we define a linear operator $S(t)$ by means of Dunford integral as follows

$$S(t) = \frac{1}{2\pi i} \int_{\Gamma_{\delta, \theta}} e^{zt} H(z) dz, \tag{3.9}$$

where

$$H(z) = \frac{g(z)}{z} R(g(z), -A), \quad g(z) = \frac{z}{1 + z^\alpha}.$$

Remark 3.1. It is worth noting that the paper [40] applied the technique of the subordination principle to study the solution operator in (3.9) of problem (3.8), that is,

$$S(t) = \int_0^\infty \phi(t, \tau) T(\tau) d\tau,$$

where $T(t)$ is a bounded C_0-semigroup and function $\phi(t, \tau)$ is a probability density function with respect to both variables t and τ, that is $\int_0^\infty \phi(t, \tau)d\tau = \int_0^\infty \phi(t, \tau)dt = 1$. Nevertheless, we also find that it is hard to get the estimate of $AS(t)$ on a Banach space unless this estimate $\|AT(t)\|_{\mathcal{L}(X)} \leq M$ is valid for all $t \geq 0$, constant $M > 0$ under the bounded analytic semigroup of $T(t)$. From this point of view, we can not apply the subordination principle to estimate the operator defined as in (3.9) in the current section.

Recall that for any $\theta > 0$,

$$\Sigma_\theta = \{z \in \mathbb{C} : z \neq 0, |\mathrm{arg} z| < \theta\}.$$

It should be noticed that estimates of the operator $S(t)$ are standard in the theory of analytic semigroups as follows.

Lemma 3.5. *[15, Lemma 4.1.1] Given $\theta \in (0, \pi/2)$, let C be an arbitrary piecewise smooth simple curve in $\Sigma_{\theta+\pi/2}$ running from $\infty e^{-i(\theta+\pi/2)}$ to $\infty e^{i(\theta+\pi/2)}$, and let X be a Banach space. Suppose that the map $f : \Sigma_{\theta+\pi/2} \times X \times \mathbb{R}^+ \to X$ has the following properties:*

(i) *$f(\cdot, x, t) : \Sigma_{\theta+\pi/2} \to X$ is holomorphic for $(x, t) \in X \times \mathbb{R}^+$.*
(ii) *$f(z, \cdot, \cdot) \in C(X \times \mathbb{R}^+, X)$ for $z \in \Sigma_{\theta+\pi/2}$.*
(iii) *There are constants $\varrho \in \mathbb{R}$ and $M > 0$ such that*

$$|f(z, x, t)| \leq M|z|^{\varrho-1}e^{t Re(z)}, \quad (z, x, t) \in \Sigma_{\theta+\pi/2} \times X \times \mathbb{R}^+.$$

Then

$$(x, t) \mapsto \int_C f(z, x, t)dz \in C(X \times \mathbb{R}^+, X)$$

and

$$\left| \int_C f(z, x, t)dz \right| \leq M t^{-\varrho}, \quad (x, t) \in X \times \mathbb{R}^+.$$

Lemma 3.6. *Let $\alpha \in (0, 1)$. The operator $S(t)$ defined in (3.9) is well defined, $S(t)x \in C([0, \infty); X)$ and $S(t)x \to x$ as $t \to 0$ for any $x \in X$. Furthermore, there exists a constant $M > 0$ such that*

$$\|S(t)\|_{\mathcal{L}(X)} \leq M, \quad for \ t \geq 0.$$

Moreover, $AS(t)x \in C((0, \infty); X)$ and there exists $C > 0$ such that

$$\|AS(t)\|_{\mathcal{L}(X)} \leq C t^{\alpha-1}, \quad for \ t > 0. \tag{3.10}$$

Proof. Let $t > 0$, $\theta \in (0, \pi/2)$, $\delta > 0$. We choose $\delta = 1/t$, since operator $-A$ generates a bounded analytic semigroup of the spectral angle is less than or equal to $\pi/2$, i.e., for any $\theta \in (0, \pi/2)$,

$$\|(zI + A)^{-1}\|_{\mathcal{L}(X)} \leq M/|z|, \quad z \in \Sigma_{\theta+\pi/2}, \tag{3.11}$$

where $\Sigma_{\theta+\pi/2} \subset \rho(-A)$. As a similarly approach in [39, Lemma 2.1], we conclude that $g(z) \in \Sigma_{\theta+\pi/2}$ and $|g(z)| \le M|z|^{1-\alpha}$ for any $z \in \Sigma_{\theta+\pi/2}$, and thus from Lemma 3.5, we get $\|S(t)\|_{\mathcal{L}(X)} \le M$ and $S(t)x \in C((0,\infty); X)$, which shows the well-defined part for $t > 0$. Additionally, for the Laplace transform of $S(t)$, by virtue of the Fubini's theorem and the Cauchy's integral formula, we obtain for $\lambda > 0$,

$$\overline{S}(\lambda) = \int_0^\infty e^{-\lambda t} S(t)dt = \frac{1}{2\pi i} \int_0^\infty \int_{\Gamma_{\delta,\theta}} e^{-\lambda t} e^{\mu t} H(\mu)d\mu dt$$

$$= \frac{1}{2\pi i} \int_{\Gamma_{\delta,\theta}} (\lambda - \mu)^{-1} H(\mu)d\mu$$

$$= H(\lambda).$$

Consequently, in order to get the limit of $S(t)x$ at $t = 0$ for $x \in X$, the similar technique as in [307, Corollary 2.2] deduces that $S(z) \to I$ strongly as $z \to 0$, $z \in \Sigma_{\theta+\pi/2}$. This means that $S(t)x \in C([0,\infty); X)$. Moreover, by using the identity

$$AH(z) = \left(-H(z) + z^{-1}I\right)g(z),$$

we see from (3.11) that

$$\|AH(z)\|_{\mathcal{L}(X)} \le M(M+1)|z|^{-\alpha}, \quad \text{for any } z \in \Sigma_{\theta+\pi/2}.$$

Thus, the estimate (3.10) of $AS(t)$ for $t > 0$ is given by

$$\|AS(t)\|_{\mathcal{L}(X)} \le \int_{\Gamma_{1/t,\pi-\theta_1}} e^{\mathrm{Re}(z)t} \|AH(z)\|_{\mathcal{L}(X)} |dz|$$

$$\le M(M+1) \left(2 \int_{1/t}^\infty r^{-\alpha} e^{-rt\cos(\theta_1)} dr + \int_{-\pi+\theta_1}^{\pi-\theta_1} t^{\alpha-1} e^{\cos(\psi)} d\psi\right)$$

$$\le M_\alpha t^{\alpha-1},$$

where $\theta_1 = \pi/2 - \theta$, M_α is a positive constant and it may depend on M, α and θ. Consequently, it follows that $AS(t)x \in X$ for $x \in X$, $t > 0$. Moreover, $AS(t)x \in C((0,\infty); X)$ according to Lemma 3.5 for $x \in X$. The proof is completed.

□

The inequality in (3.10) enables us to get another estimate about $S(t)x$ in fractional power spaces. To do this, we need the following inequality.

Lemma 3.7. *[291] Let $\omega \in (0,1)$. Then there exists a constant $C > 0$ such that*

$$|A^\omega x| \le C|x|^{1-\omega} |Ax|^\omega, \quad x \in D(A).$$

Lemma 3.8. *Let $\alpha \in (0,1)$ and $\omega \in (0,1)$. Then there exists a constant $C > 0$ such that $S(t)x \in D(A^\omega)$ for all $x \in X$ and for every $t > 0$, moreover,*

$$|A^\omega S(t)x| \le Ct^{-(1-\alpha)\omega} |x|, \quad \text{for } t > 0$$

and $A^\omega S(t)x \in C((0,\infty); X)$ for $x \in X$.

Proof. This conclusion is an immediate result of Lemma 3.6 and Lemma 3.7. So, we omit it. □

Now, we introduce the concept of admissible triplet.

Definition 3.3. We call (p, q, μ) an admissible triplet with respect to $\alpha \in (0, 1)$ if

$$\frac{1}{\mu} = \frac{N(1-\alpha)}{2}\left(\frac{1}{p} - \frac{1}{q}\right), \quad \text{for } N \geq 1, \ 1 \leq p \leq q \leq +\infty.$$

Remark 3.2. For the definition of admissible triplet, it is inspired by [172] where they concerned with space-time estimates to a fractional integro-differential equation, in the meantime, one finds that this concept matches the L^p-L^q estimates of operator $S(t)$ appropriately, see below lemma 3.9. Furthermore, one can find that $\mu = \mu(p, q)$ is completely determined by p and q.

Thenceforth, we give some useful L^p-L^q estimates about the linear operator $S(t)$.

Lemma 3.9. *The operator $S(t)$ has the following properties.*

(i) *Let $N \geq 1$, if $v \in L^p(\mathbb{R}^N)$ for $1 < p \leq q \leq +\infty$, then there exists a constant $C > 0$ such that*

$$\|S(t)v\|_{L^q(\mathbb{R}^N)} \leq Ct^{-\frac{N(1-\alpha)}{2}\left(\frac{1}{p} - \frac{1}{q}\right)}\|v\|_{L^p(\mathbb{R}^N)}, \quad t > 0.$$

(ii) *Let $N > 1$, if $v \in L^p(\mathbb{R}^N)$ for $1 < q < +\infty$, $1 < p < N$, then there exists a constant $C > 0$ such that*

$$\|A^{1/2}S(t)v\|_{L^q(\mathbb{R}^N)} \leq Ct^{-\frac{1-\alpha}{2}\left(1+N\left(\frac{1}{p} - \frac{1}{q}\right)\right)}\|v\|_{L^p(\mathbb{R}^N)}, \quad t > 0.$$

Proof. Using the classical Gagliardo-Nirenberg's inequality, we know that there exists a constant $C > 0$ such that

$$\|S(t)v\|_{L^q(\mathbb{R}^N)} \leq C\|AS(t)v\|^{\theta}_{L^p(\mathbb{R}^N)}\|S(t)v\|^{1-\theta}_{L^p(\mathbb{R}^N)},$$

where $\frac{1}{q} = \theta\left(\frac{1}{p} - \frac{2}{N}\right) + (1 - \theta)\frac{1}{p}$ for any $\theta \in [0, 1]$ with respect to each $N \geq 1$. Thus, from (3.10) in Lemma 3.6, we have

$$\|S(t)v\|_{L^q(\mathbb{R}^N)} \leq Ct^{(\alpha-1)\theta}\|v\|^{\theta}_{L^p(\mathbb{R}^N)}\|S(t)v\|^{1-\theta}_{L^p(\mathbb{R}^N)},$$

it follows that

$$\|S(t)v\|_{L^q(\mathbb{R}^N)} \leq Ct^{(\alpha-1)\theta}\|v\|_{L^p(\mathbb{R}^N)},$$

taking the exponents of p, q into above inequality, we immediately obtain the $L^p - L^q$ estimate of operator $S(t)$. Hence, we have showed (i).

On the other hand, similarly, by virtue of the Gagliardo-Nirenberg's inequality of fractional version, (see e.g. [158, Corollary 2.3.]), in view of Lemma 3.8, there exists a constant $C > 0$ such that

$$\|A^{\frac{1}{2}}S(t)v\|_{L^q(\mathbb{R}^N)} \leq C\|AS(t)v\|^{\theta}_{L^p(\mathbb{R}^N)}\|A^{\frac{1}{2}}S(t)v\|^{1-\theta}_{L^p(\mathbb{R}^N)},$$

where $\frac{1}{q} = \theta\left(\frac{1}{p} - \frac{1}{N}\right) + (1 - \theta)\frac{1}{p}$ for any $\theta \in (0, 1)$. Thus, the L^p-L^q estimate of $A^{\frac{1}{2}}S(t)v$ follows. The proof of (ii) is completed. □

In the sequel, we set a function W_f with respect to $f \in L^1(0,T;X)$ with any $T > 0$ (or $T = +\infty$), given by

$$W_f(t) = \int_0^t S(t-s)f(s)ds,$$

in which we will prove some properties of this function.

Lemma 3.10. *Let $0 < T < +\infty$. If $f \in L^1(0,T;X)$, then $W_f(\cdot) \in C([0,T];X)$. If $f \in L^p(0,T;X)$ with $p > \frac{1}{1-(1-\alpha)\omega}$ for some $0 < \omega \leq 1$, then $A^\omega W_f(\cdot) \in C([0,T];X)$. Furthermore, let $1 \leq p < +\infty$, if $r \in [p,+\infty]$ satisfies $\frac{N(1-\alpha)}{2}\left(\frac{1}{p} - \frac{1}{r}\right) < 1$ and there is $\xi \in [0,1)$ such that $\sup_{t \in [0,T]} t^\xi \|f(t)\|_{L^p(\mathbb{R}^N)} < +\infty$, then $W_f(\cdot) \in C((0,T];L^r(\mathbb{R}^N))$ and $W_f(\cdot) \in C([0,T];L^r(\mathbb{R}^N))$ provided with $\xi < 1 - \frac{N(1-\alpha)}{2}\left(\frac{1}{p} - \frac{1}{r}\right)$.*

Proof. Observe that for any $t_1, t_2 \in [0,T]$ with $t_1 < t_2$, we have

$$W_f(t_2) - W_f(t_1) = \int_{t_1}^{t_2} S(t_2 - s)f(s)ds + \int_0^{t_1} (S(t_2 - s) - S(t_1 - s))f(s)ds.$$

Since $f \in L^1(0,T;X)$ and from Lemma 3.6, it follows that

$$\left| \int_{t_1}^{t_2} S(t_2 - s)f(s)ds \right| \leq M \int_{t_1}^{t_2} |f(s)|ds \to 0, \quad \text{as } t_2 \to t_1.$$

Additionally, we have

$$|(S(t_2 - s) - S(t_1 - s))f(s)| \leq 2M|f(s)|, \quad s \in [0,t],$$

which is integrable in $L^1(0,t;X)$. By virtue of $S(t)f(\cdot) \in C([0,T];X)$, we thus conclude that $W_f(\cdot) \in C([0,T];X)$ by the Lebesgue's dominated convergence theorem. Next, for $\omega \in (0,1]$, we know $(1-\alpha)\omega \in (0,1)$. By Lemma 3.8 we obtain

$$\left| \int_{t_1}^{t_2} A^\omega S(t_2 - s)f(s)ds \right|$$

$$\leq C \int_{t_1}^{t_2} (t_2 - s)^{-(1-\alpha)\omega}|f(s)|ds$$

$$\leq C \left(\frac{p-1}{-(1-\alpha)\omega p + p - 1} \right)^{1-\frac{1}{p}} (t_2 - t_1)^{-(1-\alpha)\omega + 1 - \frac{1}{p}} \|f\|_{L^p(0,T;X)},$$

which tends to zero as $t_2 \to t_1$. On the other hand, we have

$$|A^\omega(S(t_2 - s) - S(t_1 - s))f(s)| \leq 2C(t_1 - s)^{-(1-\alpha)\omega}|f(s)|, \quad s \in [0,t),$$

which is integrable in $L^1(0,T;X)$. By virtue of $A^\omega S(t)f(\cdot) \in C((0,T];X)$ and the Lebesgue's dominated convergence theorem, we thus conclude that $A^\omega W_f(\cdot) \in C([0,T];X)$.

By Lemma 3.9, for $t_1, t_2 \in (0,T]$ with $t_1 < t_2$, we see from

$$\sup_{t \in [0,T]} t^\xi \|f(t)\|_{L^p(\mathbb{R}^N)} < +\infty, \quad \text{for } \xi \in [0,1)$$

that

$$\left\| \int_{t_1}^{t_2} S(t_2 - s) f(s) ds \right\|_{L^r(\mathbb{R}^N)}$$

$$\leq C \int_{t_1}^{t_2} (t_2 - s)^{-\frac{N(1-\alpha)}{2}\left(\frac{1}{p} - \frac{1}{r}\right)} \|f(s)\|_{L^p(\mathbb{R}^N)} ds$$

$$\leq C t_2^{1-\xi-\frac{N(1-\alpha)}{2}\left(\frac{1}{p}-\frac{1}{r}\right)} \int_{t_1/t_2}^{1} (1-s)^{-\frac{N(1-\alpha)}{2}\left(\frac{1}{p}-\frac{1}{r}\right)} s^{-\xi} ds \sup_{s \in [0,T]} s^\xi \|f(s)\|_{L^p(\mathbb{R}^N)},$$

which tends to zero as $t_2 \to t_1$ by the properties of incomplete Beta function. Moreover,

$$\|(S(t_2 - s) - S(t_1 - s)) f(s)\|_{L^r(\mathbb{R}^N)}$$

$$\leq 2C(t_1 - s)^{-\frac{N(1-\alpha)}{2}\left(\frac{1}{p}-\frac{1}{r}\right)} \|f(s)\|_{L^p(\mathbb{R}^N)}$$

$$\leq 2C(t_1 - s)^{-\frac{N(1-\alpha)}{2}\left(\frac{1}{p}-\frac{1}{r}\right)} s^{-\xi} \sup_{s \in [0,T]} s^\xi \|f(s)\|_{L^p(\mathbb{R}^N)},$$

which is integrable in $L^1(0, t_1)$. Therefore, it follows from the similar method that $W_f(\cdot) \in C((0, T]; L^r(\mathbb{R}^N))$. In addition, if $\xi < 1 - \frac{N(1-\alpha)}{2}\left(\frac{1}{p} - \frac{1}{r}\right)$, then it is easy to check that there exists a constant $C > 0$ such that

$$\|W_f(t)\|_{L^r(\mathbb{R}^N)} \leq C t^{1-\xi-\frac{N(1-\alpha)}{2}\left(\frac{1}{p}-\frac{1}{r}\right)}.$$

This implies that $W_f(t)$ tends to zero as $t \to 0$ in $L^r(\mathbb{R}^N)$. Thus, $W_f(\cdot) \in C([0, T]; L^r(\mathbb{R}^N))$. The proof is completed. $\qquad \square$

3.2.3 *Well-posedness*

Let u be a solution of problem (3.8), taking the Laplace transform of (3.8) yields

$$\bar{u}(z) = H(z)\varphi + H(z)\bar{f}(u)(z),$$

by means of the inverse Laplace transform, we thus derive an integral representation of problem (3.8) by

$$u(t) = S(t)\varphi + \int_0^t S(t - s) f(u(s)) ds, \tag{3.12}$$

following this, we regard $S(t)$ defined in (3.9) as the solution operator of problem (3.8). Next, we introduce the definitions of global/local mild solutions to the problem (3.8).

Definition 3.4. Let $p \geq 1$.

(i) A continuous function $u : [0, +\infty) \to L^p(\mathbb{R}^N)$ satisfying (3.12) for $t \in [0, +\infty)$ is called a global mild solution to problem (3.8) in $L^p(\mathbb{R}^N)$.

(ii) If there exists $0 < T < +\infty$ such that a continuous function $u : [0, T] \to L^p(\mathbb{R}^N)$ satisfies (3.12) for $t \in [0, T]$, we say that u is a local mild solution to problem (3.8) in $L^p(\mathbb{R}^N)$.

In the sequel, the well-posedness of problem (3.8) will be considered, in order to achieve this goal, the following general hypothesis of the semilinear function introduced by [65] will be also considered. Let r', r be the conjugate indices.

(Hf) We suppose that $f(0) = 0$ and $f : L^r(\mathbb{R}^N) \to L^{r'}(\mathbb{R}^N)$, for some

$$r \in \left[2, \frac{2N}{N-2}\right), \quad \text{if } N \geq 2, \quad (r \in [2, \infty], \text{ if } N = 1).$$

Additionally, we suppose that there exist constants $\sigma \geq 0$ and $0 < K < +\infty$ such that

$$\|f(u) - f(v)\|_{L^{r'}(\mathbb{R}^N)} \leq K\big(\|u\|_{L^r(\mathbb{R}^N)}^{\sigma} + \|v\|_{L^r(\mathbb{R}^N)}^{\sigma}\big)\|u - v\|_{L^r(\mathbb{R}^N)},$$

for all $u, v \in L^r(\mathbb{R}^N)$.

We first consider the case $T = +\infty$, i.e., the global well-posedness of the problem for mild solutions. For any $\alpha \in (0, 1)$, let (p, r, μ) be an admissible triplet such that $1 \leq r' \leq p < r \leq +\infty$, consider the Banach space X_{pr} of continuous functions $v : [0, \infty) \to L^p(\mathbb{R}^N)$ equipped with its natural norm

$$\|v\|_{X_{pr}} = \sup_{t \geq 0} \|v(t)\|_{L^p(\mathbb{R}^N)} + \sup_{t \geq 0} t^{\frac{1}{\mu}} \|v(t)\|_{L^r(\mathbb{R}^N)}.$$

Theorem 3.2. *Let* $1 \leq N < \frac{2}{(1-\alpha)(1-2/r)}$ *and* $\frac{1}{\mu} = \frac{1}{\sigma}\left(1 - \frac{N(1-\alpha)}{2}\left(1 - \frac{2}{r}\right)\right)$. *If* (Hf) *holds and there exists* $\lambda > 0$ *for all* $\varphi \in L^p(\mathbb{R}^N)$ *such that* $\|\varphi\|_{L^p(\mathbb{R}^N)} \leq \lambda$. *Then the problem* (3.8) *admits a unique global mild solution* $u \in X_{pr}$. *If* u *and* $\tilde{u}$ *are solutions starting at* φ *and* ψ, *both values are on* $L^p(\mathbb{R}^N)$ *respectively, then there exists a constant* $C > 0$ *such that*

$$\|u - \tilde{u}\|_{X_{pr}} \leq C\|\varphi - \psi\|_{L^p(\mathbb{R}^N)}.$$

Proof. Let $\varepsilon > 0$ and set

$$\Omega_\varepsilon = \{u \in X_{pr} : \|u\|_{X_{pr}} \leq 2\varepsilon\}.$$

It is easy to see that Ω_ε is a closed ball of X_{pr} with center 0 and radius 2ε. Define the operator Φ in Ω_ε as

$$\Phi(u)(t) = S(t)\varphi + \int_0^t S(t-s)f(u(s))ds. \tag{3.13}$$

The proof of the existence of unique global solution to problem (3.8) is based on the contraction mapping technique. From this point, we shall need some estimates which comes from this argument, we recall Lemma 3.6 and Lemma 3.9, it follows that there exists a constant $C > 0$ such that

$$\|S(t)\varphi\|_{L^p(\mathbb{R}^N)} \leq C\|\varphi\|_{L^p(\mathbb{R}^N)} \quad \text{and} \quad t^{\frac{1}{\mu}}\|S(t)\varphi\|_{L^r(\mathbb{R}^N)} \leq C\|\varphi\|_{L^p(\mathbb{R}^N)}.$$

Let us define $\lambda := \varepsilon/(2C)$ and observe that $\varphi \in L^p(\mathbb{R}^N)$ with $\|\varphi\|_{L^p(\mathbb{R}^N)} \leq \lambda$, we thus get $S(t)\varphi \in X_{pr}$. Moreover, for any $u \in \Omega_\varepsilon$, from the hypothesis (Hf) we deduce that

$$\|f(u(t))\|_{L^{r'}(\mathbb{R}^N)} \leq K\|u(t)\|_{L^r(\mathbb{R}^N)}^{\sigma+1}, \quad \text{for } t \geq 0.$$

From the choice of μ, we know that it is also determined by (p, r), we get

$$\frac{N(1-\alpha)}{2}\left(\frac{1}{r'}-\frac{1}{p}\right)+\frac{N(1-\alpha)}{2}\left(\frac{1}{p}-\frac{1}{r}\right)(\sigma+1)=1,$$

which implies that $\frac{N(1-\alpha)}{2}\left(\frac{1}{r'}-\frac{1}{p}\right)<1$ and $\sigma+1<\mu$, combined with Lemma 3.9, we can easily derive the following estimates

$$\|W_{f(u)}(t)\|_{L^p(\mathbb{R}^N)}\leq\int_0^t\|S(t-s)f(u(s))\|_{L^p(\mathbb{R}^N)}ds$$

$$\leq C\int_0^t(t-s)^{-\frac{N(1-\alpha)}{2}\left(\frac{1}{r'}-\frac{1}{p}\right)}\|f(u(s))\|_{L^{r'}(\mathbb{R}^N)}ds$$

$$\leq CK\int_0^t(t-s)^{-\frac{N(1-\alpha)}{2}\left(\frac{1}{r'}-\frac{1}{p}\right)}s^{-\vartheta}\left(\sup_{s\geq0}s^{\frac{1}{\mu}}\|u(s)\|_{L^r(\mathbb{R}^N)}\right)^{\sigma+1}ds$$

$$\leq CKB(\vartheta,1-\vartheta)\|u\|_{X_{pr}}^{\sigma+1},$$

where $B(\cdot,\cdot)$ stands for the Beta function, $\vartheta=(\sigma+1)/\mu\in(0,1)$ and

$$1-\frac{N(1-\alpha)}{2}\left(\frac{1}{r'}-\frac{1}{p}\right)=\vartheta.$$

Consequently, one derives

$$\|W_{f(u)}(t)\|_{L^p(\mathbb{R}^N)}\leq CKB(\vartheta,1-\vartheta)(2\varepsilon)^{\sigma+1}.$$

On the other hand, the choice of μ implies that $\frac{N(1-\alpha)}{2}\left(\frac{1}{r'}-\frac{1}{r}\right)<1$, as the same way in above arguments, we have

$$\|W_{f(u)}(t)\|_{L^r(\mathbb{R}^N)}\leq\int_0^t\|S(t-s)f(u(s))\|_{L^r(\mathbb{R}^N)}ds$$

$$\leq C\int_0^t(t-s)^{-\frac{N(1-\alpha)}{2}\left(\frac{1}{r'}-\frac{1}{r}\right)}\|f(u(s))\|_{L^{r'}(\mathbb{R}^N)}ds$$

$$\leq CK\int_0^t(t-s)^{-\frac{N(1-\alpha)}{2}\left(\frac{1}{r'}-\frac{1}{r}\right)}s^{-\vartheta}\left(\sup_{s\geq0}s^{\frac{1}{\mu}}\|u(s)\|_{L^r(\mathbb{R}^N)}\right)^{\sigma+1}ds$$

$$\leq CKB\left(1-\frac{N(1-\alpha)}{2}\left(\frac{1}{r'}-\frac{1}{r}\right),1-\vartheta\right)\|u\|_{X_{pr}}^{\sigma+1}t^{-\frac{1}{\mu}}$$

$$=CKB\left(\frac{\sigma}{\mu},1-\vartheta\right)\|u\|_{X_{pr}}^{\sigma+1}t^{-\frac{1}{\mu}}.$$

It follows that

$$t^{\frac{1}{\mu}}\|W_{f(u)}(t)\|_{L^r(\mathbb{R}^N)}\leq CKB\left(\frac{\sigma}{\mu},1-\vartheta\right)(2\varepsilon)^{\sigma+1}.$$

Noting that for $\vartheta=(\sigma+1)/\mu>0$,

$$B(\vartheta,1-\vartheta)=\int_0^1 z^{\vartheta-1}(1-z)^{-\vartheta}dz\leq\int_0^1 z^{\frac{\sigma}{\mu}-1}(1-z)^{-\vartheta}dz=B(\sigma/\mu,1-\vartheta),$$

for choosing $\varepsilon \leq \left(\frac{1}{2^{\sigma+3}CKB(\sigma/\mu,1-\vartheta)}\right)^{1/\sigma}$, we thus get $\|W_{f(u)}\|_{X_{pr}} \leq \varepsilon$ for $u \in \Omega_\varepsilon$. Hence, by the same choice of ε, it yields

$$\|S(t)\varphi\|_{X_{pr}} + \|W_{f(u)}\|_{X_{pr}} \leq 2C\|\varphi\|_{L^p(\mathbb{R}^N)} + 2CKB(\sigma/\mu,1-\vartheta)(2\varepsilon)^{\sigma+1}$$
$$\leq 2C\lambda + \varepsilon/2 < 2\varepsilon, \quad \text{for } u \in \Omega_\varepsilon.$$

In addition, we now concern with continuity properties of (3.13). By virtue of Lemma 3.6 and similarly to Lemma 3.10, we know that $\Phi(u) \in C([0,\infty); L^p(\mathbb{R}^N)) \cap C((0,\infty); L^r(\mathbb{R}^N))$. Consequently, operator Φ maps Ω_ε into itself.

From the assumption of f and Lemma 3.9, for any $u, v \in \Omega_\varepsilon$, we further obtain that

$$\|\Phi(u)(t) - \Phi(v)(t)\|_{L^p(\mathbb{R}^N)}$$
$$< \int_0^t \|S(t-s)(f(u(s)) - f(v(s)))\|_{L^p(\mathbb{R}^N)}ds$$
$$\leq C \int_0^t (t-s)^{\vartheta-1}\|f(u(s)) - f(v(s))\|_{L^{r'}(\mathbb{R}^N)}ds$$
$$\leq CK \int_0^t (t-s)^{\vartheta-1}\left(\|u(s)\|_{L^r(\mathbb{R}^N)}^\sigma + \|v(s)\|_{L^r(\mathbb{R}^N)}^\sigma\right)\|u(s) - v(s)\|_{L^r}ds$$
$$\leq 2CKB(\vartheta,1-\vartheta)(2\varepsilon)^\sigma\|u-v\|_{X_{pr}},$$

with a similar argument, we get

$$t^{\frac{1}{\mu}}\|\Phi(u)(t) - \Phi(v)(t)\|_{L^r(\mathbb{R}^N)} \leq 2CKB(\sigma/\mu,1-\vartheta)(2\varepsilon)^\sigma\|u-v\|_{X_{pr}}.$$

For the choice of ε, we have

$$\|\Phi(u) - \Phi(v)\|_{X_{pr}} \leq \frac{1}{2}\|u-v\|_{X_{pr}}, \tag{3.14}$$

which shows that Φ is a strict contraction on Ω_ε. Thus Φ has a fixed point u, and it is the unique solution of the problem (3.8).

Next, it just remains to prove the continuous dependence upon the initial data. Let $\tilde{u}$ be another mild solution of the problem (3.8) associated with initial data $\psi \in L^p(\mathbb{R}^N)$. We perform as in (3.14) to get

$$\|u - \tilde{u}\|_{X_{pr}} \leq C\|\varphi - \psi\|_{L^p(\mathbb{R}^N)} + \frac{1}{2}\|u - \tilde{u}\|_{X_{pr}},$$

and then the continuous dependence follows. The proof is completed. $\square$

Remark 3.3. It is notice that if the solution takes value in $L^p(\mathbb{R}^N)$ of $p = 2$, the choice of μ in $\frac{1}{\mu} = \frac{1}{\sigma}\left(1 - \frac{N(1-\alpha)}{2}\left(1 - \frac{2}{r}\right)\right)$ for $1 \leq N < \frac{2}{(1-\alpha)(1-2/r)}$ combined with the admissible triplet $(2, r, \mu)$ can reduce to $r = \frac{2(\sigma+2)N(1-\alpha)}{(\sigma+2)N(1-\alpha)-4}$ provided with $N \geq \frac{4}{(\sigma+2)(1-\alpha)}$, and we find the explicit value $\mu = \sigma + 2$. Furthermore, let us take the semilinear function $f(u) = \varrho|u|^\sigma u$ to the current problem for $\varrho \in \mathbb{R}$, if $N \geq 1$, $\sigma = \frac{4}{N(1-\alpha)}$ and $r = \sigma + 2$, or if $\sqrt{\frac{5-\alpha}{1-\alpha}} - 1 \leq N < \frac{1+\sqrt{4(1-\alpha)+1}}{1-\alpha}$, $\sigma = N$ and $r = \sigma + 2$, then from Theorem 3.2, we know that there exists a unique global solution.

Corollary 3.1. *Let* $\sqrt{\frac{5-\alpha}{1-\alpha}} - 1 \leq N < \frac{1+\sqrt{4(1-\alpha)+1}}{1-\alpha}$. *For the problem*

$$\partial_t u - \partial_t^\alpha \Delta u - \Delta u = \varrho |u|^N u, \quad \varrho \in \mathbb{R},$$

associated with initial value condition $u(0, x) = \varphi(x)$ *for* $x \in \mathbb{R}^N$, *if there exists* $\lambda > 0$ *such that* $\|\varphi\|_{L^2(\mathbb{R}^N)} \leq \lambda$. *Then there admits a unique global mild solution on space* X *of continuous functions* $v : [0, \infty) \to L^2(\mathbb{R}^N)$ *equipped with its norm*

$$|v| = \sup_{t \geq 0} \|v(t)\|_{L^2(\mathbb{R}^N)} + \sup_{t \geq 0} t^{\frac{1}{N+2}} \|v(t)\|_{L^{N+2}(\mathbb{R}^N)}.$$

In particular, from the restriction on μ in Theorem 3.2 and the admissible triplet $(2, N + 2, \mu)$, the dimension N in Corollary 3.1 is $N = \sqrt{\frac{4}{1-\alpha}}$ for some suitable $\alpha \in (0, 1)$.

Now, let us turn to the case $T > 0$ and discuss the local well-posedness of the problem. Let $\alpha \in (0, 1)$ and (p, r, μ) be an admissible triplet such that $1 < p \leq r \leq +\infty$, $p = r' < N$ and

$$\begin{cases} 2 \leq r \leq +\infty, & \text{if } 1 < N \leq \dfrac{2}{1-\alpha}, \\[2mm] 2 \leq r < \dfrac{2N(1-\alpha)}{N(1-\alpha)-2}, & \text{if } N > \dfrac{2}{1-\alpha}. \end{cases}$$

Consider the Banach space $Y_{pr}[T]$ of continuous functions $v : [0, T] \to L^p(\mathbb{R}^N)$ under this admissible triplet which satisfy

$$t^{\frac{1}{\mu}} v \in C_b([0, T]; L^r(\mathbb{R}^N)), \quad \lim_{t \to 0} t^{\frac{1}{\mu}} v(t) = 0,$$

$$t^{\frac{1-\alpha}{2}} (-\Delta)^{\frac{1}{2}} v \in C_b([0, T]; L^p(\mathbb{R}^N)), \quad \lim_{t \to 0} t^{\frac{1-\alpha}{2}} (-\Delta)^{\frac{1}{2}} v(t) = 0,$$

equipped with its natural norm

$$\|v\|_{Y_{pr}[T]} = \sup_{t \in (0, T]} t^{\frac{1}{\mu}} \|v(t)\|_{L^r(\mathbb{R}^N)} + \sup_{t \in (0, T]} t^{\frac{1-\alpha}{2}} \|(-\Delta)^{\frac{1}{2}} v(t)\|_{L^p(\mathbb{R}^N)}.$$

Theorem 3.3. *Let* $\mu > \sigma + 1$ *and (Hf) hold, then there exists* $T_* > 0$ *such that the problem (3.8) has a unique local mild solution* u *in* $Y_{pr}[T_*]$. *Moreover,*

$$t^{\frac{1}{\mu}} u \in C_b([0, T_*]; L^r(\mathbb{R}^N)), \quad t^{\frac{1-\alpha}{2}} (-\Delta)^{\frac{1}{2}} u \in C_b([0, T_*]; L^p(\mathbb{R}^N))$$

both values are zero at $t = 0$ *except for* $r = 2$ *in the first term, in which* $u(0) = \varphi$. *And if* u *and* $\tilde{u}$ *are solutions starting at* φ *and* ψ, *both values are on* $L^p(\mathbb{R}^N)$ *respectively, there exists a constant* $C > 0$ *such that*

$$\|u - \tilde{u}\|_{Y_{pr}[T_*]} \leq C \|\varphi - \psi\|_{L^p(\mathbb{R}^N)}.$$

Proof. Define the operator Φ in $B(R)$ a closed ball of $Y_{pr}[T]$ with center 0 and radius $R > 0$ as in (3.13) and we take $4C \|\varphi\|_{L^p(\mathbb{R}^N)} = R$. From (Hf), assumption $\mu > \sigma + 1$ and the proof in Theorem 3.2, it is easy to check that $\Phi(u)$ belongs to $L^p(\mathbb{R}^N)$ for $u \in B(R)$. Moreover, there exists a constant $C > 0$ such that

$$t^{\frac{1}{\mu}} \|S(t)\varphi\|_{L^r(\mathbb{R}^N)} \leq C \|\varphi\|_{L^p(\mathbb{R}^N)} \quad \text{and} \quad t^{\frac{1-\alpha}{2}} \|(-\Delta)^{\frac{1}{2}} S(t)\varphi\|_{L^p(\mathbb{R}^N)} \leq C \|\varphi\|_{L^p(\mathbb{R}^N)}.$$

Moreover, the assumption $\mu > \sigma + 1$ derives that $\mu > 1$ for $\sigma \geq 0$, that is,

$$\frac{N(1-\alpha)}{2}\left(\frac{1}{r'} - \frac{1}{r}\right)(\sigma + 1) < 1 \quad \text{and} \quad \frac{N(1-\alpha)}{2}\left(\frac{1}{r'} - \frac{1}{r}\right) < 1.$$

From Lemma 3.9 (i) we have

$$\|W_{f(u)}(t)\|_{L^r(\mathbb{R}^N)} \leq C\int_0^t (t-s)^{-\frac{1}{\mu}}\|f(u(s))\|_{L^{r'}(\mathbb{R}^N)}ds$$

$$\leq CK\int_0^t (t-s)^{-\frac{1}{\mu}}s^{-\vartheta}\left(\sup_{s\in[0,T]} s^{\frac{1}{\mu}}\|u(s)\|_{L^r(\mathbb{R}^N)}\right)^{\sigma+1}ds$$

$$\leq CKB\left(1-\frac{1}{\mu}, 1-\vartheta\right)t^{1-\frac{\sigma+2}{\mu}}\|u\|_{Y_{pr}[T]}^{\sigma+1},$$

for $\vartheta = (\sigma+1)/\mu \in (0,1)$, which implies that

$$t^{\frac{1}{\mu}}\|W_{f(u)}(t)\|_{L^r(\mathbb{R}^N)} \leq CKB\left(1-\frac{1}{\mu}, 1-\vartheta\right)T^{1-\vartheta}R^{\sigma+1}.$$

On the other hand, from Lemma 3.9 (ii) we have

$$\|(-\Delta)^{\frac{1}{2}}W_{f(u)}(t)\|_{L^p(\mathbb{R}^N)} \leq \int_0^t \|(-\Delta)^{\frac{1}{2}}S(t-s)f(u(s))\|_{L^p(\mathbb{R}^N)}ds$$

$$\leq C\int_0^t (t-s)^{-\frac{1-\alpha}{2}}\|f(u(s))\|_{L^{r'}(\mathbb{R}^N)}ds$$

$$\leq CK\int_0^t (t-s)^{-\frac{1-\alpha}{2}}s^{-\vartheta}\left(\sup_{s\in[0,T]} s^{\frac{1}{\mu}}\|u(s)\|_{L^r(\mathbb{R}^N)}\right)^{\sigma+1}ds$$

$$\leq CKB\left(\frac{1+\alpha}{2}, 1-\vartheta\right)\|u\|_{Y_{pr}[T]}^{\sigma+1}t^{\frac{1+\alpha}{2}-\vartheta},$$

it follows that

$$t^{\frac{1-\alpha}{2}}\|(-\Delta)^{\frac{1}{2}}W_{f(u)}(t)\|_{L^p(\mathbb{R}^N)} \leq CKB\left(\frac{1+\alpha}{2}, 1-\vartheta\right)T^{1-\vartheta}R^{\sigma+1}.$$

Let

$$M_\sigma = \max\left\{B\left(1-\frac{1}{\mu}, 1-\vartheta\right), B\left(\frac{1+\alpha}{2}, 1-\vartheta\right)\right\},$$

then there exists $T_* > 0$ small enough such that $4CKM_\sigma T_*^{1-(\sigma+1)/\mu}R^\sigma < 1$, we thus get $\|W_{f(u)}\|_{Y_{pr}[T_*]} \leq R/2$ for $u \in B(R)$, that is,

$$\|S(t)\varphi\|_{Y_{pr}[T_*]} + \|W_{f(u)}\|_{Y_{pr}[T_*]} \leq 2C\|\varphi\|_{L^p(\mathbb{R}^N)} + 2CKM_\sigma T_*^{1-\vartheta}R^{\sigma+1} \leq R. \tag{3.15}$$

Additionally, Lemma 3.6 and Lemma 3.8 ensure the continuity of $\Phi(u)$. Hence, the operator Φ maps $B(R)$ into itself. Proceeding as in the proof of Theorem 3.2, we can conclude that Φ has a fixed point u, which is the unique solution of problem (3.8) in $[0, T_*]$. To complete the proof, it just remains to prove $t^{\frac{1}{\mu}}u \in C_b([0,T_*]; L^r(\mathbb{R}^N))$ and $t^{\frac{1-\alpha}{2}}(-\Delta)^{\frac{1}{2}}u \in C_b([0,T_*]; L^p(\mathbb{R}^N))$ both vanishing at $t = 0$.

For this purpose, we first claim that for $1 < p \leq r$ and $\varphi \in L^p(\mathbb{R}^N)$, then

$$\lim_{t \to 0} t^{\frac{1}{\mu}} \|S(t)\varphi\|_{L^r(\mathbb{R}^N)} = 0.$$

To see this, note that by Lemma 3.9 (i) the maps $t^{\frac{1}{\mu}} S(t) : L^p(\mathbb{R}^N) \to L^r(\mathbb{R}^N)$, $t \in (0,T]$, are uniformly bounded and converge strongly to zero on the subset $L^r(\mathbb{R}^N)$ in $C_0^\infty(\mathbb{R}^N)$, where $C_0^\infty(\mathbb{R}^N)$ is dense in $L^p(\mathbb{R}^N)$. Hence, we get the desired argument. Similarly, Lemma 3.9 (ii) shows that the maps $t^{\frac{1-\alpha}{2}}(-\Delta)^{\frac{1}{2}} S(t)$ are uniformly bounded from $L^p(\mathbb{R}^N)$ to itself and converge to zero strongly as $t \to 0$. This means that $t^{\frac{1-\alpha}{2}}\|(-\Delta)^{\frac{1}{2}} S(t)\varphi\|_{L^p(\mathbb{R}^N)} \to 0$ as $t \to 0$.

Now, for $t \in (0, T_0]$ with T_0 enough small, let $u \in B(R)$, we get

$$\sup_{0 < t \leq T_0} t^{\frac{1}{\mu}} \|W_{f(u)}(t)\|_{L^r(\mathbb{R}^N)} \leq \sup_{0 < t \leq T_0} t^{\frac{1}{\mu}} \int_0^t \|S(t-s)f(u(s))\|_{L^r(\mathbb{R}^N)} ds$$

$$\leq CKM_\sigma T_0^{1-\vartheta} \|u\|_{Y_{pr}[T_0]}^{\sigma+1}$$

and

$$\sup_{0 < t \leq T_0} t^{\frac{1-\alpha}{2}} \|(-\Delta)^{\frac{1}{2}} W_{f(u)}(t)\|_{L^p(\mathbb{R}^N)}$$

$$\leq \sup_{0 < t \leq T_0} t^{\frac{1-\alpha}{2}} \int_0^t \|(-\Delta)^{\frac{1}{2}} S(t-s)f(u(s))\|_{L^r(\mathbb{R}^N)} ds$$

$$\leq CKM_\sigma T_0^{1-\vartheta} \|u\|_{Y_{pr}[T_0]}^{\sigma+1}.$$

For the choice of T_0, we get

$$t^{\frac{1}{\mu}} \|W_{f(u)}(t)\|_{L^r(\mathbb{R}^N)} + t^{\frac{1-\alpha}{2}} \|(-\Delta)^{\frac{1}{2}} W_{f(u)}(t)\|_{L^p(\mathbb{R}^N)} \to 0, \quad \text{as } t \to 0.$$

It follows that $\lim_{t \to 0} t^{\frac{1}{\mu}} u(t) = 0$ in $L^r(\mathbb{R}^N)$ and $\lim_{t \to 0} t^{\frac{1-\alpha}{2}}(-\Delta)^{\frac{1}{2}} u(t) = 0$ in $L^p(\mathbb{R}^N)$. Similar to the proof of the Theorem 3.2, the continuous dependence is clearly true. The proof is completed. $\qquad\square$

Remark 3.4. Notice that if $p = r$ in Theorem 3.2, it may not exist a global continuous and bounded solution in X_{pr}, because a singular integral term is not globally essentially bounded for $t \in [0, +\infty)$ at the showing estimate of $W_{f(u)}(\cdot)$ in $L^p(\mathbb{R}^N)$, that is,

$$\int_0^t (t-s)^{-\frac{N(1-\alpha)}{2}\left(\frac{1}{r'} - \frac{1}{p}\right)} \|u(s)\|_{L^r(\mathbb{R}^N)}^{\sigma+1} ds \leq \|u\|_{L^p(\mathbb{R}^N)}^{\sigma+1} \int_0^t (t-s)^{-\frac{N(1-\alpha)}{2}\left(\frac{1}{r'} - \frac{1}{p}\right)} ds,$$

for the assumption (Hf) and admissible triplet $(p, p, +\infty)$, where X_{pr} reduces to $L^p(\mathbb{R}^N)$. Obviously, the above right-hand side inequality of integral term is not integrable for all $t \in [0, +\infty)$. Thus, the case $p = r$ is not valid on this argument. However, let $p = r$, it is easy to show that the local solution will belong to $L^\infty(0, T; L^2(\mathbb{R}^N))$ in Theorem 3.3 for some $T > 0$.

In the sequel, we establish the continuation and a blow-up alternative.

Theorem 3.4. *Let the assumptions of Theorem 3.3 hold and u be the mild solution. Then u can be extended to a maximal interval $[0, T_{max})$. If $T_{max} < +\infty$, then $\|u(t)\|_{L^r(\mathbb{R}^N)} \to +\infty$ as $t \to T_{max}$. Furthermore, there exists a constant $C > 0$ such that*

$$\|u(t)\|_{L^2(\mathbb{R}^N)} \geq C(T_{max} - t)^{-\frac{1}{\sigma}}.$$

Proof. Since the assumptions of Theorem 3.3 hold, let $u \in Y_{pr}[T_*]$ be the solution, we proceed in a similar way to the proof of continuation as in Theorem 3.3, we just point out the differences of the proof. In fact, define $\Phi : \widetilde{Y_{pr}}[T] \to \widetilde{Y_{pr}}[T]$ by (3.13), where $\widetilde{Y_{pr}}[T]$ is the space of the functions $v \in Y_{pr}[T_*]$ equipped with the same norm such that $u \equiv v$ on $t \in [0, T_*]$ and for $t \in [T_*, T]$,

$$\sup_{t \in [T_*, T]} t^{\frac{1}{\mu}} \|v(t) - u(T)\|_{L^r(\mathbb{R}^N)} + \sup_{t \in [T_*, T]} t^{\frac{1-\alpha}{2}} \|(-\Delta)^{\frac{1}{2}} (v(t) - u(T))\|_{L^p(\mathbb{R}^N)} \leq R.$$

Given $v \in \widetilde{Y_{pr}}[T]$, the continuity of $\Phi(v) : (0, T] \to \widetilde{Y_{pr}}[T]$ follows as in Theorem 3.3. Clearly, one finds that $\Phi(v)(t) = u(t)$, for every $t \in [0, T_*]$. For $t \in [T_*, T]$, we have

$$\Phi(v)(t) - u(T_*) = S(t)\varphi - S(T_*)\varphi + \int_{T_*}^t S(t - s)f(v(s))ds \tag{3.16}$$
$$+ \int_0^{T_*} (S(t - s) - S(T_* - s))f(u(s))ds.$$

Lemma 3.9 implies that the first term of the right hand side of (3.16) is in $\widetilde{Y_{pr}}[T]$, and then it goes to zero as $t \to T_*$. For this reason, we can choose T_a so close to T_* such that $\|S(\cdot)\varphi - S(T_*)\varphi\|_{\widetilde{Y_{pr}}[T]} \leq R/3$. For the second term, we have

$$t^{\frac{1}{\mu}} \left\| \int_{T_*}^t S(t - s)f(v(s))ds \right\|_{L^r(\mathbb{R}^N)} \leq Ct^{\frac{1}{\mu}} \int_{T_*}^t (t - s)^{-\frac{1}{\mu}} \|f(v(s))\|_{L^p(\mathbb{R}^N)}ds$$

$$\leq CKt^{1-\vartheta} \int_{T_*/t}^1 (1 - s)^{-\frac{1}{\mu}} s^{-\vartheta} ds \|v\|_{\widetilde{Y_{pr}}[T]}^{\sigma+1} \to 0,$$

as $t \to T_*$ by the property of incomplete Beta function for $\vartheta = (\sigma + 1)/\mu$. Similarly,

$$t^{\frac{1-\alpha}{2}} \left\| \int_{T_*}^t (-\Delta)^{\frac{1}{2}} S(t - s)f(v(s))ds \right\|_{L^p(\mathbb{R}^N)}$$

$$\leq Ct^{\frac{1-\alpha}{2}} \int_{T_*}^t (t - s)^{-\frac{1-\alpha}{2}} \|f(v(s))\|_{L^p(\mathbb{R}^N)}ds$$

$$\leq CKt^{1-\vartheta} \int_{T_*/t}^1 (1 - s)^{-\frac{1-\alpha}{2}} s^{-\vartheta} ds \|v\|_{\widetilde{Y_{pr}}[T]}^{\sigma+1} \to 0,$$

as $t \to T_*$. Therefore, as for the second term, we know that it belongs $\widetilde{Y_{pr}}[T]$ and we can choose T_b so close to T_* such that its norm is less than $R/3$. As similar arguments, the last term of the right hand side of (3.16) belongs $\widetilde{Y_{pr}}[T]$, moreover, Lemma 3.9 and the Lebesgue's dominated convergence theorem can be applied to prove that its norm is less than $R/3$ when we choose T_c so close to T_*. Now, let $T = \min\{T_a, T_b, T_c\}$, it follows that

$$\sup_{t \in [T_*, T]} t^{\frac{1}{\mu}} \|\Phi(v)(t) - u(T)\|_{L^r(\mathbb{R}^N)}$$

$$+ \sup_{t \in [T_*, T]} t^{\frac{1-\alpha}{2}} \|(-\Delta)^{\frac{1}{2}}(\Phi(v)(t) - u(T))\|_{L^p(\mathbb{R}^N)} \leq R.$$

In the same way, we can prove that Φ is a contraction on $\widetilde{Y_{pr}}[T]$. Thus, Φ has a unique fixed point by the Banach fixed point theorem, which is a mild solution that extends to $[0, T]$.

Next, set

$$T_{max} := \sup\{T \in (0, +\infty) : \exists \text{ unique local solution}$$

$$u \text{ to the problem (3.8) on } [0, T]\}.$$

Suppose by contradiction that $T_{max} < \infty$ and there exists a constant $\widetilde{C} > 0$ such that $t^{\frac{1}{\mu}} \|u(t)\|_{L^r(\mathbb{R}^N)} \leq \widetilde{C}$ and $t^{\frac{1-\alpha}{2}} \|(-\Delta)^{\frac{1}{2}} u(t)\|_{L^p(\mathbb{R}^N)} \leq \widetilde{C}$ for all $t \in (0, T_{max})$. Next, consider a sequence of positive real numbers $\{t_n\}_{n=1}^{\infty}$ satisfying $t_n \to T_{max}$ as $n \to \infty$, we will verify that the sequence $\{u(t_n)\}_{n=1}^{\infty}$ belongs to $L^r(\mathbb{R}^N)$. Let us show that the sequence $\{u(t_n)\}_{n=1}^{\infty}$ is a Cauchy sequence in $L^r(\mathbb{R}^N)$. Indeed, for $0 < t_i < t_j < T_{max}$, we get

$$\|u(t_j) - u(t_i)\|_{L^r(\mathbb{R}^N)} \leq \|S(t_j)\varphi - S(t_i)\varphi\|_{L^r(\mathbb{R}^N)} + \int_{t_i}^{t_j} \|S(t_j - s)f(u(s))\|_{L^r(\mathbb{R}^N)} ds$$

$$+ \int_0^{t_i} \|S(t_j - s) - S(t_i - s))f(u(s))\|_{L^r(\mathbb{R}^N)} ds.$$

Therefore, the same reasoning used to estimate (3.16) gives that

$$\|t_j^{\frac{1}{\mu}} u(t_j) - t_i^{\frac{1}{\mu}} u(t_i)\|_{L^r(\mathbb{R}^N)} \to 0, \quad \text{as } t_i \to t_j.$$

Hence, the limit $\lim_{n \to \infty} u(t_n) := u(T_{max})$ exists in $L^r(\mathbb{R}^N)$. Similarly, we can check the limit $\lim_{n \to \infty}(-\Delta)^{\frac{1}{2}} u(t_n)$ in $L^p(\mathbb{R}^N)$. Therefore, $u(T_{max})$ and $(-\Delta)^{\frac{1}{2}} u(T_{max})$ exist in $\widetilde{Y_{pr}}[T]$. As the before results in this theorem, the mild solution of the problem (3.8) contradicts the maximality of T_{max}. So $\lim_{t \to T_{max}} \|u(t)\|_{L^r(\mathbb{R}^N)} = +\infty$, as $T_{max} < +\infty$. Thus, we can find the maximal mild solution u of the problem (3.8) on the interval $[0, T_{max})$.

If we consider $u(t_0)$ as the initial value for some $0 < t_0 < T_{max}$, so $\|u(t_0)\|_{L^r(\mathbb{R}^N)} < +\infty$ for $p = r = 2$, by above arguments we can prolong this solution u at least on $[t_0, t_1]$, it follows from (3.15) and the fixed point argument, we have for some $C > 0$,

$$2C\|u(t_0)\|_{L^r(\mathbb{R}^N)} + 2CKM_\sigma(t_1 - t_0)^{1-\vartheta} R^{\sigma+1} \leq R,$$

for some $t_1 < T_{max}$. Observe that if $0 \leq t_0 < T_{max}$ and $2C\|u(t_0)\|_{L^r(\mathbb{R}^N)} < R$, then

$$(T_{max} - t_0)^{1-\vartheta} > \frac{R - 2C\|u(t_0)\|_{L^r(\mathbb{R}^N)}}{2CKM_\sigma R^{\sigma+1}}.$$

In fact, otherwise for some $R > 2C\|u(t_0)\|_{L^r(\mathbb{R}^N)}$ and all $t \in (t_0, T_{max})$ we would have

$$(t - t_0)^{1-\vartheta} \leq \frac{R - 2C\|u(t_0)\|_{L^r(\mathbb{R}^N)}}{2CKM_\sigma R^{\sigma+1}},$$

which implies $2C\|u(t)\|_{L^r(\mathbb{R}^N)} < R$ for all $t \in (t_0, T_{max})$ by the previous arguments. However, this is impossible since $\|u(t)\|_{L^r(\mathbb{R}^N)} \to +\infty$ as $t \to T_{max}$. Hence, choosing for example, $R = 4C\|u(t_0)\|_{L^r(\mathbb{R}^N)}$, we see that for $0 < t_0 < T_{max}$,

$$(T_{max} - t_0)^{1-\vartheta}\|u(t_0)\|_{L^r(\mathbb{R}^N)}^{\sigma} > \widehat{C},$$

where $\widehat{C} > 0$ is a new fixed constant. Therefore, for the arbitrariness of t_0 we obtain the desired blow-up rate estimate. $\qquad\qquad\qquad\qquad\qquad\qquad\qquad\square$

3.2.4   *Integrability of Solution*

In this subsection, we will present the integrability of the global mild solution for current problem. For this purpose, we first discuss the properties of the solution operator in $L^r(0, \infty; L^q(\mathbb{R}^N))$.

Lemma 3.11. *Let $\alpha \in (0,1)$ and (p, q, r) be an admissible triplet such that $1 \leq p < +\infty$. Assume that*

(i) $q \in (p, +\infty)$ *if* $1 \leq N \leq \frac{2}{1-\alpha}$ *and*

(ii) $q \in \left(p, \frac{(1-\alpha)Np}{(1-\alpha)N-2}\right)$ *if* $N > \frac{2}{1-\alpha}$.

It holds that $S(t)v \in L^r(0, \infty; L^q(\mathbb{R}^N))$ for any $v \in L^p(\mathbb{R}^N)$. Moreover, there exists $M_\alpha = M(\alpha, q, N) > 0$ such that

$$\int_0^\infty \|S(t)v\|_{L^q(\mathbb{R}^N)}^r dt \leq M_\alpha \|v\|_{L^p(\mathbb{R}^N)}^r, \quad v \in L^p(\mathbb{R}^N).$$

Proof. Fix q, ϱ with $p < q < \infty$ and $1 \leq \varrho < p$. Consider the operator $\mathscr{T}$ defined in $L^\varrho(\mathbb{R}^N)$ given by

$$\mathscr{T}(v)(t) := \|S(t)v\|_{L^q(\mathbb{R}^N)}, \quad v \in L^\varrho(\mathbb{R}^N), \quad t > 0.$$

Observe that for each $v \in L^\varrho(\mathbb{R}^N)$, Lemma 3.9 (i) guarantees the inclusion

$$\{t > 0 : |\mathscr{T}(v)(t)| > \lambda\} \subset \{t > 0 : Ct^{-\frac{N(1-\alpha)}{2}\left(\frac{1}{\varrho}-\frac{1}{q}\right)}\|v\|_{L^\varrho(\mathbb{R}^N)} > \lambda\}, \quad \text{for } \lambda > 0,$$

which ensures the inequality

$$\left|\{t > 0 : |\mathscr{T}(v)(t)| > \lambda\}\right| \leq \left|\left\{t > 0 : Ct^{-\frac{N(1-\alpha)}{2}\left(\frac{1}{\varrho}-\frac{1}{q}\right)}\|v\|_{L^p(\mathbb{R}^N)} > \lambda\right\}\right|$$

$$= \left| \left\{ t > 0 : \ t < \left(\frac{C\|v\|_{L^\varrho(\mathbb{R}^N)}}{\lambda} \right)^{\frac{2\varrho q}{N(1-\alpha)(q-\varrho)}} \right\} \right|$$

$$\leq \left(\frac{C\|v\|_{L^\varrho(\mathbb{R}^N)}}{\lambda} \right)^{\frac{2\varrho q}{N(1-\alpha)(q-\varrho)}}.$$

Therefore, operator $\mathscr{T}$ is of the weak-type $\left(\varrho, \frac{2\varrho q}{N(1-\alpha)(q-\varrho)} \right)$. For considering the operator $\mathscr{T}$ defined in $L^q(\mathbb{R}^N)$. Thus, we also have that

$$\{ t > 0 : \ |\mathscr{T}(v)(t)| > \lambda \} \subset \{ t > 0 : \ C\|v\|_{L^q(\mathbb{R}^N)} > \lambda \}, \quad \lambda > 0,$$

which guarantees $\mathscr{T}$ is of the weak-type (q, ∞).

In the sequel, we divide the proof into two steps.

Step 1. Since $N \leq \frac{2}{1-\alpha}$, we see that $\varrho \leq \frac{2\varrho q}{(1-\alpha)N(q-\varrho)}$ with $\varrho < q$. The Marcinkiewicz interpolation theorem implies that $\mathscr{T}$ is of type (a_θ, b_θ), where

$$\frac{1}{a_\theta} = \frac{1-\theta}{\varrho} + \frac{\theta}{q} \quad \text{and} \quad \frac{1}{b_\theta} = \frac{(1-\theta)(1-\alpha)N(q-\varrho)}{2\varrho q},$$

for any $\theta \in (0,1)$. Therefore, there exists $M_\theta > 0$ such that for any $v \in L^{a_\theta}(\mathbb{R}^N)$,

$$\|S(t)v\|_{L^{b_\theta}(0,\infty;L^q(\mathbb{R}^N))} \leq M_\theta \|v\|_{L^{a_\theta}(\mathbb{R}^N)}.$$

Particularly, taking $\theta = \frac{q(p-\varrho)}{p(q-\varrho)}$, we obtain $a_\theta = p$ and $b_\theta = r$, this means that $S(t)v \in L^r(0,\infty;L^q(\mathbb{R}^N))$ for any $v \in L^p(\mathbb{R}^N)$.

Step 2. Assume that $N > \frac{2}{1-\alpha}$. Fix q such that $p < q < \frac{(1-\alpha)Np}{(1-\alpha)N-2}$ and choose ϱ satisfying $\frac{((1-\alpha)N-2)q}{(1-\alpha)N} < \varrho < p$. It follows that the operator $\mathscr{T}$ is the weak-type $\left(\varrho, \frac{2\varrho q}{(1-\alpha)N(q-\varrho)} \right)$ as well as the weak-type (q, ∞). The proof is similar to that of Step 1 for $\varrho < q$ and $\varrho \leq \frac{2\varrho q}{(1-\alpha)N(q-\varrho)}$, so we omit it. $\qquad \square$

The next is the last main theorem concerning integrability for the solutions of the problem (3.8).

Theorem 3.5. *For $\alpha \in (0,1)$ and $\varphi \in L^p(\mathbb{R}^N)$. Let (p, r, μ) be the admissible triplet given in X_{pr} and let $\frac{1}{\mu} = \frac{1}{\sigma}\left(1 - \frac{N(1-\alpha)}{2}\left(1 - \frac{2}{r} \right) \right)$. If (Hf) holds and there exists $\lambda > 0$ such that $\|\varphi\|_{L^p(\mathbb{R}^N)} \leq \lambda$, then we get the following conclusions*

(i) *for $1 \leq N \leq \frac{2}{1-\alpha}$, the problem (3.8) has a unique global mild solution which belongs to $L^\mu(0,\infty;L^r(\mathbb{R}^N))$, for $1 < p < r < +\infty$;*

(ii) *for $\frac{2}{1-\alpha} < N \leq \frac{2}{(1-\alpha)(1-2/r)}$, the problem (3.8) has a unique global mild solution which belongs to $L^\mu(0,\infty;L^r(\mathbb{R}^N))$, for $1 < p < r < \frac{(1-\alpha)Np}{(1-\alpha)N-2}$.*

Proof. We first prove (i). Define a Banach space by $Z_{pr} := X_{pr} \cap L^\mu(0,\infty;L^r(\mathbb{R}^N))$, where X_{pr} is the Banach space considered in Theorem 3.2. Then the norm of space Z_{pr} given by

$$\|v\|_{Z_{pr}} := \|v\|_{X_{pr}} + \|v\|_{L^\mu(0,\infty;L^r(\mathbb{R}^N))}.$$

By Theorem 3.2 and Lemma 3.11, there exists $M = M(N, p, r, \alpha) > 0$ such that

$$\|S(t)\varphi\|_{Z_{pr}} \leq M\|\varphi\|_{L^p(\mathbb{R}^N)}.$$

It remains to prove the another term of (3.13) in Z_{pr}. It follows that

$$\left\| \int_0^t S(t-s)f(u(s))ds \right\|_{L^r(\mathbb{R}^N)}$$

$$\leq C \int_0^t (t-s)^{-\frac{N(1-\alpha)}{2}\left(\frac{1}{r'}-\frac{1}{r}\right)} \|f(u(s))\|_{L^{r'}(\mathbb{R}^N)} ds$$

$$\leq CK \int_0^t (t-s)^{-\frac{N(1-\alpha)}{2}\left(\frac{1}{r'}-\frac{1}{r}\right)} \|u(s)\|_{L^r(\mathbb{R}^N)}^{\sigma+1} ds$$

$$\leq CK\|u\|_{X_{pr}}^{\sigma} \int_0^t (t-s)^{-\frac{N(1-\alpha)}{2}\left(\frac{1}{r'}-\frac{1}{r}\right)} s^{-\frac{\sigma}{\mu}} \|u(s)\|_{L^r(\mathbb{R}^N)} ds.$$

The assumption $\frac{1}{\mu} = \frac{1}{\sigma}\left(1 - \frac{N(1-\alpha)}{2}\left(1 - \frac{2}{r}\right)\right)$ and the doubly weighted Hardy-Littlewood-Sobolev's inequality imply that

$$\left\| \int_0^t (t-s)^{-\frac{N(1-\alpha)}{2}\left(\frac{1}{r'}-\frac{1}{r}\right)} s^{-\frac{\sigma}{\mu}} \|u(s)\|_{L^r(\mathbb{R}^N)} ds \right\|_{L^\mu(0,\infty)} \leq L_\alpha \|u\|_{L^\mu(0,\infty;L^r(\mathbb{R}^N))},$$

for a constant $L_\alpha = L(\alpha, p, r) > 0$. Hence, there exists constant $C = C(N, p, \alpha, r) > 0$ such that

$$\left\| \int_0^t S(t-s)f(u(s))ds \right\|_{L^\mu(0,\infty;L^r(\mathbb{R}^N))} \leq C\|u\|_{X_{pr}}^\sigma \|u\|_{L^\mu(0,\infty;L^r(\mathbb{R}^N))} \leq C\|u\|_{Z_{pr}}^{\sigma+1}.$$

Together with above arguments, we conclude that there is a unique global solution which belongs to $L^\mu(0, \infty; L^r(\mathbb{R}^N))$. The proof of (ii) is analogous, so we omit it. The proof is completed. $\qquad\square$

3.3 Well-posedness and Blow-up

3.3.1 *Introduction*

In this section, we introduce the fractional Rayleigh-Stokes problem

$$\begin{cases} \partial_t u - (1 + \gamma \partial_t^\alpha)\Delta u = f(t, u), & t > 0, \ x \in \Omega, \\ u(t, x) = 0, & t > 0, \ x \in \partial\Omega, \\ u(0, x) = u_0(x), & x \in \Omega, \end{cases} \tag{3.17}$$

where $\gamma > 0$ is a given constant, ∂_t^α is the Riemann-Liouville fractional partial derivative of order $\alpha \in (0, 1)$. Δ is the Laplace operator, $\Omega \subset \mathbb{R}^d (d \geq 1)$ is a bounded domain with smooth boundary $\partial\Omega$, $u_0(x)$ is the initial data for u in $L^2(\Omega)$, $f : [0, \infty] \times \mathbb{R} \to \mathbb{R}$ is an ϵ-regular map defined in the later subsection.

The section is organized as follows. In Subsection 3.3.2, the interpolation-extrapolation scales and ϵ-regular map are briefly introduced, some concepts and lemmas which will be used in this section are given. Subsection 3.3.3 begins with the definition of ϵ-regular mild solutions of the problem (3.17), and then, the properties of solution operators are discussed and the well-posedness results are obtained of the problem (3.17) in the case that the nonlinearity term is an ϵ-regular map. Furthermore, the continuation and blow-up alternative results are given in Subsection 3.3.4.

3.3.2 *Preliminaries*

In this subsection, we recall some concepts and lemmas which are useful in the next.

$\mathbb{R}_+$ is the set of all non-negative real numbers, $\mathbb{N}^+$ denotes the set of all positive integer numbers, $L^2(\Omega)$ denotes the Banach space of all measurable functions on $\Omega \subset \mathbb{R}^d (d \geq 1)$ with the inner product $(\cdot, \cdot)$ and the norm $\|v\|_{L^2(\Omega)} = |(v,v)|^{\frac{1}{2}}$. For any given Banach space Y and a set $J \subset \mathbb{R}_+$, we denote by $C(J,Y)$ the space of all continuous functions from J into Y equipped with the norm

$$\|u\|_{C(J,Y)} = \sup_{t \in J} \|u(t)\|_Y.$$

$H_0^1(\Omega)$ is the closure of $C_0^\infty(\Omega)$ in $H^1(\Omega)$, and equipped with the norm

$$\|v\|_{H_0^1(\Omega)} := \|\nabla v\|_{L^2(\Omega)},$$

here $C_0^\infty(\Omega)$ is the space of all infinitely differentiable functions with compact support in Ω, $H^1(\Omega)$ is a Hilbert space.

The left Riemann-Liouville fractional partial derivative of order $\alpha \in (0,1)$ is defined by

$$(\partial_t^\alpha u)(t,x) = \partial_t({}_0D_t^{\alpha-1}u(t,x)) = \frac{1}{\Gamma(1-\alpha)}\frac{\partial}{\partial t}\int_0^t (t-s)^{-\alpha}u(s,x)ds, \quad t > 0.$$

In the following, we introduce the spectral problem

$$\begin{cases} -\Delta\varphi_j(x) = \lambda_j\varphi_j(x), & x \in \Omega, \\ \varphi_j(x) = 0, & x \in \partial\Omega, \end{cases} \tag{3.18}$$

where $\{\varphi_j(x)\}_{j=1}^\infty$ is an orthogonal basis of $H_0^1(\Omega)$ and an orthonormal basis of $L^2(\Omega)$, λ_j is the eigenvalue of $-\Delta$ corresponding to φ_j which satisfies $0 < \lambda_1 \leq \lambda_2 \leq \ldots \lambda_j \leq \ldots$ and has the property that $\lambda_j \to \infty$ as $j \to \infty$.

In the following, denote $A = -\Delta$ for the convenience of writing, we define the fractional power operator A^β for $\beta \geq 0$ as

$$A^\beta v := \sum_{j=1}^\infty \lambda_j^\beta(v, \varphi_j)\varphi_j(x), \quad v \in D(A^\beta).$$

Obviously, $D(A^\beta)$ is a Banach space equipped with the norm

$$\|v\|_{D(A^\beta)} = \left(\sum_{j=1}^\infty \lambda_j^{2\beta}|(v,\varphi_j)|^2\right)^{\frac{1}{2}}.$$

Now we introduce the construction of abstract interpolation-extrapolation scales, for more information, we refer readers to the monograph [15]. For Banach spaces E^0 and E^1, if $E^1 \hookrightarrow E^0$ densely, we say the pair (E^0, E^1) is a densely injected Banach couple. Denote by $(\cdot, \cdot)_\theta$ the admissible interpolation functor of exponent $\theta \in (0,1)$, we mean that an interpolation functor of exponent θ for the category of densely injected Banach couples such that $E^1 \hookrightarrow (E^0, E^1)_\theta$ densely. Let $E^1 = D(A)$ with the graph norm $\|\cdot\|_{E^1} = \|A \cdot\|_{E^0}$ and $A_1 : D(A_1) \subset E^1 \to E^1$ be the realization of A in E^1, we can define $E^2 = (D(A_1), \|A_1 \cdot\|_{E^1})$.

Similarly, for $k \in \mathbb{N}^+$, we can define $E^{k+1} = (D(A_k), \|A_k \cdot\|_{E^k})$ and $A_{k+1} = E^{k+1}$ – the realization of A_k. On the other side, there is no loss of assuming $0 \in \rho(A)$. Consider the space $(E^0, \|A^{-1} \cdot \|)$ and define E^{-1} to be the completion of E^0 with the norm $\|A^{-1} \cdot \|$, we know that $E^0 \hookrightarrow E^{-1}$. Then, for $k \in \mathbb{N}^+ \cup \{-1\}$ and $\theta \in (0,1)$, we denote $E^{k+\theta} = (E^k, E^{k+1})_\theta$ and $A_{k+\theta} = E^{k+\theta}$ – the realization of A_k. We say $\{(E^\beta, A_\beta) : -1 \le \beta < \infty\}$ the interpolation-extrapolation scale over $[-1, \infty)$ associated to A and $(\cdot, \cdot)_\theta$.

In this section, for the reason that the scale of fractional power spaces may not suitable to treat critical problems, we consider the interpolation space $X^\beta := E^{\frac{\beta}{2}}(\beta \ge 0)$, thus the norm of X^β is given as follows

$$\|v\|_{X^\beta} := \left(\sum_{j=1}^\infty \lambda_j^\beta |(v, \varphi_j)|^2 \right)^{\frac{1}{2}}.$$

It is easy to see that $X^0 = L^2(\Omega)$.

In the following, we introduce a definition given in [64].

Definition 3.5. For $\epsilon \ge 0$, we say a map g is an ϵ-regular map relative to the pair (X^0, X^1) if there exist $\rho > 1$, $\delta(\epsilon) \in [\rho\epsilon, 1)$ and a positive constant M such that $g : X^{1+\epsilon} \to X^{\delta(\epsilon)}$ satisfies

$$\|g(u) - g(v)\|_{X^{\delta(\epsilon)}} \le M\|u - v\|_{X^{1+\epsilon}} \left(\|u\|_{X^{1+\epsilon}}^{\rho-1} + \|v\|_{X^{1+\epsilon}}^{\rho-1} + 1 \right), \quad \forall u, v \in X^{1+\epsilon}.$$

Denote by $\mathcal{F}$ the family of all ϵ-regular maps relative to (X^0, X^1) which satisfy the following conditions:

(i) for $u, v \in X^{1+\epsilon}$, $\rho > 1$ and $\rho\epsilon \le \delta(\epsilon) < 1$, $f(t, \cdot)$ satisfies

$$\|f(t, u) - f(t, v)\|_{X^{\delta(\epsilon)}} \le C\|u - v\|_{X^{1+\epsilon}} (\|u\|_{X^{1+\epsilon}}^{\rho-1} + \|v\|_{X^{1+\epsilon}}^{\rho-1} + w(t)t^\varsigma), \quad (3.19)$$

where C is a positive constant, $w(t)$ is a non-decreasing function satisfies $0 \le w(t) \le a$, $-\delta(\epsilon) + \epsilon \le \varsigma \le 0$, $t > 0$;

(ii) for $v \in X^{1+\epsilon}$, $-\delta(\epsilon) \le \tilde{\varsigma} \le 0$, $f(t, \cdot)$ satisfies

$$\|f(t, v)\|_{X^{\delta(\epsilon)}} \le C(\|v\|_{X^{1+\epsilon}}^\rho + w(t)t^{\tilde{\varsigma}}). \quad (3.20)$$

It is easy to verify that there exist functions belong to $\mathcal{F}$, which implies that $\mathcal{F}$ is not empty, for example, $f(t, u) = u|u|^{\rho-1}$, $f(t, u) = P(u \cdot \nabla)u$, where P is the orthogonal projection, for more details, we refer readers to [18, 64] and other references cited therein.

For any $b > 0$ and $\eta > 0$, we denote by $C^\eta([0, b]; X^{1+\epsilon})$ the space of all functions $v \in C((0, b]; X^{1+\epsilon})$ with the property $t^\eta v(t) \in C([0, b]; X^{1+\epsilon})$.

It is easy to see that $C^\eta([0, b]; X^{1+\epsilon})$ is a Banach space under the norm

$$\|v\|_{C^\eta([0,b];X^{1+\epsilon})} := \sup_{t \in [0,b]} t^\eta \|v(t)\|_{X^{1+\epsilon}}.$$

Let $S_{\alpha,j}(t) = \mathcal{L}^{-1}((z + (1 + \gamma z^\alpha)\lambda_j)^{-1})(t) = \frac{1}{2\pi i} \int_B e^{zt}(z + (1 + \gamma z^\alpha)\lambda_j)^{-1}dz$, where $t \geq 0$, $j = 1, 2, \ldots$, $\mathcal{L}^{-1}(\cdot)$ denotes the inverse Laplace transform and $B = \{z : Re(z) = \varpi, \varpi > 0\}$ is the Brownwich path. From [39, Theorem 2.2], we know that

$$S_{\alpha,j}(t) = \int_0^\infty e^{-\xi t} K_j(\xi)d\xi, \tag{3.21}$$

where

$$K_j(\xi) = \frac{\gamma}{\pi} \frac{\lambda_j \xi^\alpha \sin(\alpha\pi)}{(-\xi + \lambda_j \gamma \xi^\alpha \cos(\alpha\pi) + \lambda_j)^2 + (\lambda_j \gamma \xi^\alpha \sin(\alpha\pi))^2}.$$

We introduce some properties of $S_{\alpha,j}(t)$ in the following.

Lemma 3.12. *[39, Theorem 2.2] The functions $S_{\alpha,j}(t), j = 1, 2, \ldots$ have the following properties:*

(i) $S_{\alpha,j}(0) = 1$, $0 < S_{\alpha,j}(t) \leq 1$ *for* $t \geq 0$;
(ii) *for any* $j \in \mathbb{N}^+$, $S_{\alpha,j}(t)$ *is completely monotone for* $t \geq 0$.

Lemma 3.13. *[278, Lemma 2.2] For $\alpha \in (0, 1)$ and $t \geq 0$, $S_{\alpha,j}(t)$ satisfies*

$$S_{\alpha,j}(t) \leq \frac{M_1}{1 + \lambda_j t^{1-\alpha}},$$

where $j \in \mathbb{N}^+$ and

$$M_1 = \frac{\Gamma(1 - \alpha)}{\gamma\pi \sin(\alpha\pi)} + 1.$$

3.3.3 Existence of Mild Solutions

At the beginning, we give the definition of ϵ-regular mild solutions of problem (3.17). In fact, multiplying both sides of the first equation in (3.17) by $\varphi_j(x)$ and taking the Laplace transform, we obtain that

$$z\overline{u_j}(z) - u_j(0) + (1 + \gamma z^\alpha)\lambda_j \overline{u_j}(z) = \overline{f_j}(z),$$

that is,

$$\overline{u_j}(z) = (z + (1 + \gamma z^\alpha)\lambda_j)^{-1}(u_j(0) + \overline{f_j}(z)),$$

where $u_j = (u, \varphi_j)$, $f_j = (f, \varphi_j)$, $\overline{u_j}$ and $\overline{f_j}$ stand for the Laplace transform of u_j and f_j, respectively. And then, from the uniqueness of the inverse Laplace transform, we get

$$u_j(t) = S_{\alpha,j}(t)u_{0,j} + \int_0^t S_{\alpha,j}(t-s)f_j(s, u(s))ds,$$

where $u_{0,j} = (u_0, \varphi_j)$. Denote

$$S_\alpha(t)v = \sum_{j=1}^\infty S_{\alpha,j}(t)(v, \varphi_j)\varphi_j(x), \quad \text{for} \quad v \in L^2(\Omega),$$

then we have

$$u(t, x) = S_\alpha(t)u_0 + \int_0^t S_\alpha(t-s)f(s, u(s))ds. \tag{3.22}$$

Now we can give the definition of ϵ-regular mild solutions.

Definition 3.6. By an ϵ-regular mild solution of the problem (3.17), we mean a function $u \in C^\eta([0, b]; X^{1+\epsilon}) \cap C([0, b]; X^1)$ and satisfies (3.22).

In the following, we give some properties of $S_\alpha(t)$, which will be useful for us to get the main results.

Lemma 3.14. *Assume* $0 \le p < \infty$ *and* $0 \le q - p \le 2$, *for* $t > 0$ *and any* $v \in X^p$, *the following inequality holds*

$$t^{(1-\alpha)(q-p)/2}\|S_\alpha(t)v\|_{X^q} \le M_1\|v\|_{X^p}, \tag{3.23}$$

where M_1 *is a positive constant given in Lemma 3.13.*

Proof. From Lemma 3.13, for $t > 0$ and any $v \in X^p$, we have

$$
\begin{aligned}
\|S_\alpha(t)v\|_{X^q}^2 &= \sum_{j=1}^\infty \lambda_j^q S_{\alpha,j}^2(t)|(v, \varphi_j)|^2 \\
&\le M_1^2 \sum_{j=1}^\infty \frac{\lambda_j^q}{(1 + \lambda_j t^{1-\alpha})^2}|(v, \varphi_j)|^2 \\
&\le M_1^2 \sum_{j=1}^\infty \frac{\lambda_j^q}{(1 + \lambda_j t^{1-\alpha})^{q-p}}|(v, \varphi_j)|^2 \\
&\le M_1^2 t^{-(1-\alpha)(q-p)}\|v\|_{X^p}^2,
\end{aligned}
$$

which implies

$$t^{(1-\alpha)(q-p)/2}\|S_\alpha(t)v\|_{X^q} \le M_1\|v\|_{X^p}.$$

This concludes the result. $\square$

Remark 3.5. From the proof process of Lemma 3.14, we can see that (3.23) holds at $t = 0$ if and only if $p = q$. When $p \ne q$, the property $\lambda_j \to \infty$ as $j \to \infty$ makes the conclusion (3.23) invalid at $t = 0$.

Lemma 3.15. *For $0 \le p < \infty$ and $0 \le q - p \le 2$, $S_\alpha(t)$ is strongly continuous in terms of $t > 0$ on X^q. Moreover, for $t_2 > t_1 > 0$ and any $v \in X^p$, the following inequality holds*

$$\|S_\alpha(t_2)v - S_\alpha(t_1)v\|_{X^q} \le M_2 \lambda_1^{(q-p-2)/2}(t_1^{\alpha-1} - t_2^{\alpha-1})\|v\|_{X^p}, \qquad (3.24)$$

where $M_2 = \dfrac{\Gamma(1-\alpha)}{\pi\gamma\sin(\alpha\pi)}$.

Proof. For any $t_2 > t_1 > 0$ and $v \in X^p$, by virtue of (3.21), we have

$$\|S_\alpha(t_2)v - S_\alpha(t_1)v\|_{X^q}^2 = \sum_{j=1}^{\infty} \lambda_j^q |S_{\alpha,j}(t_2) - S_{\alpha,j}(t_1)|^2 |(v, \varphi_j)|^2$$

$$= \sum_{j=1}^{\infty} \lambda_j^q \left(\int_0^{\infty} (e^{-\tau t_1} - e^{-\tau t_2}) K_j(\tau) d\tau \right)^2 |(v, \varphi_j)|^2 \quad (3.25)$$

$$= \sum_{j=1}^{\infty} \lambda_j^q \left(\int_0^{\infty} \int_{t_1}^{t_2} e^{-\tau t} \tau K_j(\tau) dt d\tau \right)^2 |(v, \varphi_j)|^2,$$

and then, from the expression of $K_j(\tau)$, we can easily obtain that

$$0 < K_j(\tau) \le \frac{\tau^{-\alpha}}{\pi\gamma\sin(\alpha\pi)\lambda_j},$$

which implies

$$\left| \int_0^{\infty} \int_{t_1}^{t_2} e^{-\tau t} \tau K_j(\tau) dt d\tau \right| \le \frac{1}{\pi\gamma\sin(\alpha\pi)\lambda_j} \int_{t_1}^{t_2} \int_0^{\infty} e^{-\tau t} \tau^{1-\alpha} d\tau dt$$

$$= \frac{\Gamma(2-\alpha)}{\pi\gamma\sin(\alpha\pi)\lambda_j} \int_{t_1}^{t_2} t^{\alpha-2} dt \qquad (3.26)$$

$$= \frac{\Gamma(1-\alpha)}{\pi\gamma\sin(\alpha\pi)\lambda_j} (t_1^{\alpha-1} - t_2^{\alpha-1}).$$

Substituting (3.26) into (3.25), we get that

$$\|S_\alpha(t_2)v - S_\alpha(t_1)v\|_{X^q}^2 \le \left(\frac{\Gamma(1-\alpha)(t_1^{\alpha-1} - t_2^{\alpha-1})}{\pi\gamma\sin(\alpha\pi)} \right)^2 \lambda_1^{q-p-2}\|v\|_{X^p}^2.$$

It shows that in the case $t_2 \to t_1$, $\|S_\alpha(t_2)v - S_\alpha(t_1)v\|_{X^q} \to 0$, which implies that $S_\alpha(t)v$ is strongly continuous for $t > 0$. This completes the proof. $\qquad \square$

Remark 3.6. In the case $p = q$, the strong continuity of $S_\alpha(t)$ still holds at $t = 0$ on X^p. In fact, for any $v \in X^p$, taking $t_1 = 0$, $t_2 > 0$, from the fact that $S_{\alpha,j}(t) = 1$ at $t = 0$ in Lemma 3.12, we have $S_\alpha(t_1)v = v$. By virtue of Lemma 3.13, we get

$$\|S_\alpha(t_2)v - v\|_{X^p}^2 = \sum_{j=1}^{\infty} \lambda_j^p (1 - S_{\alpha,j}(t_2))^2 |(v, \varphi_j)|^2 \le 2(1 + M_1^2)\|v\|_{X^p}^2.$$

The Weierstrass discriminance implies that $\|S_\alpha(t_2)v - v\|_{X^p}$ is uniformly convergent. Then we have

$$\lim_{t_2 \to 0^+} \|S_\alpha(t_2)v - v\|_{X^p} = \left(\sum_{j=1}^{\infty} \lim_{t_2 \to 0^+} \lambda_j^p (1 - S_{\alpha,j}(t_2))^2 |(v, \varphi_j)|^2 \right)^{\frac{1}{2}} = 0,$$

which implies that $S_\alpha(t)$ is strongly continuous for all $t \ge 0$ on X^p.

In the following, we give some estimations when the source term is $\mathcal{F}$ type, which will be useful for verifying the main results. For the convenience in our writing, we use the notation σ as $\sigma = (1 - \alpha)(1 + \epsilon - \delta(\epsilon))$. Obviously, $0 < \sigma < 1$.

Lemma 3.16. *Assume* $f \in \mathcal{F}$ *and* $0 < \eta < \frac{1}{\rho}$, *then for* $t \in (0, b]$ *and any* $u \in C^\eta([0, b]; X^{1+\epsilon})$, *we have*

$$\left\| \int_0^t S_\alpha(t - s) f(s, u(s)) ds \right\|_{X^{1+\epsilon}} \leq M_1 C B_1 t^{1-\sigma/2} \left(t^{-\rho\eta} \|u\|^\rho_{C^\eta([0,b]; X^{1+\epsilon})} + w(t) t^{\tilde{\varsigma}} \right),$$
(3.27)

where $B_1 = \max\{B(1 - \sigma/2, 1 - \rho\eta), B(1 - \sigma/2, 1 + \tilde{\varsigma})\}$.

Proof. For $t > 0$ and any $u(t, x) \in C^\eta([0, b]; X^{1+\epsilon})$, from Lemma 3.14, we have

$$\left\| \int_0^t S_\alpha(t - s) f(s, u(s)) ds \right\|_{X^{1+\epsilon}} \leq \int_0^t \|S_\alpha(t - s) f(s, u(s))\|_{X^{1+\epsilon}} ds$$

$$\leq M_1 \int_0^t (t - s)^{-\sigma/2} \|f(s, u(s))\|_{X^{\delta(\epsilon)}} ds.$$

Substituting the condition (3.20) into above, we have

$$\left\| \int_0^t S_\alpha(t - s) f(s, u(s)) ds \right\|_{X^{1+\epsilon}}$$

$$\leq M_1 \int_0^t (t - s)^{-\sigma/2} C \left(\|u(s)\|^\rho_{X^{1+\epsilon}} + w(s) s^{\tilde{\varsigma}} \right) ds$$

$$\leq M_1 C \int_0^t (t - s)^{-\sigma/2} s^{-\rho\eta} ds \|u\|^\rho_{C^\eta([0,b]; X^{1+\epsilon})} + M_1 C w(t) \int_0^t (t - s)^{-\sigma/2} s^{\tilde{\varsigma}} ds$$

$$= M_1 C B(1 - \sigma/2, 1 - \rho\eta) t^{1-\sigma/2-\rho\eta} \|u\|^\rho_{C^\eta([0,b]; X^{1+\epsilon})}$$

$$+ M_1 C w(t) B(1 - \sigma/2, 1 + \tilde{\varsigma}) t^{1-\sigma/2+\tilde{\varsigma}},$$

where $B(\cdot, \cdot)$ denotes the Beta function. From the definition of σ, we know that $1 - \sigma/2 > 0$, and then, combined with $\eta < \frac{1}{\rho}$ and $\tilde{\varsigma} \geq -\delta(\epsilon) > -1$, we can easily check that the Beta functions in the last inequality of above are meaningful. The proof is completed. $\square$

Lemma 3.17. *Assume* $f \in \mathcal{F}$ *and* $0 < \eta < \min\{\frac{1}{\rho}, 1 + \varsigma\}$, *for any* $u, v \in C^\eta([0, b]; X^{1+\epsilon})$, *denote* $U = \max\{\|u\|_{C^\eta([0,b]; X^{1+\epsilon})}, \|v\|_{C^\eta([0,b]; X^{1+\epsilon})}\}$, *then we have*

$$\left\| \int_0^t S_\alpha(t - s)(f(s, u(s)) - f(s, v(s))) ds \right\|_{X^{1+\epsilon}}$$

$$\leq M_1 B_2 C t^{1-\sigma/2} \left(2 U^{\rho-1} t^{-\eta\rho} + w(t) t^{-\eta+\varsigma} \right) \|u - v\|_{C^\eta([0,b]; X^{1+\epsilon})}, \quad 0 < t \leq b,$$
(3.28)

where $B_2 = \max\{B(1 - \sigma/2, 1 - \eta\rho), B(1 - \sigma/2, 1 - \eta + \varsigma)\}$.

Proof. For any $u, v \in C^\eta([0, b]; X^{1+\epsilon})$, from Lemma 3.14, we have

$$\left\| \int_0^t S_\alpha(t - s)(f(s, u(s)) - f(s, v(s))) ds \right\|_{X^{1+\epsilon}}$$

$$\leq \int_0^t \|S_\alpha(t-s)(f(s,u(s)) - f(s,v(s)))\|_{X^{1+\epsilon}} ds$$

$$\leq M_1 \int_0^t (t-s)^{-\sigma/2} \|f(s,u(s)) - f(s,v(s))\|_{X^{\delta(\epsilon)}} ds.$$

And then, the condition (3.19) implies that

$$\left\| \int_0^t S_\alpha(t-s)(f(s,u(s)) - f(s,v(s))) ds \right\|_{X^{1+\epsilon}}$$

$$\leq M_1 \int_0^t (t-s)^{-\sigma/2} C \|u(s) - v(s)\|_{X^{1+\epsilon}} (\|u(s)\|_{X^{1+\epsilon}}^{\rho-1} + \|v(s)\|_{X^{1+\epsilon}}^{\rho-1} + w(s)s^\varsigma) ds$$

$$\leq 2M_1 C U^{\rho-1} \int_0^t (t-s)^{-\sigma/2} s^{-\eta\rho} ds \|u - v\|_{C^\eta([0,b];X^{1+\epsilon})}$$

$$+ M_1 C w(t) \int_0^t (t-s)^{-\sigma/2} s^{-\eta+\varsigma} ds \|u - v\|_{C^\eta([0,b];X^{1+\epsilon})}$$

$$= 2M_1 C U^{\rho-1} B(1-\sigma/2, 1-\eta\rho)) t^{1-\sigma/2-\eta\rho} \|u - v\|_{C^\eta([0,b];X^{1+\epsilon})}$$

$$+ M_1 C w(t) B(1-\sigma/2, 1-\eta+\varsigma) t^{1-\sigma/2-\eta+\varsigma} \|u - v\|_{C^\eta([0,b];X^{1+\epsilon})},$$

where $B(\cdot, \cdot)$ denotes the Beta function. Similarly, combined with the fact $\sigma \in (0,1)$ and $\eta < \min\{\frac{1}{\rho}, 1+\varsigma\}$, we know that $1 - \sigma/2 > 0$, $\rho\eta < 1$ and $1 - \eta + \varsigma > 0$, which imply that the Beta functions in the last inequality of above are meaningful. The proof is completed. □

In the following, we present an existence result and investigate the behaviors of ϵ-regular mild solutions of problem (3.17) at $t = 0$. Furthermore, the dependence of ϵ-regular mild solutions on initial conditions is also obtained. Before giving our main results, let's verify the relationships between the parameters at first. From the constraint on ς and the relation that $\delta(\epsilon) < 1$, we have that

$$1 + \varsigma - (1-\alpha)\epsilon/2 \geq 1 - \delta(\epsilon) + (1+\alpha)\epsilon/2 > 0$$

and

$$1 - \sigma/2 + \varsigma \geq 1 - \sigma/2 - \delta(\epsilon) + \epsilon = (1 - \delta(\epsilon) + \epsilon)(1+\alpha)/2 > 0.$$

By virtue of the fact that $\delta(\epsilon) \geq \rho\epsilon$, we get that

$$\frac{1 - (1-\alpha)(1-\delta(\epsilon))/2}{\rho} - (1-\alpha)\epsilon/2 = \frac{1+\alpha}{2\rho} + \frac{1-\alpha}{2\rho}(\delta(\epsilon) - \rho\epsilon) > 0.$$

Theorem 3.6. *Assume $f \in \mathcal{F}$ and $u_0 \in X^1$, let*

$$(1-\alpha)\epsilon/2 < \eta < \min\left\{1+\varsigma, \frac{1-(1-\alpha)(1-\delta(\epsilon))/2}{\rho}\right\},$$

then there exists $b > 0$ such that the problem (3.17) has a unique ϵ-regular mild solution in $[0,b]$, and the solution $u(t,x)$ satisfies

$$t^\eta \|u(t)\|_{X^{1+\epsilon}} \to 0, \quad \text{as } t \to 0^+.$$

Moreover, if $u(t,x)$ and $z(t,x)$ are two ϵ-regular mild solutions of the problem (3.17) corresponding to the initial value conditions $u_0, z_0 \in X^1$, respectively, then for any $t \in [0, b]$, we have

$$t^\eta \|u(t) - z(t)\|_{X^{1+\epsilon}} \le C_1 \|u_0 - z_0\|_{X^1},$$

where $C_1 = 4M_1 b^{\eta - (1-\alpha)\epsilon/2}$.

Proof. From the condition $\eta < \frac{1 - (1-\alpha)(1-\delta(\epsilon))/2}{\rho}$, we have $\eta < \frac{1}{\rho}$, which implies that B_1 and B_2 exist and are well-defined, we denote $B_{max} = \max\{B_1, B_2\}$. For $u_0 \in X^1$, we take $0 < r \le 1$ and choose $b > 0$ such that for any $t \in [0, b]$, the following inequalities hold:

$$M_1 t^{\eta - (1-\alpha)\epsilon/2} \|u_0\|_{X^1} \le \frac{r}{2},$$

$$M_1 C B_{max} t^{1+\eta - \sigma/2 - \rho\eta} r^{\rho - 1} \le \frac{1}{4} \tag{3.29}$$

and

$$\max\{M_1 C B_1 t^{1+\eta - \sigma/2 + \varsigma} w(t), M_1 C B_2 t^{1 - \sigma/2 + \varsigma} w(t)\} \le \frac{r}{4}. \tag{3.30}$$

For any $u \in C^\eta([0, b]; X^{1+\epsilon})$, we define an operator T as follows:

$$(Tu)(t) = S_\alpha(t)u_0 + \int_0^t S_\alpha(t - s)f(s, u(s))ds, \quad t \ge 0. \tag{3.31}$$

The main idea of our proof here is to prove that T exists a unique fixed point in $C^\eta([0, b]; X^{1+\epsilon})$, and then to check that this fixed point belongs to $C([0, b]; X^1)$.

Let

$$\mathfrak{B}(r, b) = \left\{ u \in C^\eta([0, b]; X^{1+\epsilon}) : \sup_{t \in [0, b]} t^\eta \|u(t)\|_{X^{1+\epsilon}} \le r \right\}.$$

Now we shall prove that T maps $\mathfrak{B}(r, b)$ into itself and T is a contraction mapping.

Step I. T maps $\mathfrak{B}(r, b)$ into $\mathfrak{B}(r, b)$.

We start by showing that for any $u \in \mathfrak{B}(r, b)$, $Tu \in C((0, b]; X^{1+\epsilon})$. In fact, for $0 < t_1 < t_2 \le b$, we have

$$\|(Tu)(t_2) - (Tu)(t_1)\|_{X^{1+\epsilon}}$$

$$\le \|S_\alpha(t_2)u_0 - S_\alpha(t_1)u_0\|_{X^{1+\epsilon}} + \left\| \int_{t_1}^{t_2} S_\alpha(t_2 - s)f(s, u(s))ds \right\|_{X^{1+\epsilon}}$$

$$+ \left\| \int_0^{t_1} (S_\alpha(t_2 - s) - S_\alpha(t_1 - s))f(s, u(s))ds \right\|_{X^{1+\epsilon}}$$

$$=: \mathcal{I}_1 + \mathcal{I}_2 + \mathcal{I}_3.$$

Since $t_1 > 0$, according to the strong continuity of $S_\alpha(t)v$ for $v \in X^1$ and combining with the inequality (3.24), we have

$$\mathcal{I}_1 \le M_2 \lambda_1^{(\epsilon - 2)/2} (t_1^{\alpha - 1} - t_2^{\alpha - 1}) \|u_0\|_{X^1},$$

which implies that $\mathcal{I}_1$ goes to zero as $t_2 \to t_1$.

And then, from the process of proof of Lemma 3.16, we obtain that

$$\mathcal{I}_2 \leq M_1 C \int_{t_1}^{t_2} (t_2 - s)^{-\sigma/2}(\|u(s)\|_{X^{1+\epsilon}}^{\rho} + w(s)s^{\tilde{\varsigma}})ds$$

$$\leq M_1 Cr^{\rho} \int_{t_1}^{t_2} (t_2 - s)^{-\sigma/2}s^{-\rho\eta}ds + M_1 Cw(t_2) \int_{t_1}^{t_2} (t_2 - s)^{-\sigma/2}s^{\tilde{\varsigma}}ds$$

$$= M_1 Cr^{\rho}t_2^{1-\sigma/2-\rho\eta} \int_{\frac{t_1}{t_2}}^{1} (1 - s)^{-\sigma/2}s^{-\rho\eta}ds \qquad (3.32)$$

$$+ M_1 Cw(t_2)t_2^{1-\sigma/2+\tilde{\varsigma}} \int_{\frac{t_1}{t_2}}^{1} (1 - s)^{-\sigma/2}s^{\tilde{\varsigma}}ds.$$

The ranges of σ, η, ρ and $\tilde{\varsigma}$ show that $-\sigma/2 > -1$, $-\rho\eta > -1$ and $\tilde{\varsigma} \geq -\delta(\epsilon) > -1$, which imply that $(1 - s)^{-\sigma/2}s^{-\rho\eta}$ and $(1 - s)^{-\sigma/2}s^{\tilde{\varsigma}}$ are integrable for $s \in (\frac{t_1}{t_2}, 1)$, thus the last inequality of (3.32) converges to zero as $t_2 \to t_1$, which implies that $\mathcal{I}_2$ converges to zero as $t_2 \to t_1$.

For any $0 < \xi < t_1$, we can estimate $\mathcal{I}_3$ as follows

$$\mathcal{I}_3 \leq \int_0^{t_1} \|(S_\alpha(t_2 - s) - S_\alpha(t_1 - s))f(s, u(s))\|_{X^{1+\epsilon}}ds$$

$$= \int_0^{t_1-\xi} \|(S_\alpha(t_2 - s) - S_\alpha(t_1 - s))f(s, u(s))\|_{X^{1+\epsilon}}ds$$

$$+ \int_{t_1-\xi}^{t_1} \|(S_\alpha(t_2 - s) - S_\alpha(t_1 - s))f(s, u(s))\|_{X^{1+\epsilon}}ds$$

$$=: \mathcal{I}_{31} + \mathcal{I}_{32}.$$

From Lemma 3.14 and (3.20), we have

$$\|(S_\alpha(t_2 - s) - S_\alpha(t_1 - s))f(s, u(s))\|_{X^{1+\epsilon}} \leq 2M_1(t_1 - s)^{-\sigma/2}\|f(s, u(s))\|_{X^{\delta(\epsilon)}}$$

$$\leq 2M_1 C(t_1 - s)^{-\sigma/2}(s^{-\eta\rho}r^{\rho} + w(s)s^{\tilde{\varsigma}}),$$

and then, from $-\sigma/2 > -1$, $-\rho\eta > -1$ and $\tilde{\varsigma} \geq -\delta(\epsilon) > -1$, we know that $2M_1 C(t_1 - s)^{-\sigma/2}(s^{-\eta\rho}r^{\rho} + w(s)s^{\tilde{\varsigma}})$ is integrable for $s \in (0, t_1)$. Consequently, in the case $s \in (0, t_1 - \xi)$, by virtue of the Lebesgue's dominated convergence theorem and Lemma 3.15, we get $\mathcal{I}_{31} \to 0$ as $t_2 \to t_1$. When $s \in (t_1 - \xi, t_1)$, the above inequality implies that

$$\mathcal{I}_{32} \leq 2M_1 Cr^{\rho} \int_{t_1-\xi}^{t_1} (t_1 - s)^{-\sigma/2}s^{-\rho\eta}ds + 2M_1 Cw(t_1) \int_{t_1-\xi}^{t_1} (t_1 - s)^{-\sigma/2}s^{\tilde{\varsigma}}ds,$$

in light of the integrability of the integrand terms, we obtain that $\mathcal{I}_{32} \to 0$ as $\xi \to 0$. Thus we obtain that $\mathcal{I}_3 \to 0$ as $t_2 \to t_1$.

Therefore, combined with above discussion, we get that $\|(Tu)(t_2) - (Tu)(t_1)\|_{X^{1+\epsilon}} \to 0$ as $t_2 \to t_1$ for $0 < t_1 < t_2 \leq b$, which implies that $(Tu)(t)$ is continuous for $t \in (0, b]$.

Now we show that $t^{\eta}(Tu)(t) \in C([0, b]; X^{1+\epsilon})$.

From the above discussion, we can easily get that $t^\eta (Tu)(t) \in C((0,b]; X^{1+\epsilon})$, there only left to prove the continuity of $t^\eta (Tu)(t)$ at $t = 0$. For any $u \in \mathfrak{B}(r,b)$ and $t \in (0,b]$, from Lemma 3.14 and Lemma 3.16, we can see that

$$t^\eta \|(Tu)(t)\|_{X^{1+\epsilon}}$$

$$\leq t^\eta \|S_\alpha(t)u_0\|_{X^{1+\epsilon}} + t^\eta \left\| \int_0^t S_\alpha(t-s)f(s,u(s))ds \right\|_{X^{1+\epsilon}}$$

$$\leq M_1 t^{\eta - (1-\alpha)\epsilon/2} \|u_0\|_{X^1} + M_1 C B_1 t^{1+\eta-\sigma/2-\rho\eta} r^\rho + M_1 C B_1 w(t) t^{1+\eta-\sigma/2+\tilde\varsigma},$$
(3.33)

where B_1 is given in Lemma 3.16. According to the facts that $(1-\alpha)\epsilon/2 < \eta < \frac{1-(1-\alpha)(1-\delta(\epsilon))/2}{\rho}$ and $\tilde\varsigma \geq -\delta(\epsilon)$, we get $1+\eta-\sigma/2-\rho\eta > 1+\eta-\sigma/2-(1-(1-\alpha)(1-\delta(\epsilon))/2) = \eta - (1-\alpha)\epsilon/2 > 0$ and $1+\eta-\sigma/2+\tilde\varsigma \geq 1+\eta-\sigma/2-\delta(\epsilon) = (1-\delta(\epsilon))(1+\alpha)/2 + \eta - (1-\alpha)\epsilon/2 > 0$, which imply that $t^\eta \|(Tu)(t)\|_{X^{1+\epsilon}} \to 0$ as $t \to 0^+$, thus we get that $t^\eta (Tu)(t) \in C([0,b]; X^{1+\epsilon})$.

Now we show that for any $u \in \mathfrak{B}(r,b)$ and all $t \in [0,b]$, $t^\eta \|(Tu)(t)\|_{X^{1+\epsilon}} \leq r$.

In fact, substituting (3.29) and (3.30) into (3.33), we can get that

$$t^\eta \|(Tu)(t)\|_{X^{1+\epsilon}} \leq M_1 t^{\eta - (1-\alpha)\epsilon/2} \|u_0\|_{X^1} + M_1 C B_1 t^{1+\eta-\sigma/2} \left(t^{-\rho\eta} r^\rho + w(t)t^{\tilde\varsigma} \right)$$

$$\leq \frac{r}{2} + \frac{r}{4} + \frac{r}{4}$$

$$= r.$$

Therefore, we obtain that T maps $\mathfrak{B}(r,b)$ into itself.

Step II. T is a contraction mapping.

For any $u, v \in \mathfrak{B}(r,b)$, from Lemma 3.17, (3.29) and (3.30), we have

$$t^\eta \|(Tu)(t) - (Tv)(t)\|_{X^{1+\epsilon}}$$

$$= t^\eta \left\| \int_0^t S_\alpha(t-s)(f(s,u(s)) - f(s,v(s)))ds \right\|_{X^{1+\epsilon}}$$

$$\leq M_1 B_2 C(2r^{\rho-1} t^{1+\eta-\sigma/2-\eta\rho} + w(t)t^{1-\sigma/2+\tilde\varsigma}) \|u-v\|_{C^\eta([0,b];X^{1+\epsilon})}$$

$$\leq \left(\frac{1}{2} + \frac{r}{4} \right) \|u-v\|_{C^\eta([0,b];X^{1+\epsilon})}$$

$$\leq \frac{3}{4} \|u-v\|_{C^\eta([0,b];X^{1+\epsilon})}.$$

Thus we get that T is a contraction mapping. The Banach fixed point theorem implies that T exists a fixed point $u(t,x)$ in $\mathfrak{B}(r,b)$.

Step III. The fixed point $u(t,x) \in C([0,b]; X^1)$.

From Step I, we know that the fixed point $u \in C((0,b]; X^{1+\epsilon})$, and then, combined with the fact that $X^{1+\epsilon} \hookrightarrow X^1$, analogous to the process in proving the continuity of $(Tu)(t)$ in Step I, we can get that $u \in C((0,b]; X^1)$, here we omit the proof process. There only left to prove the continuity of $u(t)$ at $t = 0$ in X^1.

Actually, for $t > 0$, from the fact that u is the fixed point of T in $\mathfrak{B}(r,b)$, we have

$$\|u(t) - u_0\|_{X^1} \leq \|S_\alpha(t)u_0 - u_0\|_{X^1} + \left\| \int_0^t S_\alpha(t-s)f(s,u(s))ds \right\|_{X^1}$$

$$=: \mathcal{O}_1 + \mathcal{O}_2.$$

From Remark 3.6, we can easily get that $\mathcal{O}_1 \to 0$ as $t \to 0^+$.

Now we estimate $\mathcal{O}_2$.

In fact, by using an analogous argument presented in the process of proof of Lemma 3.16, we have

$$\mathcal{O}_2 \le \int_0^t \|S_\alpha(t-s)f(s,u(s))\|_{X^1} ds$$

$$\le M_1 C \int_0^t (t-s)^{-(1-\alpha)(1-\delta(\epsilon))/2}(\|u(s)\|_{X^{1+\epsilon}}^\rho + w(s)s^{\tilde{\varsigma}}) ds$$

$$\le M_1 C \int_0^t (t-s)^{-(1-\alpha)(1-\delta(\epsilon))/2} s^{-\rho\eta} ds(\sup_{s\in[0,t]} s^\eta\|u(s)\|_{X^{1+\epsilon}})^\rho$$

$$+ M_1 C \int_0^t (t-s)^{-(1-\alpha)(1-\delta(\epsilon))/2} w(s)s^{\tilde{\varsigma}} ds$$

$$\le M_1 CB(1-(1-\alpha)(1-\delta(\epsilon))/2, 1-\eta\rho)$$

$$\times t^{1-(1-\alpha)(1-\delta(\epsilon))/2-\eta\rho}(\sup_{s\in[0,t]}(s^\eta\|u(s)\|_{X^{1+\epsilon}}))^\rho$$

$$+ M_1 Cw(t)B(1-(1-\alpha)(1-\delta(\epsilon))/2, 1+\tilde{\varsigma})t^{1-(1-\alpha)(1-\delta(\epsilon))/2+\tilde{\varsigma}},$$

where the facts $0 < \alpha < 1$ and $-\tilde{\varsigma} \le \delta(\epsilon) < 1$ are used, which show that the Beta functions in the last inequality of above are well-defined, and $1-(1-\alpha)(1-\delta(\epsilon))/2+\tilde{\varsigma} \ge (1-\delta(\epsilon))(1+\alpha)/2 > 0$ can be checked as well. According to the constraint of η, we can get that the exponents of t in the last inequality of above are positive. Thus, the right-hand side of above inequalities converges to zero as $t \to 0^+$, which implies $u(t,x) \in C([0,b]; X^1)$.

Based on the discussion of above steps, we conclude that $u(t,x)$ is an ϵ-regular mild solution of the problem (3.17).

Now we show that $t^\eta\|u(t)\|_{X^{1+\epsilon}} \to 0$ as $t \to 0^+$.

From Lemma 3.14, Lemma 3.16 and (3.29), we have

$$t^\eta\|u(t)\|_{X^{1+\epsilon}}$$

$$\le M_1 t^{\eta-(1-\alpha)\epsilon/2}\|u_0\|_{X^1}$$

$$+ M_1 CB(1-\sigma/2, 1-\rho\eta)t^{1+\eta-\sigma/2-\rho\eta} \sup_{s\in[0,t]}(s^\eta\|u(s)\|_{X^{1+\epsilon}})^\rho$$

$$+ M_1 Cw(t)B(1-\sigma/2, 1+\tilde{\varsigma})t^{1+\eta-\sigma/2+\tilde{\varsigma}}$$

$$\le M_1 t^{\eta-(1-\alpha)\epsilon/2}\|u_0\|_{X^1} + M_1 CB_1 t^{1+\eta-\sigma/2-\rho\eta}r^{\rho-1} \sup_{s\in[0,t]}(s^\eta\|u(s)\|_{X^{1+\epsilon}})$$

$$+ M_1 Cw(t)B_1 t^{1+\eta-\sigma/2+\tilde{\varsigma}}$$

$$\le M_1 t^{\eta-(1-\alpha)\epsilon/2}\|u_0\|_{X^1} + \frac{1}{4}\sup_{s\in[0,t]}(s^\eta\|u(s)\|_{X^{1+\epsilon}}) + M_1 Cw(t)B_1 t^{1+\eta-\sigma/2+\tilde{\varsigma}},$$

which implies

$$\sup_{s\in[0,t]}(s^\eta\|u(s)\|_{X^{1+\epsilon}}) \le \frac{4}{3}\left(M_1 t^{\eta-(1-\alpha)\epsilon/2}\|u_0\|_{X^1} + M_1 Cw(t)B_1 t^{1+\eta-\sigma/2+\tilde{\varsigma}}\right)$$

$$\to 0, \quad \text{as } t \to 0^+.$$

Step IV. The continuous dependence of ϵ-regular mild solutions on the initial value conditions.

Assume that $u(t, x), z(t, x) \in C^\eta([0, b]; X^{1+\epsilon}) \cap C([0, b]; X^1)$ are two ϵ-regular mild solutions of the problem (3.17) obtained from above under the given initial value conditions u_0 and z_0, respectively, thus they have the following forms

$$u(t, x) = S_\alpha(t)u_0 + \int_0^t S_\alpha(t - s)f(s, u(s, x))ds,$$

$$z(t, x) = S_\alpha(t)z_0 + \int_0^t S_\alpha(t - s)f(s, z(s, x))ds.$$

For $u_0, z_0 \in X^1$, from Lemma 3.14, Lemma 3.17, (3.29) and (3.30), we can estimate $t^\eta \|u(t) - z(t)\|_{X^{1+\epsilon}}$ as follows

$$t^\eta \|u(t) - z(t)\|_{X^{1+\epsilon}}$$

$$\leq t^\eta \|S_\alpha(t)(u_0 - z_0)\|_{X^{1+\epsilon}} + t^\eta \left\| \int_0^t S_\alpha(t - s)(f(s, u(s)) - f(s, z(s)))ds \right\|_{X^{1+\epsilon}}$$

$$\leq M_1 t^{\eta-(1-\alpha)\epsilon/2} \|u_0 - z_0\|_{X^1}$$

$$\quad + 2M_1 B_2 C t^{\eta+1-\sigma/2-\eta\rho} r^{\rho-1} \sup_{s\in[0,t]} (s^\eta \|u(s) - z(s)\|_{X^{1+\epsilon}})$$

$$\quad + M_1 B_2 C t^{1-\sigma/2+\varsigma} w(t) \sup_{s\in[0,t]} (s^\eta \|u(s) - z(s)\|_{X^{1+\epsilon}})$$

$$\leq M_1 t^{\eta-(1-\alpha)\epsilon/2} \|u_0 - z_0\|_{X^1} + \frac{3}{4} \sup_{s\in[0,t]} (s^\eta \|u(s) - z(s)\|_{X^{1+\epsilon}}),$$

which implies

$$t^\eta \|u(t) - z(t)\|_{X^{1+\epsilon}} \leq 4M_1 t^{\eta-(1-\alpha)\epsilon/2} \|u_0 - z_0\|_{X^1}$$

$$\leq 4M_1 b^{\eta-(1-\alpha)\epsilon/2} \|u_0 - z_0\|_{X^1}.$$

Thus the continuous dependence of ϵ-regular mild solutions on the initial value conditions is proved. The proof is completed. $\square$

3.3.4 *Continuation and Blow-up Alternative*

In this subsection, we consider the continuation and blow-up alternative for the ϵ-regular mild solution obtained by Theorem 3.6. We give the definition of continuation of ϵ-regular mild solutions at the beginning.

Definition 3.7. For the ϵ-regular mild solution $u \in C^\eta([0, b]; X^{1+\epsilon}) \cap C([0, b]; X^1)$ of the problem (3.17), we say that v is a continuation of u in $[0, \tilde{b}]$ if $v \in C^\eta([0, \tilde{b}]; X^{1+\epsilon}) \cap C([0, \tilde{b}]; X^1)$ is an ϵ-regular mild solution for $\tilde{b} > b$ and $v(t) = u(t)$ whenever $t \in [0, b]$.

Theorem 3.7. *Assume the conditions in Theorem 3.6 hold and let u be an ϵ-regular mild solution of the problem (3.17) in $[0, b]$, then there exists a unique continuation $\tilde{u}$ of u in $[0, \tilde{b}]$ for some $\tilde{b} > b$.*

Proof. Take $\tilde{b} > b$ and choose $\tilde{r} > 0$ such that for $u_0 \in X^1$ and any $t \in [b, \tilde{b}]$, the following inequalities hold:

$$M_2 \lambda_1^{(\epsilon-2)/2} t^\eta (b^{\alpha-1} - t^{\alpha-1}) \|u_0\|_{X^1} \leq \frac{\tilde{r}}{3},$$

$$M_1 C w(t) t^{1+\eta-\sigma/2+\varsigma} \int_{\frac{b}{t}}^1 (1-s)^{-\sigma/2} s^{\tilde{\varsigma}} ds \leq \frac{\tilde{r}}{6},$$

$$M_2 \lambda_1^{(\epsilon-\delta(\epsilon)-1)/2} C m^\rho t^\eta \left(B(\alpha, 1 - \rho\eta)(b^{\alpha-\rho\eta} - t^{\alpha-\rho\eta}) \right.$$
$$\left. + t^{\alpha-\rho\eta} \int_{\frac{b}{t}}^1 (1-s)^{\alpha-1} s^{-\rho\eta} ds \right) \leq \frac{\tilde{r}}{6}, \tag{3.34}$$

$$M_2 \lambda_1^{(\epsilon-\delta(\epsilon)-1)/2} C w(b) t^\eta \left(B(\alpha, 1 + \tilde{\varsigma})(b^{\alpha+\tilde{\varsigma}} - t^{\alpha+\tilde{\varsigma}}) \right.$$
$$\left. + t^{\alpha+\tilde{\varsigma}} \int_{\frac{b}{t}}^1 (1-s)^{\alpha-1} s^{\tilde{\varsigma}} ds \right) \leq \frac{\tilde{r}}{6}$$

and

$$M_1 C m^\rho t^{1+\eta-\sigma/2-\eta\rho} \int_{\frac{b}{t}}^1 (1-s)^{-\sigma/2} s^{-\eta\rho} ds \leq \frac{\tilde{r}}{6},$$

$$M_1 C w(t) t^{1-\sigma/2+\varsigma} \int_{\frac{b}{t}}^1 (1-s)^{-\sigma/2} s^{-\eta+\varsigma} ds \leq \frac{1}{3}, \tag{3.35}$$

where $m = \max\{r, \tilde{r} + \tilde{b}^\eta \|u(b)\|_{X^{1+\epsilon}}\}$. From the condition $(1-\alpha)\epsilon/2 < \eta < \min\{1 + \varsigma, \frac{1-(1-\alpha)(1-\delta(\epsilon))/2}{\rho}\}$, we can easily check that $1 + \eta - \sigma/2 - \rho\eta > 0$, $1 + \eta - \sigma/2 + \varsigma \geq 1 + \eta - \sigma/2 - \delta(\epsilon) > 0$. The fact $\rho\eta < 1$ and $1 + \varsigma \geq 1 - \delta(\epsilon) > 0$ imply that the Beta functions in (3.34) are well-defined. We can also verify functions $(1-s)^{-\sigma/2} s^{\varsigma}$, $(1-s)^{-\sigma/2} s^{-\eta\rho}$ and $(1-s)^{-\sigma/2} s^{-\eta+\varsigma}$ are integrable for $s \in (\frac{b}{t}, 1)$ and $t \in (b, \tilde{b}]$.

We define the set $\mathfrak{B}(\tilde{r}, \tilde{b})$ of all functions $v \in C^\eta([0, \tilde{b}]; X^{1+\epsilon}) \cap C([0, \tilde{b}]; X^1)$ such that $v(t) = u(t)$ for $t \in [0, b]$ and $\sup_{s \in [b, t]} s^\eta \|v(s) - u(b)\|_{X^{1+\epsilon}} \leq \tilde{r}$ for $t \in [b, \tilde{b}]$. For any $v \in \mathfrak{B}(\tilde{r}, \tilde{b})$ and $t \in [0, \tilde{b}]$, we still use the definition of T in (3.31).

Now we prove that T maps $\mathfrak{B}(\tilde{r}, \tilde{b})$ into $\mathfrak{B}(\tilde{r}, \tilde{b})$.

For any $v \in \mathfrak{B}(\tilde{r}, \tilde{b})$, in the case $t \in [0, b]$, we know that $v(t) = u(t)$, then from Theorem 3.6, we get that T maps $\mathfrak{B}(\tilde{r}, \tilde{b})$ into itself and $(Tv)(t) = (Tu)(t) = u(t)$. Now we prove that $(Tv)(t) \in C^\eta([0, \tilde{b}]; X^{1+\epsilon}) \cap C([0, \tilde{b}]; X^1)$. In fact, using a similar process of proof provided in Theorem 3.6, we can show that for any $v \in \mathfrak{B}(\tilde{r}, \tilde{b})$, $(Tv)(t) \in C^\eta([0, \tilde{b}]; X^{1+\epsilon}) \cap C([0, \tilde{b}]; X^1)$, here we omit the details.

There left to prove that $\sup_{s \in [b, t]} s^\eta \|(Tv)(s) - u(b)\|_{X^{1+\epsilon}} \leq \tilde{r}$ holds for $t \in [b, \tilde{b}]$.

In fact, for any $v \in \mathfrak{B}(\tilde{r}, \tilde{b})$, from Theorem 3.6 and the definition of $\mathfrak{B}(\tilde{r}, \tilde{b})$, we have $u(b) = S_\alpha(b)u_0 + \int_0^b S_\alpha(b-s)f(s, u(s))ds = S_\alpha(b)u_0 + \int_0^b S_\alpha(b-s)f(s, v(s))ds$, then for any $t \in [b, \tilde{b}]$, we obtain

$$\|(Tv)(t) - u(b)\|_{X^{1+\epsilon}} \leq \|(S_\alpha(t) - S_\alpha(b))u_0\|_{X^{1+\epsilon}} + \left\| \int_b^t S_\alpha(t-s)f(s,v(s))ds \right\|_{X^{1+\epsilon}}$$

$$+ \left\| \int_0^b (S_\alpha(t-s) - S_\alpha(b-s))f(s,v(s))ds \right\|_{X^{1+\epsilon}}$$

$$=: \mathcal{Q}_1 + \mathcal{Q}_2 + \mathcal{Q}_3.$$

From Lemma 3.15, we get that

$$\mathcal{Q}_1 \leq M_2 \lambda_1^{(\epsilon-2)/2}(b^{\alpha-1} - t^{\alpha-1})\|u_0\|_{X^1}.$$

And then, similar as (3.32), we use Lemma 3.14 to estimate the second term, that is

$$\mathcal{Q}_2 \leq \int_b^t \|S_\alpha(t-s)f(s,v(s))\|_{X^{1+\epsilon}}ds$$

$$\leq M_1 C \int_b^t (t-s)^{-\sigma/2}s^{-\rho\eta}ds\big(\sup_{s\in[b,t]} s^\eta\|v(s)\|_{X^{1+\epsilon}}\big)^\rho$$

$$+ M_1 Cw(t) \int_b^t (t-s)^{-\sigma/2}s^{\tilde{\varsigma}}ds$$

$$\leq M_1 Ct^{1-\sigma/2-\rho\eta} \int_{\frac{b}{t}}^1 (1-s)^{-\sigma/2}s^{-\rho\eta}ds\big(\sup_{s\in[b,t]} s^\eta\|v(s)\|_{X^{1+\epsilon}}\big)^\rho$$

$$+ M_1 Cw(t)t^{1-\sigma/2+\tilde{\varsigma}} \int_{\frac{b}{t}}^1 (1-s)^{-\sigma/2}s^{\tilde{\varsigma}}ds.$$

From Lemma 3.15 and (3.20), we can estimate the third term as follows

$$\mathcal{Q}_3 \leq \int_0^b \|(S_\alpha(t-s) - S_\alpha(b-s))f(s,v(s))\|_{X^{1+\epsilon}}ds$$

$$\leq M_2 \lambda_1^{(\epsilon-\delta(\epsilon)-1)/2} \int_0^b ((b-s)^{\alpha-1} - (t-s)^{\alpha-1}) \|f(s,v(s))\|_{X^{\delta(\epsilon)}}ds$$

$$\leq M_2 \lambda_1^{(\epsilon-\delta(\epsilon)-1)/2}C \int_0^b ((b-s)^{\alpha-1} - (t-s)^{\alpha-1}) (\|v(s)\|_{X^{1+\epsilon}}^\rho + w(s)s^{\tilde{\varsigma}})ds$$

$$\leq M_2 \lambda_1^{(\epsilon-\delta(\epsilon)-1)/2}C\big(\sup_{s\in[0,b]} s^\eta\|v(s)\|_{X^{1+\epsilon}}\big)^\rho \int_0^b ((b-s)^{\alpha-1} - (t-s)^{\alpha-1}) s^{-\rho\eta}ds$$

$$+ M_2 \lambda_1^{(\epsilon-\delta(\epsilon)-1)/2}Cw(b) \int_0^b ((b-s)^{\alpha-1} - (t-s)^{\alpha-1}) s^{\tilde{\varsigma}}ds$$

$$= M_2 \lambda_1^{(\epsilon-\delta(\epsilon)-1)/2}C\big(\sup_{s\in[0,b]} s^\eta\|v(s)\|_{X^{1+\epsilon}}\big)^\rho$$

$$\times \left(\int_0^b (b-s)^{\alpha-1}s^{-\rho\eta}ds - \int_0^t (t-s)^{\alpha-1}s^{-\rho\eta}ds \right)$$

$$+ M_2\lambda_1^{(\epsilon-\delta(\epsilon)-1)/2}C\Big(\sup_{s\in[0,b]} s^\eta\|v(s)\|_{X^{1+\epsilon}}\Big)^\rho \int_b^t (t-s)^{\alpha-1}s^{-\rho\eta}ds$$

$$+ M_2\lambda_1^{(\epsilon-\delta(\epsilon)-1)/2}Cw(b)\left(\int_0^b (b-s)^{\alpha-1}s^{\tilde\varsigma}ds - \int_0^t (t-s)^{\alpha-1}s^{\tilde\varsigma}ds\right)$$

$$+ M_2\lambda_1^{(\epsilon-\delta(\epsilon)-1)/2}Cw(b)\int_b^t (t-s)^{\alpha-1}s^{\tilde\varsigma}ds.$$

From the definition of Beta function, and by virtue of the facts $\rho\eta < 1$, $\tilde\varsigma \geq -\delta(\epsilon) > -1$, we can compute the integrals of above inequality as follows

$$\int_0^b (b-s)^{\alpha-1}s^{-\rho\eta}ds = b^{\alpha-\rho\eta}B(\alpha, 1-\rho\eta),$$

$$\int_0^b (b-s)^{\alpha-1}s^{\tilde\varsigma}ds = b^{\alpha+\tilde\varsigma}B(\alpha, 1+\tilde\varsigma),$$

similarly,

$$\int_0^t (t-s)^{\alpha-1}s^{-\rho\eta}ds = t^{\alpha-\rho\eta}B(\alpha, 1-\rho\eta),$$

$$\int_0^t (t-s)^{\alpha-1}s^{\tilde\varsigma}ds = t^{\alpha+\tilde\varsigma}B(\alpha, 1+\tilde\varsigma).$$

Substituting the above results into the original inequality, we obtain that

$$\mathcal{Q}_3 \leq M_2\lambda_1^{(\epsilon-\delta(\epsilon)-1)/2}CB(\alpha, 1-\rho\eta)\Big(\sup_{s\in[0,b]} s^\eta\|v(s)\|_{X^{1+\epsilon}}\Big)^\rho(b^{\alpha-\rho\eta}-t^{\alpha-\rho\eta})$$

$$+ M_2\lambda_1^{(\epsilon-\delta(\epsilon)-1)/2}C\Big(\sup_{s\in[0,b]} s^\eta\|v(s)\|_{X^{1+\epsilon}}\Big)^\rho t^{\alpha-\rho\eta}\int_{\frac{b}{t}}^1 (1-s)^{\alpha-1}s^{-\rho\eta}ds$$

$$+ M_2\lambda_1^{(\epsilon-\delta(\epsilon)-1)/2}Cw(b)\Big(B(\alpha, 1+\tilde\varsigma)(b^{\alpha+\tilde\varsigma}-t^{\alpha+\tilde\varsigma})$$

$$+ t^{\alpha+\tilde\varsigma}\int_{\frac{b}{t}}^1 (1-s)^{\alpha-1}s^{\tilde\varsigma}ds\Big).$$

Observing that

$$\sup_{s\in[0,t]} s^\eta\|v(s)\|_{X^{1+\epsilon}} \leq \sup_{s\in[0,t]} s^\eta\|v(s)-u(b)\|_{X^{1+\epsilon}} + t^\eta\|u(b)\|_{X^{1+\epsilon}} \leq m,$$

and in virtue of the conditions (3.34) and (3.35), we obtain that

$$t^\eta\|Tv(t) - u(b)\|_{X^{1+\epsilon}}$$

$$\leq t^\eta M_2\lambda_1^{(\epsilon-2)/2}(b^{\alpha-1}-t^{\alpha-1})\|u_0\|_{X^1}$$

$$+ M_1Ct^{1+\eta-\sigma/2-\rho\eta}m^\rho\int_{\frac{b}{t}}^1 (1-s)^{-\sigma/2}s^{-\rho\eta}ds$$

$$+ M_1Cw(t)t^{1+\eta-\sigma/2+\tilde\varsigma}\int_{\frac{b}{t}}^1 (1-s)^{-\sigma/2}s^{\tilde\varsigma}ds$$

$$+ M_2\lambda_1^{(\epsilon-\delta(\epsilon)-1)/2}Cm^\rho t^\eta\Big(B(\alpha, 1-\rho\eta)(b^{\alpha-\rho\eta}-t^{\alpha-\rho\eta})$$

$$+ t^{\alpha - \rho\eta} \int_{\frac{b}{t}}^{1} (1-s)^{\alpha-1} s^{-\rho\eta} ds \Big)$$

$$+ M_2 \lambda_1^{(\epsilon - \delta(\epsilon) - 1)/2} C w(b) t^\eta \Big(B(\alpha, 1 + \tilde{\varsigma})(b^{\alpha + \tilde{\varsigma}} - t^{\alpha + \tilde{\varsigma}})$$

$$+ t^{\alpha + \tilde{\varsigma}} \int_{\frac{b}{t}}^{1} (1-s)^{\alpha-1} s^{\tilde{\varsigma}} ds \Big)$$

$$\leq \frac{\tilde{r}}{3} + \frac{\tilde{r}}{6} + \frac{\tilde{r}}{6} + \frac{\tilde{r}}{6} + \frac{\tilde{r}}{6} = \tilde{r}.$$

Thus we conclude that T maps $\mathfrak{B}(\tilde{r}, \tilde{b})$ into $\mathfrak{B}(\tilde{r}, \tilde{b})$.

In the following, we prove that T is a contraction mapping on $\mathfrak{B}(\tilde{r}, \tilde{b})$.

For any $v_1, v_2 \in \mathfrak{B}(\tilde{r}, \tilde{b})$, it is easy to see that $(Tv_1)(t) - (Tv_2)(t) = 0$ for $t \in [0, b]$, we just have to discuss the case when $t \in [b, \tilde{b}]$, in this case, we have

$$(Tv_1)(t) - (Tv_2)(t) = \int_{b}^{t} S_\alpha(t - s)(f(s, v_1(s)) - f(s, v_2(s))) ds.$$

It follows from the proof process of Lemma 3.17 that

$$t^\eta \|(Tv_1)(t) - (Tv_2)(t)\|_{X^{1+\epsilon}}$$

$$\leq 2 M_1 C (V_{max})^{\rho-1} t^\eta \int_{b}^{t} (t-s)^{-\sigma/2} s^{-\eta\rho} ds \|v_1 - v_2\|_{C^\eta([0,\tilde{b}];X^{1+\epsilon})}$$

$$+ M_1 C w(t) t^\eta \int_{b}^{t} (t-s)^{-\sigma/2} s^{-\eta+\varsigma} ds \|v_1 - v_2\|_{C^\eta([0,\tilde{b}];X^{1+\epsilon})}$$

$$\leq 2 M_1 C m^{\rho-1} t^{1+\eta-\sigma/2-\eta\rho} \int_{\frac{b}{t}}^{1} (1-s)^{-\sigma/2} s^{-\eta\rho} ds \|v_1 - v_2\|_{C^\eta([0,\tilde{b}];X^{1+\epsilon})}$$

$$+ M_1 C w(t) t^{1-\sigma/2+\varsigma} \int_{\frac{b}{t}}^{1} (1-s)^{-\sigma/2} s^{-\eta+\varsigma} ds \|v_1 - v_2\|_{C^\eta([0,\tilde{b}];X^{1+\epsilon})},$$

where $V_{max} = \max \left\{ \|v_1\|_{C^\eta([0,\tilde{b}];X^{1+\epsilon})}, \|v_2\|_{C^\eta([0,\tilde{b}];X^{1+\epsilon})} \right\}$.

Applying the fact $\frac{\tilde{r}}{m} \leq 1$ into the first condition of (3.35), we get that

$$M_1 C m^{\rho-1} t^{1+\eta-\sigma/2-\eta\rho} \int_{\frac{b}{t}}^{1} (1-s)^{-\sigma/2} s^{-\eta\rho} ds \leq \frac{1}{6} \frac{\tilde{r}}{m} \leq \frac{1}{6}. \qquad (3.36)$$

Combined with the above discussion, by virtue of (3.36) and the second condition of (3.35), we obtain that

$$t^\eta \|(Tv_1)(t) - (Tv_2)(t)\|_{X^{1+\epsilon}} \leq \frac{2}{3} \|v_1 - v_2\|_{C^\eta([0,\tilde{b}];X^{1+\epsilon})},$$

which implies that T is a strictly contraction mapping. By virtue of the Banach fixed point theorem, we obtain that there exists a unique fixed point $\tilde{u} \in \mathfrak{B}(\tilde{r}, \tilde{b})$, which implies that $\tilde{u}$ is a continuation of u in $[0, \tilde{b}]$.

Now we prove the continuation $\tilde{u}$ is unique in $[0, \tilde{b}]$.

Assume that u_1 and u_2 are two continuations of the ϵ-regular mild solution u obtained by Theorem 3.6 in $[0, \tilde{b}]$. From the definition of continuation, it is obviously that $u_1(t) = u_2(t) = u(t)$ for $t \in [0, b]$. Thus, we only need to check that $u_1(t) = u_2(t)$ for $t \in [b, \tilde{b}]$ in the sequel. From the proof process of Lemma 3.17, we have

$$\|u_1(t) - u_2(t)\|_{X^{1+\epsilon}}$$

$$= \left\| \int_0^t S_\alpha(t - s)(f(s, u_1(s)) - f(s, u_2(s)))ds \right\|_{X^{1+\epsilon}}$$

$$= \left\| \int_b^t S_\alpha(t - s)(f(s, u_1(s)) - f(s, u_2(s)))ds \right\|_{X^{1+\epsilon}}$$

$$\leq M_1 \int_b^t (t - s)^{-\sigma/2} C \|u_1(s) - u_2(s)\|_{X^{1+\epsilon}} (\|u_1(s)\|_{X^{1+\epsilon}}^{\rho-1} + \|u_2(s)\|_{X^{1+\epsilon}}^{\rho-1} + w(s)s^\varsigma) ds$$

$$\leq M_1 C M_3 \int_b^t (t - s)^{-\sigma/2} \|u_1(s) - u_2(s)\|_{X^{1+\epsilon}} ds,$$

where $M_3 = (\sup_{s \in [b, \tilde{b}]} \|u_1(s)\|_{X^{1+\epsilon}})^{\rho-1} + (\sup_{s \in [b, \tilde{b}]} \|u_2(s)\|_{X^{1+\epsilon}})^{\rho-1} + w(\tilde{b})b^\varsigma$. For the reason that $(t - s)^{-\sigma/2}$ is continuous and nonnegative for $s \in (b, t)$, the singular Gronwall's inequality implies that the relation $u_1(t) = u_2(t)$ also holds for $t \in [b, \tilde{b}]$. Thus we conclude that $\tilde{u}$ is the unique continuation of u in $[0, \tilde{b}]$. This completes the proof. $\qquad\square$

Remark 3.7. We note that the conclusion in Theorem 3.7 still holds if we change the existence time b of $u(t)$ into the maximal time we can get from Theorem 3.6.

Theorem 3.8. *Assume the conditions in Theorem 3.6 hold and u is an ϵ-regular mild solution of the problem (3.17) with maximal time of existence b_{max}. Then either $b_{max} = \infty$ or $\limsup_{t \to b_{max}^-} \|u(t)\|_{X^{1+\epsilon}} = \infty$.*

Proof. Assume that the maximal time of existence $b_{max} < \infty$ and there exists a positive constant R such that the solution $u(t)$ satisfies $\sup_{t \in (0, b_{max})} t^\eta \|u(t)\|_{X^{1+\epsilon}} \leq R$. In the following, we shall prove that this assumption is a contradiction.

Taking a sequence $\{t_n\}_{n \in \mathbb{N}^+} (t_n < b_{max})$ which satisfies $t_n \to b_{max}^-$ as $n \to \infty$. Now we show that $\{u(t_n)\}_{n \in \mathbb{N}^+}$ is a Cauchy sequence in $X^{1+\epsilon}$.

In fact, for $0 < t_m < t_n < b_{max}$, we have

$$u(t_n) - u(t_m) = (S_\alpha(t_n) - S_\alpha(t_m))u_0 + \int_0^{t_n} S_\alpha(t_n - s)f(s, u(s))ds$$

$$- \int_0^{t_m} S_\alpha(t_m - s)f(s, u(s))ds$$

$$= (S_\alpha(t_n) - S_\alpha(t_m))u_0 + \int_{t_m}^{t_n} S_\alpha(t_n - s)f(s, u(s))ds$$

$$+ \int_0^{t_m} (S_\alpha(t_n - s) - S_\alpha(t_m - s))f(s, u(s))ds.$$

By using an analogous argument presented in the proof process of Theorem 3.6 (Step I) and Theorem 3.7 ($\mathcal{Q}_3$), we can easily get that

$$\|u(t_n) - u(t_m)\|_{X^{1+\epsilon}}$$

$$\leq \|(S_\alpha(t_n) - S_\alpha(t_m))u_0\|_{X^{1+\epsilon}} + \left\| \int_{t_m}^{t_n} S_\alpha(t_n - s)f(s, u(s))ds \right\|_{X^{1+\epsilon}}$$

$$+ \left\| \int_0^{t_m} (S_\alpha(t_n - s) - S_\alpha(t_m - s))f(s, u(s))ds \right\|_{X^{1+\epsilon}}$$

$$\leq M_2 \lambda_1^{(\epsilon-2)/2}(t_m^{\alpha-1} - t_n^{\alpha-1})\|u_0\|_{X^1}$$

$$+ M_1 C R^\rho t_n^{1-\sigma/2-\rho\eta} \int_{\frac{t_m}{t_n}}^1 (1-s)^{-\sigma/2}s^{-\rho\eta}ds$$

$$+ M_1 C w(t_n) t_n^{1-\sigma/2+\tilde\xi} \int_{\frac{t_m}{t_n}}^1 (1-s)^{-\sigma/2}s^{\tilde\xi}ds$$

$$+ M_2 \lambda_1^{(\epsilon-\delta(\epsilon)-1)/2} C \int_0^{t_m} ((t_m - s)^{\alpha-1} - (t_n - s)^{\alpha-1})(R^\rho s^{-\rho\eta} + w(t_m)s^{\tilde\xi})ds,$$

where $(1-s)^{-\sigma/2}s^{-\rho\eta}$, $(1-s)^{-\sigma/2}s^{\tilde\xi}$ and $((t_m - s)^{\alpha-1} - (t_n - s)^{\alpha-1})(R^\rho s^{-\rho\eta} + w(t_m)s^{\tilde\xi})$ are integrable according to our earlier discussion. Moreover, for given $\theta > 0$, since $\{t_n\}_{n \in \mathbb{N}^+}(t_n < b_{max})$ is a sequence which converges to b_{max}^-, then there exists an $N \in \mathbb{N}^+$ such that for any $n, m \geq N$, $|t_n - t_m|$ can be small enough so that

$$M_2 \lambda_1^{(\epsilon-2)/2}(t_m^{\alpha-1} - t_n^{\alpha-1})\|u_0\|_{X^1} < \frac{\theta}{3},$$

$$M_1 C R^\rho t_n^{1-\sigma/2-\rho\eta} \int_{\frac{t_m}{t_n}}^1 (1-s)^{-\sigma/2}s^{-\rho\eta}ds + M_1 C w(t_n) t_n^{1-\sigma/2+\tilde\xi} \int_{\frac{t_m}{t_n}}^1 (1-s)^{-\sigma/2}s^{\tilde\xi}ds$$

$$< \frac{\theta}{3},$$

$$M_2 \lambda_1^{(\epsilon-\delta(\epsilon)-1)/2} C \int_0^{t_m} ((t_m - s)^{\alpha-1} - (t_n - s)^{\alpha-1})(R^\rho s^{-\rho\eta} + w(t_m)s^{\tilde\xi})ds < \frac{\theta}{3}.$$

Thus, for $n, m \geq N$, we have

$$\|u(t_n) - u(t_m)\|_{X^{1+\epsilon}} < \theta,$$

which implies that $\{u(t_n)\}_{n \in \mathbb{N}^+}$ is a Cauchy sequence in $X^{1+\epsilon}$, it follows that $\lim_{t \to b_{max}^-} u(t)$ exists and is finite, thus we can extend $u(t)$ from $(0, b_{max})$ to $(0, b_{max}]$ in $X^{1+\epsilon}$. Furthermore, Theorem 3.7 implies that $u(t)$ can be extended to a bigger interval than $(0, b_{max}]$, which contradicts the assumption that b_{max} is the maximal time of existence. Therefore, we conclude that either $\limsup_{t \to b_{max}^-} \|u(t)\|_{X^{1+\epsilon}} = \infty$ or $b_{max} = \infty$. The proof is completed. $\qquad\square$

3.4 Weak Solutions

3.4.1 *Introduction*

To the best of our knowledge, there are few results of the existence, uniqueness and regularity of weak solutions for the Rayleigh-Stokes problem. As we know, the research on weak solutions for partial differential equations is of great theoretical and physical significance, for the reason that some partial differential equations are not easy to find the classical solutions. In consequence, researchers can obtain weak solutions of certain partial differential equations, and the regularity of weak solutions can be obtained by using theories of Sobolev spaces. In this section, we investigate the following fractional Rayleigh-Stokes problem for a heated second grade fluid

$$\begin{cases} \partial_t u - (1 + \gamma \partial_t^\alpha)\Delta u = f(t,x), & t \in (0,b), \ x \in \Omega, \\ u(t,x) = 0, & t \in (0,b), \ x \in \partial\Omega, \\ u(0,x) = h(x), & x \in \Omega, \end{cases} \tag{3.37}$$

where ∂_t^α is the Riemann-Liouville fractional partial derivative of order $\alpha \in (0,1)$, $\Omega \subset \mathbb{R}^d (d \geq 1)$ is a bounded domain with smooth boundary $\partial\Omega$, Δ is the Laplacian operator, $b > 0$ is a given time, $\gamma > 0$ is a constant, $h(x)$ is the state of u at the initial time, $f : [0,b] \times \Omega \to \mathbb{R}$ is the source term of our problem.

The outline of this section is as follows. In Subsection 3.4.2, we give some notations, concepts and lemmas which will be used in this section. In Subsection 3.4.3, firstly, we give the definition of weak solutions of the fractional Rayleigh-Stokes problem, and then, by using the Galerkin method, we obtain the existence, uniqueness and the energy estimate of approximate solutions, afterwards, the existence, uniqueness and regularity results of weak solutions are obtained with the initial data $h(x)$ in $H_0^1(\Omega)$ and the source term $f(t,x)$ in $H^{-1}(\Omega)$. Finally, under conditions that $h \in H^2(\Omega)$ and $f \in L^2(0,b; L^2(\Omega))$, the spacial regularity of weak solutions of the fractional Rayleigh-Stokes problem is improved.

3.4.2 Preliminaries

In this subsection, we recall some notations, concepts and lemmas which are useful in this section.

Denote $\mathbb{R}^d (d \geq 1)$ is the d-dimensional Euclidean space, and $\Omega \subset \mathbb{R}^d$ is a bounded domain with smooth boundary $\partial\Omega$.

In order to obtain our main results, we introduce the spectral problem: $-\Delta\varphi_k = \lambda_k\varphi_k$ in Ω and $\varphi_k = 0$ on $\partial\Omega$, where $\{\varphi_k\}_{k=1}^\infty$ is both an orthogonal basis of $H_0^1(\Omega)$ and an orthonormal basis of $L^2(\Omega)$, λ_k is the eigenvalue of $-\Delta$ corresponding to φ_k and satisfies $0 < \lambda_1 \leq \lambda_2 \leq \dots \lambda_k \leq \dots$ with the property that $\lambda_k \to \infty$ as $k \to \infty$.

Some functional spaces are introduced in the following.

We denote by $L^2(\Omega)$ the Banach space of all measurable functions on Ω endowed with the inner product $(\cdot,\cdot)$ and the norm

$$\|v\|_{L^2(\Omega)} := (v,v)^{\frac{1}{2}} = \left(\sum_{k=1}^\infty |(v,\varphi_k)|^2 \right)^{\frac{1}{2}}.$$

$H_0^1(\Omega)$ denotes the closure of $C_0^\infty(\Omega)$ in $H^1(\Omega)$, and equipped with the norm

$$\|v\|_{H_0^1(\Omega)} := \|\nabla v\|_{L^2(\Omega)},$$

here $C_0^\infty(\Omega)$ is the space of all infinitely differentiable functions with compact support, and it is easy to see that $\|v\|^2_{H_0^1(\Omega)} = \sum_{k=1}^\infty \lambda_k |(v, \varphi_k)|^2$.

Denote by $H^{-1}(\Omega)$ the dual space of $H_0^1(\Omega)$ with the norm

$$\|v\|_{H^{-1}(\Omega)} := \left(\sum_{k=1}^\infty \lambda_k^{-1} |(v, \varphi_k)|^2 \right)^{\frac{1}{2}}.$$

We denote by $L^p(0, b; X)(p \geq 1)$ the space of all strongly measurable functions v with the norm

$$\|v\|_{L^p(0,b;X)} := \left(\int_0^b \|v(t)\|^p_X dt \right)^{\frac{1}{p}} < \infty, \quad \text{for} \quad 1 \leq p < \infty$$

and

$$\|v\|_{L^\infty(0,b;X)} := \operatorname{esssup}_{t \in [0,b]} \|v(t)\|_X < \infty,$$

where X mainly takes $L^2(\Omega)$, $H_0^1(\Omega)$, $H^{-1}(\Omega)$ or $H^2(\Omega)$ in this section.

Let $C([0, b]; X)$ be the Banach space of continuous functions v from $[0, b]$ into X equipped with the norm

$$\|v\|_{C([0,b];X)} := \max_{t \in [0,b]} \|v(t)\|_X < \infty.$$

For $u : [0, \infty) \times \mathbb{R}^n \to \mathbb{R}$, the left Riemann-Liouville fractional derivative and the left Caputo fractional derivative with respect to time t of u for order $\alpha \in (0, 1)$ are defined by

$$\partial_t^\alpha u(t, x) = \frac{1}{\Gamma(1 - \alpha)} \frac{\partial}{\partial t} \int_0^t (t - s)^{-\alpha} u(s, x) ds, \quad t > 0$$

and

$$^C \partial_t^\alpha u(t, x) = \partial_t^\alpha (u(t, x) - u(0, x)), \quad t > 0,$$

respectively.

In the following, we introduce some inequalities which will be useful in proving our main results.

Lemma 3.18. *[233] Let $a(t) \in \mathbb{R}$ be a continuous and differentiable function. Then the following inequality holds*

$$\frac{1}{2} {_0}D_t^\alpha a^2(t) \leq a(t) {_0}D_t^\alpha a(t), \quad \forall \alpha \in (0, 1), \quad \forall t \geq 0.$$

Lemma 3.19. *[273] If $\alpha \in (0, 1)$ and $v(\cdot, x) \in L^2(0, b)$ for each $x \in \Omega$, then*

$$\int_0^t ({_0}D_s^{-\alpha} v(s), v(s)) ds \geq \cos\left(\frac{\alpha\pi}{2}\right) \int_0^t \|{_0}D_s^{-\frac{\alpha}{2}} v(s)\|^2_{L^2(\Omega)} ds, \quad \text{for } t \in [0, b].$$

Now we give an existence result under a special case of Proposition 5.5 and Theorem 5.5 in [197].

Lemma 3.20. *Let $0 < \alpha < 1$, $\lambda, \mu \in \mathbb{R}$ and let $g(t)$ be a given real function defined on $\mathbb{R}^+$. Then the Cauchy type problem*

$$
\begin{cases}
y'(t) - \lambda_0 D_t^\alpha y(t) - \mu y(t) = g(t), & t > 0, \\
y(0) = w
\end{cases}
$$

is solvable, and its solution has the form

$$
y(t) = P_{\alpha,\lambda,\mu}(t)w + \int_0^t P_{\alpha,\lambda,\mu}(t-s)g(s)ds, \tag{3.38}
$$

provided that the right-hand side in (3.38) is convergent, where

$$
P_{\alpha,\lambda,\mu}(t) = \mathcal{L}^{-1}\left(\frac{1}{s - \lambda s^\alpha - \mu}\right)(t),
$$

$\mathcal{L}^{-1}(v(s))(t)$ stands for the inverse Laplace transform of v.

It is easy to see that in the case $\lambda = -\lambda_k \gamma$ and $\mu = -\lambda_k (k = 1, 2, \ldots)$, the function $P_{\alpha,\lambda,\mu}(t)$ is equivalent to the function $u_k(t)$ in [39], in this section, for the convenience of writing, we denote

$$
P_{\alpha,k}(t) = \mathcal{L}^{-1}\left(\frac{1}{s + \gamma\lambda_k s^\alpha + \lambda_k}\right)(t), \tag{3.39}
$$

and it has the following properties.

Lemma 3.21. *[39] The functions $P_{\alpha,k}(t), k = 1, 2, \ldots$, have the following properties:*

(i) $P_{\alpha,k}(0) = 1$, $0 < P_{\alpha,k}(t) \leq 1$, $t \geq 0$;
(ii) $P_{\alpha,k}(t)$ *are completely monotone for $t \geq 0$.*

3.4.3 Existence, Uniqueness and Regularity

In this subsection, we give the definition of weak solutions of problem (3.37) and estimate the existence, uniqueness and regularity of weak solutions by virtue of Galerkin method. Furthermore, an improved regularity result of weak solutions is obtained.

Definition 3.8. We say a function $u \in L^2(0, b; H_0^1(\Omega)) \cap L^\infty(0, b; L^2(\Omega))$ with $\partial_t^\alpha u \in L^2(0, b; H_0^1(\Omega))$, $\partial_t u \in L^2(0, b; H^{-1}(\Omega))$ is a weak solution of problem (3.37) provided

(i) $\langle \partial_t u, v \rangle + ((1 + \gamma\partial_t^\alpha)\nabla u, \nabla v) = (f, v)$, for each $v \in H_0^1(\Omega)$ and a.e. $t \in [0, b]$;
(ii) $u(0) = h$.

For a fixed positive integer m, the work on finding the existence of weak solutions of problem (3.37) is turned to looking for a function u_m of the following form

$$u_m(t) = \sum_{k=1}^{m} d_m^k(t)\varphi_k, \qquad (3.40)$$

where u_m is required to satisfy

$$\begin{cases} (\partial_t u_m(t), \varphi_k) + ((1 + \gamma\partial_t^\alpha)\nabla u_m(t), \nabla\varphi_k) = (f(t), \varphi_k), \\ (u_m(0), \varphi_k) = (h, \varphi_k), \end{cases} \qquad (3.41)$$

here, for the convenience of writing, we omit the spatial variable in u_m and f .

Lemma 3.22. *Assume $h \in H_0^1(\Omega)$ and $f \in L^2(0, b; H^{-1}(\Omega))$. Then for every fixed integer $m = 1, 2, \ldots$ and any fixed integer $k = 1, \ldots, m$, the problem (3.41) exists a unique solution u_m of the form*

$$u_m(t) = \sum_{k=1}^{m} d_m^k(t)\varphi_k = P_\alpha^m(t)h + \int_0^t P_\alpha^m(t - s)f(s)ds, \qquad (3.42)$$

where $P_\alpha^m(t)v = \sum_{k=1}^{m} P_{\alpha,k}(t)(v, \varphi_k)$ for any $v \in L^2(\Omega)$ and $P_{\alpha,k}(t)$ is defined in (3.39).

Proof. Assume that $u_m(t)$ is given by (3.40). From the assumptions of $\{\varphi_k\}_{k=1}^\infty$, we obtain that

$$(\partial_t u_m(t), \varphi_k) = d_m^{k\,\prime}(t)$$

and

$$((1 + \gamma\partial_t^\alpha)\nabla u_m(t), \nabla\varphi_k) = (1 + \gamma_0 D_t^\alpha)\lambda_k d_m^k(t).$$

Then the problem (3.41) is equivalent to the following fractional ordinary differential equations

$$\begin{cases} d_m^{k\,\prime}(t) + (1 + \gamma_0 D_t^\alpha)\lambda_k d_m^k(t) = f^k(t), \\ d_m^k(0) = h^k, \end{cases} \qquad (3.43)$$

where $f^k(t) := (f(t), \varphi_k)$ and $h^k := (h, \varphi_k)$.

From Lemma 3.20, we can easy to check that the problem (3.43) is solvable, which implies that there exists a unique solution $d_m^k(t)$ for a.e. $t \in [0, b]$ of the following form

$$d_m^k(t) = P_{\alpha,k}(t)h^k + \int_0^t P_{\alpha,k}(t - s)f^k(s)ds, \qquad (3.44)$$

where $P_{\alpha,k}(t)$ is defined in (3.39). Lemma 3.21 indicates that the solution $d_m^k(t)$ is continuous for $t \in [0, b]$. Thus we conclude that the problem (3.41) exists a unique continuous solution u_m of the form

$$u_m(t) = \sum_{k=1}^{m} d_m^k(t)\varphi_k = P_\alpha^m(t)h + \int_0^t P_\alpha^m(t - s)f(s)ds,$$

where the operator $P_\alpha^m(t)v = \sum_{k=1}^{m} P_{\alpha,k}(t)(v, \varphi_k)$ for any $v \in L^2(\Omega)$. $\qquad \square$

Before giving the main results, we propose to use the energy estimate method to show the priori estimate of u_m in the following.

Lemma 3.23. *Let $0 < \alpha < \frac{1}{2}$, and assume $h \in H_0^1(\Omega)$, $f \in L^2(0, b; H^{-1}(\Omega))$. Then there exists a constant $C > 0$ depending only on α, γ and b such that*

$$\|u_m\|_{L^\infty(0,b;L^2(\Omega))} + \|u_m\|_{L^2(0,b;H_0^1(\Omega))}$$

$$+ \|\partial_t u_m\|_{L^2(0,b;H^{-1}(\Omega))} + \|\partial_t^\alpha u_m\|_{L^2(0,b;H_0^1(\Omega))} \tag{3.45}$$

$$\leq C \left(\|h\|_{L^2(\Omega)} + \|h\|_{H_0^1(\Omega)} + \|f\|_{L^2(0,b;H^{-1}(\Omega))} \right).$$

Proof. Multiplying both sides of the first equation in (3.41) by $d_m^k(t)$ and summing it from $k = 1$ to m, we obtain the following equation

$$(\partial_t u_m(t), u_m(t)) + ((1 + \gamma \partial_t^\alpha) \nabla u_m(t), \nabla u_m(t)) = (f(t), u_m(t)). \tag{3.46}$$

From Lemma 3.18 and the relation $(\partial_t v(t), v(t)) = \frac{1}{2} \frac{d}{dt}(v(t), v(t))$, we can estimate the left side of (3.46) as follows

$$(\partial_t u_m(t), u_m(t)) + ((1 + \gamma \partial_t^\alpha) \nabla u_m(t), \nabla u_m(t))$$

$$\geq \frac{1}{2} \frac{d}{dt} \|u_m(t)\|_{L^2(\Omega)}^2 + \frac{\gamma}{2} {}_0 D_t^\alpha \|\nabla u_m(t)\|_{L^2(\Omega)}^2 + \|\nabla u_m(t)\|_{L^2(\Omega)}^2.$$

Substituting the above result into (3.46), and then, applying the Young's inequality and the inequality

$$|\langle g, v \rangle| \leq \|g\|_{H^{-1}(\Omega)} \|v\|_{H_0^1(\Omega)}, \quad \text{for} \quad g \in H^{-1}(\Omega), v \in H_0^1(\Omega), \tag{3.47}$$

we get that

$$\frac{1}{2} \frac{d}{dt} \|u_m(t)\|_{L^2(\Omega)}^2 + \frac{\gamma}{2} {}_0 D_t^\alpha \|\nabla u_m(t)\|_{L^2(\Omega)}^2 + \|\nabla u_m(t)\|_{L^2(\Omega)}^2$$

$$\leq |(f(t), u_m(t))| \leq \|f(t)\|_{H^{-1}(\Omega)} \|u_m(t)\|_{H_0^1(\Omega)} \leq \frac{1}{2} \|f(t)\|_{H^{-1}(\Omega)}^2 + \frac{1}{2} \|u_m(t)\|_{H_0^1(\Omega)}^2,$$

which implies

$$\frac{d}{dt} \|u_m(t)\|_{L^2(\Omega)}^2 + \gamma {}_0 D_t^\alpha \|\nabla u_m(t)\|_{L^2(\Omega)}^2 + \|\nabla u_m(t)\|_{L^2(\Omega)}^2 \leq \|f(t)\|_{H^{-1}(\Omega)}^2.$$

Integrating both sides of the above inequality from 0 to t, from $\|u_m(0)\|_{L^2(\Omega)} = (\sum_{k=1}^m |(h, \varphi_k)|^2)^{\frac{1}{2}} \leq \|h\|_{L^2(\Omega)}$, we obtain

$$\|u_m(t)\|_{L^2(\Omega)}^2 + \frac{\gamma}{\Gamma(1-\alpha)} \int_0^t (t-s)^{-\alpha} \|\nabla u_m(s)\|_{L^2(\Omega)}^2 ds + \int_0^t \|\nabla u_m(s)\|_{L^2(\Omega)}^2 ds$$

$$\leq \int_0^t \|f(s)\|_{H^{-1}(\Omega)}^2 ds + \|h\|_{L^2(\Omega)}^2,$$

and then, the fact that $(t-s)^{-\alpha} \geq t^{-\alpha} \geq b^{-\alpha}$ for $0 \leq s < t \leq b$ implies

$$\|u_m(t)\|_{L^2(\Omega)}^2 + \left(\frac{\gamma b^{-\alpha}}{\Gamma(1-\alpha)} + 1 \right) \int_0^t \|\nabla u_m(s)\|_{L^2(\Omega)}^2 ds$$

$$\leq \int_0^t \|f(s)\|_{H^{-1}(\Omega)}^2 ds + \|h\|_{L^2(\Omega)}^2.$$

Taking the maximum for $t \in [0, b]$ of both sides of the above inequality yields that

$$\|u_m\|^2_{L^\infty(0,b;L^2(\Omega))} + \left(\frac{\gamma b^{-\alpha}}{\Gamma(1-\alpha)} + 1\right) \|u_m\|^2_{L^2(0,b;H^1_0(\Omega))}$$

$$\leq \|f\|^2_{L^2(0,b;H^{-1}(\Omega))} + \|h\|^2_{L^2(\Omega)}. \tag{3.48}$$

Thus we obtain $u_m \in L^\infty(0, b; L^2(\Omega)) \cap L^2(0, b; H^1_0(\Omega))$.

In the following, we will estimate that $\partial_t u_m \in L^2(0, b; H^{-1}(\Omega))$.

At the beginning, we multiply both sides of the first equation of (3.43) by $\lambda_k^{-1} d_m^{k\,'}(t)$, it follows that

$$\lambda_k^{-1}(d_m^{k\,'}(t))^2 + \gamma d_m^{k\,'}(t) {}_0 D_t^\alpha d_m^k(t) + d_m^{k\,'}(t)d_m^k(t) = \lambda_k^{-1}f^k(t)d_m^{k\,'}(t),$$

and then, let $g(t, x) = \sum_{k=1}^\infty \lambda_k^{-1} f^k(t)\varphi_k(x)$ and sum the above equation from $k = 1$ to m, we have

$$\|\partial_t u_m(t)\|^2_{H^{-1}(\Omega)} + \gamma(\partial_t^\alpha u_m(t), \partial_t u_m(t)) + (u_m(t), \partial_t u_m(t)) = (g(t), \partial_t u_m(t)).$$

More precisely, denote $\omega_\alpha(t) = \frac{t^{\alpha-1}}{\Gamma(\alpha)}$, from the facts that $\partial_t^\alpha u_m(t) = {}_0 D_t^{\alpha-1}\partial_t u_m(t) + \omega_{1-\alpha}(t)u_m(0)$, $(\partial_t u_m(t), u_m(t)) = \frac{1}{2}\frac{d}{dt}(u_m(t), u_m(t))$, and on account of the Young's inequality and (3.47), we get that

$$\|\partial_t u_m(t)\|^2_{H^{-1}(\Omega)} + \gamma({}_0 D_t^{\alpha-1}\partial_t u_m(t), \partial_t u_m(t)) + \frac{1}{2}\frac{d}{dt}(u_m(t), u_m(t))$$

$$=(g(t), \partial_t u_m(t)) - \gamma(\omega_{1-\alpha}(t)u_m(0), \partial_t u_m(t))$$

$$\leq \|g(t)\|_{H^1_0(\Omega)}\|\partial_t u_m(t)\|_{H^{-1}(\Omega)} + \gamma\|\omega_{1-\alpha}(t)u_m(0)\|_{H^1_0(\Omega)}\|\partial_t u_m(t)\|_{H^{-1}(\Omega)}$$

$$=\|f(t)\|_{H^{-1}(\Omega)}\|\partial_t u_m(t)\|_{H^{-1}(\Omega)} + \gamma\|\omega_{1-\alpha}(t)u_m(0)\|_{H^1_0(\Omega)}\|\partial_t u_m(t)\|_{H^{-1}(\Omega)}$$

$$\leq \|f(t)\|^2_{H^{-1}(\Omega)} + \gamma^2\omega^2_{1-\alpha}(t)\|u_m(0)\|^2_{H^1_0(\Omega)} + \frac{1}{2}\|\partial_t u_m(t)\|^2_{H^{-1}(\Omega)},$$

that is,

$$\|\partial_t u_m(t)\|^2_{H^{-1}(\Omega)} + 2\gamma({}_0 D_t^{\alpha-1}\partial_t u_m(t), \partial_t u_m(t)) + \frac{d}{dt}(u_m(t), u_m(t))$$

$$\leq 2\|f(t)\|^2_{H^{-1}(\Omega)} + 2\gamma^2\omega^2_{1-\alpha}(t)\|u_m(0)\|^2_{H^1_0(\Omega)}.$$

Integrating the above inequality from 0 to t, then we conclude from Lemma 3.19 that

$$\int_0^t \|\partial_s u_m(s)\|^2_{H^{-1}(\Omega)}ds$$

$$+ 2\gamma\cos\left(\frac{1-\alpha}{2}\pi\right)\int_0^t \|{}_0 D_s^{\frac{\alpha-1}{2}}\partial_s u_m(s)\|^2_{L^2(\Omega)}ds + \|u_m(t)\|^2_{L^2(\Omega)}$$

$$\leq 2\int_0^t \|f(s)\|^2_{H^{-1}(\Omega)}ds + \frac{2\gamma^2 t^{1-2\alpha}}{(1-2\alpha)\Gamma^2(1-\alpha)}\|h\|^2_{H^1_0(\Omega)} + \|h\|^2_{L^2(\Omega)}.$$

Taking the maximum of both sides of the above inequality for $t \in [0, b]$, we get

$$\|\partial_t u_m\|^2_{L^2(0,b;H^{-1}(\Omega))} + 2\gamma\cos\left(\frac{1-\alpha}{2}\pi\right)\|C\partial_t^{\frac{1+\alpha}{2}}u_m\|^2_{L^2(0,b;L^2(\Omega))}$$

$$+ \|u_m\|^2_{L^\infty(0,b;L^2(\Omega))} \tag{3.49}$$

$$\leq 2\|f\|^2_{L^2(0,b;H^{-1}(\Omega))} + \frac{2\gamma^2 b^{1-2\alpha}}{(1-2\alpha)\Gamma^2(1-\alpha)}\|h\|^2_{H^1_0(\Omega)} + \|h\|^2_{L^2(\Omega)}.$$

Thus we conclude that $\partial_t u_m \in L^2(0,b; H^{-1}(\Omega))$.

In the sequel, we shall prove that $\partial_t^\alpha u_m \in L^2(0,b; H_0^1(\Omega))$.

We multiply both sides of the first equation in (3.41) by $_0 D_t^\alpha d_m^k(t)$ and sum it from $k=1$ to m, that is,

$$\gamma(\partial_t^\alpha \nabla u_m(t), \partial_t^\alpha \nabla u_m(t)) + (\nabla u_m(t), \partial_t^\alpha \nabla u_m(t))$$
$$= (f(t), \partial_t^\alpha u_m(t)) - (\partial_t u_m(t), \partial_t^\alpha u_m(t)),$$

and then, on account of (3.47), the Young's inequality and Lemma 3.18, we have

$$\gamma \|\partial_t^\alpha \nabla u_m(t)\|_{L^2(\Omega)}^2 + \frac{1}{2} {_0 D_t^\alpha} \|\nabla u_m(t)\|_{L^2(\Omega)}^2$$
$$\leq \|f(t)\|_{H^{-1}(\Omega)} \|\partial_t^\alpha u_m(t)\|_{H_0^1(\Omega)} + \|\partial_t u_m(t)\|_{H^{-1}(\Omega)} \|\partial_t^\alpha u_m(t)\|_{H_0^1(\Omega)}$$
$$\leq \frac{1}{\gamma} \|f(t)\|_{H^{-1}(\Omega)}^2 + \frac{1}{\gamma} \|\partial_t u_m(t)\|_{H^{-1}(\Omega)}^2 + \frac{\gamma}{2} \|\partial_t^\alpha u_m(t)\|_{H_0^1(\Omega)}^2,$$

which shows that

$$\gamma \|\partial_t^\alpha \nabla u_m(t)\|_{L^2(\Omega)}^2 + {_0 D_t^\alpha} \|\nabla u_m(t)\|_{L^2(\Omega)}^2 \leq \frac{2}{\gamma} \|f(t)\|_{H^{-1}(\Omega)}^2 + \frac{2}{\gamma} \|\partial_t u_m(t)\|_{H^{-1}(\Omega)}^2.$$

Integrating the above inequality from 0 to t, we get that

$$\gamma \int_0^t \|\partial_s^\alpha \nabla u_m(s)\|_{L^2(\Omega)}^2 ds + \frac{1}{\Gamma(1-\alpha)} \int_0^t (t-s)^{-\alpha} \|\nabla u_m(s)\|_{L^2(\Omega)}^2 ds$$
$$\leq \frac{2}{\gamma} \int_0^t \|f(s)\|_{H^{-1}(\Omega)}^2 ds + \frac{2}{\gamma} \int_0^t \|\partial_s u_m(s)\|_{H^{-1}(\Omega)}^2 ds.$$

Taking the maximum for $t \in [0,b]$ of both sides of the above inequality and using $(t-s)^{-\alpha} \geq t^{-\alpha} \geq b^{-\alpha}$ for $0 \leq s < t \leq b$, we conclude from (3.49) that

$$\gamma \|\partial_t^\alpha u_m\|_{L^2(0,b;H_0^1(\Omega))}^2 + \frac{b^{-\alpha}}{\Gamma(1-\alpha)} \|u_m\|_{L^2(0,b;H_0^1(\Omega))}^2$$
$$\leq \frac{2}{\gamma} \|f\|_{L^2(0,b;H^{-1}(\Omega))}^2 + \frac{2}{\gamma} \|\partial_t u_m\|_{L^2(0,b;H^{-1}(\Omega))}^2$$
$$\leq \frac{6}{\gamma} \|f\|_{L^2(0,b;H^{-1}(\Omega))}^2 + \frac{4\gamma b^{1-2\alpha}}{(1-2\alpha)\Gamma^2(1-\alpha)} \|h\|_{H_0^1(\Omega)}^2 + \frac{2}{\gamma} \|h\|_{L^2(\Omega)}^2,$$

which implies that $\partial_t^\alpha u_m \in L^2(0,b; H_0^1(\Omega))$.

Therefore, there exists a positive constant C such that (3.45) holds. This completes the proof. $\qquad\square$

Theorem 3.9. *Assume* $h \in H_0^1(\Omega)$ *and* $f \in L^2(0,b; H^{-1}(\Omega))$. *Then for* $0 < \alpha < \frac{1}{2}$, *the problem* (3.37) *exists at least one weak solution* $u \in L^2(0,b; H_0^1(\Omega)) \cap L^\infty(0,b; L^2(\Omega))$ *with* $\partial_t^\alpha u \in L^2(0,b; H_0^1(\Omega))$ *and* $\partial_t u \in L^2(0,b; H^{-1}(\Omega))$.

Proof. From Lemma 3.23, we know that the sequence $\{u_m\}_{m=1}^\infty$ is bounded in $L^2(0,b; H_0^1(\Omega)) \cap L^\infty(0,b; L^2(\Omega))$, $\{\partial_t u_m\}_{m=1}^\infty$ is bounded in $L^2(0,b; H^{-1}(\Omega))$, and $\{\partial_t^\alpha u_m\}_{m=1}^\infty$ is bounded in $L^2(0,b; H_0^1(\Omega))$, and then, from the weak compactness theorem in Hilbert space, there exist a subsequence $\{u_{\hat{m}}\}_{\hat{m}=1}^\infty \subset \{u_m\}_{m=1}^\infty$ and a function $u \in L^2(0, b; H_0^1(\Omega)) \cap L^\infty(0, b; L^2(\Omega))$ with $\partial_t u \in L^2(0, b; H^{-1}(\Omega))$, $\partial_t^\alpha u \in L^2(0, b; H_0^1(\Omega))$ such that

$$u_{\hat{m}} \rightharpoonup u \text{ in } L^2(0, b; H_0^1(\Omega)), \quad u_{\hat{m}} \rightharpoonup^* u \text{ in } L^\infty(0, b; L^2(\Omega)), \tag{3.50}$$

$$\partial_t u_{\hat{m}} \rightharpoonup \partial_t u \text{ in } L^2(0, b; H^{-1}(\Omega)) \tag{3.51}$$

and

$$\partial_t^\alpha u_{\hat{m}} \rightharpoonup \partial_t^\alpha u \text{ in } L^2(0, b; H_0^1(\Omega)). \tag{3.52}$$

We choose a function $v \in C^1([0, b]; H_0^1(\Omega))$ with $v(b) = 0$, and substitute φ_k in the first equation of (3.41) by v, which implies

$$(\partial_t u_m(t), v(t)) + ((1 + \gamma \partial_t^\alpha)\nabla u_m(t), \nabla v(t)) = (f(t), v(t)).$$

Integrating the above equation with respect to t from $t = 0$ to b, we have

$$\int_0^b (\partial_t u_m(t), v(t))dt + \int_0^b ((1 + \gamma \partial_t^\alpha)\nabla u_m(t), \nabla v(t))dt = \int_0^b (f(t), v(t))dt. \tag{3.53}$$

Choose $m = \hat{m}$ and on account of (3.50)-(3.52), in the limiting case $\hat{m} \to \infty$ we get the following equation

$$\int_0^b (\partial_t u(t), v(t))dt + \int_0^b ((1 + \gamma \partial_t^\alpha)\nabla u(t), \nabla v(t))dt = \int_0^b (f(t), v(t))dt. \tag{3.54}$$

And then, from the fact that $C^1([0, b]; H_0^1(\Omega))$ is dense in $L^2(0, b; H_0^1(\Omega))$, we conclude that (3.54) holds for all functions in $L^2(0, b; H_0^1(\Omega))$. Particularly, for each $v \in H_0^1(\Omega)$, the following equation holds:

$$(\partial_t u(t), v) + ((1 + \gamma \partial_t^\alpha)\nabla u(t), \nabla v) = (f(t), v), \quad \text{a.e. } 0 \le t \le b. \tag{3.55}$$

It remains to prove that $u(0) = h$. From the facts that $u \in L^2(0, b; H_0^1(\Omega))$ and $\partial_t u \in L^2(0, b; H^{-1}(\Omega))$, we have $u \in C([0, b]; L^2(\Omega))$, which implies that the function $u(t)$ is well-defined at $t = 0$.

Integrating (3.53) by parts gives

$$\int_0^b -(u_m(t), \partial_t v(t))dt + \int_0^b ((1 + \gamma \partial_t^\alpha)\nabla u_m(t), \nabla v(t))dt$$

$$= \int_0^b (f(t), v(t))dt + (u_m(0), v(0)).$$

Let $m = \hat{m}$, from (3.50)-(3.52), in the case that $\hat{m} \to \infty$, we conclude from the fact $u_{\hat{m}}(0) \to h$ in $L^2(\Omega)$ that

$$\int_0^b -(u(t), \partial_t v(t))dt + \int_0^b ((1 + \gamma \partial_t^\alpha)\nabla u(t), \nabla v(t))dt = \int_0^b (f(t), v(t))dt + (h, v(0)). \tag{3.56}$$

Similarly, (3.54) gives that

$$\int_0^b -(u(t), \partial_t v(t))dt + \int_0^b ((1 + \gamma\partial_t^\alpha)\nabla u(t), \nabla v(t))dt$$
$$= \int_0^b (f(t), v(t))dt + (u(0), v(0)). \tag{3.57}$$

Comparing (3.56) with (3.57), we can easy to get that

$$(u(0), v(0)) = (h, v(0)),$$

for the reason that $v(0)$ is arbitrarily selected, we conclude

$$u(0) = h.$$

Therefore, u satisfies Definition 3.8, which implies that u is a weak solution of the problem (3.37). The proof is completed. $\qquad\square$

In the following, we prove the uniqueness of weak solutions of the problem (3.37).

Theorem 3.10. *Assume the conditions in Theorem 3.9 hold. Then the weak solutions of the problem (3.37) is unique.*

Proof. Suppose that u_1, u_2 are two weak solutions of (3.37). If we denote by $u = u_1 - u_2$, then u satisfies

$$\begin{cases} (\partial_t u, v) + ((1 + \gamma\partial_t^\alpha)\nabla u, \nabla v) = 0, \\ u(0) = 0. \end{cases} \tag{3.58}$$

Thus the purpose of getting the uniqueness of weak solutions of the problem (3.37) is converted to check that $u \equiv 0$.

We choose $v = u$ in (3.58), that is,

$$(\partial_t u(t), u(t)) + (\nabla u(t), \nabla u(t)) + \gamma(\partial_t^\alpha \nabla u(t), \nabla u(t)) = 0.$$

According to the fact $2(\partial_t u(t), u(t)) = \frac{d}{dt}(u(t), u(t))$ and by virtue of Lemma 3.18, we get

$$\frac{1}{2}\frac{d}{dt}(u(t), u(t)) + (\nabla u(t), \nabla u(t)) = -\gamma(\partial_t^\alpha \nabla u(t), \nabla u(t))$$
$$\le -\frac{\gamma}{2}{}_0 D_t^\alpha(\nabla u(t), \nabla u(t)),$$

which implies that

$$\frac{d}{dt}(u(t), u(t)) \le -\gamma_0 D_t^\alpha(\nabla u(t), \nabla u(t)).$$

Integrating both sides of the above inequality from 0 to t, we have

$$(u(t), u(t)) \le -\gamma_0 D_t^{\alpha-1}(\nabla u(t), \nabla u(t)).$$

It is easy to see that the right-hand side of the above inequality is less than 0, that is,

$$(u(t), u(t)) \le 0,$$

which implies that $u \equiv 0$.

Therefore, the uniqueness of weak solutions of the problem (3.37) is concluded. The proof is completed. $\qquad\square$

In the following, under the improved regularity conditions, we shall discuss the spacial regularity of weak solutions of problem (3.37).

Theorem 3.11. *Assume the conditions in Theorem 3.9 hold, and u is the weak solution of the problem* (3.37). *If $h \in H^2(\Omega)$ and $f \in L^2(0, b; L^2(\Omega))$, then we have*

$$u \in L^\infty(0, b; H_0^1(\Omega)) \bigcap L^2(0, b; H^2(\Omega)),$$

$$\partial_t^\alpha u \in L^2(0, b; H^2(\Omega)), \ \partial_t u \in L^2(0, b; L^2(\Omega)),$$

and the following estimation holds

$$\begin{aligned}
&\|u\|_{L^\infty(0,b;H_0^1(\Omega))} + \|u\|_{L^2(0,b;H^2(\Omega))} \\
&+ \|\partial_t u\|_{L^2(0,b;L^2(\Omega))} + \|\partial_t^\alpha u\|_{L^2(0,b;H^2(\Omega))} \\
&\leq \tilde{C}(\|f\|_{L^2(0,b;L^2(\Omega))} + \|h\|_{H^2(\Omega)}),
\end{aligned} \tag{3.59}$$

where $\tilde{C}$ is a positive constant depending only on γ, b and α.

Proof. From the fact that u is the weak solution of the problem (3.37), we know that u_m satisfies (3.41) under the form (3.40). Now we estimate that $u \in L^2(0, b; H^2(\Omega)) \cap L^\infty(0, b; H_0^1(\Omega))$.

Multiplying the first equation in (3.41) by $\lambda_k d_m^k(t)$ and summing it from $k = 1$ to m, we have

$$\begin{aligned}
&(\partial_t u_m(t), \Delta u_m(t)) + (\nabla u_m(t), \nabla \Delta u_m(t)) + \gamma(\partial_t^\alpha \nabla u_m(t), \nabla \Delta u_m(t)) \\
&= (f(t), \Delta u_m(t)),
\end{aligned}$$

which implies

$$\begin{aligned}
&(\partial_t \nabla u_m(t), \nabla u_m(t)) + (\Delta u_m(t), \Delta u_m(t)) + \gamma(\partial_t^\alpha \Delta u_m(t), \Delta u_m(t)) \\
&= -(f(t), \Delta u_m(t)).
\end{aligned} \tag{3.60}$$

From Lemma 3.18 and the fact $(\partial_t v(t), v(t)) = \frac{1}{2}\frac{d}{dt}(v(t), v(t))$, we can estimate the left-hand side of (3.60) as following

$$\begin{aligned}
&(\partial_t \nabla u_m(t), \nabla u_m(t)) + (\Delta u_m(t), \Delta u_m(t)) + \gamma(\partial_t^\alpha \Delta u_m(t), \Delta u_m(t)) \\
&\geq \frac{1}{2}\frac{d}{dt}\|\nabla u_m(t)\|^2_{L^2(\Omega)} + \|\Delta u_m(t)\|^2_{L^2(\Omega)} + \frac{\gamma}{2}{}_0 D_t^\alpha \|\Delta u_m(t)\|^2_{L^2(\Omega)}.
\end{aligned}$$

Applying the Hölder's inequality and the Young's inequality to the right-hand side of (3.60), we can get that

$$\begin{aligned}
|-(f(t), \Delta u_m(t))| &\leq \|f(t)\|_{L^2(\Omega)}\|\Delta u_m(t)\|_{L^2(\Omega)} \\
&\leq \frac{1}{2}\|f(t)\|^2_{L^2(\Omega)} + \frac{1}{2}\|\Delta u_m(t)\|^2_{L^2(\Omega)}.
\end{aligned}$$

Indeed, the above discussion and (3.60) imply that

$$\frac{d}{dt}\|\nabla u_m(t)\|^2_{L^2(\Omega)} + \|\Delta u_m(t)\|^2_{L^2(\Omega)} + \gamma\, {}_0 D_t^\alpha \|\Delta u_m(t)\|^2_{L^2(\Omega)} \leq \|f(t)\|^2_{L^2(\Omega)},$$

and then, integrating the above inequality with respect to the time variable from 0 to t, we obtain that

$$\|\nabla u_m(t)\|^2_{L^2(\Omega)} + \int_0^t \|\Delta u_m(s)\|^2_{L^2(\Omega)} ds + \frac{\gamma}{\Gamma(1-\alpha)} \int_0^t (t-s)^{-\alpha} \|\Delta u_m(s)\|^2_{L^2(\Omega)} ds$$
$$\leq \int_0^t \|f(s)\|^2_{L^2(\Omega)} ds + \|\nabla h\|^2_{L^2(\Omega)}.$$

Taking the maximum of $t \in [0, b]$ and applying the fact that $(t-s)^{-\alpha} \geq t^{-\alpha} \geq b^{-\alpha}$ for $0 \leq s < t \leq b$, we have

$$\|\nabla u_m\|^2_{L^\infty(0,b;L^2(\Omega))} + \left(1 + \frac{\gamma b^{-\alpha}}{\Gamma(1-\alpha)}\right) \|\Delta u_m\|^2_{L^2(0,b;L^2(\Omega))}$$
$$\leq \|f\|^2_{L^2(0,b;L^2(\Omega))} + \|h\|^2_{H^1_0(\Omega)}.$$

Taking $m = \hat{m}$ in the above inequality, and passing $\hat{m} \to \infty$, we can easy to get the same bounds of u and the conclusion

$$u \in L^\infty(0, b; H^1_0(\Omega)) \cap L^2(0, b; H^2(\Omega)).$$

We estimate that $\partial_t u \in L^2(0, b; L^2(\Omega))$ in the sequel.

Multiplying both sides of the first equation of (3.41) by $(d_m^k)'$ and summing it from $k = 1$ to m show that

$$(\partial_t u_m(t), \partial_t u_m(t)) + \gamma(\partial_t^\alpha \nabla u_m(t), \nabla \partial_t u_m(t)) + (\nabla u_m(t), \nabla \partial_t u_m(t))$$
$$= (f(t), \partial_t u_m(t)).$$

In view of the Hölder's inequality, the Young's inequality and the fact that $2(\partial_t v(t), v(t)) = \frac{d}{dt}(v(t), v(t))$, we have

$$\|\partial_t u_m(t)\|^2_{L^2(\Omega)} + \gamma({}_0 D_t^{\alpha-1} \partial_t \nabla u_m(t), \nabla \partial_t u_m(t)) + \frac{1}{2} \frac{d}{dt} \|\nabla u_m(t)\|^2_{L^2(\Omega)}$$
$$\leq \|f(t)\|_{L^2(\Omega)} \|\partial_t u_m(t)\|_{L^2(\Omega)} - \gamma(\omega_{1-\alpha}(t) \nabla u_m(0), \nabla \partial_t u_m(t))$$
$$\leq \|f(t)\|^2_{L^2(\Omega)} + \frac{1}{4} \|\partial_t u_m(t)\|^2_{L^2(\Omega)} + \gamma \omega_{1-\alpha}(t) \|\Delta u_m(0)\|_{L^2(\Omega)} \|\partial_t u_m(t)\|_{L^2(\Omega)}$$
$$\leq \|f(t)\|^2_{L^2(\Omega)} + \frac{1}{2} \|\partial_t u_m(t)\|^2_{L^2(\Omega)} + \gamma^2 \omega^2_{1-\alpha}(t) \|\Delta u_m(0)\|^2_{L^2(\Omega)},$$

which implies

$$\|\partial_t u_m(t)\|^2_{L^2(\Omega)} + 2\gamma({}_0 D_t^{\alpha-1} \partial_t \nabla u_m(t), \nabla \partial_t u_m(t)) + \frac{d}{dt} \|\nabla u_m(t)\|^2_{L^2(\Omega)}$$
$$\leq 2\|f(t)\|^2_{L^2(\Omega)} + 2\gamma^2 \omega^2_{1-\alpha}(t) \|\Delta u_m(0)\|^2_{L^2(\Omega)}. \tag{3.61}$$

Integrating (3.61) for both sides from 0 to t, we conclude from Lemma 3.19 that

$$\int_0^t \|\partial_s u_m(s)\|^2_{L^2(\Omega)} ds + 2\gamma \cos\left(\frac{1-\alpha}{2}\pi\right) \|{}_0 D_t^{\frac{\alpha-1}{2}} \partial_t \nabla u_m(t)\|^2_{L^2(\Omega)}$$
$$+ \|\nabla u_m(t)\|^2_{L^2(\Omega)}$$

$$\leq 2 \int_0^t \|f(s)\|^2_{L^2(\Omega)} ds + \frac{2\gamma^2 t^{1-2\alpha}}{(1-2\alpha)\Gamma^2(1-\alpha)} \|\Delta u_m(0)\|^2_{L^2(\Omega)} + \|\nabla u_m(0)\|^2_{L^2(\Omega)}.$$

Then taking the maximum of $t \in [0, b]$ of the above inequality implies that

$$\|\partial_t u_m\|^2_{L^2(0,b;L^2(\Omega))} + 2\gamma \cos\left(\frac{1-\alpha}{2}\pi\right) \|{}^C \partial_t^{\frac{1+\alpha}{2}} \nabla u_m\|^2_{L^2(0,b;L^2(\Omega))}$$

$$+ \|\nabla u_m\|^2_{L^\infty(0,b;L^2(\Omega))}$$

$$\leq 2\|f\|^2_{L^2(0,b;L^2(\Omega))} + \frac{2\gamma^2 b^{1-2\alpha}}{(1-2\alpha)\Gamma^2(1-\alpha)} \|\Delta u_m(0)\|^2_{L^2(\Omega)} + \|\nabla u_m(0)\|^2_{L^2(\Omega)}$$

$$\leq 2\|f\|^2_{L^2(0,b;L^2(\Omega))} + \frac{2\gamma^2 b^{1-2\alpha}}{(1-2\alpha)\Gamma^2(1-\alpha)} \|h\|^2_{H^2(\Omega)} + \|h\|^2_{H_0^1(\Omega)}.$$

From the above discussion, choose $m = \hat{m}$ and let $\hat{m} \to \infty$, we conclude that $\partial_t u \in L^2(0, b; L^2(\Omega))$.

In the following, we estimate $\partial_t^\alpha u \in L^2(0, b; H^2(\Omega))$.

Multiplying the first equation in (3.41) by $\lambda_k \, {}_0 D_t^\alpha d_m^k(t)$ and summing it from $k = 1$ to m, we have

$$\gamma(\partial_t^\alpha \Delta u_m(t), \partial_t^\alpha \Delta u_m(t)) + (\Delta u_m(t), \partial_t^\alpha \Delta u_m(t))$$
$$= (\partial_t u_m(t), \partial_t^\alpha \Delta u_m(t)) - (f(t), \partial_t^\alpha \Delta u_m(t)).$$

By virtue of Lemma 3.18, the Hölder's inequality and the Young's inequality, we have

$$\gamma\|\partial_t^\alpha \Delta u_m(t)\|^2_{L^2(\Omega)} + \frac{1}{2}{}_0 D_t^\alpha \|\Delta u_m(t)\|^2_{L^2(\Omega)}$$

$$\leq \|\partial_t u_m(t)\|_{L^2(\Omega)} \|\partial_t^\alpha \Delta u_m(t)\|_{L^2(\Omega)} + \|f(t)\|_{L^2(\Omega)} \|\partial_t^\alpha \Delta u_m(t)\|_{L^2(\Omega)}$$

$$\leq \frac{1}{\gamma}\|\partial_t u_m(t)\|^2_{L^2(\Omega)} + \frac{1}{\gamma}\|f(t)\|^2_{L^2(\Omega)} + \frac{\gamma}{2}\|\partial_t^\alpha \Delta u_m(t)\|^2_{L^2(\Omega)},$$

that is,

$$\gamma\|\partial_t^\alpha \Delta u_m(t)\|^2_{L^2(\Omega)} + {}_0 D_t^\alpha \|\Delta u_m(t)\|^2_{L^2(\Omega)} \leq \frac{2}{\gamma}\|\partial_t u_m(t)\|^2_{L^2(\Omega)} + \frac{2}{\gamma}\|f(t)\|^2_{L^2(\Omega)}.$$

Integrating both sides of the above inequality for $s \in [0, t]$, and taking the maximum of $t \in [0, b]$, we have

$$\gamma\|\partial_t^\alpha \Delta u_m\|^2_{L^2(0,b;L^2(\Omega))} + \frac{1}{\Gamma(1-\alpha)}\int_0^b (b-s)^{-\alpha}\|\Delta u_m(s)\|^2_{L^2(\Omega)} ds$$

$$\leq \frac{2}{\gamma}\|\partial_t u_m\|^2_{L^2(0,b;L^2(\Omega))} + \frac{2}{\gamma}\|f\|^2_{L^2(0,b;L^2(\Omega))},$$

let $m = \hat{m}$, in the limiting case $\hat{m} \to \infty$, we deduce that $\partial_t^\alpha u \in L^2(0, b; H^2(\Omega))$.

Combined with the above discussion, we conclude that (3.59) holds. This completes the proof of the theorem. $\qquad\square$

3.5 Global Solutions in Besov-Morrey Spaces

3.5.1 *Introduction*

The two of the most important mathematical and physical models among non-Newtonian fluids are the Stokes's first problem for a flat plate as well as the Rayleigh-Stokes problem for an edge. We refer to [122–124, 275, 314] for details about velocity and temperature of the Rayleigh-Stokes problems.

In this section, we study the nonlinear Rayleigh-Stokes equations which involve the Riemann-Liouville fractional derivative operator (in time) as follows:

$$\begin{cases} \partial_t u - (1 + \gamma_0 \partial_t^\alpha)\Delta u = f(u), & t > 0, \ x \in \mathbb{R}^N, \\ u(0, x) = u_0(x), & x \in \mathbb{R}^N, \end{cases} \quad (3.62)$$

where $\gamma_0 > 0$ is a constant, Δ is the Laplacian operator, ∂_t^α is the Riemann-Liouville fractional derivative of order $\alpha \in (0, 1)$ which is defined by

$$\partial_t^\alpha u(t, x) = \frac{1}{\Gamma(1 - \alpha)} \frac{\partial}{\partial t} \int_0^t (t - s)^{-\alpha} u(s, x) ds, \quad t > 0, \ x \in \mathbb{R}^N,$$

provided the right-handed side of the equality is pointwise defined.

In order to gain insights into the behavior of solutions of the model (3.62), there has been substantial interest in a form solution and its existence (see e.g., [39, 40, 248, 280, 335, 338] and references therein). For instance, some recent results on local existence in time in a bounded domain $\Omega \subset \mathbb{R}^N$ have been obtained in [210, 409], whose solutions obtained in these studies involve infinite series and special functions, and thus are inconvenient for deriving the properties of solutions in unbounded domains. To the best of the authors' knowledge, there are limited articles treating. with solvability of fractional Rayleigh-Stokes equations in the whole space $\mathbb{R}^N$. We would like to mention that He et al. [163] studied the qualitative properties relied on the Gagliardo-Nirenberg's inequalities, like global/local well-posedness, blow-up results and integrability of mild solutions for small initial value in L^q. Very recently, our proceeding paper [297] obtained two global existence results in the framework of Besov type with the initial value $u_0 \in \dot{B}_{p,\infty}^{N/p-2/(\rho(1-\alpha))}$. However, it remains to take more efforts to find a solution space as large as possible.

Theoretical studies on well-posedness analyses for differential equations involving fractional derivatives have received considerable attention in the last decade, see [155, 298, 322, 360, 383]. Recent results on global solvability of fractional diffusion cases [64, 284, 334] tell us that solutions consisting of C_0-semigroup and the Wright-type function is a key point in deriving the properties of solutions operators in different spaces such as L^q, Besov spaces and Besov-Morrey spaces. Though the solution operator of the considered equation (3.62) can be established in the form:

$$S(t) = \int_0^\infty \phi(t, \tau) T(\tau) d\tau, \quad t > 0,$$

where $T(t)$ is a C_0-semigroup and $\phi(t, \tau)$ is a probability density, it is a difficulty for discriminating convergence of the integral $\int_0^\infty \phi(t, \tau) \tau^{-\varrho} d\tau$ for $\varrho \in (0, 1)$ and $t > 0$.

Therefore, it seems to be infeasible for the equation (3.62) along the way as we studied fractional diffusion case based on C_0-semigroup and thus it is imperative to develop several suitable approaches deriving the properties of its solution operator.

For $\rho > 0$, $\lambda = N - \frac{2p}{\rho(1-\alpha)}$ and $\mu = N - \frac{2}{\rho(1-\alpha)}$ one has the continuous inclusions

$$L^q \subsetneq \mathcal{M}_{p,\lambda} \subsetneq \mathcal{N}^{s_0}_{p,\mu,\infty} \quad \text{and} \quad \dot{B}^{k_0}_{r,\infty} \subset \mathcal{N}^{s_0}_{p,\mu,\infty}$$

provided $\frac{N}{q} = \frac{N-\lambda}{p} = -s_0 + \frac{N-\mu}{p} = -k_0 + \frac{N}{r}$, where $s_0 = \frac{N-\mu}{p} - \frac{2}{\rho(1-\alpha)}$, $k_0 = \frac{N}{r} - \frac{2}{\rho(1-\alpha)}$ and $1 \le p < r \le q \le \frac{\rho N(1-\alpha)}{2}$. Moreover, there are no results on the global solvability of fractional Rayleigh-Stokes equations in a class such that $Z \supsetneq \mathcal{N}^{s_0}_{p,\mu,\infty}$. In this sense, we provide a long time existence of solutions for the equation (3.62). The aims of this section is to establish the global well-posedness and asymptotic behavior of solutions for the equation (3.62) in the framework of Besov-Morrey spaces. The main features of our work are three aspects. The first is that our results are improved compared with the existence results obtained in [297]. Furthermore, the approach is also different which depends on real interpolation, a multiplier and some inequalities such as the Gagliardo-Nirenberg's inequalities and Sobolev-type embedding. The last but not the least is the selection of the initial value that allows for a larger range than the previous works.

The rest of the section is organized as follows. In the next subsection, we introduce some notations, function spaces that will be used throughout the section. In Subsection 3.5.3, the several estimates of resolvent operators and solution operators in Besov-Morrey spaces are presented, respectively. In Subsection 3.5.4, the existence, uniqueness and continuous dependence on initial value of global solutions for nonlinear fractional Rayleigh-Stokes equations are investigated in Besov-Morrey spaces. Further, we also establish the asymptotic behavior of solutions at infinity.

3.5.2 Notations and Function Spaces

Here, we recall definitions of function spaces which are main tools for estimates in the following sections.

Throughout this section we use the following notations. $\mathcal{S}$ and $\mathcal{S}'$ denote the Schwartz spaces of rapidly decreasing smooth functions and tempered distributions, respectively. Let $(X, |\cdot|)$ be a Banach space. We denote by $C_b(I, X)$ the space of bounded continuous operators from $I \subset \mathbb{R}_+$ to X, equipped with the norm $\sup_{t \in I} |\cdot|$ and

$$C_*(I, X) = \{v \in C(I, X) : \|v\|_{C_*(I,X)} = \sup_{t \in I} t^{\frac{1-\alpha}{\vartheta}} |v(t)| < \infty\}$$

for $\vartheta > 1$.

Now we recall the definition of Sobolev-Morrey spaces. Let $U_\varepsilon(x_0)$ be the open ball in $\mathbb{R}^N$ centered at x_0 and with radius $\varepsilon > 0$ and let $1 \le p < \infty$ and $0 \le \mu < N$, the Morrey spaces $\mathcal{M}_{p,\mu} = \mathcal{M}_{p,\mu}(\mathbb{R}^N)$ is defined to be the set of functions $v \in L_p(U_\varepsilon(x_0))$ such that

$$\|v\|_{\mathcal{M}_{p,\mu}} = \sup_{x_0 \in \mathbb{R}^N} \sup_{\varepsilon > 0} \varepsilon^{-\frac{\mu}{p}} \|v\|_{L_p(U_\varepsilon(x_0))} < \infty.$$

For $s \in \mathbb{R}$ and $1 \leq p < \infty$, the homogeneous Sobolev-Morrey space $\mathcal{M}_{p,\mu}^s = (-\Delta)^{-\frac{s}{2}}\mathcal{M}_{p,\mu}$ with norm $\|v\|_{\mathcal{M}_{p,\mu}^s} = \|(-\Delta)^{\frac{s}{2}}v\|_{\mathcal{M}_{p,\mu}}$. Especially, we have $\mathcal{M}_{p,0} = L^p$ and $\mathcal{M}_{p,0}^s = \dot{H}_p^s$ is the well known Sobolev space for $p > 1$. Moreover, $(-\Delta)^{\frac{l}{2}}\mathcal{M}_{p,\mu}^s = \mathcal{M}_{p,\mu}^{s-l}$ for $0 \leq l \leq s$.

Let $\{\varphi_j\}_{j\in\mathbb{Z}}$ be the Littlewood-Paley decomposition. We take a function $\phi \in C_0^\infty(\mathbb{R}^N)$ with $supp(\phi) = \{\xi \in \mathbb{R}^N : \frac{1}{2} \leq |\xi| \leq 2\}$ such that $\sum_{j\in\mathbb{Z}} \phi(2^{-j}\xi) = 1$ for all $\xi \neq 0$. The functions φ_j and ψ_j are defined by

$$\widehat{\varphi_j}(\xi) = \phi(2^{-j}\xi), \quad \widehat{\psi_j}(\xi) = 1 - \sum_{j=1}^\infty \phi(2^{-j}\xi).$$

For $s \in \mathbb{R}$, $1 \leq p < \infty$, $1 \leq r \leq \infty$ and $0 \leq \mu < N$, the homogeneous Besov-Morrey space $\mathcal{N}_{p,\mu,r}^s$ is defined by

$$\mathcal{N}_{p,\mu,r}^s = \{v \in \mathcal{S}'/\mathcal{P} : \|v\|_{\mathcal{N}_{p,\mu,r}^s} < \infty\},$$

where the set $\mathcal{S}'/\mathcal{P}$ consists in equivalence classes in $\mathcal{S}'$ modulo polynomials, and

$$\|v\|_{\mathcal{N}_{p,\mu,r}^s} = \begin{cases} \left(\sum_{j\in\mathbb{Z}}(2^{sj}\|\varphi_j * v\|_{\mathcal{M}_{p,\mu}})^r\right)^{1/r}, & 1 \leq r < \infty, \\ \sup_{j\in\mathbb{Z}}(2^{sj}\|\varphi_j * v\|_{\mathcal{M}_{p,\mu}}), & r = \infty. \end{cases}$$

In particular, $\mathcal{N}_{p,0,r}^s = \dot{B}_{p,r}^s$ is the homogeneous Besov space when $\mu = 0$.

Let $0 \leq \mu < N$ and $1 \leq p < \infty$. We recall inclusion relations between Morrey spaces and Besov-Morrey spaces (see [207, 328])

$$\mathcal{N}_{p,\mu,1}^0 \subset \mathcal{M}_{p,\mu} \subset \mathcal{N}_{p,\mu,\infty}^0. \tag{3.63}$$

Let $p_j, s_j, j = 1, 2$ such that $p_2 \leq p_1$, $s_1 \leq s_2$ and $s_2 - \frac{N-\mu}{p_2} = s_1 - \frac{N-\mu}{p_1}$, we obtain the Sobolev-type embedding as follows

$$\mathcal{M}_{p_2,\mu}^{s_2} \subset \mathcal{M}_{p_1,\mu}^{s_1} \quad \text{and} \quad \mathcal{N}_{p_2,\mu,\infty}^{s_2} \subset \mathcal{N}_{p_1,\mu,\infty}^{s_1}. \tag{3.64}$$

For $1 < p, q < \infty$, we define the space based on Besov-Morrey type spaces as

$$Z_{p,q} = C_b((0,\infty); \mathcal{N}_{p,\mu,\infty}^{s_0}) \cap C_*((0,\infty); \mathcal{M}_{q,\mu}),$$

which is a Banach space endowed with the norm

$$\|v\|_{Z_{p,q}} = \sup_{t\in(0,\infty)} \|v(t)\|_{\mathcal{N}_{p,\mu,\infty}^{s_0}} + \sup_{t\in(0,\infty)} t^{\frac{1-\alpha}{\vartheta}}\|v(t)\|_{\mathcal{M}_{q,\mu}}.$$

Here $\vartheta > 1$ and s_0 are given by

$$\frac{1}{\vartheta} = \frac{1}{\rho(1-\alpha)} - \frac{N-\mu}{2q} \quad \text{and} \quad s_0 = \frac{N-\mu}{p} - \frac{2}{\rho(1-\alpha)}.$$

We present an estimate for a multiplier operator on $\mathcal{M}_{p,\mu}^s$ (see [207]).

Lemma 3.24. *Let $m, s \in \mathbb{R}$ and $0 \leq \mu < N$ and $P(\xi) \in C^{[N/2]+1}(\mathbb{R}^N \setminus \{0\})$. Assume that there is $C > 0$ such that*

$$\left|\frac{\partial^k}{\partial\xi^k}P(\xi)\right| \leq C|\xi|^{m-|k|},$$

for all $k \in (\mathbb{N}_0)^N$ with $|k| \leq [N/2] + 1$ and for all $\xi \neq 0$. Then the multiplier operator $P(D)v = \mathcal{F}^{-1}[P(\xi)\widehat{v}]$ on $\mathcal{S}'/\mathcal{P}$ satisfies the estimate

$$\|P(D)v\|_{\mathcal{M}_{p,\mu}^{s-m}} \leq MC\|v\|_{\mathcal{M}_{p,\mu}^s},$$

for $1 < p < \infty$, where $M > 0$ is a constant independent of v.

We finish this subsection recalling an elementary fixed point lemma whose proof can be found in [121].

Lemma 3.25. *Let $(X, |\cdot|)$ be a Banach space and $1 < \rho < \infty$. Suppose that $W : X \to X$ satisfies $W(0) = 0$ and*

$$|W(u) - W(v)| \leq L|u - v|(|u|^\rho + |v|^\rho).$$

Let $R > 0$ be the unique positive root of $2^{\rho+1}LR^\rho - 1 = 0$. Given $0 < \varepsilon < R$ and $\varphi \in X$ such that $|\varphi| \leq \varepsilon$, there exists a solution $u \in X$ for the equation $u = \varphi + W(u)$ which is the unique one in the closed ball $\{v \in X : |v| \leq 2\varepsilon\}$. Moreover, if $|\bar{\varphi}| \leq \varepsilon$ and $|\bar{u}| \leq 2\varepsilon$ satisfies the equation $\bar{u} = \bar{\varphi} + W(\bar{u})$ then

$$|u - \bar{u}| \leq \frac{1}{1 - 2^{\rho+1}L\varepsilon^\rho}(|\varphi - \bar{\varphi}|).$$

3.5.3 *Properties of Solution Operators*

The goal of this subsection is to derive several estimates of solution operators for the fractional Rayleigh-Stokes equations on Sobolev-Morrey, Besov-Morrey and Besov spaces, respectively.

Let $1 < p < \infty$. We consider the operator A in $L^p(\mathbb{R}^N)$ defined by $A = -\Delta$ with the domain

$$D(A) = \{u \in L^p(\mathbb{R}^N) : \Delta u \in L^p(\mathbb{R}^N)\}.$$

It follows from [63, Theorem 2.3.2] that $-A$ generates a bounded analytic semigroup $T(t)$ and the spectrum is equal to $(-\infty, 0)$. Let $\Sigma_{\pi-\sigma} = \{z \in \mathbb{C} : z \neq 0, |\text{arg}z| < \pi - \sigma\}$ for $\sigma \in (0, \pi/2)$. Define $G_z(x)$ by $\widehat{G_z}(\xi) := \frac{1}{z+|\xi|^2}$. Then we have that for $v \in L^p(\mathbb{R}^N)$,

$$(z + A)^{-1}v = (4\pi)^{-\frac{N}{2}} \int_{\mathbb{R}^N} e^{ix\cdot\xi}\widehat{G_z}(\xi)\hat{v}(\xi)d\xi, \quad \text{for } x \in \mathbb{R}^N \text{ and } z \in \Sigma_{\pi-\sigma}.$$

We specify several estimates of the operator $(z + A)^{-1}$ on $\mathcal{M}^s_{p,\mu}$ and $\mathcal{N}^s_{p,\mu,r}$, which are playing an important role in proving properties of the solution operators. Here and below the letter M_1 will denote constants which can change from line to line.

Lemma 3.26. *Let $s, \beta \in \mathbb{R}$ with $s \leq \beta$, $1 < p \leq q < \infty$ and $0 \leq \mu < N$. Assume $z \in \Sigma_{\pi-\sigma}$.*

(i) *Let $\beta - s + (\frac{N-\mu}{p} - \frac{N-\mu}{q}) \leq 2$. There exists a constant $M_1 > 0$ such that*

$$\|(z + A)^{-1}v\|_{\mathcal{M}^\beta_{q,\mu}} \leq M_1|z|^{\frac{\beta-s}{2}+\frac{1}{2}(\frac{N-\mu}{p}-\frac{N-\mu}{q})-1}\|v\|_{\mathcal{M}^s_{p,\mu}}, \tag{3.65}$$

  for $v \in \mathcal{M}^s_{p,\mu}$.

(ii) *Let $1 \leq r \leq \infty$ and $\beta - s + (\frac{N-\mu}{p} - \frac{N-\mu}{q}) \leq 2$. There exists a constant $M_1 > 0$ such that*

$$\|(z + A)^{-1}v\|_{\mathcal{N}^\beta_{q,\mu,r}} \leq M_1|z|^{\frac{\beta-s}{2}+\frac{1}{2}(\frac{N-\mu}{p}-\frac{N-\mu}{q})-1}\|v\|_{\mathcal{N}^s_{p,\mu,r}}, \tag{3.66}$$

  for $v \in \mathcal{S}'/\mathcal{P}$.

(iii) *Let $1 \leq r \leq \infty$ and $s < \beta$ be such that $\beta - s + (\frac{N-\mu}{p} - \frac{N-\mu}{q}) < 2$. There exists a constant $M_1 > 0$ such that*

$$\|(z+A)^{-1}v\|_{\mathcal{N}^{\beta}_{q,\mu,1}} \leq M_1 |z|^{\frac{\beta-s}{2}+\frac{1}{2}(\frac{N-\mu}{p}-\frac{N-\mu}{q})-1}\|v\|_{\mathcal{N}^s_{p,\mu,r}}$$

for $v \in \mathcal{S}'/\mathcal{P}$.

Proof. (i) Let $1 < q < \infty$ and $\theta \in [0,2]$. Since

$$\frac{\partial^k}{\partial\xi^k}\frac{|\xi|^\theta}{1+|\xi|^2} \leq C|\xi|^{-k}, \quad \xi \neq 0,$$

for $k \in (\mathbb{N}_0)^N$ with $|k| \leq [N/2]+1$. Using Lemma 3.24, it follows that $A^{\frac{\theta}{2}}G_1 v = \mathcal{F}^{-1}[\frac{|\xi|^\theta}{1+|\xi|^2}\hat{v}(\xi)]$ satisfies the estimate

$$\|A^{\frac{\theta}{2}}G_1 v\|_{\mathcal{M}^s_{q,\mu}} \leq MC\|v\|_{\mathcal{M}^s_{q,\mu}},$$

where M may depend on q.

Let $h(\xi,z) := \widehat{A^{\frac{\theta}{2}}G_z} = \frac{|\xi|^\theta}{z+|\xi|^2}$, we have by changing variable $\xi \to \eta = \frac{\xi}{\sqrt{|z|}}$ that

$$h(\xi,|z|) = |z|^{\frac{\theta}{2}-1}\frac{|\eta|^\theta}{1+|\eta|^2} = |z|^{\frac{\theta}{2}-1}h(\frac{\xi}{\sqrt{|z|}},1).$$

We can calculate immediately to derive the following relation

$$
\begin{aligned}
A^{\frac{\theta}{2}}(z+A)^{-1}v &= \mathcal{F}^{-1}[h(\xi,|z|)\hat{v}(\xi)](x)\\
&= |z|^{\frac{\theta}{2}+\frac{N}{2}-1}\mathcal{F}^{-1}[h(\eta,1)\hat{v}(\sqrt{|z|}\eta)](\sqrt{|z|}x)\\
&= |z|^{\frac{\theta}{2}-1}\mathcal{F}^{-1}[h(\cdot,1)\widehat{v_{|z|^{-1/2}}}(\cdot)](\sqrt{|z|}x)\\
&= |z|^{\frac{\theta}{2}-1}[A^{\frac{\theta}{2}}(1+A)^{-1}v_{|z|^{1/2}}]_{|z|^{-1/2}}(x),
\end{aligned}
$$

where we have used $\widehat{v_\lambda}(\xi) = \lambda^{-N}\hat{v}(\frac{\xi}{\lambda})$ for $v_\lambda(x) := v(\lambda x)$. Recalling that $\|v(\lambda\cdot)\|_{\mathcal{M}^s_{q,\mu}} = \lambda^{s-\frac{N-\mu}{p}}\|v(\cdot)\|_{\mathcal{M}^s_{q,\mu}}$, it follows that

$$
\begin{aligned}
\|A^{\frac{\theta}{2}}(z+A)^{-1}v\|_{\mathcal{M}^s_{q,\mu}} &= |z|^{\frac{\theta}{2}-1-\frac{1}{2}(s-\frac{N-\mu}{p})}\|A^{\frac{\theta}{2}}(1+A)^{-1}v_{|z|^{1/2}}\|_{\mathcal{M}^s_{q,\mu}}\\
&\leq M_1 |z|^{\frac{\theta}{2}-1-\frac{1}{2}(s-\frac{N-\mu}{p})}\|v_{|z|^{1/2}}\|_{\mathcal{M}^s_{q,\mu}} \qquad (3.67)\\
&= M_1 |z|^{\frac{\theta}{2}-1}\|v\|_{\mathcal{M}^s_{q,\mu}},
\end{aligned}
$$

where $M_1 = MC$. From the Sobolev-type embedding for $l = \frac{N-\mu}{p} - \frac{N-\mu}{q}$ and (3.67) for $\theta = \beta - s + l$ one can see that

$$
\begin{aligned}
\|(z+A)^{-1}v\|_{\mathcal{M}^\beta_{q,\mu}} &= \|A^{\frac{\beta}{2}}(z+A)^{-1}v\|_{\mathcal{M}_{q,\mu}}\\
&\leq \|A^{\frac{\beta}{2}}(z+A)^{-1}v\|_{\mathcal{M}^l_{p,\mu}}\\
&= |A^{\frac{\beta-s}{2}+\frac{l}{2}}(z+A)^{-1}v\|_{\mathcal{M}^s_{p,\mu}}\\
&\leq M_1 |z|^{\frac{\beta-s}{2}+\frac{1}{2}(\frac{N-\mu}{p}-\frac{N-\mu}{q})-1}\|v\|_{\mathcal{M}^s_{p,\mu}}.
\end{aligned}
$$

Thus (3.65) holds.

(ii) By real interpolation, we see that $\mathcal{N}^s_{p,\mu,r} = (\mathcal{M}^{s_1}_{p,\mu}, \mathcal{M}^{s_2}_{p,\mu})_{\theta_1,r}$ for $0 \leq s_0 < s < s_1$ and $0 < \theta_1 < 1$ with $s = (1-\theta_1)s_0 + \theta_1 s_1$. This together with the interpolation theorem on operators shows that

$$\|(z+A)^{-1}v\|_{\mathcal{N}^{\beta}_{q,\mu,r}} \leq m_0^{1-\theta_1} m_1^{\theta_1} |z|^{\frac{N}{2}(\frac{1}{p}-\frac{1}{q})+\frac{\theta}{2}-1} \|v\|_{\mathcal{N}^s_{p,\mu,r}},$$

where $m_i = \|(z+A)^{-1}\|_{\mathcal{M}^{s_i}_{p,\mu} \to \mathcal{M}^{\beta_i}_{p,\mu}}$ for $i = 0, 1$. Using the relation (3.65), we obtain the inequality (3.66).

(iii) We choose $\varepsilon > 0$ so small that $\beta + \varepsilon - s + \frac{1}{2}(\frac{N-\mu}{p} - \frac{N-\mu}{q}) \leq 2$. Thus we can take $\theta_2 = \frac{\varepsilon}{\varepsilon+\beta-s}$. It ensures that $\theta_2 \in (0,1)$ and $\beta = (\beta+\varepsilon)(1-\theta_2) + s\theta_2$. In virtue of the real interpolation, we have $\mathcal{N}^{\beta}_{q,\mu,1} = (\mathcal{N}^{\beta+\varepsilon}_{q,\mu,\infty}, \mathcal{N}^s_{q,\mu,\infty})_{\theta_2,1}$, it follows from (3.66) that

$$\|(z+A)^{-1}v\|_{\mathcal{N}^{\beta+\varepsilon}_{q,\mu,\infty}} \leq M_1 |z|^{\frac{\beta+\varepsilon-s}{2}+\frac{1}{2}(\frac{N-\mu}{p}-\frac{N-\mu}{q})-1} \|v\|_{\mathcal{N}^s_{p,\mu,\infty}},$$

$$\|(z+A)^{-1}v\|_{\mathcal{N}^s_{q,\mu,\infty}} \leq M_1 |z|^{\frac{1}{2}(\frac{N-\mu}{p}-\frac{N-\mu}{q})-1} \|v\|_{\mathcal{N}^s_{p,\mu,\infty}}.$$

This yields that

$$\|(z+A)^{-1}v\|_{\mathcal{N}^{\beta}_{q,\mu,1}} \leq M_1 |z|^{\frac{\beta-s}{2}+\frac{1}{2}(\frac{N-\mu}{p}-\frac{N-\mu}{q})-1} \|v\|_{\mathcal{N}^s_{p,\mu,\infty}}.$$

The proof is completed. $\qquad\square$

Despite $\mathcal{M}^s_{p,0} = \dot{H}^s_p$ and $\mathcal{N}^s_{p,0,\infty} = \dot{B}^s_{p,r}$, the restriction $\beta - s + (\frac{N-\mu}{p} - \frac{N-\mu}{q}) < 2$ can be weaken to $\beta - s \leq 2$ when we consider the corresponding estimates of the operator $(z+A)^{-1}$ on Soblev and Besov spaces.

Remark 3.8. Let $N \geq 1$ and $1 < p < +\infty$. From (3.67), we know that $\|A^{\frac{\theta}{2}}(z+A)^{-1}v\|_{L_p} \leq M_1 |z|^{\frac{\theta}{2}-1} \|v\|_{L_p}$ for $\theta \in [0,2]$ and $v \in L^p(\mathbb{R}^N)$.

(i) Let $s \in \mathbb{R}$, $\gamma \in [0,2]$ and $1 < p \leq q < +\infty$. Similarly, by virtue of the Gagliardo-Nirenberg's inequality with fractional derivatives (see [344, Corollary 1.3]), we know that for $v \in L^p(\mathbb{R}^N)$,

$$\|(z+A)^{-1}v\|_{\dot{H}^{\gamma}_q} \leq \|(z+A)^{-1}v\|_{L^p}^{1-\nu} \|(z+A)^{-1}v\|_{\dot{H}^2_p}^{\nu} \leq M_1 |z|^{\nu-1} \|v\|_{L^p},$$

where $N/q - \gamma = \nu(N/p - 2) + (1-\nu)N/p$ for $\nu \in [0,1]$ and $M_1 > 0$ is a constant. Therefore, one can see that for $v \in L^p(\mathbb{R}^N)$,

$$\|(z+A)^{-1}v\|_{\dot{H}^{\gamma}_q} \leq M_1 |z|^{\frac{N}{2}(\frac{1}{p}-\frac{1}{q})+\frac{\gamma}{2}-1} \|v\|_{L^p},$$

for $\gamma \in [0,2]$ and $1 < p \leq q < +\infty$. This also implies that

$$\|(z+A)^{-1}v\|_{\dot{H}^{\beta}_q} \leq M_1 |z|^{\frac{N}{2}(\frac{1}{p}-\frac{1}{q})+\frac{\beta-s}{2}-1} \|v\|_{\dot{H}^s_p},$$

for $v \in \dot{H}^s_p$, $\gamma = \beta - s \in [0,2]$ and $1 < p \leq q < +\infty$.

(ii) Let $N \geq 1$, $s, \beta \in \mathbb{R}$ with $\beta - s \in [0,2]$, $1 \leq r \leq +\infty$ and $1 < p \leq q < +\infty$. By real interpolation, we see that $\dot{B}^s_{p,r} = (\dot{H}^{s_0}_p, \dot{H}^{s_1}_p)_{\nu_1,r}$ for $0 \leq s_0 < s < s_1$ and $0 < \nu_1 < 1$ with $s = (1-\nu_1)s_0 + \nu_1 s_1$. This together with the interpolation theorem on operators shows that

$$\|(z+A)^{-1}v\|_{\dot{B}^{\beta}_{q,r}} \leq M_1 |z|^{\frac{N}{2}(\frac{1}{p}-\frac{1}{q})+\frac{\theta}{2}-1} \|v\|_{\dot{B}^s_{p,r}},$$

for $v \in \dot{B}^s_p$, $\beta - s \in [0, 2]$ and $1 < p \le q < +\infty$.

(iii) Let $1 < p < \infty$ and $\beta - s \in (0, 2)$. If $v \in \dot{B}^s_{p,\infty}$, then there exists a constant $M_1 > 0$ such that

$$\|(z + A)^{-1}v\|_{\dot{B}^\beta_{p,1}} \le M_1 |z|^{\frac{\theta}{2} - 1} \|v\|_{\dot{B}^s_{p,\infty}}.$$

For the proof of (iii), we can see [208].

For $\delta > 0$ and $\sigma \in (0, \pi/2)$ we introduce the contour $\Gamma_{\delta,\sigma}$ defined by

$$\Gamma_{\delta,\sigma} = \{re^{-i\sigma} : r \ge \delta\} \cup \{\delta e^{i\psi} : |\psi| \le \sigma\} \cup \{re^{i\sigma} : r \ge \delta\},$$

where the circular arc is oriented counterclockwise, and the two rays are oriented with an increasing imaginary part. In the sequel, we introduce an operator $S(t)$ as

$$S(t) = \frac{1}{2\pi i} \int_{\Gamma_{\delta,\pi-\sigma}} e^{zt} H(z) dz, \quad \text{for } t > 0, \tag{3.68}$$

where

$$H(z) = \frac{\omega(z)}{z} (\omega(z) + A)^{-1}, \quad \omega(z) = \frac{z}{1 + \gamma_0 z^\alpha}.$$

Note that if $\gamma_0 = 0$, then we recover the classical heat equation

$$\partial_t u - \Delta u = f(u), \quad t > 0.$$

Thus following the notation above we have $\omega(z) = z$, $H(z) = (z - \Delta)^{-1}$. From the Cauchy's theorem, we have the identity

$$S(t)\phi(x) = \frac{1}{2\pi i} \int_{\Gamma_{\delta,\pi-\sigma}} e^{zt} (z - \Delta)^{-1} dz\phi(x) = e^{\Delta t}\phi(x) = g_t * \phi(x),$$

where g_t is the classical heat (Gaussian) kernel.

From [163], we give a concept of mild solutions of the problem (3.62).

Definition 3.9. Let $1 < p < \infty$, $0 \le \mu < N$ and $s \in \mathbb{R}$. A continuous function $u : (0, \infty) \to \mathcal{N}^s_{p,\mu,\infty}$ satisfies

$$u(t) = S(t)u_0 + \int_0^t S(t - \tau)f(u(\tau))d\tau, \quad t > 0,$$

we say that u is a global mild solution to the problem (3.62) on $\mathcal{N}^s_{p,\mu,\infty}$.

Our next lemma describes several estimates of the solution operator $S(t)v$ for $t > 0$ on $\mathcal{M}^\beta_{q,\mu}$ and $\mathcal{N}^\beta_{q,\mu,r}$.

Lemma 3.27. *Let* $s, \beta \in \mathbb{R}$ *with* $s \le \beta$, $1 < p \le q < \infty$, $0 \le \mu < N$ *be such that* $\beta - s + (\frac{N-\mu}{p} - \frac{N-\mu}{q}) < 2$.

(i) *Then there exists a constant* $M_1 > 0$ *such that*

$$\|S(t)v\|_{\mathcal{M}^\beta_{q,\mu}} \le M_1 t^{-(1-\alpha)\frac{(\beta-s)}{2} - \frac{1-\alpha}{2}(\frac{N-\mu}{p} - \frac{N-\mu}{q})} \|v\|_{\mathcal{M}^s_{p,\mu}},$$

for $t > 0$ *and* $v \in \mathcal{M}^s_{p,\mu}$.

(ii) *Let $1 \leq r \leq \infty$. There exists a constant $M_1 > 0$ such that*

$$\|S(t)v\|_{\mathcal{N}^{\beta}_{q,\mu,r}} \leq M_1 t^{-(1-\alpha)\frac{(\beta-s)}{2}-\frac{1-\alpha}{2}(\frac{N-\mu}{p}-\frac{N-\mu}{q})}\|v\|_{\mathcal{N}^{s}_{p,\mu,r}},$$

for $t > 0$ and $v \in \mathcal{S}'/\mathcal{P}$.

(iii) *Let $1 \leq r \leq \infty$ and $s < \beta$. There exists a constant $M_1 > 0$ such that*

$$\|S(t)v\|_{\mathcal{N}^{\beta}_{q,\mu,1}} \leq M_1 t^{-(1-\alpha)\frac{(\beta-s)}{2}-\frac{1-\alpha}{2}(\frac{N-\mu}{p}-\frac{N-\mu}{q})}\|v\|_{\mathcal{N}^{s}_{p,\mu,r}}, \qquad (3.69)$$

for $t > 0$ and $v \in \mathcal{S}'/\mathcal{P}$.

Proof. (i) Let $\sigma \in (0, \pi/2)$. From Lemma 2.1 in [39], we have that $w(z) \in \Sigma_{\pi-\sigma}$ and $|w(z)| \leq C_0|z|^{1-\alpha}$ for any $z \in \Sigma_{\pi-\sigma}$. Then we deduce from (3.65) that

$$\|(\omega(z)+A)^{-1}v\|_{\mathcal{M}^{\beta}_{q,\mu}} \leq M_1|\omega(z)|^{\frac{\beta-s}{2}+\frac{1}{2}(\frac{N-\mu}{p}-\frac{N-\mu}{q})-1}\|v\|_{\mathcal{M}^{s}_{p,\mu}},$$

for $\beta - s + (\frac{N-\mu}{p} - \frac{N-\mu}{q}) < 2$, this also shows

$$\|H(z)v\|_{\mathcal{M}^{\beta}_{q,\mu}} \leq \frac{M_1}{|z|}|\omega(z)|^{\frac{\beta-s}{2}+\frac{1}{2}(\frac{N-\mu}{p}-\frac{N-\mu}{q})}\|v\|_{\mathcal{M}^{s}_{p,\mu}}$$

$$\leq M_1 C_0|z|^{(1-\alpha)[\frac{\beta-s}{2}+\frac{1}{2}(\frac{N-\mu}{p}-\frac{N-\mu}{q})]-1}\|v\|_{\mathcal{M}^{s}_{p,\mu}}.$$

Since $1 < p \leq q < \infty$ satisfy $\beta - s + (\frac{N-\mu}{p} - \frac{N-\mu}{q}) < 2$, it implies $\frac{(1-\alpha)}{2}[\beta - s + (\frac{N-\mu}{p} - \frac{N-\mu}{q})] < 1$ for $\alpha \in (0,1)$. Let $t > 0$ and $\delta > 0$. We choose $\delta = 1/t$, it follows from the property of the Gamma function that

$$\|H(z)v\|_{\mathcal{M}^{\beta}_{q,\mu}} \leq \frac{1}{2\pi}\int_{\Gamma_{1/t,\pi-\sigma}} e^{Re(z)t}\|H(z)v\|_{\mathcal{M}^{\beta}_{q,\mu}}|dz|$$

$$\leq \frac{M_1 C_0}{2\pi}\|v\|_{\mathcal{M}^{s}_{p,\mu}}\left(2\int_{t^{-1}}^{\infty}\tau^{\frac{(1-\alpha)}{2}[\beta-s+(\frac{N-\mu}{p}-\frac{N-\mu}{q})]-1}e^{-\tau t\cos(\sigma)}d\tau\right.$$

$$\left.+\int_{-\pi+\theta}^{\pi-\theta}t^{-\frac{(1-\alpha)}{2}[\beta-s+(\frac{N-\mu}{p}-\frac{N-\mu}{q})]}e^{\cos(\psi)}d\psi\right)$$

$$\leq M_1 t^{-\frac{(1-\alpha)}{2}[\beta-s+(\frac{N-\mu}{p}-\frac{N-\mu}{q})]}\|v\|_{\mathcal{M}^{s}_{p,\mu}},$$

for $v \in \mathcal{M}^{s}_{p,\mu}$, where $M_1 = M_1(C_0, N, p, q, \sigma, \beta, s, \mu)$.

(ii) By the same arguments as we derived the result of (i), we can obtain from (3.66) that

$$\|S(t)v\|_{\mathcal{N}^{\beta}_{q,\mu,r}} \leq \frac{1}{2\pi}\int_{\Gamma_{1/t,\pi-\sigma}} e^{Re(z)t}\|H(z)v\|_{\mathcal{N}^{\beta}_{q,\mu,r}}|dz|$$

$$\leq M_1 t^{-\frac{(1-\alpha)}{2}[\beta-s+(\frac{N-\mu}{p}-\frac{N-\mu}{q})]}\|v\|_{\mathcal{N}^{s}_{p,\mu,r}}.$$

(iii) Similar arguments follow the inequality (3.69). Therefore, we omit the proof. $\qquad \square$

Now we present the continuity of $S(t)v$ for $t > 0$ in $\mathcal{N}^{\beta}_{q,\mu,r}$ and $\mathcal{M}^{\beta}_{q,\mu}$ as follows.

Remark 3.9. Let $\beta, s \in \mathbb{R}$, $1 < p \leq q < \infty$ and $0 \leq \mu < N$ be such that $s \leq \beta$ and $\beta - s + (\frac{N-\mu}{p} - \frac{N-\mu}{q}) < 2$. By differentiating $S(t)v$ we have

$$S'(t)v = \frac{1}{2\pi i} \int_{\Gamma_{\delta, \pi - \sigma}} e^{zt} z H(z) v \, dz, \qquad \text{for } t > 0.$$

Using the similar arguments as we derived Lemma 3.27, we arrive at

$$\|S'(t)v\|_{\mathcal{N}^{\beta}_{q,\mu,r}} \leq \frac{1}{2\pi} \int_{\Gamma_{1/t, \pi - \sigma}} e^{Re(z)t} \|z H(z) v\|_{\mathcal{N}^{\beta}_{q,\mu,r}} |dz|$$

$$\leq M_1 t^{-\frac{(1-\alpha)}{2}[\beta - s + (\frac{N-\mu}{p} - \frac{N-\mu}{q})] - 1} \|v\|_{\mathcal{N}^{s}_{p,\mu,r}}.$$

Therefore for any $0 < t_1 < t_2 < \infty$ and $\beta - s + (\frac{N-\mu}{p} - \frac{N-\mu}{q}) \neq 0$, we have

$$\|S(t_2)v - S(t_1)v\|_{\mathcal{N}^{\beta}_{q,\mu,r}}$$

$$\leq \int_{t_1}^{t_2} \|S'(t)v\|_{\mathcal{N}^{\beta}_{q,\mu,r}} dt$$

$$\leq M_1 \|v\|_{\mathcal{N}^{s}_{p,\mu,r}} \int_{t_1}^{t_2} t^{-\frac{(1-\alpha)}{2}[\beta - s + (\frac{N-\mu}{p} - \frac{N-\mu}{q})] - 1} dt$$

$$= \widetilde{M_1} \|v\|_{\mathcal{N}^{s}_{p,\mu,r}} \left(t_1^{-\frac{(1-\alpha)}{2}[\beta - s + (\frac{N-\mu}{p} - \frac{N-\mu}{q})]} - t_2^{-\frac{(1-\alpha)}{2}[\beta - s + (\frac{N-\mu}{p} - \frac{N-\mu}{q})]} \right),$$

where $\widetilde{M_1} = \frac{M_1}{\frac{(1-\alpha)}{2}[\beta - s + (\frac{N-\mu}{p} - \frac{N-\mu}{q})]}$. In fact, from the above arguments, we can also derive the continuity of $S(t)v$ for $t > 0$ in $\mathcal{N}^{\beta}_{q,\mu,r}$ when $\beta - s + (\frac{N-\mu}{p} - \frac{N-\mu}{q}) = 0$. Therefore, $S(\cdot)v \in C((0, \infty); \mathcal{N}^{\beta}_{q,\mu,r})$.

On the other hand, by the similar arguments as above, the continuity of the operator $S(t)v$ in $\mathcal{M}^{\beta}_{q,\mu}$ also holds for $t > 0$.

In virtue of Remark 3.8, the restriction $\beta - s + (\frac{N-\mu}{p} - \frac{N-\mu}{q}) < 2$ that is indispensable for Lemma 3.2 in [297] can be weaken to the restriction $(1 - \alpha)(\beta - s + \frac{N}{p} - \frac{N}{q}) < 2$ when we study the estimates about the operator $S(t)$ for $t > 0$ on Besov spaces.

Remark 3.10. Let $1 \leq r \leq +\infty$ and $s, \beta \in \mathbb{R}$. It holds that

(i) Let $\beta - s \in [0, 2]$ and $1 < p \leq q < \infty$ be such that $\frac{(1-\alpha)(\beta - s)}{2} + \frac{N(1-\alpha)}{2}(\frac{1}{p} - \frac{1}{q}) < 1$. If $v \in \dot{B}^{s}_{p,r}(\mathbb{R}^N)$, then there exists a constant $M_1 > 0$ such that

$$\|S(t)v\|_{\dot{B}^{\beta}_{q,r}} \leq M_1 t^{-\frac{N(1-\alpha)}{2}(\frac{1}{p} - \frac{1}{q}) - \frac{(1-\alpha)(\beta - s)}{2}} \|v\|_{\dot{B}^{s}_{p,r}}, \qquad \text{for } t > 0.$$

(ii) Let $1 < p < \infty$ and $\beta - s \in (0, 2)$. If $v \in \dot{B}^{s}_{p,\infty}(\mathbb{R}^N)$, then there exists a constant $M_1 > 0$ such that

$$\|S(t)v\|_{\dot{B}^{\beta}_{p,1}} \leq M_1 t^{-\frac{(1-\alpha)(\beta - s)}{2}} \|v\|_{\dot{B}^{s}_{p,\infty}}, \qquad \text{for } t > 0.$$

3.5.4   *Well-posedness of Global Solutions*

In this subsection we shall study the well-posedness and asymptotic behavior of global solutions for the semi-linear fractional Rayleigh-Stokes equations.

Let $N \geq 1$ and $0 \leq \mu < N$, we recall that a triplet (ϑ, q, q_0) is admissible if

$$\frac{1}{\vartheta} = \frac{N}{2}\left(\frac{1}{q_0} - \frac{1}{q}\right),$$

where

$$1 < q_0 \leq q < \begin{cases} \dfrac{Nq_0}{N-2}, & N > 2, \\ \infty, & N \leq 2. \end{cases}$$

We denote (ϑ, q, q_0) as a generalized admissible triplet if

$$\frac{1}{\vartheta} = \frac{N-\mu}{2}\left(\frac{1}{q_0} - \frac{1}{q}\right),$$

where

$$1 < q_0 \leq q < \begin{cases} \dfrac{q_0(N-\mu)}{N-\mu-2q_0}, & N-\mu > 2q_0, \\ \infty, & N-\mu \leq 2q_0. \end{cases}$$

Clearly, $q < \vartheta \leq \infty$ if (ϑ, q, q_0) is an admissible triplet, and $1 < \vartheta \leq \infty$ if (ϑ, q, q_0) is a generalized admissible triplet.

Next, the hypothesis of the semilinear term is introduced.

(Hf) Let $f \in C(\mathbb{R}, \mathbb{R})$. We suppose that $f(0) = 0$ and that there exist $\rho > 0$ and $K > 0$ such that

$$|f(u) - f(v)| \leq K(|u|^\rho + |v|^\rho)|u - v|, \quad \text{for } u, v \in \mathbb{R}.$$

For $1 < q < \infty$ and $0 \leq \mu < N$, from the hypothesis (Hf) we deduce that

$$\|f(u) - f(v)\|_{\mathcal{M}_{\frac{q}{1+\rho},\mu}} \leq (\|u\|_{\mathcal{M}_{q,\mu}}^\rho + \|v\|_{\mathcal{M}_{q,\mu}}^\rho)\|u - v\|_{\mathcal{M}_{q,\mu}}. \tag{3.70}$$

The following lemma implies strong continuity of the operator $S(t)u_0$ for $t \in (0, \infty)$ with values in $\mathcal{M}_{q,\mu} \cap \mathcal{N}_{p,\mu,\infty}^{s_0}$ for the initial data $u_0 \in \mathcal{N}_{p,\mu,\infty}^{s_0}$.

Lemma 3.28. *Let* $0 \leq \mu < N$, $1 < q_0 := \frac{(N-\mu)\rho(1-\alpha)}{2} < p \leq q$ *and* (ϑ, q, q_0) *be a generalized admissible triplet. Assume that* $u_0 \in \mathcal{N}_{p,\mu,\infty}^{s_0}$. *Then it holds that* $S(t)u_0 \in C_b((0, \infty); \mathcal{N}_{p,\mu,\infty}^{s_0})$ *and* $t^{\frac{1-\alpha}{\vartheta}}S(t)u_0 \in C_b((0, \infty); \mathcal{M}_{q,\mu})$ *with the estimate*

$$\sup_{t \in (0,\infty)} \|S(t)u_0\|_{\mathcal{N}_{p,\mu,\infty}^{s_0}} + \sup_{t \in (0,\infty)} t^{\frac{1-\alpha}{\vartheta}}\|S(t)u_0\|_{\mathcal{M}_{q,\mu}} \leq L_0\|u_0\|_{\mathcal{N}_{p,\mu,\infty}^{s_0}}, \tag{3.71}$$

where L_0 *is a constant.*

Proof. Consider $p > \frac{(N-\mu)\rho(1-\alpha)}{2} > 1$, we have $s_0 = \frac{N-\mu}{p} - \frac{2}{\rho(1-\alpha)} < 0$. Notice that $N(\frac{1}{p} - \frac{1}{q}) - s_0 = \frac{2}{\vartheta} < 2$ for a generalized admissible triplet (μ, q, q_0), and which gives by Lemma 3.27(iii) and (3.63) that

$$\sup_{t\in(0,\infty)} \|S(t)u_0\|_{\mathcal{N}^{s_0}_{p,\mu,\infty}} + \sup_{t\in(0,\infty)} t^{\frac{1-\alpha}{\vartheta}} \|S(t)u_0\|_{\mathcal{M}_{q,\mu}}$$

$$\leq M_1 \|u_0\|_{\mathcal{N}^{s_0}_{p,\mu,\infty}} + \sup_{t\in(0,\infty)} t^{\frac{1-\alpha}{\vartheta}} \|S(t)u_0\|_{\mathcal{N}^{0}_{q,\mu,1}}$$

$$\leq M_1 \|u_0\|_{\mathcal{N}^{s_0}_{p,\mu,\infty}} + M_1 \sup_{t\in(0,\infty)} t^{\frac{1-\alpha}{\vartheta} + \frac{(N-\mu)(1-\alpha)}{2q} - \frac{1}{p}} \|u_0\|_{\mathcal{N}^{s_0}_{p,\mu,\infty}}$$

$$= L_0 |u_0\|_{\mathcal{N}^{s_0}_{p,\mu,\infty}}$$

for all $u_0 \in \mathcal{N}^{s_0}_{p,\mu,\infty}$ and $t > 0$, where $L_0 = 2M_1$. This yields (3.71).

Next we prove that $S(t)u_0$ is strongly continuous with values in $\mathcal{M}_{q,\mu} \cap \mathcal{N}^{s_0}_{p,\mu,\infty}$ for all $t > 0$. Indeed, Remark 3.9 implies the continuity of $S(t)u_0$ in $\mathcal{N}^{s_0}_{p,\mu,\infty}$ for $t > 0$. It remains to prove the continuity of $S(t)u_0$ in $\mathcal{M}_{q,\mu}$. For any $0 < t_1 < t_2 < \infty$, we can obtain from (3.63) that

$$\left\| t_2^{\frac{1-\alpha}{\vartheta}} S(t_2)u_0 - t_1^{\frac{1-\alpha}{\vartheta}} S(t_1)u_0 \right\|_{\mathcal{M}_{q,\mu}}$$

$$\leq \left\| (t_2^{\frac{1-\alpha}{\vartheta}} - t_1^{\frac{1-\alpha}{\vartheta}}) S(t_2)u_0 \right\|_{\mathcal{N}^{0}_{q,\mu,1}} + t_1^{\frac{1-\alpha}{\vartheta}} \|S(t_2)u_0 - S(t_1)u_0\|_{\mathcal{N}^{0}_{q,\mu,1}}.$$

The first term for the right-hand side of above inequality is bounded by $M_1 \|u_0\|_{\mathcal{N}^{s_0}_{p,\mu,\infty}} [1 - (t_1/t_2)^{\frac{1-\alpha}{\vartheta}}]$ due to Lemma 3.27(iii). Using Remark 3.9 again, the second term tends to 0 as $t_2 \to t_1$. So one can see that $t^{\frac{1-\alpha}{\vartheta}} S(t)u_0 \in C_b((0,\infty); \mathcal{M}_{q,\mu})$. The proof is completed. $\qquad\square$

We state the well-posedness results of global solutions for the equation (3.62).

Theorem 3.12. *Let* $0 \leq \mu < N$, $1 < q_0 := \frac{(N-\mu)\rho(1-\alpha)}{2} < p \leq q$ *and* (ϑ, q, q_0) *be a generalized admissible triplet satisfying*

$$\max\{q_0, 1 + \rho\} < q < q_0(1 + \rho). \tag{3.72}$$

Let $u_0 \in \mathcal{N}^{s_0}_{p,\mu,\infty}$ *and (Hf) holds. If there exist* $\varepsilon > 0$ *and* $\kappa := \kappa(\varepsilon)$ *such that* $\|u_0\|_{\mathcal{N}^{s_0}_{p,\mu,\infty}} \leq \kappa$, *then the equation (3.62) has a unique solution* $u \in Z_{p,q}$ *with* $\|u\|_{Z_{p,q}} \leq 2\varepsilon$. *Moreover, the map* $u_0 \in U_\kappa \mapsto u \in Z_{p,q}$ *is Lipschitz continuous, where* U_κ *is denoted by*

$$U_\kappa = \{u_0 \in \mathcal{N}^{s_0}_{p,\mu,\infty} : \|u_0\|_{\mathcal{N}^{s_0}_{p,\mu,\infty}} \leq \kappa\}.$$

Proof. Let $u \in Z_{p,q}$ and $t > 0$. Define the operator Φ in $Z_{p,q}$ as

$$\Phi(u)(t) = S(t)u_0 + W(u)(t), \tag{3.73}$$

where

$$W(u)(t) = \int_0^t S(t-\tau)f(u(\tau))d\tau.$$

Now we estimate the nonlinear part. For any generalized admissible triplet (ϑ, q, q_0), we get

$$\frac{\rho(N-\mu)(1-\alpha)}{2q} + \frac{\rho(1-\alpha)}{\vartheta} = 1, \tag{3.74}$$

which implies that $\frac{\rho(N-\mu)(1-\alpha)}{2q} < 1$. Let $u, v \in Z_{p,q}$. Notice that $\mathcal{M}_{q_0,\mu} \subset \mathcal{N}^0_{q_0,\mu,\infty} \subset \mathcal{N}^{s_0}_{p,\mu,\infty}$ due to (3.63) and (3.64) for $0 - \frac{N-\mu}{q_0} = s_0 - \frac{N-\mu}{p}$, we obtain from Lemma 3.27(i), (3.70) and (3.74) that

$$\|W(u)(t) - W(v)(t)\|_{\mathcal{N}^{s_0}_{p,\mu,\infty}}$$

$$\leq \|W(u)(t) - W(v)(t)\|_{\mathcal{M}_{q_0,\mu}}$$

$$\leq M_1 \int_0^t (t-\tau)^{-\frac{(N-\mu)(1-\alpha)}{2}(\frac{1+\rho}{q} - \frac{1}{q_0})} \|f(u(\tau) - f(v(\tau)))\|_{\mathcal{M}_{q/(1+\rho),\mu}} d\tau$$

$$\leq M_1 K \int_0^t (t-\tau)^{-\frac{(N-\mu)(1-\alpha)}{2}(\frac{1+\rho}{q} - \frac{1}{q_0})} \|u(\tau) - v(\tau)\|_{\mathcal{M}_{q,\mu}} (\|u(\tau)\|^\rho_{\mathcal{M}_{q,\mu}}$$

$$+ \|v(\tau)\|^\rho_{\mathcal{M}_{q,\mu}}) d\tau$$

$$\leq M_1 K \|u - v\|_{C_*((0,\infty);\mathcal{M}_{q,\mu})} (\|u\|^\rho_{C_*((0,\infty);\mathcal{M}_{q,\mu})} + \|v\|^\rho_{C_*((0,\infty);\mathcal{M}_{q,\mu})})$$

$$\times \int_0^t (t-\tau)^{\frac{1-\alpha}{\vartheta} - \frac{\rho(N-\mu)(1-\alpha)}{2q}} \tau^{-\frac{(\rho+1)}{\vartheta}(1-\alpha)} d\tau$$

$$\leq M_1 K B(\chi, 1-\chi) \|u - v\|_{C_*((0,\infty);\mathcal{M}_{q,\mu})} (\|u\|^\rho_{C_*((0,\infty);\mathcal{M}_{q,\mu})} + \|v\|^\rho_{C_*((0,\infty);\mathcal{M}_{q,\mu})}), \tag{3.75}$$

where $B(\cdot,\cdot)$ stands for the Beta function, $\chi = \frac{(\rho+1)}{\vartheta}(1-\alpha) \in (0,1)$, here we have used (3.72). Further, let $\beta = s = 0$, it follows from Lemma 3.27(i), (3.70) and (3.74) again that

$$\|W(u)(t) - W(v)(t)\|_{\mathcal{M}_{q,\mu}}$$

$$\leq M_1 \int_0^t (t-\tau)^{-\frac{\rho(N-\mu)(1-\alpha)}{2q}} \|f(u(\tau) - f(v(\tau)))\|_{\mathcal{M}_{q/(1+\rho),\mu}} d\tau$$

$$\leq M_1 K \|u - v\|_{C_*((0,\infty);\mathcal{M}_{q,\mu})} (\|u\|^\rho_{C_*((0,\infty);\mathcal{M}_{q,\mu})} + \|v\|^\rho_{C_*((0,\infty);\mathcal{M}_{q,\mu})})$$

$$\times \int_0^t (t-\tau)^{-\frac{\rho(N-\mu)(1-\alpha)}{2q}} \tau^{-\frac{(\rho+1)}{\vartheta}(1-\alpha)} d\tau$$

$$\leq M_1 K B(\frac{\chi\rho}{1+\rho}, 1-\chi) \|u - v\|_{C_*((0,\infty);\mathcal{M}_{q,\mu})} (\|u\|^\rho_{C_*((0,\infty);\mathcal{M}_{q,\mu})} + \|v\|^\rho_{C_*((0,\infty);\mathcal{M}_{q,\mu})}). \tag{3.76}$$

We observe that

$$B(\frac{\chi\rho}{1+\rho}, 1-\chi) = \int_0^1 \tau^{\frac{\chi\rho}{1+\rho}-1}(1-\tau)^{\chi-1} d\tau \geq \int_0^1 \tau^{\chi-1}(1-\tau)^{\chi-1} d\tau = B(\chi, 1-\chi).$$

Thus we deduce the relation

$$\|W(u) - W(v)\|_{Z_{p,q}} \leq L_1(\|u\|^\rho_{Z_{p,q}} + \|v\|^\rho_{Z_{p,q}})\|u - v\|_{Z_{p,q}}, \tag{3.77}$$

where $L_1 = 2M_1 K B(\frac{\chi\rho}{1+\rho}, 1-\chi)$.

It remains to prove that $W(u) \in Z_{p,q}$ for $u \in Z_{p,q}$. Now we show the continuity of $W(u)(t)$ in $\mathcal{M}_{q,\mu} \cap \mathcal{N}_{p,\mu,\infty}^{s_0}$ for all $t \in (0,\infty)$ and $u \in Z_{p,q}$. For $0 < t_1 < t_2 < \infty$, it holds that

$$W(u)(t_2) - W(u)(t_1)$$
$$= \int_{t_1}^{t_2} S(t_2 - \tau) f(u(\tau)) d\tau + \int_0^{t_1} [S(t_2 - \tau) - S(t_1 - \tau)] f(u(\tau)) d\tau$$
$$=: I_1(t_1, t_2) + I_2(t_1, t_2).$$

Similar to the arguments on (3.75), we can calculate $\|I_1(t_1, t_2)\|_{\mathcal{N}_{p,\mu,\infty}^{s_0}}$ as follows

$$\|I_1(t_1, t_2)\|_{\mathcal{N}_{p,\mu,\infty}^{s_0}}$$
$$\leq M_1 K \|u\|_{C_*((0,\infty);\mathcal{M}_{q,\mu})}^{\rho+1} \int_{t_1}^{t_2} (t_2 - \tau)^{-\frac{(N-\mu)(1-\alpha)}{2}(\frac{1+\rho}{q} - \frac{1}{q_0})} \tau^{-\frac{(\rho+1)}{\vartheta}(1-\alpha)} d\tau$$
$$\leq M_1 K \|u\|_{C_*((0,\infty);\mathcal{M}_{q,\mu})}^{\rho+1} \int_{t_1/t_2}^1 (1-\tau)^{\frac{1-\alpha}{\vartheta} - \frac{\rho(N-\mu)(1-\alpha)}{2q}} \tau^{-\chi} d\tau$$
$$\to 0, \quad \text{as } t_2 \to t_1,$$

where we have used the properties of the Beta function. Using the continuity of $S(t)v$ on $\mathcal{M}_{q_0,\mu}$ in Remark 3.9 for $v \in \mathcal{M}_{\frac{q}{1+\rho},\mu}$, it holds that

$$\|[S(t_2 - \tau) - S(t_1 - \tau)] f(u(\tau))\|_{\mathcal{M}_{q_0,\mu}}$$
$$\leq [(t_2 - \tau)^{-\frac{(N-\mu)(1-\alpha)}{2}(\frac{1+\rho}{q} - \frac{1}{q_0})} - (t_1 - \tau)^{-\frac{(N-\mu)(1-\alpha)}{2}(\frac{1+\rho}{q} - \frac{1}{q_0})}] \|f(u(\tau))\|_{\mathcal{M}_{\frac{q}{1+\rho},\mu}}.$$

It yields that

$$\|I_2(t_1, t_2)\|_{\mathcal{N}_{p,\mu,\infty}^{s_0}}$$
$$\leq M_1 K \|u\|_{C_*((0,\infty);\mathcal{M}_{q,\mu})}^{\rho+1} \left[\int_0^{t_1} (t_1 - \tau)^{-\frac{(N-\mu)(1-\alpha)}{2}(\frac{1+\rho}{q} - \frac{1}{q_0})} \tau^{-\frac{(\rho+1)}{\vartheta}(1-\alpha)} d\tau \right.$$
$$\left. - \int_0^{t_1} (t_2 - \tau)^{-\frac{(N-\mu)(1-\alpha)}{2}(\frac{1+\rho}{q} - \frac{1}{q_0})} \tau^{-\frac{(\rho+1)}{\vartheta}(1-\alpha)} d\tau \right]$$
$$\leq M_1 K \|u\|_{C_*((0,\infty);\mathcal{M}_{q,\mu})}^{\rho+1} \left[\int_0^1 (1-\tau)^{\frac{1-\alpha}{\vartheta} - \frac{\rho(N-\mu)(1-\alpha)}{2q}} \tau^{-\chi} d\tau \right.$$
$$\left. - \int_0^{t_1/t_2} (1-\tau)^{\frac{1-\alpha}{\vartheta} - \frac{\rho(N-\mu)(1-\alpha)}{2q}} \tau^{-\chi} d\tau \right]$$
$$\to 0, \quad \text{as } t_2 \to t_1.$$

This shows $W(u) \in C((0,\infty); \mathcal{N}_{p,\mu,\infty}^{s_0})$. Moreover, in the same way as above we have that

$$\|I_1(t_1, t_2)\|_{\mathcal{M}_{q,\mu}}$$
$$\leq M_1 K \|u\|_{C_*((0,\infty);\mathcal{M}_{q,\mu})}^{\rho+1} t_2^{-\frac{1-\alpha}{\vartheta}} \int_{t_1/t_2}^1 (1-\tau)^{-\frac{\rho(N-\mu)(1-\alpha)}{2q}} \tau^{-\frac{\rho+1}{\mu}(1-\alpha)} d\tau,$$

$$\|I_2(t_1, t_2)\|_{\mathcal{M}_{q,\mu}}$$

$$\leq M_1 K \|u\|_{C_*((0,\infty);\mathcal{M}_{q,\mu})}^{\rho+1} \left[t_1^{-\frac{1-\alpha}{\vartheta}} \int_0^1 (1-\tau)^{-\frac{\rho(N-\mu)(1-\alpha)}{2q}} \tau^{-\frac{\rho+1}{\mu}(1-\alpha)} d\tau \right.$$

$$\left. - t_2^{-\frac{1-\alpha}{\vartheta}} \int_0^{t_1/t_2} (1-\tau)^{-\frac{\rho(N-\mu)(1-\alpha)}{2q}} \tau^{-\frac{\rho+1}{\mu}(1-\alpha)} d\tau \right].$$

The above inequalities also tend to zero as $t_2 \to t_1$, and then $W(u) \in C((0,\infty); \mathcal{M}_{q,\mu})$. Consequently, this together with the relation (3.77) for $v = 0$ shows that $W(u) \in Z_{p,q}$ for $u \in Z_{p,q}$.

We choose $0 < \varepsilon < R = (\frac{1}{2^{\rho+1}L_1})^\rho$, where $L_1 > 0$ is the constant obtained in (3.77). Let $\kappa = \frac{\varepsilon}{L_0}$, where $L_0 > 0$ is the constant obtained in Lemma 3.28. Lemma 3.25 with $X = Z_{p,q}$ and $\varphi = S(t)u_0$ yields the existence of a unique global mild solution $u \in Z_{p,q}$ such that $\|u\|_{Z_{p,q}} \leq 2\varepsilon$.

Finally, we show the continuous dependence on the initial value. Let $\overline{u_0} \in U_\kappa$ and $\bar{u}$ be the solution of the equation $\bar{u}(t) = S(t)\overline{u_0} + W(\bar{u})(t)$. From Lemma 3.25 and Lemma 3.28 for $\overline{u_0} \in \mathcal{N}_{p,\mu,\infty}^{s_0}$, we obtain that

$$\|u - \bar{u}\|_{Z_{p,q}} \leq \frac{1}{1 - 2^{\rho+1}L_1\varepsilon^\rho} \|S(t)(u_0 - \overline{u_0})\|_{Z_{p,q}} \leq \frac{L_0}{1 - 2^{\rho+1}L_1\varepsilon^\rho} \|u_0 - \overline{u_0}\|_{\mathcal{N}_{p,\mu,\infty}^{s_0}}.$$

The proof is completed. □

The following corollary means that the equation (3.62) is solvable without the restrictions $q < q_0(1 + \rho)$.

Corollary 3.2. *Let the conditions of Theorem 3.12 hold except for $q < q_0(1 + \rho)$. Then the equation (3.62) has a unique solution $u \in Z_{p,q}$ satisfying*

$$\|u(t)\|_{\mathcal{M}_{\tilde{q},\mu}} \leq M_1 t^{-\frac{1}{\rho} + \frac{(N-\mu)(1-\alpha)}{2q}}. \tag{3.78}$$

Proof. Let $q \geq q_0(1 + \rho)$. Then we can choose $\zeta > 0$ so small that

$$\frac{N - \mu}{2} \left(\frac{1 + \rho}{\tilde{q}} - \frac{1}{q} \right) < 1,$$

where we denote $\tilde{q} = q_0(1 + \rho) - \zeta$. Observe that the above inequality can be guaranteed by the fact that $q < (N - \mu)q_0/(N - \mu - 2q_0)$ if $N - \mu > 2q_0$ and $q < \infty$ if $N - \mu \leq 2q_0$. Let

$$\frac{1}{\vartheta} = \frac{N - \mu}{2} \left(\frac{1}{q_0} - \frac{1}{\tilde{q}} \right).$$

Then $(\vartheta, \tilde{q}, q_0)$ is a generalized admissible triplet and $\vartheta > (1 + \rho)(1 - \alpha)$. Using Theorem 3.12 we know that the equation (3.62) has a unique solution $u \in C_b((0,\infty); \mathcal{N}_{p,\mu,\infty}^{s_0}) \cap C_*((0,\infty); \mathcal{M}_{\tilde{q},\mu})$. A direct calculation yields that

$$\|W(u)(t)\|_{\mathcal{M}_{q,\mu}}$$

$$\leq M_1 \int_0^t (t-\tau)^{-\frac{(N-\mu)(1-\alpha)}{2}\left(\frac{1+\rho}{\tilde{q}} - \frac{1}{q}\right)} \|f(u(\tau))\|_{\mathcal{M}_{\tilde{q}/(1+\rho),\mu}} d\tau$$

$$\leq M_1 K \|u\|_{C_*((0,\infty);\mathcal{M}_{\tilde{q},\mu})}^{\rho+1} \int_0^t (t-\tau)^{-\frac{(N-\mu)(1-\alpha)}{2}(\frac{1+\rho}{\tilde{q}}-\frac{1}{q})} \tau^{-\frac{(\rho+1)}{\vartheta}(1-\alpha)} d\tau$$

$$\leq M_1 K B_1 t^{-\frac{1-\alpha}{\vartheta}} \|u\|_{C_*((0,\infty);\mathcal{M}_{\tilde{q},\mu})}^{\rho+1},$$

where $B_1 = B(1 - \frac{N(1-\alpha)}{2}(\frac{1+\rho}{\tilde{q}} - \frac{1}{q}), 1 - \frac{\rho+1}{\vartheta}(1-\alpha))$. Furthermore, by the similar arguments, the continuity of $W(u)(t)$ on $\mathcal{M}_{q,\mu}$ is obvious for $t > 0$. This together with Lemma 3.28 implies that $u \in C_*((0,\infty);\mathcal{M}_{q,\mu})$. This implies $u \in Z_{p,q}$ and (3.78). The proof is completed. $\qquad\square$

Finally we state an asymptotic behavior result of the solution of the equation equation (3.62) as $t \to \infty$.

Theorem 3.13. *Let the conditions of Theorem 3.12 hold. Assume that u and v be two global mild solutions for the equation (3.62) obtained in Theorem 3.12, with the initial values u_0 and v_0, respectively. Then*

$$\lim_{t\to\infty} \|u(t) - v(t)\|_{\mathcal{N}_{p,\mu,\infty}^{s_0}} + \lim_{t\to\infty} t^{\frac{1-\alpha}{\vartheta}} \|u(t) - v(t)\|_{\mathcal{M}_{q,\mu}} = 0$$

if and only if

$$\lim_{t\to\infty} \|S(t)(u_0 - v_0)\|_{\mathcal{N}_{p,\mu,\infty}^{s_0}} + \lim_{t\to\infty} t^{\frac{1-\alpha}{\vartheta}} \|S(t)(u_0 - v_0)\|_{\mathcal{M}_{q,\mu}} = 0.$$

Proof. ($\Leftarrow$) We denote

$$J(u,v)(t) = W(u)(t) - W(v)(t).$$

Using (3.75), $\|u\|_{Z_{p,q}} \leq 2\varepsilon$ and $\|v\|_{Z_{p,q}} \leq 2\varepsilon$, we obtain that

$$\|J(u,v)(t)\|_{\mathcal{N}_{p,\mu,\infty}^{s_0}}$$

$$\leq M_1 K \int_0^t (t-\tau)^{-\frac{(N-\mu)(1-\alpha)}{2}(\frac{1+\rho}{q}-\frac{1}{q_0})} \|u(\tau) - v(\tau)\|_{\mathcal{M}_{q,\mu}} (\|u(\tau)\|_{\mathcal{M}_{q,\mu}}^{\rho}$$

$$+ \|v(\tau)\|_{\mathcal{M}_{q,\mu}}^{\rho}) d\tau$$

$$\leq 2 M_1 K (2\varepsilon)^{\rho} \int_0^t (t-\tau)^{-\frac{(N-\mu)(1-\alpha)}{2}(\frac{1+\rho}{q}-\frac{1}{q_0})} \tau^{-\frac{\rho(1-\alpha)}{\vartheta}} \|u(\tau) - v(\tau)\|_{\mathcal{M}_{q,\mu}} d\tau.$$

For $\|J(u,v)(t)\|_{\mathcal{M}_{q,\mu}}$, in view of (3.76), one can see that

$$\|J(u,v)(t)\|_{\mathcal{M}_{q,\mu}} \leq 2 M_1 K (2\varepsilon)^{\rho} \int_0^t (t-\tau)^{-\frac{\rho(N-\mu)(1-\alpha)}{2q}} \tau^{-\frac{\rho(1-\alpha)}{\vartheta}} \|u(\tau) - v(\tau)\|_{\mathcal{M}_{q,\mu}} d\tau.$$

Let $\Pi(t) = \|u(t) - v(t)\|_{\mathcal{N}_{p,\mu,\infty}^{s_0}} + t^{\frac{1-\alpha}{\vartheta}} \|u(t) - v(t)\|_{\mathcal{M}_{q,\mu}}$. By performing a change of variables in $\|J(u,v)(t)\|_{\mathcal{N}_{p,\mu,\infty}^{s_0}}$ and $\|J(u,v)(t)\|_{\mathcal{M}_{q,\mu}}$, we get that

$$\|J(u,v)(t)\|_{\mathcal{N}_{p,\mu,\infty}^{s_0}} + t^{\frac{1-\alpha}{\vartheta}} \|J(u,v)(t)\|_{\mathcal{M}_{q,\mu}}$$

$$\leq 2 M_1 K (2\varepsilon)^{\rho} \int_0^1 (1-\tau)^{-\frac{(N-\mu)(1-\alpha)}{2}(\frac{1+\rho}{q}-\frac{1}{q_0})} \tau^{-\frac{(\rho+1)}{\vartheta}(1-\alpha)} \Pi(t\tau) d\tau$$

$$+ 2 M_1 K (2\varepsilon)^{\rho} \int_0^1 (1-\tau)^{-\frac{\rho(N-\mu)(1-\alpha)}{2q}} \tau^{-\frac{(\rho+1)}{\vartheta}(1-\alpha)} \Pi(t\tau) d\tau.$$

Now we claim that $\limsup_{t\to\infty} \Pi(t) = 0$. Indeed, we take $\limsup_{t\to\infty}$ in above inequality in order to obtain

$$\limsup_{t\to\infty}[\||J(u,v)(t)\|_{\mathcal{N}^{s_0}_{p,\mu,\infty}} + t^{\frac{\rho(1-\alpha)}{\vartheta}}\|J(u,v)(t)\|_{\mathcal{M}_{q,\mu}}]$$

$$\leq 2M_1 K(2\varepsilon)^\rho \int_0^1 (1-\tau)^{-\frac{(N-\mu)(1-\alpha)}{2}(\frac{1+\rho}{q}-\frac{1}{q_0})}\tau^{-\frac{(\rho+1)}{\vartheta}(1-\alpha)}\limsup_{t\to\infty}\Pi(t\tau)d\tau$$

$$+ 2M_1 K(2\varepsilon)^\rho \int_0^1 (1-\tau)^{-\frac{\rho(N-\mu)(1-\alpha)}{2q}}\tau^{-\frac{(\rho+1)}{\vartheta}(1-\alpha)}\limsup_{t\to\infty}\Pi(t\tau)d\tau \qquad (3.79)$$

$$\leq L_1 2^{\rho+1}\varepsilon^\rho \limsup_{t\to\infty}\Pi(t).$$

It follows that

$$\limsup_{t\to\infty}\Pi(t) \leq \limsup_{t\to\infty}[\|S(t)(u_0-v_0)\|_{\mathcal{N}^{s_0}_{p,\mu,\infty}} + t^{\frac{1-\alpha}{\vartheta}}\|S(t)(u_0-v_0)\|_{\mathcal{M}_{q,\mu}}]$$

$$+ \limsup_{t\to\infty}[\||J(u,v)(t)\|_{\mathcal{N}^{s_0}_{p,\mu,\infty}} + t^{\frac{1-\alpha}{\vartheta}}\|J(u,v)(t)\|_{\mathcal{M}_{q,\mu}}]$$

$$\leq L_1 2^{\rho+1}\varepsilon^\rho \limsup_{t\to\infty}\Pi(t).$$

Since $L_1 2^{\rho+1}\varepsilon^\rho < 1$, we get $\limsup_{t\to\infty}\Pi(t) = 0$, as required.

($\Rightarrow$) According to the relation $S(t)(u_0-v_0) = u(t) - v(t) - (J(u,v)(t))$ and (3.79), we have that

$$\limsup_{t\to\infty}[\|S(t)(u_0-v_0)\|_{\mathcal{N}^{s_0}_{p,\mu,\infty}} + t^{\frac{1-\alpha}{\vartheta}}\|S(t)(u_0-v_0)\|_{\mathcal{M}_{q,\mu}}]$$

$$\leq \limsup_{t\to\infty}\Pi(t) + \limsup_{t\to\infty}[\||J(u,v)(t)\|_{\mathcal{N}^{s_0}_{p,\mu,\infty}} + t^{\frac{1-\alpha}{\vartheta}}\|J(u,v)(t)\|_{\mathcal{M}_{q,\mu}}]$$

$$\leq (1 + L_1 2^{\rho+1}\varepsilon^\rho)\limsup_{t\to\infty}\Pi(t) = 0.$$

The proof is completed. $\square$

3.6   Final Value Problem

3.6.1   *Introduction*

The aim of this section is to obtain the existence result of mild solutions for the nonlinear fractional Rayleigh-Stokes problem under a weaker condition on the nonlinear source term than Lipschitz condition. And the quasi-boundary value method will be used to regularize the problem, which is mainly by perturbing the final condition or boundary condition to a new regularization condition depending on a small parameter.

Consider the nonlinear fractional Rayleigh-Stokes problem with final condition

$$\begin{cases} \partial_t u - (1+\gamma\partial_t^\alpha)\Delta u = f(t,x,u), & t \in (0,b), \ x \in \Omega, \\ u(t,x) = 0, & t \in (0,b), \ x \in \partial\Omega, \\ u(b,x) = g(x), & x \in \Omega, \end{cases} \qquad (3.80)$$

where $\Omega \subset \mathbb{R}^d (d \geq 1)$ is a bounded domain with smooth boundary $\partial\Omega$, Δ is the Laplacian operator, $b > 0$ is a given time, $\gamma > 0$ is a fixed constant, $g \in L^2(\Omega)$ is the final value, $f : [0, b] \times \Omega \times \mathbb{R} \to \mathbb{R}$ is a given function satisfying some assumptions, and ∂_t^α is the Riemann-Liouville fractional partial derivative of order $\alpha \in (0, 1)$ which is defined as follows

$$\partial_t^\alpha v(t, x) = \frac{1}{\Gamma(1-\alpha)} \frac{\partial}{\partial t} \int_0^t (t-s)^{-\alpha} v(s, x) ds, \quad t > 0,$$

where $\Gamma(\cdot)$ is the Gamma function.

In Subsection 3.6.2, some concepts and lemmas which will be used in this section are given. In Subsection 3.6.3, the compactness and continuity in the uniform operator topology of solution operator for problem (3.80) are discussed at first, then we obtain the existence result of mild solutions for problem (3.80). Furthermore, a regularization method is used in Subsection 3.6.4, the well-posedness and the convergence rate for regularized solutions are obtained.

3.6.2 *Preliminaries*

Let $b > 0$ be a given number, $\mathbb{N}^+$ denotes the set of positive integer numbers, $L^2(\Omega)$ is the Banach space of all measurable functions on $\Omega \subset \mathbb{R}^d (d \geq 1)$ with the inner product $(\cdot, \cdot)$ and norm $\|\cdot\|$. Denote by $\mathcal{B}(X)$ the space of all bounded linear operators from normed linear space X into X. $L^\infty(0, b; L^2(\Omega))$ is the Banach space of all essentially bounded and measurable functions v equipped with the norm

$$\|v\|_\infty = \text{esssup}_{t \in [0, b]} \|v(t)\|.$$

We recall the following spectral problem

$$\begin{cases} -\Delta\varphi_n(x) = \lambda_n \varphi_n(x), & x \in \Omega, \\ \varphi_n(x) = 0, & x \in \partial\Omega, \end{cases} \tag{3.81}$$

where $n \in \mathbb{N}^+$ and the eigenvalues satisfy

$$0 < \lambda_1 \leq \lambda_2 \leq \dots \leq \lambda_n \leq \dots,$$

with $\lambda_n \to \infty$ as $n \to \infty$. And the corresponding set of eigenfunctions $\{\varphi_n\}_{n=1}^\infty \subset H_0^1(\Omega)$. Obviously, $\{\varphi_n\}_{n=1}^\infty$ is an orthonormal basis of $L^2(\Omega)$.

Let $C([0, b]; L^2(\Omega))$ stand for the Banach space of all continuous functions defined on $[0, b]$ with the norm

$$\|v\|_{C([0,b];L^2(\Omega))} = \sup_{t \in [0,b]} \|v(t)\|.$$

In the following, for $0 < \mu < 1$, we introduce the weighted continuous function space of all continuous functions defined on $(0, b]$, which has the following form

$$C^\mu([0, b]; L^2(\Omega)) := \{v \in C((0, b]; L^2(\Omega)) | t^\mu v(t) \in C([0, b]; L^2(\Omega))\},$$

and it is easy to see that $C^\mu([0, b]; L^2(\Omega))$ is a Banach space equipped with the norm

$$\|v\|_\mu = \sup_{t \in [0,b]} t^\mu \|v(t)\|.$$

For $\sigma \geq 0$, we denote the space $\mathcal{H}^\sigma(\Omega)$ as follows:

$$\mathcal{H}^\sigma(\Omega) = \left\{ v \in L^2(\Omega) : \sum_{j=1}^\infty \lambda_j^{2\sigma} |(v, \varphi_j)|^2 < \infty \right\},$$

and the norm of $\mathcal{H}^\sigma(\Omega)$ is defined by

$$\|v\|_{\mathcal{H}^\sigma(\Omega)} = \left(\sum_{j=1}^\infty \lambda_j^{2\sigma} |(v, \varphi_j)|^2 \right)^{\frac{1}{2}}.$$

Lemma 3.29. *[279, Lemma 4.2] Assume β, θ and C are positive constants, then, for $\xi > 0$, we have*

$$F(\xi) = \frac{\beta \xi^{1-\theta}}{\beta \xi + C} \leq \begin{cases} l_1(\theta, C)\beta^\theta, & 0 < \theta < 1, \\ l_2(\theta, C, \lambda_1)\beta, & \theta \geq 1, \xi \geq \lambda_1, \end{cases}$$

where $l_1(\theta, C) = (1 - \theta)^{1-\theta} \theta^\theta C^{-\theta}$ and $l_2(\theta, C, \lambda_1) = C^{-1} \lambda_1^{1-\theta}$.

3.6.3   *Existence of Solutions*

Let u be a solution of problem (3.80), and let $u(t, x)$ be expanded as follows

$$u(t, x) = \sum_{j=1}^\infty (u(t, \cdot), \varphi_j) \varphi_j(x).$$

From [39] and [278, Equation (10)], the solution of problem (3.80) can be described as

$$u(t, x) = \sum_{j=1}^\infty \frac{P_{\alpha,j}(t)}{P_{\alpha,j}(b)} \left((g, \varphi_j) - \int_0^b P_{\alpha,j}(b - s)(f(s, u(s)), \varphi_j) ds \right) \varphi_j(x)$$

$$+ \sum_{j=1}^\infty \int_0^t P_{\alpha,j}(t - s)(f(s, u(s)), \varphi_j) ds \varphi_j(x), \tag{3.82}$$

where

$$P_{\alpha,j}(t) = \int_0^\infty e^{-\xi t} K_j(\xi) d\xi, \tag{3.83}$$

and

$$K_j(\xi) = \frac{\gamma}{\pi} \frac{\lambda_j \xi^\alpha \sin(\alpha\pi)}{(-\xi + \lambda_j \gamma \xi^\alpha \cos(\alpha\pi) + \lambda_j)^2 + (\lambda_j \gamma \xi^\alpha \sin(\alpha\pi))^2}.$$

For any $v \in L^2(\Omega)$, we denote two operators

$$P_\alpha(t)v = \sum_{j=1}^\infty P_{\alpha,j}(t)(v, \varphi_j)\varphi_j(x), \quad S_\alpha(t)v = \sum_{j=1}^\infty \frac{P_{\alpha,j}(t)}{P_{\alpha,j}(b)}(v, \varphi_j)\varphi_j(x).$$

We use $u(t)$ to denote the spatial function $u(t, \cdot)$, then the equation (3.82) can be rewritten as

$$u(t) = S_\alpha(t)g - \int_0^b S_\alpha(t) \circ P_\alpha(b-s)f(s, u(s))ds + \int_0^t P_\alpha(t-s)f(s, u(s))ds, \quad (3.84)$$

where $\circ$ stands for the composite operation.

Noting that if u is an exact solution of problem (3.80), then u is at least once derivative with respect to time, this is a little bit constrained and higher smoothness is required for the given data. For this reason, a suitable definition of mild solutions for problem (3.80) is given below.

Definition 3.10. For $\mu \in (0, 1)$, a function u is called a mild solution of problem (3.80), if $u \in C^\mu([0, b]; L^2(\Omega))$ and it satisfies the integral equation (3.84).

Lemma 3.30. *[278, Lemma 2.2] For $\alpha \in (0, 1)$ and $t \in [0, b]$, we have the following estimates*

$$P_{\alpha,j}(t) \le \frac{C_1}{1 + \lambda_j t^{1-\alpha}}, \quad P_{\alpha,j}(b) \ge \frac{C_2}{\lambda_j},$$

where

$$C_1 = \frac{\Gamma(1-\alpha)}{\gamma\pi\sin(\alpha\pi)} + 1, \quad C_2 = \frac{\gamma\sin(\alpha\pi)e^{-b}}{3\pi(\alpha+1)(\gamma^2 + 1 + \frac{1}{\lambda_1^2})}.$$

In the following, we give some properties of operators $P_\alpha(t)$ and $S_\alpha(t)$.

Lemma 3.31. *[409, Lemma 3.1, Lemma 3.2] The operator $P_\alpha(t)$ is compact and continuous in the uniform operator topology for $t > 0$.*

For $t \in [0, b]$ and $v \in C^\mu([0, b]; L^2(\Omega))$, we define an operator as follows:

$$(Q_\alpha v)(t) = \int_0^b S_\alpha(t) \circ P_\alpha(b-s)v(s)ds.$$

Lemma 3.32. *Let $1 - \alpha < \mu < 1$. Then $Q_\alpha : C^\mu([0, b]; L^2(\Omega)) \to C^\mu([0, b]; L^2(\Omega))$ is a completely continuous operator.*

Proof. For any $n \in \mathbb{N}^+$ and $t \in (0, b]$, let $\Omega_n = \mathrm{span}\{\varphi_1, ..., \varphi_n\}$ and define $Q_\alpha^n :$ $(0, b] \times L^2(\Omega) \to \Omega_n$ as follows:

$$(Q_\alpha^n v)(t) = \int_0^b S_\alpha^n(t) \circ P_\alpha^n(b-s)v(s)ds, \quad v \in C^\mu([0, b]; L^2(\Omega)),$$

where

$$P_\alpha^n(t)v = \sum_{j=1}^n P_{\alpha,j}(t)(v, \varphi_j)\varphi_j(x), \quad S_\alpha^n(t)v = \sum_{j=1}^n \frac{P_{\alpha,j}(t)}{P_{\alpha,j}(b)}(v, \varphi_j)\varphi_j(x).$$

Clearly, Ω_n is a finite dimensional subspace of $L^2(\Omega)$. In the sequel, we choose a bounded subset $\Lambda_r = \{v \in C^\mu([0, b]; L^2(\Omega)) : \|v\|_\mu \le r\}$ of $C^\mu([0, b]; L^2(\Omega))$, here r

is a positive constant. Then we need to prove that the set $\{(Q_\alpha^n v)(t) : v \in \Lambda_r\}$ is relatively compact in $C^\mu([0,b]; L^2(\Omega))$ for every $n \in \mathbb{N}^+$.

For any $v \in \Lambda_r$, in view of Lemma 3.30, we have

$$\|S_\alpha^n(t) \circ P_\alpha^n(b-s)v\|$$

$$= \left(\sum_{j=1}^n \left(\frac{P_{\alpha,j}(t)}{P_{\alpha,j}(b)} P_{\alpha,j}(b-s) \right)^2 |(v(s), \varphi_j)|^2 \right)^{\frac{1}{2}}$$

$$\leq \left(\sum_{j=1}^n \left(\frac{C_1}{1+\lambda_j t^{1-\alpha}} \cdot \frac{\lambda_j}{C_2} \cdot \frac{C_1}{1+\lambda_j(b-s)^{1-\alpha}} \right)^2 |(v(s), \varphi_j)|^2 \right)^{\frac{1}{2}} \qquad (3.85)$$

$$\leq \frac{C_1^2}{C_2} (b-s)^{\alpha-1} \|v(s)\|.$$

It yields that

$$\|(Q_\alpha^n v)(t)\| = \left\| \int_0^b S_\alpha^n(t) \circ P_\alpha^n(b-s)v(s)ds \right\|$$

$$\leq \int_0^b \|S_\alpha^n(t) \circ P_\alpha^n(b-s)v(s)\|ds$$

$$\leq \frac{C_1^2}{C_2} \int_0^b (b-s)^{\alpha-1} \|v(s)\|ds \qquad (3.86)$$

$$\leq \frac{C_1^2}{C_2} \frac{\Gamma(\alpha)\Gamma(1-\mu)}{\Gamma(1-\mu+\alpha)} b^{\alpha-\mu} r,$$

which implies that

$$\|(Q_\alpha^n v)\|_\mu \leq \frac{C_1^2}{C_2} \frac{\Gamma(\alpha)\Gamma(1-\mu)}{\Gamma(1-\mu+\alpha)} b^\alpha r.$$

This derives the set $\{(Q_\alpha^n v)(t) : v \in \Lambda_r\}$ is uniformly bounded.

Hence, by virtue of the Arzelà-Ascoli theorem, it remains to prove that $\{t^\mu(Q_\alpha^n v)(t) : v \in \Lambda_r\}$ is equicontinuous for $t \in [0,b]$. In order to achieve this aim, we need to check that $\lim_{t\to 0^+} t^\mu(Q_\alpha^n v)(t)$ exists and is finite. Indeed, for any $v \in \Lambda_r$, when $t = 0$, we have $(Q_\alpha^n v)(0) = \int_0^b S_\alpha^n(0) \circ P_\alpha^n(b-s)v(s)ds$, from (3.86), we know that

$$\|(Q_\alpha^n v)(0)\| \leq \frac{C_1^2}{C_2} \frac{\Gamma(\alpha)\Gamma(1-\mu)}{\Gamma(1-\mu+\alpha)} b^{\alpha-\mu} r < \infty,$$

which implies

$$\lim_{t\to 0^+} t^\mu(Q_\alpha^n v)(t) = 0.$$

Thus, for $t_1 = 0, t_2 \in (0,b]$ and any $v \in \Lambda_r$, it is easy to see that

$$\|t_2^\mu(Q_\alpha^n v)(t_2) - t_1^\mu(Q_\alpha^n v)(t_1)\| \leq t_2^\mu \frac{C_1^2}{C_2} \frac{\Gamma(\alpha)\Gamma(1-\mu)}{\Gamma(1-\mu+\alpha)} b^{\alpha-\mu} r$$

$$\to 0, \quad \text{as} \quad t_2 \to 0^+.$$

Moreover, for $0 < t_1 < t_2 \le b$ and any $v \in \Lambda_r$, we have

$$\|t_2^\mu (Q_\alpha^n v)(t_2) - t_1^\mu (Q_\alpha^n v)(t_1)\|$$

$$= \left\| t_2^\mu \int_0^b S_\alpha^n(t_2) \circ P_\alpha^n(b-s)v(s)ds - t_1^\mu \int_0^b S_\alpha^n(t_1) \circ P_\alpha^n(b-s)v(s)ds \right\|$$

$$\le t_2^\mu \left\| \int_0^b (S_\alpha^n(t_2) - S_\alpha^n(t_1)) \circ P_\alpha^n(b-s)v(s)ds \right\|$$

$$+ (t_2^\mu - t_1^\mu) \left\| \int_0^b S_\alpha^n(t_1) \circ P_\alpha^n(b-s)v(s)ds \right\|$$

$$=: I_1 + I_2.$$

By using the inequalities $1 - e^{-t} \le t$, $e^{-t} \le \frac{1}{t}$ for $t > 0$, from (3.83), we have

$$\|(S_\alpha^n(t_2) - S_\alpha^n(t_1)) \circ P_\alpha^n(b-s)v(s)\|^2$$

$$= \sum_{j=1}^n \left(\frac{|P_{\alpha,j}(t_2) - P_{\alpha,j}(t_1)|}{P_{\alpha,j}(b)} P_{\alpha,j}(b-s) \right)^2 |(v(s), \varphi_j)|^2$$

$$\le \sum_{j=1}^n \left(\frac{C_1}{C_2} \int_0^\infty |e^{-\xi t_2} - e^{-\xi t_1}| K_j(\xi)d\xi(b-s)^{\alpha-1} \right)^2 |(v(s), \varphi_j)|^2$$

$$\le \left(\frac{C_1(b-s)^{\alpha-1}}{C_2} \right)^2 \sum_{j=1}^n \left(\int_0^\infty e^{-\xi t_1}\xi(t_2 - t_1)K_j(\xi)d\xi \right)^2 |(v(s), \varphi_j)|^2$$

$$\le \left(\frac{C_1(b-s)^{\alpha-1}}{C_2} \right)^2 \sum_{j=1}^n \left(\frac{t_2 - t_1}{t_1} \int_0^\infty K_j(\xi)d\xi \right)^2 |(v(s), \varphi_j)|^2$$

$$\le \left(\frac{C_1(b-s)^{\alpha-1}}{C_2} \right)^2 \left(\frac{t_2 - t_1}{t_1} \right)^2 \|v(s)\|^2,$$

where we use the fact $\int_0^\infty K_j(\xi)d\xi = 1$. Therefore, we get

$$I_1 \le \frac{t_2^\mu(t_2 - t_1)}{t_1} \frac{C_1\Gamma(\alpha)\Gamma(1-\mu)}{C_2\Gamma(\alpha+1-\mu)} b^{\alpha-\mu} r \to 0, \quad \text{as} \quad t_2 \to t_1.$$

From (3.86), we have

$$I_2 = (t_2^\mu - t_1^\mu)\|(Q_\alpha^n v)(t_1)\|$$

$$\le \frac{C_1^2}{C_2}(t_2^\mu - t_1^\mu)\frac{\Gamma(\alpha)\Gamma(1-\mu)}{\Gamma(1-\mu+\alpha)} b^{\alpha-\mu} r \to 0, \quad \text{as} \quad t_2 \to t_1.$$

Thus, $\|t_2^\mu(Q_\alpha^n v)(t_2) - t_1^\mu(Q_\alpha^n v)(t_1)\| \to 0$ as $t_2 \to t_1$, which implies that the set $\{t^\mu(Q_\alpha^n v)(t) : v \in \Lambda_r\}$ is equicontinuous for $t \in [0, b]$. Consequently, for every $n \in \mathbb{N}^+$, Q_α^n is a compact operator.

In the sequel, we prove the compactness of operator Q_α. From the fact that Q_α^n is compact, it requires to verify $Q_\alpha^n \to Q_\alpha$ as $n \to \infty$ in $C^\mu([0,b]; L^2(\Omega))$. Indeed, for any $v \in C^\mu([0,b]; L^2(\Omega))$, we have

$$\|(Q_\alpha v)(t) - (Q_\alpha^n v)(t)\| = \left\| \int_0^b (S_\alpha(t) \circ P_\alpha(b-s) - S_\alpha^n(t) \circ P_\alpha^n(b-s)) v(s) ds \right\|$$

$$\leq \int_0^b \left(\sum_{j=n+1}^\infty \left(\frac{P_{\alpha,j}(t)}{P_{\alpha,j}(b)} P_{\alpha,j}(b-s) \right)^2 |(v(s), \varphi_j)|^2 \right)^{\frac{1}{2}} ds$$

$$\leq \frac{C_1^2}{C_2} \frac{1}{1 + \lambda_{n+1} t^{1-\alpha}} \int_0^b (b-s)^{\alpha-1} \|v(s)\| ds$$

$$\leq \frac{C_1^2}{C_2} \frac{1}{1 + \lambda_{n+1} t^{1-\alpha}} \frac{\Gamma(\alpha)\Gamma(1-\mu)}{\Gamma(\alpha+1-\mu)} b^{\alpha-\mu} \|v\|_\mu.$$

Therefore, from the fact that $\lambda_{n+1} \to \infty$ as $n \to \infty$, we have

$$\|Q_\alpha v - Q_\alpha^n v\|_\mu \leq \frac{C_1^2}{C_2} \frac{\Gamma(\alpha)\Gamma(1-\mu)}{\Gamma(\alpha+1-\mu)} b^{2\alpha-1} \frac{1}{\lambda_{n+1}} \|v\|_\mu$$

$$\to 0, \quad \text{as} \quad n \to \infty.$$

This implies the operator Q_α is compact in $C^\mu([0,b]; L^2(\Omega))$.

Finally, we prove the continuity of Q_α. In fact, let $\{v_m\}_{m=1}^\infty \subset C^\mu([0,b]; L^2(\Omega))$ and $v \in C^\mu([0,b]; L^2(\Omega))$, which satisfy $\lim_{m\to\infty} v_m = v$. Similar to the process of (3.85), we have

$$\|S_\alpha(t) \circ P_\alpha(b-s)(v_m(s) - v(s))\| \leq \frac{C_1^2}{C_2} (b-s)^{\alpha-1} \|v_m(s) - v(s)\|, \qquad (3.87)$$

which implies that

$$\|(Q_\alpha v_m)(t) - (Q_\alpha v)(t)\| = \left\| \int_0^b S_\alpha(t) \circ P_\alpha(b-s)(v_m(s) - v(s)) ds \right\|$$

$$\leq \frac{C_1^2}{C_2} \int_0^b (b-s)^{\alpha-1} \|v_m(s) - v(s)\| ds$$

$$\leq \frac{C_1^2}{C_2} \frac{\Gamma(\alpha)\Gamma(1-\mu)}{\Gamma(\alpha+1-\mu)} b^{\alpha-\mu} \|v_m - v\|_\mu.$$

Hence

$$\|Q_\alpha v_m - Q_\alpha v\|_\mu \leq \frac{C_1^2}{C_2} \frac{\Gamma(\alpha)\Gamma(1-\mu)}{\Gamma(\alpha+1-\mu)} b^\alpha \|v_m - v\|_\mu \to 0, \quad \text{as} \quad m \to \infty.$$

This means that Q_α is continuous on $C^\mu([0,b]; L^2(\Omega))$. Combined with the above arguments, we conclude that Q_α is completely continuous. The proof is completed. $\square$

Before giving our main result, we need the following hypothesis:

(H1) $f(t, v)$ is measurable for $t \in (0, b]$, $f(t, v)$ is continuous for $v \in L^2(\Omega)$ and there exists a constant $C_f > 0$ such that

$$\|f(t, v)\| \leq C_f \|v\|.$$

Theorem 3.14. *Assume $g \in L^2(\Omega)$ and (H1) holds. If $\mu \in (1 - \alpha, 1)$, then the problem (3.80) has at least one mild solution in $C^\mu([0, b]; L^2(\Omega))$ provided that*

$$\frac{C_1^2 C_f \Gamma(\alpha) \Gamma(1 - \mu)}{C_2 \Gamma(\alpha + 1 - \mu)} b^\alpha + \frac{C_1 C_f}{1 - \mu} b < 1. \tag{3.88}$$

Proof. For any $u \in C^\mu([0, b]; L^2(\Omega))$, we define an operator as follows

$$(Tu)(t) = S_\alpha(t)g - (Q_\alpha f)(t) + (G_\alpha f)(t),$$

where

$$(G_\alpha f)(t) = \int_0^t P_\alpha(t - s) f(s, u(s)) ds.$$

Clearly, proving the existence of solutions to the problem (3.80) is equivalent to proving that the operator equation $Tu = u$ has at least one fixed point in $C^\mu([0, b]; L^2(\Omega))$. In view of inequality (3.88), let us choose a positive constant r such that

$$\left(1 - \frac{C_1^2 C_f \Gamma(\alpha) \Gamma(1 - \mu)}{C_2 \Gamma(\alpha + 1 - \mu)} b^\alpha - \frac{C_1 C_f}{1 - \mu} b \right)^{-1} \frac{C_1}{C_2} b^{\alpha - 1 + \mu} \|g\| \leq r,$$

and we denote a bounded closed and convex subset $B_r = \{v \in C^\mu([0, b]; L^2(\Omega)) : \|v\|_\mu \leq r\}$ of $C^\mu([0, b]; L^2(\Omega))$. Next, we will check that T is a completely continuous operator and T maps B_r into B_r.

From Lemma 3.30, for any $v \in C^\mu([0, b]; L^2(\Omega))$, we first estimate the operator G_α as follows:

$$\|(G_\alpha v)(t)\| \leq C_1 \int_0^t \|v(s)\| ds \leq \frac{C_1}{1 - \mu} t^{1 - \mu} \|v\|_\mu. \tag{3.89}$$

Claim I. The set $\{(Tu)(t) : u \in B_r\}$ is relatively compact in $C^\mu([0, b]; L^2(\Omega))$.

From Lemma 3.32 and the assumption of f, Q_α is completely continuous from $C^\mu([0, b]; L^2(\Omega))$ into $C^\mu([0, b]; L^2(\Omega))$, and for any $u \in C^\mu([0, b]; L^2(\Omega))$, it is easy to see that $f(t, u) \in C^\mu([0, b]; L^2(\Omega))$, which implies that $\{(Q_\alpha f)(t) : u \in B_r\}$ is relatively compact in $C^\mu([0, b]; L^2(\Omega))$. We only need to prove that $\{(G_\alpha f)(t) : u \in B_r\}$ is relatively compact in $C^\mu([0, b]; L^2(\Omega))$.

From (3.89) and (H1), we have

$$t^\mu \|(G_\alpha f)(t)\| \leq C_1 t^\mu \int_0^t \|f(s, u(s))\| ds$$

$$\leq C_1 C_f t^\mu \int_0^t \|u(s)\| ds$$

$$\leq \frac{C_1 C_f}{1 - \mu} tr,$$

it yields that

$$\|G_\alpha f\|_\mu \le \frac{C_1 C_f}{1-\mu} br,$$

which implies that G_α is a uniformly bounded operator.

Now we prove the set $\{t^\mu (G_\alpha f)(t) : u \in B_r\}$ is equicontinuous for $t \in [0, b]$. For $t_1 = 0, 0 < t_2 \le b$, from (H1), we have

$$\|t_2^\mu (G_\alpha f)(t_2) - t_1^\mu (G_\alpha f)(t_1)\| = \left\| t_2^\mu \int_0^{t_2} P_\alpha (t_2 - s) f(s, u(s)) ds \right\|$$

$$\le \frac{C_1 C_f}{1-\mu} t_2 r \to 0, \quad \text{as } t_2 \to 0^+.$$

For $0 < t_1 < t_2 \le b$, we have

$$\|t_2^\mu (G_\alpha f)(t_2) - t_1^\mu (G_\alpha f)(t_1)\|$$

$$= \left\| t_2^\mu \int_0^{t_2} P_\alpha (t_2 - s) f(s, u(s)) ds - t_1^\mu \int_0^{t_1} P_\alpha (t_1 - s) f(s, u(s)) ds \right\|$$

$$\le \left\| t_2^\mu \int_0^{t_1 - \delta} (P_\alpha (t_2 - s) - P_\alpha (t_1 - s)) f(s, u(s)) ds \right\|$$

$$+ \left\| t_2^\mu \int_{t_1 - \delta}^{t_1} (P_\alpha (t_2 - s) - P_\alpha (t_1 - s)) f(s, u(s)) ds \right\|$$

$$+ \left\| t_2^\mu \int_{t_1}^{t_2} P_\alpha (t_2 - s) f(s, u(s)) ds \right\| + \left\| (t_2^\mu - t_1^\mu) \int_0^{t_1} P_\alpha (t_1 - s) f(s, u(s)) ds \right\|$$

$$=: I_1 + I_2 + I_3 + I_4.$$

Obviously, from Lemma 3.30, (H1) and Lemma 3.31, we have

$$I_1 \le t_2^\mu \int_0^{t_1 - \delta} \|P_\alpha (t_2 - s) - P_\alpha (t_1 - s)\|_{\mathcal{B}(L^2(\Omega))} \|f(s, u(s))\| ds$$

$$\le \frac{C_f t_2^\mu (t_1 - \delta)^{1-\mu}}{1-\mu} \sup_{s \in [0, t_1 - \delta]} \|P_\alpha (t_2 - s) - P_\alpha (t_1 - s)\|_{\mathcal{B}(L^2(\Omega))} r$$

$$\to 0, \quad \text{as } t_2 \to t_1.$$

From Lemma 3.30 and (H1), in the case $\delta \to 0$, we have

$$I_2 \le t_2^\mu \int_{t_1 - \delta}^{t_1} \|P_\alpha (t_2 - s) - P_\alpha (t_1 - s)\|_{\mathcal{B}(L^2(\Omega))} \|f(s, u(s))\| ds$$

$$\le 2 C_1 C_f t_2^\mu \int_{t_1 - \delta}^{t_1} \|u(s)\| ds$$

$$\le \frac{2 C_1 C_f t_2^\mu}{1-\mu} \left(t_1^{1-\mu} - (t_1 - \delta)^{1-\mu} \right) r \to 0.$$

As for I_3, one has

$$I_3 \le t_2^\mu \int_{t_1}^{t_2} \|P_\alpha (t_2 - s)\|_{\mathcal{B}(L^2(\Omega))} \|f(s, u(s))\| ds$$

$$\leq C_1 C_f t_2^\mu \int_{t_1}^{t_2} \|u(s)\| ds$$

$$\leq \frac{C_1 C_f t_2^\mu}{1-\mu} (t_2^{1-\mu} - t_1^{1-\mu}) r \to 0, \quad \text{as} \quad t_2 \to t_1.$$

And it is easy to see that

$$I_4 \leq (t_2^\mu - t_1^\mu) \int_0^{t_1} \|P_\alpha(t_1 - s)\|_{\mathcal{B}(L^2(\Omega))} \|f(s, u(s))\| ds$$

$$\leq C_1 C_f (t_2^\mu - t_1^\mu) \int_0^{t_1} \|u(s)\| ds$$

$$\leq \frac{C_1 C_f t_1^{1-\mu}}{1-\mu} (t_2^\mu - t_1^\mu) r \to 0, \quad \text{as } t_2 \to t_1.$$

Therefore,

$$\|t_2^\mu (G_\alpha f)(t_2) - t_1^\mu (G_\alpha f)(t_1)\| \to 0, \quad \text{as } t_2 \to t_1,$$

which implies that the set $\{t^\mu (G_\alpha f)(t) : u \in B_r\}$ is equicontinuous on $[0, b]$.

From Lemma 3.31, we know that $\{P_\alpha(\epsilon) \int_0^{t-\epsilon} P_\alpha(t-s) f(s, u(s)) ds : u \in B_r\}$ is relatively compact, then we have

$$\left\| P_\alpha(\epsilon) \int_0^{t-\epsilon} P_\alpha(t-s) f(s, u(s)) ds - (G_\alpha f)(t, u) \right\|$$

$$\leq \|P_\alpha(\epsilon) - I\|_{\mathcal{B}(L^2(\Omega))} \left\| \int_0^{t-\epsilon} P_\alpha(t-s) f(s, u(s)) ds \right\| + \left\| \int_{t-\epsilon}^{t} P_\alpha(t-s) f(s, u(s)) ds \right\|$$

$$\leq \|P_\alpha(\epsilon) - I\|_{\mathcal{B}(L^2(\Omega))} \frac{C_1 C_f}{1-\mu} (t-\epsilon)^{1-\mu} r + \frac{C_1 C_f}{1-\mu} \left(t^{1-\mu} - (t-\epsilon)^{1-\mu} \right) r$$

$$\to 0, \quad \text{as } \epsilon \to 0.$$

Thus, the set $\{(G_\alpha f)(t) : u \in B_r\}$ is relatively compact in $C^\mu([0, b]; L^2(\Omega))$.

Therefore, the set $\{(Tu)(t) : u \in B_r\}$ is relatively compact in $C^\mu([0, b]; L^2(\Omega))$.

Claim II. T maps B_r into B_r.

For any $u \in B_r$, from the above discussion, we get that $t^\mu (Q_\alpha f)(t)$ and $t^\mu (G_\alpha f)(t)$ is continuous for $t \in [0, b]$. It remains to show the continuity of $t^\mu S_\alpha(t) g$. At first, from the expression of $S_\alpha(t)$ and Lemma 3.30, we can estimate $\|S_\alpha(t) g\|$ as follows

$$\|S_\alpha(t) g\| \leq \frac{C_1}{C_2} t^{\alpha-1} \|g\|, \tag{3.90}$$

it yields that

$$t^\mu \|S_\alpha(t) g\| \leq \frac{C_1}{C_2} t^{\alpha-1+\mu} \|g\| \to 0, \quad \text{as } t \to 0^+,$$

which implies that $\lim_{t \to 0^+} t^\mu S_\alpha(t) g$ exists and is finite. Thus, for $t_1 = 0, 0 < t_2 \leq b$, we have

$$\|t_2^\mu S_\alpha(t_2) g - t_1^\mu S_\alpha(t_1) g\| \to 0, \quad \text{as } t_2 \to t_1.$$

From the expression of $K_j(\xi)$, we can easy to check that $K_j(\xi) \leq \frac{1}{\pi\lambda_j\gamma\xi^\alpha\sin(\alpha\pi)}$. For $0 < t_1 < t_2 \leq b$, we estimate $P_{\alpha,j}(t)$ in the following

$$|P_{\alpha,j}(t_2) - P_{\alpha,j}(t_1)| = \int_{t_1}^{t_2}\int_0^\infty e^{-\xi t}\xi K_j(\xi)d\xi dt$$

$$\leq \frac{1}{\pi\lambda_j\gamma\sin(\alpha\pi)}\int_{t_1}^{t_2}\int_0^\infty e^{-\xi t}\xi^{1-\alpha}d\xi dt$$

$$= \frac{\Gamma(2-\alpha)}{(1-\alpha)\pi\lambda_j\gamma\sin(\alpha\pi)}(t_1^{\alpha-1} - t_2^{\alpha-1}).$$

By virtue of Lemma 3.30, (3.90) and combines with the above results, we have

$$\|t_2^\mu S_\alpha(t_2)g - t_1^\mu S_\alpha(t_1)g\|$$

$$\leq t_2^\mu\|S_\alpha(t_2)g - S_\alpha(t_1)g\| + (t_2^\mu - t_1^\mu)\|S_\alpha(t_1)g\|$$

$$\leq \frac{\Gamma(2-\alpha)}{C_2(1-\alpha)\pi\gamma\sin(\alpha\pi)}(t_1^{\alpha-1} - t_2^{\alpha-1})\|g\| + \frac{C_1}{C_2}t_1^{\alpha-1}(t_2^\mu - t_1^\mu)\|g\|$$

$$\to 0, \quad \text{as } t_2 \to t_1.$$

Thus we obtain that $t^\mu S_\alpha(t)g$ is continuous for $t \in [0,b]$. This implies that $Tu \in C^\mu([0,b]; L^2(\Omega))$.

For any $u \in B_r$, from (3.87), (3.89) and (3.90), we have

$$\|(Tu)(t)\| \leq \|S_\alpha(t)g\| + \|(Q_\alpha f)(t)\| + \|(G_\alpha f)(t)\|$$

$$\leq \frac{C_1}{C_2}t^{\alpha-1}\|g\| + \frac{C_1^2 C_f\Gamma(\alpha)\Gamma(1-\mu)}{C_2\Gamma(\alpha+1-\mu)}b^{\alpha-\mu}\|u\|_\mu + \frac{C_1 C_f}{1-\mu}t^{1-\mu}\|u\|_\mu$$

$$\leq \frac{C_1}{C_2}t^{\alpha-1}\|g\| + \frac{C_1^2 C_f\Gamma(\alpha)\Gamma(1-\mu)}{C_2\Gamma(\alpha+1-\mu)}b^{\alpha-\mu}r + \frac{C_1 C_f}{1-\mu}t^{1-\mu}r,$$

which implies that

$$t^\mu\|(Tu)(t)\| \leq \frac{C_1}{C_2}t^{\alpha-1+\mu}\|g\| + \frac{C_1^2 C_f\Gamma(\alpha)\Gamma(1-\mu)}{C_2\Gamma(\alpha+1-\mu)}b^{\alpha-\mu}rt^\mu + \frac{C_1 C_f}{1-\mu}tr.$$

It yields that

$$\|Tu\|_\mu \leq \frac{C_1}{C_2}b^{\alpha-1+\mu}\|g\| + \frac{C_1^2 C_f\Gamma(\alpha)\Gamma(1-\mu)}{C_2\Gamma(\alpha+1-\mu)}b^\alpha r + \frac{C_1 C_f}{1-\mu}br \leq r.$$

Thus, we conclude that T maps B_r into B_r.

Claim III. T is a continuous operator of $u \in B_r$.

From Lemma 3.32, it is easy to see that Q_α is continuous in B_r. In the sequel, we only need to prove the continuity of G_α. Indeed, let $\{u_n\}_{n=1}^\infty \subset B_r$ and $u \in B_r$, assume that $\lim_{n\to\infty} u_n = u$, from hypothesis (H1), for $t \in (0, b]$, we have

$$\|f(t, u_n(t)) - f(t, u(t)))\| \to 0, \quad \text{as } n \to \infty$$

and

$$\|f(s, u_n(s)) - f(s, u(s)))\| \leq 2\|f(s, u(s)))\| \leq 2C_f s^{-\mu}\|u\|_\mu,$$

it is easy to see that $s^{-\mu}$ is integrable on $(0, t]$, thus by the Lebesgue's dominated convergence theorem, we have

$$\int_0^t \|f(s, u_n(s)) - f(s, u(s)))\| ds \to 0, \quad \text{as } n \to \infty.$$

On the other hand, we get

$$\int_0^t \|P_\alpha(t - s)(f(s, u_n(s)) - f(s, u(s)))\| ds \leq C_1 \int_0^t \|(f(s, u_n(s)) - f(s, u(s)))\| ds,$$

which implies that G_α is continuous in B_r. Thus T is continuous of u in B_r.

From the above arguments, we conclude that T is completely continuous, by applying the Schauder fixed point theorem, we conclude that the operator T has a fixed point on $C^\mu([0, b]; L^2(\Omega))$, which implies that the problem (3.80) has at least one mild solution. $\qquad\square$

3.6.4 *Quasi-boundary Value Method*

In this subsection, in view of inequality (3.90), one can check that $S_\alpha(\cdot)g$ does not belong to $L^\infty(0, b; L^2(\Omega))$, and then the solution u of the problem (3.80) does not belong to $L^\infty(0, b; L^2(\Omega))$, it implies that the problem (3.80) is ill-posed in the sense of Hadamard. In order to solve this puzzle, the quasi-boundary value method is used to regularize the problem (3.80), and the corresponding regularized problem is given in the following:

$$\begin{cases} \partial_t v - (1 + \gamma \partial_t^\alpha)\Delta v = f(t, x, v), & x \in \Omega, \ t \in (0, b), \\ v(t, x) = 0, & x \in \partial\Omega, \ t \in (0, b), \\ v(b, x) + \beta(\delta)v(0, x) = g^\delta(x), & x \in \Omega, \end{cases} \quad (3.91)$$

where $\beta(\delta) > 0$ is a regularization parameter and the noise data g^δ satisfies

$$\|g^\delta - g\| \leq \delta. \quad (3.92)$$

In what follows, we give the definition of mild solutions for the problem (3.91), and then, the well-posed results are obtained. Assume that u_β^δ is a solution of the problem (3.91), from [409, Definition 3.1], under the initial condition $u_\beta^\delta(0)$, u_β^δ can be described as

$$u_\beta^\delta(t, x) = \sum_{j=1}^\infty P_{\alpha,j}(t)(u_\beta^\delta(0), \varphi_j)\varphi_j(x) + \sum_{j=1}^\infty \int_0^t P_{\alpha,j}(t - s)(f(s, u_\beta^\delta(s)), \varphi_j)ds\varphi_j(x),$$

$$(3.93)$$

substituting $t = b$ into (3.93), we have

$$u_\beta^\delta(b, x) = \sum_{j=1}^\infty P_{\alpha,j}(b)(u_\beta^\delta(0), \varphi_j)\varphi_j(x) + \sum_{j=1}^\infty \int_0^b P_{\alpha,j}(b - s)(f(s, u_\beta^\delta(s)), \varphi_j)ds\varphi_j(x).$$

By virtue of $u_\beta^\delta(b, x) + \beta(\delta)u_\beta^\delta(0, x) = g^\delta(x)$ in (3.91), we obtain

$$(u_\beta^\delta(0), \varphi_j) = \frac{1}{\beta(\delta) + P_{\alpha,j}(b)} \left((g^\delta, \varphi_j) - \int_0^b P_{\alpha,j}(b - s)(f(s, u_\beta^\delta(s)), \varphi_j)ds \right),$$

$$(3.94)$$

which implies

$$u_\beta^\delta(t,x) = \sum_{j=1}^{\infty} \frac{P_{\alpha,j}(t)}{\beta(\delta) + P_{\alpha,j}(b)} \left((g^\delta, \varphi_j) - \int_0^b P_{\alpha,j}(b-s)(f(s, u_\beta^\delta(s)), \varphi_j)ds \right) \varphi_j(x)$$

$$+ \sum_{j=1}^{\infty} \int_0^t P_{\alpha,j}(t-s)(f(s, u_\beta^\delta(s)), \varphi_j)ds\varphi_j(x).$$

(3.95)

Let

$$S_\alpha^\beta(t)v = \sum_{j=1}^{\infty} \frac{P_{\alpha,j}(t)}{\beta(\delta) + P_{\alpha,j}(b)}(v, \varphi_j)\varphi_j, \quad \text{for} \quad v \in L^2(\Omega),$$

and let $u_\beta^\delta(t)$ stand for $u_\beta^\delta(t, \cdot)$, then (3.95) can be rewritten as

$$u_\beta^\delta(t) = S_\alpha^\beta(t)g^\delta - \int_0^b S_\alpha^\beta(t) \circ P_\alpha(b-s)f(s, u_\beta^\delta(s))ds + \int_0^t P_\alpha(t-s)f(s, u_\beta^\delta(s))ds.$$

(3.96)

According to the above discussion, we introduce the following definition of mild solutions of the problem (3.91).

Definition 3.11. A function u_β^δ is called a mild solution of the problem (3.91), if u_β^δ belongs to $L^\infty(0, b; L^2(\Omega))$ and it satisfies the integral equation (3.96).

Before giving our main results, we provide the following constraint of f.

(H2) For any $u, v \in L^2(\Omega)$, there exists a constant $C_f' > 0$ such that

$$\|f(t, u) - f(t, v)\| \le C_f'\|u - v\|, \quad \text{for} \quad t \in (0, b].$$

Theorem 3.15. *Assume that (H2) holds. For every $\delta > 0$ and $g^\delta \in L^2(\Omega)$, the problem (3.91) exists a unique solution in $L^\infty(0, b; L^2(\Omega))$ provided*

$$\frac{C_1^2 C_f' b^\alpha}{C_2 \alpha} + C_1 C_f' b < 1.$$

(3.97)

In addition, the solution depends continuously on g^δ.

Proof. For any $v \in L^\infty(0, b; L^2(\Omega))$, let

$$(\mathcal{T}v)(t) = S_\alpha^\beta(t)g^\delta - \int_0^b S_\alpha^\beta(t) \circ P_\alpha(b-s)f(s, v(s))ds + \int_0^t P_\alpha(t-s)f(s, v(s))ds.$$

Clearly, the operator $\mathcal{T}$ has a unique fixed point which means that the problem (3.91) has a unique solution. At the beginning, we prove that $\mathcal{T}$ maps $L^\infty(0, b; L^2(\Omega))$ into $L^\infty(0, b; L^2(\Omega))$.

For any $v \in L^\infty(0, b; L^2(\Omega))$, from Lemma 3.30 and (H2), we have

$$\|(\mathcal{T}v)(t)\| \le \frac{C_1}{\beta(\delta)}\|g^\delta\| + \frac{C_1^2 C_f'}{C_2} \int_0^b (b-s)^{\alpha-1}\|v(s)\|ds + C_1 C_f' \int_0^t \|v(s)\|ds$$

$$\le \frac{C_1}{\beta(\delta)}\|g^\delta\| + \frac{C_1^2 C_f' b^\alpha}{C_2 \alpha}\|v\|_\infty + C_1 C_f' t\|v\|_\infty,$$

which implies that

$$\|\mathcal{T}v\|_\infty \le \frac{C_1}{\beta(\delta)}\|g^\delta\| + \left(\frac{C_1^2 C_f' b^\alpha}{C_2 \alpha} + C_1 C_f' b\right)\|v\|_\infty.$$

Thus we get that $\mathcal{T}$ maps $L^\infty(0, b; L^2(\Omega))$ into itself.

For any $v, w \in L^\infty(0, b; L^2(\Omega))$, in virtue of (H2) and (3.97), we prove that $\mathcal{T}$ is a contraction operator.

$$\|(\mathcal{T}v)(t) - (\mathcal{T}w)(t)\|$$

$$\le \left\|\int_0^b S_\alpha^\beta(t) \circ P_\alpha(b - s)\left(f(s, v(s)) - f(s, w(s))\right) ds\right\|$$

$$+ \left\|\int_0^t P_\alpha(t - s)\left(f(s, v(s)) - f(s, w(s))\right) ds\right\|$$

$$\le \frac{C_1^2 C_f'}{C_2}\int_0^b (b - s)^{\alpha-1}\|v(s) - w(s)\| ds + C_1 C_f'\int_0^t \|v(s) - w(s)\| ds$$

$$\le \frac{C_1^2 C_f' b^\alpha}{C_2 \alpha}\|v - w\|_\infty + C_1 C_f' t\|v - w\|_\infty,$$

thus we have

$$\|\mathcal{T}v - \mathcal{T}w\|_\infty \le \left(\frac{C_1^2 C_f' b^\alpha}{C_2 \alpha} + C_1 C_f' b\right)\|v - w\|_\infty.$$

From the fact that $\frac{C_1^2 C_f' b^\alpha}{C_2 \alpha} + C_1 C_f' b < 1$, by using the contraction mapping principle, we conclude that the problem (3.91) exists a unique solution in $L^\infty(0, b; L^2(\Omega))$.

In the sequel, we prove the continuous dependence of the solution u_β^δ on the data g^δ. Denote u_β^δ and $\tilde{u}_\beta^\delta$ are two regularization solutions corresponding to the conditions g^δ and $\tilde{g}^\delta$, respectively. From Lemma 3.30 and (H2), for every $\delta > 0$, we have

$$\|u_\beta^\delta(t) - \tilde{u}_\beta^\delta(t)\|$$

$$\le \|S_\alpha^\beta(t)(g^\delta - \tilde{g}^\delta)\| + \left\|\int_0^b S_\alpha^\beta(t) \circ P_\alpha(b - s)(f(s, u_\beta^\delta(s)) - f(s, \tilde{u}_\beta^\delta(s))) ds\right\|$$

$$+ \left\|\int_0^t P_\alpha(t - s)(f(s, u_\beta^\delta(s)) - f(s, \tilde{u}_\beta^\delta(s))) ds\right\|$$

$$\le \frac{C_1}{\beta(\delta)}\|g^\delta - \tilde{g}^\delta\| + \frac{C_1^2 C_f'}{C_2}\int_0^b (b - s)^{\alpha-1}\|u_\beta^\delta(s) - \tilde{u}_\beta^\delta(s)\| ds$$

$$+ C_1 C_f'\int_0^t \|u_\beta^\delta(s) - \tilde{u}_\beta^\delta(s)\| ds$$

$$\le \frac{C_1}{\beta(\delta)}\|g^\delta - \tilde{g}^\delta\| + \left(\frac{C_1^2 C_f' b^\alpha}{C_2 \alpha} + C_1 C_f' b\right)\|u_\beta^\delta - \tilde{u}_\beta^\delta\|_\infty,$$

which implies that

$$\|u_\beta^\delta - \tilde{u}_\beta^\delta\|_\infty \le \left(1 - \frac{C_1^2 C_f' b^\alpha}{C_2\alpha} - C_1 C_f' b\right)^{-1} \frac{C_1}{\beta(\delta)}\|g^\delta - \tilde{g}^\delta\|.$$

Particularly,

$$\|u_\beta^\delta(0) - \tilde{u}_\beta^\delta(0)\|$$

$$\le \|S_\alpha^\beta(0)(g^\delta - \tilde{g}^\delta)\| + \left\|\int_0^b S_\alpha^\beta(0) \circ P_\alpha(b-s)(f(s, u_\beta^\delta(s)) - f(s, \tilde{u}_\beta^\delta(s)))ds\right\|$$

$$\le \frac{1}{\beta(\delta)}\|g^\delta - \tilde{g}^\delta\| + \frac{C_1 C_f' b^\alpha}{C_2\alpha}\|u_\beta^\delta - \tilde{u}_\beta^\delta\|_\infty$$

$$\le \frac{1}{\beta(\delta)}\left(1 + \left(1 - \frac{C_1^2 C_f' b^\alpha}{C_2\alpha} - C_1 C_f' b\right)^{-1} \frac{C_1^2 C_f' b^\alpha}{C_2\alpha}\right)\|g^\delta - \tilde{g}^\delta\|.$$

Thus we get the desired conclusion. The proof is completed. $\qquad\square$

Let

$$C_\beta = \begin{cases} l_1(\theta, C_2)\beta^\theta(\delta), & 0 < \theta < 1, \\ l_2(\theta, C_2, \lambda_1)\beta(\delta), & \theta \ge 1, \end{cases}$$

where l_1, l_2 are defined in Lemma 3.29. In the sequel, the convergence analysis result is given under a priori condition.

Theorem 3.16. *Assume the conditions in Theorem 3.15 hold, and the regularization parameter $\beta(\delta)$ satisfies the following condition:*

$$\lim_{\delta\to 0} \frac{\delta}{\beta(\delta)} = \lim_{\delta\to 0} \beta(\delta) = 0. \tag{3.98}$$

For $\theta > 0$, assume that $u(0) \in \mathcal{H}^\theta(\Omega)$. Then

$$\|u_\beta^\delta - u\|_\infty \le \left(1 - \frac{C_1^2 C_f' b^\alpha}{C_2\alpha} - C_1 C_f' b\right)^{-1} \left(\frac{C_1\delta}{\beta(\delta)} + C_\beta\|u(0)\|_{\mathcal{H}^\theta(\Omega)}\right).$$

Proof. From Theorem 3.15, we know that there exists a unique regularization solution $u_\beta^\delta \in L^\infty(0, b; L^2(\Omega))$, assume that u is a solution of the problem (3.80), for every $\delta > 0$, from (H2) and (3.92), it is not difficult to check that $u_\beta^\delta - u \in L^\infty(0, b; L^2(\Omega))$. In fact, denote

$$u_\beta(t) = S_\alpha^\beta(t)g - \int_0^b S_\alpha^\beta(t) \circ P_\alpha(b-s)f(s, u(s))ds + \int_0^t P_\alpha(t-s)f(s, u(s))ds.$$

It is easy to see that

$$\|u_\beta^\delta(t) - u(t)\| \le \|u_\beta^\delta(t) - u_\beta(t)\| + \|u_\beta(t) - u(t)\|.$$

In virtue of Lemma 3.30 and (H2), we estimate $\|u_\beta^\delta(t) - u_\beta(t)\|$ in the following:

$$\|u_\beta^\delta(t) - u_\beta(t)\|$$

$$\leq \|S_\alpha^\beta(t)(g^\delta - g)\| + \left\|\int_0^b S_\alpha^\beta(t) \circ P_\alpha(b-s)(f(s, u_\beta^\delta(s)) - f(s, u(s)))ds\right\|$$

$$+ \left\|\int_0^t P_\alpha(t-s)(f(s, u_\beta^\delta(s)) - f(s, u(s)))ds\right\|$$

$$\leq \frac{C_1}{\beta(\delta)}\|g^\delta - g\| + \frac{C_1^2 C_f'}{C_2}\int_0^b (b-s)^{\alpha-1}\|u_\beta^\delta(s) - u(s)\|ds$$

$$+ C_1 C_f' \int_0^t \|u_\beta^\delta(s) - u(s)\|ds$$

$$\leq \frac{C_1\delta}{\beta(\delta)} + \left(\frac{C_1^2 C_f' b^\alpha}{C_2 \alpha} + C_1 C_f' b\right)\|u_\beta^\delta - u\|_\infty.$$

In the sequel, we estimate $\|u_\beta(t) - u(t)\|$. From Lemma 3.29 and Lemma 3.30, we have

$$\|u_\beta(t) - u(t)\| = \left\|\sum_{j=1}^\infty \left(\frac{P_{\alpha,j}(t)}{\beta(\delta) + P_{\alpha,j}(b)} - \frac{P_{\alpha,j}(t)}{P_{\alpha,j}(b)}\right)\right.$$

$$\times \left.\left((g, \varphi_j) - \int_0^b P_{\alpha,j}(b-s)(f(s, u(s)), \varphi_j)ds\right)\varphi_j(x)\right\|$$

$$\leq \left(\sum_{j=1}^\infty \left(\frac{\beta(\delta)\lambda_j}{\beta(\delta)\lambda_j + C_2}\right)^2 |(u(0), \varphi_j)|^2\right)^{\frac{1}{2}}$$

$$\leq C_\beta \|u(0)\|_{\mathcal{H}^\theta(\Omega)},$$

which implies that

$$\|u_\beta(t) - u(t)\| \leq C_\beta \|u(0)\|_{\mathcal{H}^\theta(\Omega)}.$$

Based on the previous discussion, we obtain

$$\|u_\beta^\delta - u\|_\infty \leq \frac{C_1\delta}{\beta(\delta)} + \left(\frac{C_1^2 C_f' b^\alpha}{C_2 \alpha} + C_1 C_f' b\right)\|u_\beta^\delta - u\|_\infty + C_\beta \|u(0)\|_{\mathcal{H}^\theta(\Omega)},$$

from (3.97), we conclude that

$$\|u_\beta^\delta - u\|_\infty \leq \left(1 - \frac{C_1^2 C_f' b^\alpha}{C_2 \alpha} - C_1 C_f' b\right)^{-1}\left(\frac{C_1\delta}{\beta(\delta)} + C_\beta \|u(0)\|_{\mathcal{H}^\theta(\Omega)}\right).$$

Thus we get the desired conclusion and the convergence result is also obtained. Moreover, from (3.98), we can easy to see that $u_\beta^\delta \to u$ in $L^\infty(0, b; L^2(\Omega))$ as the regularization parameter $\delta \to 0$. The proof is completed. $\qquad\square$

3.7   Notes and Remarks

The results in Section 3.1 due to Zhou and Wang [409]. The main results in Section 3.2 are adopted from He, Zhou and Peng [163]. Section 3.3 is taken from Wang, Alsaedi, Ahmad and Zhou [346]. The contents in Section 3.4 are taken from Wang, Zhou, Alsaedi and Ahmad [348]. The results in Section 3.5 due to Peng and Zhou [296]. Section 3.6 is from Wang, Zhou and He [347].

Chapter 4

Fractional Fokker-Planck Equations

In this chapter, we study the time-fractional Fokker-Planck equations which can be used to describe the subdiffusion in an external time and space-dependent force field $F(t, x)$. In Section 4.1, we present some results on existence and uniqueness of mild solutions allowing the "working space" that may have low regularity. Secondly, we analyze the relationship between "working space" and the value range of α when investigating the problem of classical solutions. Finally, by constructing a suitable weighted Hölder continuous function space, the existence of classical solutions is derived without the restriction on $\alpha \in (\frac{1}{2}, 1)$. In Section 4.2, a time-fractional Fokker-Planck initial-boundary value problem is considered, the spatial domain $\Omega \subset \mathbb{R}^d$, where $d \geq 1$, has a smooth boundary. Existence, uniqueness and regularity of a mild solution u are proved under the hypothesis that the initial data u_0 lies in $L^2(\Omega)$. For $1/2 < \alpha < 1$ and $u_0 \in H^2(\Omega) \cap H_0^1(\Omega)$, it is shown that u becomes a classical solution of the problem. Estimates of time derivatives of the classical solution are derived.

4.1 Operator Method

4.1.1 *Introduction*

In the past few decades, the researches on anomalous diffusion problems with external force fields have been extensively investigated, and then the corresponding mathematical and physical model leads to the time-fractional Fokker-Planck equations. Metzler et al. in [264, 265] showed how fractional Fokker-Planck equations for the description of anomalous diffusion in external fields $F(x)$, can be derived from the generalized master equation, which shows that the probability density $u(t, x)$ for a particle to be at time t and position x obeys a time-fractional Fokker-Planck equation of the form

$$\partial_t u - \partial_t^{1-\alpha}(\kappa_\alpha u_{xx} - (Fu)_x) = 0,$$

where $\kappa_\alpha > 0$ is a constant, $\partial_t = \frac{\partial}{\partial_t}$ and $\partial_t^{1-\alpha}$ is the Riemann-Liouville fractional partial derivative of order $1 - \alpha \in (0, 1)$ defined by

$$\partial_t^{1-\alpha} u(t, x) = \frac{\partial}{\partial t} {}_0 D_t^{-\alpha} u(t, x), \quad t > 0,$$

with $_0D_t^{-\alpha}$ the Riemann-Liouville fractional integral given by

$$_0D_t^{-\alpha}u(t,x) = \frac{1}{\Gamma(\alpha)} \int_0^t (t-s)^{\alpha-1}u(s,x)ds, \quad t > 0,$$

where $\Gamma(\cdot)$ is the Gamma function.   It is worth mentioning that Barkai et al. [34] generalized the continuous time random walk theory to include the effect of space dependent jump probabilities. When the mean waiting time diverges they also derived a fractional Fokker-Planck equation. Subsequently, Henry et al. [166] considered the more general case when $F = F(t,x)$, which was formally equivalent to the subordinated stochastic Langevin equations for time- and space-dependent forces.

In this section, we study the following initial-boundary value problems of time-fractional Fokker-Planck equations in an open bounded domain $\Omega \subset \mathbb{R}^d (d \geq 1)$ with C^2 boundary:

$$\begin{cases} \partial_t u(t,x) - \nabla \cdot (\partial_t^{1-\alpha}\kappa_\alpha\nabla u - F\partial_t^{1-\alpha}u)(t,x) = f(t,x), & (t,x) \in (0,T) \times \Omega, \\ u(t,x) = 0, & (t,x) \in (0,T) \times \partial\Omega, \\ u(0,x) = u_0(x), & x \in \Omega. \end{cases}$$
$$(4.1)$$

It is generally known that the equation (4.1) would resolve itself into time-fractional diffusion equations when $F(t,x) = 0$, which has been the focus of many studies due to its significant application in subdiffusive model of anomalous diffusion processes.

Now, the researches on time-fractional Fokker-Planck equations were mainly focused on numerical analysis, see [173, 218]. However, its basic theoretical works are far from sufficient. McLean et al. [260] investigated the initial boundary value problem of a class of linear reaction convection-diffusion equations including the equation (4.1), and the well-posedness of weak solutions were obtained by novel energy methods in combination with a fractional Gronwall's inequality and properties of fractional integrals. Further, McLean et al. [261] obtained the regularity theory of weak solutions of a linear time-fractional, advection-diffusion-reaction equation. Recently, Le et al. [219] used the Galerkin approximation method to analyze the existence and uniqueness of mild solutions of the equations (4.1), and then the existence and regular estimates of the classical solution were obtained when $\alpha \in (\frac{1}{2}, 1)$. Compared with the diversity of the methods to time-fractional diffusion equations, the current methods of time-fractional Fokker-Planck equations are relatively single, which are basically based on the Galerkin approximation and energy estimates, combined with fractional calculus theory.

In fact, the research ideas of time-fractional diffusion equations can be basically divided into two aspects: the one is to deduce the form of mild solutions, and then analyze the properties of solutions via estimating the solution operator; the other is to establish energy estimates of approximate solutions, and then address the procedure of taking limits to obtain the properties of solutions. However, the researchers only use the second idea to make qualitative analysis of time-fractional

Fokker-Planck equations, which is insightful work. It's such a pity that the existence of classical solutions has remained elusive when $\alpha \in (0, \frac{1}{2}]$, the fundamental reason is a stubborn "working space". Considering the advantage of the first idea that the "working space" can be selected diversely, we first deduce the form of mild solutions to the objective problems expressed by Mittag-Leffler functions, and then use the properties of Mittag-Leffler functions to discuss that of solutions. The main novelties of our work are as follows.

- The representation of mild solutions without fractional derivative operators is given, which allows the "working space" of mild solutions that may have low regularity.
- The influence of the "working space" on the value range of α is analyzed when considering the existence of classical solutions.
- The existence of classical solutions is obtained without any restriction on α by constructing a suitable weighted Hölder continuous function space.

The remainder of the section is organized as follows. In the next subsection, we introduce some notations and lemmas that will be used throughout the section. In Subsection 4.1.3, firstly we give some operators and secondly we deduce the representation of mild solutions without fractional derivative operators. In Subsection 4.1.4, we derive the well-posedness and space-regularity of mild solutions. The well-posedness of classical solutions under the restriction on $\alpha \in (\frac{1}{2}, 1)$ is derived in Subsection 4.1.5. Finally, the existence and uniqueness of classical solutions are also obtained in a weighted Hölder continuous function space when $\alpha \in (0, 1)$.

4.1.2 *Preliminaries*

Throughout the section, we assume that the external force field $F = (F_1, ..., F_d)^T$, $F \in W^{1,1}(0, T; L^2(\Omega))$ denotes the space of all integrable functions $F_i : (0, T) \to L^2(\Omega)$ such that $F_i' \in L^1(0, T; L^2(\Omega))$. $F \in W^{1,\infty}(0, T; \Omega)$ denotes the space of all functions whose divergence $\nabla \cdot F$ is continuous on $[0, T] \times \bar{\Omega}$ with the norm

$$\|F\|_{W^{1,\infty}} := \|F\|_{\infty} + \max_{(t,x) \in [0,T] \times \Omega} |\nabla \cdot F(t, x)|,$$

where

$$\|F\|_{\infty} := \max_{1 \leq i \leq d} \max_{(t,x) \in [0,T] \times \Omega} |F_i(t, x)|.$$

The assumption on F will be given later.

Let $A = -\Delta$ and $\{\lambda_k\}$ be the set of eigenvalues of the operator A with a homogeneous boundary condition, let $\{e_k\}$ denote the complete orthonormal system of eigenfunctions which forms an orthogonal basis of $L^2(\Omega)$ such that

$$Ae_k = \lambda_k e_k, \quad x \in \Omega; \quad e_k|_{\partial\Omega} = 0, \quad k = 1, 2, ...,$$

where $0 < \lambda_1 \leq \lambda_2 \leq ... \leq \lambda_k \leq ...$ and $\lim_{k \to \infty} \lambda_k = +\infty$.

For all $\theta \geq 0$, the fractional power operator A^θ possesses the following representation:

$$A^\theta v = \sum_{k=1}^\infty \lambda_k^\theta (v, e_k) e_k, \quad \mathcal{D}(A^\theta) = \Big\{ v \in L^2(\Omega) : \sum_{k=1}^\infty \lambda_k^{2\theta} |(v, e_k)|^2 < \infty \Big\},$$

and that $\mathcal{D}(A^\theta)$ is a Hilbert space with the norm:

$$\|v\|_{\mathcal{D}(A^\theta)} = \Big(\sum_{k=1}^\infty \lambda_k^{2\theta} |(v, e_k)|^2 \Big)^{\frac{1}{2}}.$$

By duality, we can also set $H^{-1}(\Omega) = (H_0^1(\Omega))^*$, $\mathcal{D}(A^{-\theta}) = (\mathcal{D}(A^\theta))^*$. Then $\mathcal{D}(A^{-\theta})$ is a Hilbert space endowed with the norm $\|v\|_{\mathcal{D}(A^{-\theta})} = \Big(\sum_{k=1}^\infty \lambda_k^{-2\theta} |(v, e_k)|^2 \Big)^{\frac{1}{2}}$. We have $\mathcal{D}(A^\theta) \subset H^{2\theta}(\Omega)$ for $\theta > 0$. For the details of the Sobolev spaces with fractional powers $H^{2\theta}(\Omega)$, see Fujiwara [129] for example. In particular, $\mathcal{D}(A^{\frac{1}{2}}) = H_0^1(\Omega)$ and $\mathcal{D}(A) = H_0^1(\Omega) \cap H^2(\Omega)$, see Sakamoto and Yamamoto [313] and references therein.

Let us recall the Mittag-Leffler function

$$E_{\alpha,\beta}(z) = \sum_{k=0}^\infty \frac{z^k}{\Gamma(\alpha k + \beta)}, \quad \beta \in \mathbb{R}, \ z \in \mathbb{C}.$$

Then there exists a positive constant M_α such that

$$E_{\alpha,\beta}(-t) \leq \frac{M_\alpha}{1+t}, \quad \text{for } t \geq 0. \tag{4.2}$$

In particular, when $\beta = 1$ it is so called one parameter the Mittag-Leffler function and that for $\alpha = \beta = 1$ it is the exponential function.

Lemma 4.1. *[217] If $v : [0, T] \to H^1(\Omega)$ and $F \in W^{1,\infty}(0, T; \Omega)$, then*

$$\|\nabla \cdot (F(t) v(t))\| \leq \|F\|_{W^{1,\infty}} \|v(t)\|_{H^1(\Omega)}.$$

Let X be Banach space. The space of Hölder continuous functions on $[0, T]$ with the exponent $\theta \in (0, 1)$ is denoted by $C^\theta([0, T]; X)$ which has the representation form

$$C^\theta([0, T]; X) = \{ u \in C([0, T]; X) : \sup_{0 \leq t_1 < t_2 \leq T} \frac{|u(t_2) - u(t_1)|}{(t_2 - t_1)^\theta} < \infty \},$$

equipped with the norm

$$\|u\|_{C^\theta} = \sup_{0 \leq t_1 < t_2 \leq T} \frac{|u(t_2) - u(t_1)|}{(t_2 - t_1)^\theta}.$$

The weighted Hölder continuous function space of all continuous functions defined on $(0, T]$ for $0 < \theta < \eta < 1$ is denoted by $\mathcal{F}^{\eta,\theta}((0, T]; X)$ which satisfies:

(i) $\lim_{t \to 0} t^{1-\eta} u(t)$ exists;

(ii) u is Hölder continuous with the exponent θ and with the weighted $t_1^{1-\eta+\theta}$, i.e.,

$$\sup_{0 \le t_1 < t_2 \le T} \frac{t_1^{1-\eta+\theta}|u(t_2) - u(t_1)|}{(t_2 - t_1)^\theta} < \infty;$$

(iii) if $t \to 0$, then $\varphi_u(t) := \sup_{0 \le t_1 < t} \frac{t_1^{1-\eta+\theta}|u(t)-u(t_1)|}{(t-t_1)^\theta} \to 0$.

Then the space $\mathcal{F}^{\eta,\theta}((0,T];X)$ is a Banach space with the norm

$$\|u\|_{\mathcal{F}^{\eta,\theta}} = \sup_{t \in [0,T]} t^{1-\eta}|u(t)| + \sup_{0 \le t_1 < t_2 \le T} \frac{t_1^{1-\eta+\theta}|u(t_2) - u(t_1)|}{(t_2 - t_1)^\theta}.$$

4.1.3 The Solution Operators

Before giving the formal solution, we first consider the operators, for any $v \in L^2(\Omega)$ and a.e. $t > 0$,

$$S_\alpha(t)v = \sum_{k=1}^\infty E_{\alpha,1}(-\lambda_k \kappa_\alpha t^\alpha)(v, e_k)e_k(x),$$

$$Q_\alpha(t)v = \sum_{k=1}^\infty t^\alpha E_{\alpha,\alpha+1}(-\lambda_k \kappa_\alpha t^\alpha)(v, e_k)e_k(x),$$

$$P_\alpha(t)v = \sum_{k=1}^\infty t^{\alpha-1} E_{\alpha,\alpha}(-\lambda_k \kappa_\alpha t^\alpha)(v, e_k)e_k(x).$$

By $_0D_t^{-\alpha}[E_{\alpha,1}(\lambda t^\alpha)] = t^\alpha E_{\alpha,\alpha+1}(\lambda t^\alpha)$ and $\frac{d}{dt}[t^{\beta-1}E_{\alpha,\beta}(\lambda t^\alpha)] = t^{\beta-2}E_{\alpha,\beta-1}(\lambda t^\alpha)$ for $\beta, \lambda \in \mathbb{C}$ (see, eg, Podlubny [303]), we deduce that

$$_0D_t^{-\alpha}[S_\alpha(t)v] = Q_\alpha(t)v \quad \text{and} \quad Q'_\alpha(t)v = P_\alpha(t)v. \tag{4.3}$$

In what follows, we will also give the following estimates. They follow easily from (4.2) and some straightforward computations.

Lemma 4.2. *Let $0 \le \beta \le \gamma \le 1$ and $v \in \mathcal{D}(A^\beta)$. Then the following assertions hold:*

$$\|Q_\alpha(t)v\|_{\mathcal{D}(A^\gamma)} \le M_\alpha \kappa_\alpha^{-(\gamma-\beta)} t^{\alpha(1+\beta-\gamma)}\|v\|_{\mathcal{D}(A^\beta)},$$

$$\|P_\alpha(t)v\|_{\mathcal{D}(A^\gamma)} \le M_\alpha \kappa_\alpha^{-(\gamma-\beta)} t^{\alpha(1+\beta-\gamma)-1}\|v\|_{\mathcal{D}(A^\beta)},$$

$$\|S_\alpha(t)v\|_{\mathcal{D}(A^\gamma)} \le M_\alpha \kappa_\alpha^{-(\gamma-\beta)} t^{\alpha(\beta-\gamma)}\|v\|_{\mathcal{D}(A^\beta)},$$

for every $t > 0$.

Proof. If $0 \le \beta \le \gamma \le 1$, then for every $t > 0$ and $v \in \mathcal{D}(A^\beta)$, we have

$$\|Q_\alpha(t)v\|_{\mathcal{D}(A^\gamma)}^2 = \sum_{k=1}^\infty t^{2\alpha} \lambda_k^{2(\gamma-\beta)}[E_{\alpha,\alpha+1}(-\lambda_k \kappa_\alpha t^\alpha)]^2 \lambda_k^{2\beta}|(v, e_k)|^2$$

$$\le M_\alpha^2 t^{2\alpha} \sum_{k=1}^\infty \frac{\lambda_k^{2(\gamma-\beta)}}{(1 + \lambda_k \kappa_\alpha t^\alpha)^2} \lambda_k^{2\beta}|(v, e_k)|^2$$

$$=M_\alpha^2\kappa_\alpha^{-2(\gamma-\beta)}t^{2\alpha(1+\beta-\gamma)}\sum_{k=1}^{\infty}\left[\frac{(\lambda_k\kappa_\alpha t^\alpha)^{\gamma-\beta}}{1+\lambda_k\kappa_\alpha t^\alpha}\right]^2\lambda_k^{2\beta}|(v,e_k)|^2$$

$$\leq M_\alpha^2\kappa_\alpha^{-2(\gamma-\beta)}t^{2\alpha(1+\beta-\gamma)}\|v\|_{\mathcal{D}(A^\beta)}^2.$$

Similarly, the operators $P_\alpha(t)v$ and $S_\alpha(t)v$ can be estimated. Then the desired conclusion is obtained.   □

Next, we seek the solution $u(t,x)$ for the equation (4.1) in the form:

$$u(t,x)=\sum_{k=1}^{\infty}u_k(t)e_k(x). \qquad (4.4)$$

Take the inner product of both sides of the equation (4.1) with e_k to obtain

$$u_k'(t)+\lambda_k\kappa_\alpha\partial_t^{1-\alpha}u_k=f_k-(\nabla\cdot(F\partial_t^{1-\alpha}u),e_k)\quad\text{and}\quad u_k(0)=u_{0k}, \qquad (4.5)$$

where $u_{0k}=(u_0,e_k)$ and $f_k=(f,e_k)$, we can infer that the solution of the equation (4.5) is given as follows:

$$u_k(t)=E_{\alpha,1}(-\lambda_k\kappa_\alpha t^\alpha)u_k(0)$$

$$+\int_0^t E_{\alpha,1}(-\lambda_k\kappa_\alpha(t-s)^\alpha)[f_k-(\nabla\cdot(F\partial_s^{1-\alpha}u),e_k)]ds.$$

Thus, according to (4.4), it yields that

$$u(t,x)=\sum_{k=1}^{\infty}E_{\alpha,1}(-\lambda_k\kappa_\alpha t^\alpha)u_k(0)e_k(x)$$

$$+\sum_{k=1}^{\infty}\int_0^t E_{\alpha,1}(-\lambda_k\kappa_\alpha(t-s)^\alpha)[f_k-(\nabla\cdot(F\partial_s^{1-\alpha}u),e_k)]e_k(x)ds$$

$$=S_\alpha(t)u_0+\int_0^t S_\alpha(t-s)f(s)ds-\int_0^t S_\alpha(t-s)\nabla\cdot(F(s)\partial_s^{1-\alpha}u(s))ds.$$

$$(4.6)$$

Noting that the term $\partial_s^{1-\alpha}u$ appears in formal solutions and then the space of solutions needs more regularity. However, it is not necessary to take account of this case to the mild solutions, so we should remove it. To achieve this purpose, we integrate (4.6) by the Riemann-Liouville fractional integral of order α. From (4.3), one can immediately calculate that

$$_0D_t^{-\alpha}\left(\int_0^t S_\alpha(t-s)f(s)ds\right)=\frac{1}{\Gamma(\alpha)}\int_0^t(t-s)^{\alpha-1}\int_0^s S_\alpha(s-\tau)f(\tau)d\tau ds$$

$$=\frac{1}{\Gamma(\alpha)}\int_0^t\int_\tau^t(t-s)^{\alpha-1}S_\alpha(s-\tau)f(\tau)dsd\tau \qquad (4.7)$$

$$=\int_0^t Q_\alpha(t-s)f(s)ds.$$

On the other hand, since

$$\int_0^t \nabla \cdot (F(s)\partial_s^{1-\alpha}u(s))ds = \nabla \cdot (F(t)_0D_t^{-\alpha}u(t)) - \nabla \cdot \int_0^t (F'(s)_0D_s^{-\alpha}u(s))ds,$$

(4.8)

then using (4.7) and integrating by parts with respect to s, we can obtain from $Q_\alpha(0) = 0$ and (4.3) that

$$_0D_t^{-\alpha}\left(\int_0^t S_\alpha(t-s)\nabla \cdot (F(s)\partial_s^{1-\alpha}u(s))ds \right)$$

$$= \int_0^t Q_\alpha(t-s)\frac{\partial}{\partial s}\left[\int_0^s \nabla \cdot (F(\tau)\partial_\tau^{1-\alpha}u(\tau))d\tau \right]ds$$

$$= \int_0^t Q'_\alpha(t-s)\left[\nabla \cdot (F(s)_0D_s^{-\alpha}u(s)) - \nabla \cdot \int_0^s F'(\tau)_0D_\tau^{-\alpha}u(\tau)d\tau \right]ds$$

$$= \int_0^t P_\alpha(t-s)\nabla \cdot (F(s)_0D_s^{-\alpha}u(s)) - \int_0^t \left[\int_\tau^t P_\alpha(t-s)ds \right]\nabla \cdot (F'(\tau)_0D_\tau^{-\alpha}u(\tau))d\tau$$

$$= \int_0^t P_\alpha(t-s)\nabla \cdot (F(s)_0D_s^{-\alpha}u(s))ds - \int_0^t Q_\alpha(t-s)\nabla \cdot (F'(s)_0D_s^{-\alpha}u(s))ds.$$

This, together (4.3) and (4.7), ensures that

$$_0D_t^{-\alpha}u = Q_\alpha(t)u_0 + \int_0^t Q_\alpha(t-s)f(s)ds + \int_0^t Q_\alpha(t-s)\nabla \cdot (F'(s)_0D_s^{-\alpha}u(s))ds$$

$$- \int_0^t P_\alpha(t-s)\nabla \cdot (F(s)_0D_s^{-\alpha}u(s))ds.$$

(4.9)

From this point of view, we give the following definitions of mild solutions and classical solutions.

Definition 4.1. We say a function $u : [0,T] \to H^{-1}(\Omega)$ is a mild solution of the equation (4.1) provided $_0D_t^{-\alpha}u \in C([0,T]; L^2(\Omega))$ and u satisfies the integral equation (4.9).

Definition 4.2. We say a function $u : [0,T] \to L^2(\Omega)$ is a classical solution of the equation (4.1) provided $u(t) \in \mathcal{D}(A)$, $\partial_t^{1-\alpha}u(t) \in \mathcal{D}(A)$ and $\partial_t u \in L^2(\Omega)$ for a.e. $t \in [0,T]$ and u satisfies the equation (4.1).

4.1.4 Mild Solutions

In this subsection, we prove the well-posedness for the equation (4.1). Before this, we analysis the following integral equations:

$$w(t) = Q_\alpha(t)u_0 + \int_0^t Q_\alpha(t-s)f(s)ds + \int_0^t Q_\alpha(t-s)\nabla \cdot (F'(s)w(s))ds$$

$$- \int_0^t P_\alpha(t-s)\nabla \cdot (F(s)w(s))ds.$$

(4.10)

First, we give the existence and uniqueness of the solution to the integral equation (4.10), it also ensures that of the solution to the integral equation (4.9) provided we set $w = {}_0D_t^{-\alpha}u$.

Lemma 4.3. *Let $u_0 \in L^2(\Omega)$ and $f \in L^1(0,T;L^2(\Omega))$. Suppose that $F \in W^{1,1}(0,T; L^2(\Omega))$. Then the equation (4.10) has a unique mild solution in $C([0,T];L^2(\Omega))$.*

Proof. We define an operator $\mathcal{H}$ on $C([0,T];L^2(\Omega))$ as follows:

$$\mathcal{H}w(t) = Q_\alpha(t)u_0 + \int_0^t Q_\alpha(t-s)f(s)ds + \int_0^t Q_\alpha(t-s)\nabla \cdot (F'(s)w(s))ds$$
$$- \int_0^t P_\alpha(t-s)\nabla \cdot (F(s)w(s))ds. \tag{4.11}$$

Step 1. We shall show that $\mathcal{H}w \in C([0,T];L^2(\Omega))$ for any $w \in C([0,T];L^2(\Omega))$. For any $h > 0$, let $0 \le t < t+h \le T$, we have

$$\|\mathcal{H}w(t+h) - \mathcal{H}w(t)\|$$

$$\le \|Q_\alpha(t+h)u_0 - Q_\alpha(t)u_0\| + \left\| \int_0^{t+h} Q_\alpha(t+h-s)f(s)ds - \int_0^t Q_\alpha(t-s)f(s)ds \right\|$$

$$+ \left\| \int_0^{t+h} Q_\alpha(t+h-s)\nabla \cdot (F'(s)w(s))ds - \int_0^t Q_\alpha(t-s)\nabla \cdot (F'(s)w(s))ds \right\|$$

$$+ \left\| \int_0^{t+h} P_\alpha(t+h-s)\nabla \cdot (F(s)w(s))ds - \int_0^t P_\alpha(t-s)\nabla \cdot (F(s)w(s))ds \right\|$$

$$=: I_1(t,h) + I_2(t,h) + I_3(t,h) + I_4(t,h).$$

We estimate each term of right-handed sides of above inequality. Firstly, in view of

$$\lambda_k^\gamma \left| \tau_2^{\beta-1} E_{\alpha,\beta}(-\lambda_k\kappa_\alpha\tau_2^\alpha) - \tau_1^{\beta-1}E_{\alpha,\beta}(-\lambda_k\kappa_\alpha\tau_1^\alpha) \right|$$

$$= \lambda_k^\gamma \left| \int_{\tau_1}^{\tau_2} s^{\beta-2}E_{\alpha,\beta-1}(-\lambda_k\kappa_\alpha s^\alpha)ds \right|$$

$$\le M_\alpha \int_{\tau_1}^{\tau_2} s^{\beta-2} \frac{\lambda_k^\gamma}{1+\lambda_k\kappa_\alpha s^\alpha}ds$$

$$\le M_\alpha\kappa_\alpha^{-\gamma} \int_{\tau_1}^{\tau_2} s^{\beta-\alpha\gamma-2}ds$$

$$= \frac{M_\alpha\kappa_\alpha^{-\gamma}}{\beta-\alpha\gamma-1}[\tau_2^{\beta-\alpha\gamma-1} - \tau_1^{\beta-\alpha\gamma-1}], \quad \text{for } \gamma \in [0,1], \ \beta \ne \alpha\gamma+1, \ 0 \le \tau_1 < \tau_2, \tag{4.12}$$

we know that for $\beta = \alpha+1$ and $\gamma = 0$,

$$I_1(t,h) = \left(\sum_{k=1}^\infty |(t+h)^\alpha E_{\alpha,\alpha+1}(-\lambda_k\kappa_\alpha(t+h)^\alpha) \right.$$

$$\left. - t^\alpha E_{\alpha,\alpha+1}(-\lambda_k\kappa_\alpha t^\alpha)|^2 |(u_0,e_k)|^2 \right)^{\frac{1}{2}}$$

$$\le \frac{M_\alpha}{\alpha}[(t+h)^\alpha - t^\alpha]\|u_0\|.$$

We proceed to estimate $I_2(t, h)$. In view of Lemma 4.2, using the estimate of $I_1(t, h)$ and the inequality $\tau_1^\alpha - \tau_2^\alpha \le (\tau_1 - \tau_2)^\alpha$ for $\tau_1 \ge \tau_2 \ge 0$, we obtain

$$
I_2(t, h) \le \int_t^{t+h} \|Q_\alpha(t + h - s) f(s)\| ds + \int_0^t \|[Q_\alpha(t + h - s) - Q_\alpha(t - s)] f(s)\| ds
$$

$$
\le M_\alpha \int_t^{t+h} (t + h - s)^\alpha \|f(s)\| ds + \frac{M_\alpha}{\alpha} \int_0^t [(t + h - s)^\alpha - (t - s)^\alpha] \|f(s)\| ds
$$

$$
\le M_\alpha h^\alpha \left(1 + \frac{1}{\alpha}\right) \|f\|_{L^1(0,T;L^2(\Omega))}.
$$

For $I_3(t, h)$, we use the case $\beta = \alpha + 1, \gamma = \frac{1}{2}$ of (4.12) to derive

$$
\|A^{\frac{1}{2}} Q_\alpha(\tau_2) v - A^{\frac{1}{2}} Q_\alpha(\tau_1) v\|
$$

$$
= \left(\sum_{k=1}^\infty \lambda_k |\tau_2^\alpha E_{\alpha, \alpha+1}(-\lambda_k \kappa_\alpha \tau_2^\alpha) - \tau_1^\alpha E_{\alpha, \alpha+1}(-\lambda_k \kappa_\alpha \tau_1^\alpha)|^2 |(v, e_k)|^2\right)^{\frac{1}{2}} \tag{4.13}
$$

$$
\le \frac{2 M_\alpha \kappa_\alpha^{-\frac{1}{2}}}{\alpha} (\tau_2^{\frac{\alpha}{2}} - \tau_1^{\frac{\alpha}{2}}) \|v\|, \quad \text{for } v \in L^2(\Omega), \ 0 \le \tau_1 < \tau_2,
$$

this together with the estimate $\|A^{-\frac{1}{2}} \nabla \cdot \phi\| \le C \|\phi\|$ for $\phi \in L^2(\Omega)$ and the case $\gamma = \frac{1}{2}$ of Lemma 4.2 shows that

$$
I_3(t, h) \le \int_t^{t+h} \|A^{\frac{1}{2}} Q_\alpha(t + h - s) A^{-\frac{1}{2}} \nabla \cdot (F'(s) w(s))\| ds
$$

$$
+ \int_0^t \|[A^{\frac{1}{2}} Q_\alpha(t + h - s) - A^{\frac{1}{2}} Q_\alpha(t - s)] A^{-\frac{1}{2}} \nabla \cdot (F'(s) w(s))\| ds
$$

$$
\le C M_\alpha \kappa_\alpha^{-\frac{1}{2}} \int_t^{t+h} (t + h - s)^{\frac{\alpha}{2}} \|F'(s) w(s)\| ds
$$

$$
+ \frac{2 C M_\alpha \kappa_\alpha^{-\frac{1}{2}}}{\alpha} \int_0^t [(t + h - s)^{\frac{\alpha}{2}} - (t - s)^{\frac{\alpha}{2}}] \|F'(s) w(s)\| ds
$$

$$
\le C M_\alpha \kappa_\alpha^{-\frac{1}{2}} h^{\frac{\alpha}{2}} \left(1 + \frac{2}{\alpha}\right) \|F'\|_{L^1(0,T;L^2(\Omega))} \|w\|_{C([0,T];L^2(\Omega))}.
$$

Take the same way as we estimated $I_3(t, h)$ to deal with $I_4(t, h)$, one also needs to estimate

$$
\|A^{\frac{1}{2}} P_\alpha(\tau_2) v - A^{\frac{1}{2}} P_\alpha(\tau_1) v\|
$$

$$
= \left(\sum_{k=1}^\infty \lambda_k |\tau_2^{\alpha-1} E_{\alpha, \alpha}(-\lambda_k \kappa_\alpha \tau_2^\alpha) - \tau_1^{\alpha-1} E_{\alpha, \alpha}(-\lambda_k \kappa_\alpha \tau_1^\alpha)|^2 |(v, e_k)|^2\right)^{\frac{1}{2}} \tag{4.14}
$$

$$
\le \frac{2 M_\alpha \kappa_\alpha^{-\frac{1}{2}}}{2 - \alpha} (\tau_1^{\frac{\alpha}{2}-1} - \tau_2^{\frac{\alpha}{2}-1}) \|v\|, \quad \text{for } v \in L^2(\Omega), \ 0 \le \tau_1 < \tau_2,
$$

where we have used the case $\beta = \alpha, \gamma = \frac{1}{2}$ of (4.12). Combining with Lemma 4.2 again and $W^{1,1}(0,T;L^2(\Omega)) \hookrightarrow L^p(0,T;L^2(\Omega))$ for any $p \geq 1$, in the case $p > \frac{2}{\alpha}$, we get that

$$I_4(t,h) \leq \int_t^{t+h} \|A^{\frac{1}{2}}P_\alpha(t+h-s)A^{-\frac{1}{2}}\nabla \cdot (F(s)w(s))\|ds$$

$$+ \int_0^t \|[A^{\frac{1}{2}}P_\alpha(t+h-s) - A^{\frac{1}{2}}P_\alpha(t-s)]A^{-\frac{1}{2}}\nabla \cdot (F(s)w(s))\|ds$$

$$\leq CM_\alpha \kappa_\alpha^{-\frac{1}{2}} \int_t^{t+h} (t+h-s)^{\frac{\alpha}{2}-1}\|F(s)w(s)\|ds$$

$$+ \frac{2CM_\alpha \kappa_\alpha^{-\frac{1}{2}}}{2-\alpha} \int_0^t [(t-s)^{\frac{\alpha}{2}-1} - (t+h-s)^{\frac{\alpha}{2}-1}]\|F(s)w(s)\|ds$$

$$\leq CM_\alpha \kappa_\alpha^{-\frac{1}{2}} B_{\alpha,p} h^{\frac{\alpha p-2}{2p}}(1 + \frac{2}{2-\alpha})\|F\|_{L^p(0,T;L^2(\Omega))}\|w\|_{C([0,T];L^2(\Omega))},$$

where $B_{\alpha,p} = \left(\frac{2(p-1)}{\alpha p-2}\right)^{\frac{p-1}{p}}$ and we have used the inequality

$$\left(\int_0^t [(t-s)^{\frac{\alpha}{2}-1} - (t+h-s)^{\frac{\alpha}{2}-1}]^{\frac{p}{p-1}}ds\right)^{\frac{p-1}{p}}$$

$$\leq \left(\int_0^t [(t-s)^{(\frac{\alpha}{2}-1)\frac{p}{p-1}} - (t+h-s)^{(\frac{\alpha}{2}-1)\frac{p}{p-1}}]ds\right)^{\frac{p-1}{p}}$$

$$= B_{\alpha,p}[h^{\frac{\alpha p-2}{2(p-1)}} + t^{\frac{\alpha p-2}{2(p-1)}} - (t+h)^{\frac{\alpha p-2}{2(p-1)}}]^{\frac{p-1}{p}}.$$

Therefore, combined with the estimates of $I_i(t,h), i = 1, ..., 4$, we conclude that $\|\mathcal{H}w(t+h) - \mathcal{H}w(t)\| \to 0$ as $h \to 0^+$. Similar arguments can show that $\|\mathcal{H}w(t+h) - \mathcal{H}w(t)\| \to 0$ as $h \to 0^-$. Consequently, we obtain that $\mathcal{H}w \in C([0,T];L^2(\Omega))$ for any $w \in C([0,T];L^2(\Omega))$.

Step 2. $\mathcal{H}$ has a unique fixed point.

Indeed, for any $w_1, w_2 \in C([0,T];L^2(\Omega))$, we have

$$\|\mathcal{H}w_1(t) - \mathcal{H}w_2(t)\|$$

$$\leq \left\|\int_0^t Q_\alpha(t-s)\nabla \cdot (F'(s)(w_1(s) - w_2(s)))ds\right\|$$

$$+ \left\|\int_0^t P_\alpha(t-s)\nabla \cdot (F(s)(w_1(s) - w_2(s)))ds\right\|$$

$$\leq CM_\alpha \kappa_\alpha^{-\frac{1}{2}}[\int_0^t (t-s)^{\frac{\alpha}{2}}\|F'(s)(w_1(s) - w_2(s))\|ds$$

$$+ \int_0^t (t-s)^{\frac{\alpha}{2}-1}\|F(s)(w_1(s) - w_2(s))\|ds]$$

$$\leq CM_\alpha \kappa_\alpha^{-\frac{1}{2}}\|w_1 - w_2\|_{C([0,T];L^2(\Omega))}[t^{\frac{\alpha}{2}}\|F'\|_{L^1(0,T;L^2(\Omega))} + t^{\frac{\alpha p-2}{2(p-1)}}\|F\|_{L^p(0,T;L^2(\Omega))}].$$

Then one can choose a $T_1 \in (0,T)$ small enough which ensures that

$$CM_\alpha \kappa_\alpha^{-\frac{1}{2}}[T_1^{\frac{\alpha}{2}}\|F'\|_{L^1(0,T;L^2(\Omega))} + T_1^{\frac{\alpha p-2}{2(p-1)}}\|F\|_{L^p(0,T;L^2(\Omega))}] < 1.$$

It follows that $\mathcal{H}$ has a fixed point, thus the equation (4.10) has a unique solution in $C([0, T_1]; L^2(\Omega))$.

Now, we will deal with the continuation of the solution to the interval $[0, T]$. Let us make the assumption that we have obtained the solution $\bar{u}$ of the equation (4.10) on the interval $[0, T_l]$ for some $T_l > 0$. For $T_{l+1} > T_l$, it is tempting to prove the solution on $[T_l, T_{l+1}]$. To do this, we introduce the complete space

$$\bar{E}_{T_{l+1}} = \{u \in C([0, T_{l+1}]; L^2(\Omega)) : u(t) = \bar{u}(t) \text{ for } t \in [0, T_l]\},$$

with the distance $\|u\|_{\bar{E}_{T_{l+1}}} = \|u\|_{C([T_l, T_{l+1}]; L^2(\Omega))}$. Let $u \in \bar{E}_{T_{l+1}}$, then $u \in C([0, T_{l+1}]; L^2(\Omega))$. According to the previous proof, we have that $\mathcal{H}u \in \bar{E}_{T_{l+1}}$.

Next, we will show that the operator $\mathcal{H}$ is also a strict contraction on $\bar{E}_{T_{l+1}}$ when $T_{l+1} - T_l$ is sufficient small. We shall rewrite $\mathcal{H}$ as the following form:

$$\mathcal{H}w(t)$$

$$= Q_\alpha(t)u_0 + \int_0^{T_l} Q_\alpha(t-s)[f(s) + \nabla \cdot (F'(s)w(s))]ds$$

$$- \int_0^{T_l} P_\alpha(t-s)\nabla \cdot (F(s)w(s))ds + \int_{T_l}^t Q_\alpha(t-s)[f(s) + \nabla \cdot (F'(s)w(s))]ds$$

$$- \int_{T_l}^t P_\alpha(t-s)\nabla \cdot (F(s)w(s))ds.$$

For $u_1, u_2 \in \bar{E}_{T_{l+1}}$, we have

$$\mathcal{H}u_1(t) - \mathcal{H}u_2(t) = \int_{T_l}^t Q_\alpha(t-s)\nabla \cdot (F'(s)(u_1(s) - u_2(s)))ds$$

$$- \int_{T_l}^t P_\alpha(t-s)\nabla \cdot (F(s)(u_1(s) - u_2(s)))ds.$$

This follows from Lemma 4.2 that

$$\|\mathcal{H}u_1 - \mathcal{H}u_2\|_{C([T_l, T_{l+1}]; L^2(\Omega))}$$

$$\leq B_F C M_\alpha \kappa_\alpha^{-\frac{1}{2}} \|u_1 - u_2\|_{C([T_l, T_{l+1}]; L^2(\Omega))}[(T_{l+1} - T_l)^{\frac{\alpha}{2}} + (T_{l+1} - T_l)^{\frac{\alpha p - 2}{2(p-1)}}],$$

where $B_F = \sup\{\|F'\|_{L^1(0,T; L^2(\Omega))}, \|F\|_{L^p(0,T; L^2(\Omega))}\}$. Therefore

$$\|\mathcal{H}u_1 - \mathcal{H}u_2\|_{\bar{E}_{T_{l+1}}} \leq B_F C M_\alpha (1 + T^{\frac{2-\alpha}{2(p-1)}}) \kappa_\alpha^{-\frac{1}{2}} \|u_1 - u_2\|_{\bar{E}_{T_{l+1}}} (T_{l+1} - T_l)^{\frac{\alpha p - 2}{2(p-1)}}.$$

Moreover, we can choose one

$$T_{l+1} \in \left(T_l, T_l + \left(B_F C M_\alpha \kappa_\alpha^{-\frac{1}{2}}(1 + T^{\frac{2-\alpha}{2(p-1)}})\right)^{-\frac{2(p-1)}{\alpha p - 2}}\right),$$

which ensures that

$$0 < B_F C M_\alpha \kappa_\alpha^{-\frac{1}{2}}(1 + T^{\frac{2-\alpha}{2(p-1)}})(T_{l+1} - T_l)^{\frac{\alpha p - 2}{2(p-1)}} < 1.$$

Hence, the operator $\mathcal{H}$ is a strict contraction on $\bar{E}_{T_{l+1}}$, this also shows that the equation (4.10) has a unique solution on the interval $[T_l, T_{l+1}]$. We proceed to repeat the process on the intervals $[T_{l+1}, T_{l+2}]$, ..., until the equation (4.10) has a unique solution on the interval $[0, T]$. The claim then follows. $\qquad\square$

We now show how to transform the solution w of the integral equation (4.10) to the mild solution of the equation (4.1) as long as the former exists.

Theorem 4.1. *Under the conditions of Lemma 4.3, we proceed to obtain that* $w \in L^2(0, T; H_0^1(\Omega))$ *and the equation (4.1) has a unique mild solution* $u \in L^2(0, T; H^{-1}(\Omega))$, *which satisfies*

$$u(t) = S_\alpha(t)u_0 + \int_0^t S_\alpha(t-s)f(s)ds + \int_0^t S_\alpha(t-s)\nabla \cdot (F'(s)_0 D_s^{-\alpha}u(s))ds$$

$$- \nabla \cdot (F(t)_0 D_t^{-\alpha}u(t)) - \int_0^t S_\alpha'(t-s)\nabla \cdot (F(s)_0 D_s^{-\alpha}u(s))ds.$$

(4.15)

Proof. First, we denote

$$H(t) = \int_0^t P_\alpha(t-s)\nabla \cdot (F(s)w(s))ds.$$

Using the Young's inequality of the convolution and $\frac{d}{dt}E_{\alpha,1}(-\lambda t^\alpha) = -\lambda t^{\alpha-1}E_{\alpha,\alpha}(-\lambda t^\alpha)$ for $\lambda > 0$, it follows that

$$\left\| \int_0^t \lambda_k(t-s)^{\alpha-1}E_{\alpha,\alpha}(-\lambda_k\kappa_\alpha(t-s)^\alpha)\lambda_k^{-\frac{1}{2}}(\nabla \cdot (F(s)w(s)), e_k)ds \right\|_{L^2(0,T)}^2$$

$$\leq \frac{1}{\kappa_\alpha^2}\left(\int_0^T \lambda_k^{-1}|(\nabla \cdot (F(s)w(s)), e_k)|^2 ds\right)\left(\int_0^T |\lambda_k\kappa_\alpha s^{\alpha-1}E_{\alpha,\alpha}(-\lambda_k\kappa_\alpha s^\alpha)|ds\right)^2$$

$$= \frac{1}{\kappa_\alpha^2}[1 - E_{\alpha,1}(-\lambda_k\kappa_\alpha T^\alpha)]^2 \int_0^T \lambda_k^{-1}|(\nabla \cdot (F(s)_0 D_s^{-\alpha}u(s)), e_k)|^2 ds,$$

(4.16)

which yields

$$\|H\|_{L^2(0,T;\mathcal{D}(A^{\frac{1}{2}}))}^2 = \left\| \int_0^t AP_\alpha(t-s)A^{-\frac{1}{2}}\nabla \cdot (F(s)w(s))ds \right\|_{L^2(0,T;L^2(\Omega))}^2$$

$$\leq \frac{1}{\kappa_\alpha^2}\int_0^T \sum_{k=1}^\infty \lambda_k^{-1}|(\nabla \cdot (F(s)w(s)), e_k)|^2 ds$$

$$\leq \frac{C^2}{\kappa_\alpha^2}\|F\|_{L^2(0,T;L^2(\Omega))}^2\|w\|_{C([0,T];L^2(\Omega))}^2 < \infty.$$

Moreover, in view of Lemma 4.2, we have

$$\|w(t)\|_{H_0^1(\Omega)}$$

$$\leq M_\alpha\kappa_\alpha^{-\frac{1}{2}}\left(t^{\frac{\alpha}{2}}\|u_0\| + \int_0^t (t-s)^{\frac{\alpha}{2}}\|f(s)\|ds + \kappa_\alpha^{-\frac{1}{2}}\int_0^t \|F'(s)w(s)\|ds\right)$$

$$+ \|H(t)\|_{\mathcal{D}(A^{\frac{1}{2}})}$$

$$\leq M_\alpha\kappa_\alpha^{-\frac{1}{2}}\left(t^{\frac{\alpha}{2}}\|u_0\| + t^{\frac{\alpha}{2}}\|f\|_{L^1(0,T;L^2(\Omega))} + \kappa_\alpha^{-\frac{1}{2}}\|F'\|_{L^1(0,T;L^2(\Omega))}\|w\|_{C([0,T];L^2(\Omega))}\right)$$

$$+ \|H(t)\|_{\mathcal{D}(A^{\frac{1}{2}})}.$$

Therefore, $w \in L^2(0, T; H_0^1(\Omega))$.

We can take the Riemann-Liouville fractional derivative of order α of the equation (4.10) and invoke (4.3), it holds that

$$\partial_t^\alpha w(t) = S_\alpha(t)u_0 + \int_0^t S_\alpha(t-s)f(s)ds + \int_0^t S_\alpha(t-s)\nabla \cdot (F'(s)w(s))ds$$
$$- \nabla \cdot (F(t)w(t)) - \int_0^t S_\alpha'(t-s)\nabla \cdot (F(s)w(s))ds,$$

where we have used the following formula

$$\partial_t^\alpha \left(\int_0^t P_\alpha(t-s)\nabla \cdot (F(s)w(s))ds \right)$$
$$= \frac{1}{\Gamma(1-\alpha)} \frac{\partial}{\partial t} \int_0^t (t-s)^{-\alpha} \int_0^s P_\alpha(s-\tau)\nabla \cdot (F(\tau)w(\tau))d\tau ds$$
$$= \frac{1}{\Gamma(1-\alpha)} \frac{\partial}{\partial t} \int_0^t \int_\tau^t (t-s)^{-\alpha} P_\alpha(s-\tau)\nabla \cdot (F(\tau)w(\tau))ds d\tau$$
$$= \frac{\partial}{\partial t} \int_0^t S_\alpha(t-s)\nabla \cdot (F(s)w(s))ds$$
$$= \nabla \cdot (F(t)w(t)) + \int_0^t S_\alpha'(t-s)\nabla \cdot (F(s)w(s))ds,$$

where

$$S_\alpha'(t)v = \sum_{k=1}^\infty -\lambda_k \kappa_\alpha t^{\alpha-1} E_{\alpha,\alpha}(-\lambda_k \kappa_\alpha t^\alpha)(v, e_k)e_k(x). \tag{4.17}$$

It follows from $w = {}_0D_t^{-\alpha}u$ and Lemma 4.2 that (4.15) holds and we have

$$\|u(t)\|_{H^{-1}(\Omega)} \leq M_\alpha\|u_0\| + M_\alpha \int_0^t \|f(s)\|ds + M_\alpha \int_0^t \|F'(s){}_0D_s^{-\alpha}u(s)\|ds$$
$$+ \|F(t)\|_{L^2(\Omega)}\|{}_0D_t^{-\alpha}u(t))\|_{L^2(\Omega)}$$
$$+ \left\| \int_0^t A^{-\frac{1}{2}}S_\alpha'(t-s)\nabla \cdot (F(s){}_0D_s^{-\alpha}u(s))ds \right\|.$$

In view of (4.16), it is easy to show that $u \in L^2(0, T; H^{-1}(\Omega))$. $\square$

Next we verify under what conditions mild solutions may have improved space regularity.

Theorem 4.2.

(i) *Suppose that $F \in W^{1,1}(0, T; L^2(\Omega)) \cap L^\infty(0, T; H_0^1(\Omega))$. Then the mild solution $u \in L^2(0, T; L^2(\Omega))$ and ${}_0D_t^{-\alpha}u \in C([0, T]; H_0^1(\Omega))$.*

(ii) *Suppose that $F \in W^{1,1}(0, T; L^2(\Omega)) \cap W^{1,\infty}(0, T; \Omega)$. Then ${}_0D_t^{-\alpha}u \in L^2(0, T; H^2(\Omega))$.*

Proof. The proof of (i). It is obvious that the equation (4.10) has a unique mild solution $w \in L^2(0, T; H_0^1(\Omega)) \cap C([0, T]; L^2(\Omega))$ under the assumptions of $F \in W^{1,1}(0, T; L^2(\Omega)) \cap L^\infty(0, T; H_0^1(\Omega))$. Moreover,

$$\|{_0D_t^{-\alpha}}u(t)\|_{\mathcal{D}(A^{\frac{1}{2}})} \leq M_\alpha \kappa_\alpha^{-\frac{1}{2}} \left(t^{\frac{\alpha}{2}} \|u_0\| + t^{\frac{\alpha}{2}} \|f\|_{L^1(0,T;L^2(\Omega))} \right.$$

$$+ \kappa_\alpha^{-\frac{1}{2}} \|F'\|_{L^1(0,T;L^2(\Omega))} \|{_0D_t^{-\alpha}}u\|_{C([0,T];L^2(\Omega))} \Bigg)$$

$$+ M_\alpha \kappa_\alpha^{-\frac{1}{2}} \int_0^t (t-s)^{\frac{\alpha}{2}-1} \|F(s)\|_{H^1(\Omega)} \|{_0D_s^{-\alpha}}u(s))\|_{H_0^1(\Omega)} ds.$$

It deduces from the generalized Gronwall's inequality that

$$\|{_0D_t^{-\alpha}}u(t)\|_{H_0^1(\Omega)} \leq M_\alpha \kappa_\alpha^{-\frac{1}{2}} \left(t^{\frac{\alpha}{2}} \|u_0\| + t^{\frac{\alpha}{2}} \|f\|_{L^1(0,T;L^2(\Omega))} \right.$$

$$+ \kappa_\alpha^{-\frac{1}{2}} \|F'\|_{L^1(0,T;L^2(\Omega))} \|{_0D_t^{-\alpha}}u\|_{C([0,T];L^2(\Omega))} \Bigg) E_{\frac{\alpha}{2},1}(\mu t^{\frac{\alpha}{2}}),$$

where $\mu = M_\alpha \kappa_\alpha^{-\frac{1}{2}} \|F\|_{L^\infty(0,T;H^1(\Omega))}$, and then ${_0D_t^{-\alpha}}u \in L^\infty(0, T; H_0^1(\Omega))$. Furthermore, similar to the proof of Theorem 4.1, we also easily know that ${_0D_t^{-\alpha}}u \in C([0, T]; H_0^1(\Omega))$.

From Lemma 4.2, it follows that

$$\|u(t)\| \leq M_\alpha \|u_0\| + M_\alpha \int_0^t \|f(s)\| ds + M_\alpha \kappa_\alpha^{-\frac{1}{2}} \int_0^t (t-s)^{-\frac{\alpha}{2}} \|F'(s)_0 D_s^{-\alpha}u(s)\| ds$$

$$+ \|F(t)\|_{H^1(\Omega)} \|{_0D_t^{-\alpha}}u(t))\|_{H_0^1(\Omega)} + \left\| \int_0^t S_\alpha'(t-s) \nabla \cdot (F(s)_0 D_s^{-\alpha}u(s)) ds \right\|.$$

The similar arguments to (4.16) lead to the property that

$$\left\| \int_0^t S_\alpha'(t-s) \nabla \cdot (F(s)_0 D_s^{-\alpha}u(s)) ds \right\|_{L^2(0,T;L^2(\Omega))}^2$$

$$\leq \frac{1}{\kappa_\alpha^2} \int_0^T \sum_{k=1}^\infty |(\nabla \cdot (F(s)_0 D_s^{-\alpha}u(s)), e_k)|^2 ds$$

$$\leq \frac{1}{\kappa_\alpha^2} \|\nabla \cdot (F(\cdot)_0 D_t^{-\alpha}u(\cdot))\|_{L^2(0,T;L^2(\Omega))}^2$$

$$\leq \frac{1}{\kappa_\alpha^2} \|F\|_{L^2(0,T;H^1(\Omega))} \|{_0D_t^{-\alpha}}u\|_{C([0,T];H_0^1(\Omega))}.$$

Then we can easily derive $u \in L^2(0, T; L^2(\Omega))$.

The proof of (ii). We immediate calculate that

$$\|w(t)\|_{\mathcal{D}(A)} \leq \|Q_\alpha(t)u_0\|_{\mathcal{D}(A)} + \int_0^t \|Q_\alpha(t-s)f(s)\|_{\mathcal{D}(A)} ds$$

$$+ \int_0^t \|Q_\alpha(t-s) \nabla \cdot (F'(s)w(s))\|_{\mathcal{D}(A)} ds$$

$$+ \left\| \int_0^t P_\alpha(t-s)\nabla \cdot (F(s)w(s))ds \right\|_{\mathcal{D}(A)}$$

$$=:H_1(t) + H_2(t).$$

From Lemma 4.1 and Lemma 4.2, we can easily know that

$$H_1(t) = \|Q_\alpha(t)u_0\|_{\mathcal{D}(A)} + \int_0^t \|Q_\alpha(t-s)f(s)\|_{\mathcal{D}(A)}ds$$

$$+ \int_0^t \|Q_\alpha(t-s)\nabla \cdot (F'(s)w(s))\|_{\mathcal{D}(A)}ds$$

$$\leq M_\alpha \kappa_\alpha^{-1}\left(\|u_0\| + \int_0^t \|f(s)\|ds + \int_0^t \|F(s)\|_{W^{1,\infty}}\|w(s)\|_{H^1(\Omega)}ds\right).$$

It follows that $H_1(\cdot) \in L^2(0,T;\mathcal{D}(A))$. On the other hand, in view of (4.16) and (4.17), we obtain that $H_2(\cdot) \in L^2(0,T;\mathcal{D}(A))$. Therefore the desired conclusion follows. □

Remark 4.1. In fact, the result (i) of Theorem 4.2 is also obtained in [260, Theorem 4.1 and Theorem 4.2], which used the energy arguments combined with the generalized Gronwall's inequality.

4.1.5 *Classical Solutions in Case of $\alpha \in (\frac{1}{2}, 1)$*

In this subsection, we prove the existence of classical solutions for the equation (4.1). First, we shall show that the equation (4.10) has a unique solution which belongs to $AC([0,T];L^2(\Omega))$. To do this, we introduce the space

$$X_T = \{w \in C([0,T];H_0^1(\Omega)) : w' \in L^1(0,T;L^2(\Omega))\},$$

equipped with the norm $\|w\|_{X_T} = \|w\|_{C([0,T];H_0^1(\Omega))} + \|w'\|_{L^1(0,T;L^2(\Omega))}$.

The following result shows that the solution of the integral equation (4.10) belongs to X_T.

Lemma 4.4. *Let $u_0 \in L^2(\Omega)$ and $f \in L^1(0,T;L^2(\Omega))$. Suppose that $F \in W^{1,1}(0,T; L^2(\Omega)) \cap W^{1,\infty}(0,T;\Omega)$. Then the equation (4.10) has a unique solution $w \in X_T$.*

Proof. Consider the operator $\mathcal{H} : X_T \to X_T$ given as (4.11). Then it is well-defined. In fact, let $w \in X_T$, then $\mathcal{H}w \in C([0,T];H_0^1(\Omega))$. Further, we immediately take the first derivative of $\mathcal{H}w$ with respect to t and use (4.3) to obtain

$$(\mathcal{H}w)'(t) = P_\alpha(t)u_0 + \partial_t\left(\int_0^t Q_\alpha(t-s)f(s)ds\right)$$

$$+ \partial_t\left(\int_0^t Q_\alpha(t-s)\nabla \cdot (F'(s)w(s))ds\right)$$

$$- \partial_t\left(\int_0^t P_\alpha(t-s)\nabla \cdot (F(s)w(s))ds\right).$$

Then we can infer from $Q_\alpha(0) = 0$ that

$$\partial_t \left(\int_0^t Q_\alpha(t-s)f(s)ds \right) = \int_0^t P_\alpha(t-s)f(s)ds.$$

Besides, in view of $w(0) = 0$, it holds that

$$\partial_t \left(\int_0^t P_\alpha(t-s)\nabla \cdot (F(s)w(s))ds \right)$$

$$= P_\alpha(t)\nabla \cdot (F(0)w(0)) + \int_0^t P_\alpha(s)\partial_t[\nabla \cdot (F(t-s)w(t-s))]ds$$

$$= \int_0^t P_\alpha(t-s)\nabla \cdot (F'(s)w(s) + F(s)w'(s))ds.$$

Thus

$$(\mathcal{H}w)'(t) = P_\alpha(t)u_0 + \int_0^t P_\alpha(t-s)f(s)ds - \int_0^t P_\alpha(t-s)\nabla \cdot (F(s)w'(s))ds.$$

Using Lemma 4.2 again, one can have

$$\|(\mathcal{H}w)'(t)\| \leq M_\alpha \left(t^{\alpha-1}\|u_0\| + \int_0^t (t-s)^{\alpha-1}\|f(s)\|ds \right.$$

$$\left. + \|F\|_{W^{1,\infty}} \int_0^t (t-s)^{\frac{\alpha}{2}-1}\|w'(s)\|ds \right). \tag{4.18}$$

It follows from the Young's inequality of the convolution that $(\mathcal{H}w)' \in L^1(0,T;L^2(\Omega))$. Using the Banach fixed point method and the continuation of the solution as we proved in Lemma 4.3, one also obtains that the equation (4.10) has a unique solution in X_T. $\qquad\square$

Our next result shows the existence of the mild solution of the equation (4.1).

Theorem 4.3. *Let $u_0 \in L^2(\Omega)$ and $f \in L^1(0,T;L^2(\Omega))$. Suppose that $F \in W^{1,1}(0,T; L^2(\Omega)) \cap W^{1,\infty}(0,T;\Omega)$. Then the equation (4.1) has the mild solution $u \in L^2(0,T; L^2(\Omega)) \cap L^1(0,T;H_0^1(\Omega))$ such that $\partial_t^{1-\alpha}u \in L^1(0,T;L^2(\Omega))$, which satisfies*

$$u(t) = S_\alpha(t)u_0 + \int_0^t S_\alpha(t-s)f(s)ds - \int_0^t S_\alpha(t-s)\nabla \cdot (F(s)\partial_s^{1-\alpha}u(s))ds. \tag{4.19}$$

Proof. Similar to the proof of Theorems 4.1 and 4.2, the equation (4.1) has the mild solution $u \in L^2(0,T;L^2(\Omega))$ denoted as (4.15). In view of $\int_\tau^t S_\alpha'(t-s)vds = [S_\alpha(t-\tau) - S_\alpha(0)]v$ and $S_\alpha(0)v = v$ for $v \in L^2(\Omega)$, we obtain that

$$\int_0^t S_\alpha(t-s)\nabla \cdot (F'(s)_0 D_s^{-\alpha}u(s))ds$$

$$= \int_0^t S_\alpha'(t-s)\nabla \cdot \int_0^s (F'(\tau)_0 D_\tau^{-\alpha}u(\tau))d\tau ds + \int_0^t \nabla \cdot (F'(s)_0 D_s^{-\alpha}u(s))ds.$$

Then using the integration by parts and (4.8), one can infer from $_0D_s^{-\alpha}u(s)\big|_{s=0} = 0$ and

$$\int_0^s (F'(\tau)_0D_\tau^{-\alpha}u(\tau))d\tau\big|_{s=0} = 0$$

that

$$\int_0^t S'_\alpha(t-s)\Big[\nabla\cdot(F(s)_0D_s^{-\alpha}u(s)) - \nabla\cdot\int_0^s(F'(\tau)_0D_\tau^{-\alpha}u(\tau))d\tau\Big]ds$$

$$= \int_0^t S_\alpha(t-s)\frac{\partial}{\partial s}\Big[\int_0^s\nabla\cdot(F(\tau)\partial_\tau^{1-\alpha}u(\tau))d\tau\Big]ds$$

$$- S_\alpha(t-s)\Big[\nabla\cdot(F(s)_0D_s^{-\alpha}u(s)) - \nabla\cdot\int_0^s(F'(\tau)_0D_\tau^{-\alpha}u(\tau))d\tau\Big]\Big|_0^t$$

$$= \int_0^t S_\alpha(t-s)\nabla\cdot(F(s)\partial_s^{1-\alpha}u(s))ds - \nabla\cdot(F(t)_0D_t^{-\alpha}u(t))$$

$$+ \nabla\cdot\int_0^t(F'(\tau)_0D_\tau^{-\alpha}u(\tau))d\tau.$$

Combined with above equalities, (4.15) would be rewritten as

$$u(t) = S_\alpha(t)u_0 + \int_0^t S_\alpha(t-s)f(s)ds - \int_0^t S_\alpha(t-s)\nabla\cdot(F(s)\partial_s^{1-\alpha}u(s))ds.$$

Using Lemma 4.2 again, we can deduce that

$$\|u(t)\|_{H_0^1(\Omega)} \leq M_\alpha\kappa_\alpha^{-\frac{1}{2}}t^{-\frac{\alpha}{2}}\|u_0\| + M_\alpha\kappa_\alpha^{-\frac{1}{2}}\int_0^t(t-s)^{-\frac{\alpha}{2}}\|f(s)\|ds$$

$$+ M_\alpha\kappa_\alpha^{-1}\|F\|_{W^{1,\infty}}\int_0^t(t-s)^{-\alpha}\|\partial_s^{1-\alpha}u(s)\|ds. \tag{4.20}$$

It follows from the Young's inequality for the convolution that $u \in L^1(0,T;H_0^1(\Omega))$. $\square$

Our next result shows the precise conditions under which we have a regular solution.

Theorem 4.4. *Suppose that $f \in L^2(0,T;L^2(\Omega))$ and $F \in W^{1,1}(0,T;L^2(\Omega))\cap W^{1,\infty}(0,T;\Omega)$. If we restrict $\alpha \in (\frac{1}{2},1)$, then the mild solution u of the equation (4.1) has improved regularity.*

(i) *If $u_0 \in H_0^1(\Omega)$, then $u \in L^2(0,T;\mathcal{D}(A))$ and $\partial_t^{1-\alpha}u \in L^2(0,T;H_0^1(\Omega))$.*
(ii) *If $u_0 \in \mathcal{D}(A^{\gamma_0})$ for $\gamma_0 = \max\{\frac{1}{2},\gamma_1\}$ and $\gamma_1 \in [0,\frac{1}{2\alpha})$, then $u \in C([0,T];\mathcal{D}(A^{\gamma_1}))$.*
(iii) *If $u_0 \in \mathcal{D}(A)$, then $\partial_t^{1-\alpha}u \in L^2(0,T;\mathcal{D}(A))$.*

Proof. From Lemma 4.4, we know that

$$\partial_t^{1-\alpha}u(t) = P_\alpha(t)u_0 + \int_0^t P_\alpha(t-s)f(s)ds - \int_0^t P_\alpha(t-s)\nabla\cdot(F(s)\partial_s^{1-\alpha}u(s))ds.$$

If $\alpha \in (\frac{1}{2}, 1)$, then it follows from (4.18) and the generalized Gronwall's inequality (see e.g. [165, Lemma 7.1.1]) that

$$\|\partial_t^{1-\alpha} u(t)\| \leq C^*(\alpha, \|F\|_{W^{1,\infty}} T) t^{\alpha-1},$$

which also shows that $\partial_t^{1-\alpha} u \in L^2(0, T; L^2(\Omega))$.

The proof of (i). Since

$$\left\| \int_0^t A^{\frac{1}{2}} P_\alpha(t-s) \nabla \cdot (F(s) \partial_s^{1-\alpha} u(s)) ds \right\|$$

$$= \left\| \int_0^t A P_\alpha(t-s) A^{-\frac{1}{2}} \nabla \cdot (F(s) \partial_s^{1-\alpha} u(s)) ds \right\|,$$

by the same way as we derived (4.16), one can get

$$\left\| \int_0^t A P_\alpha(t-s) A^{-\frac{1}{2}} \nabla \cdot (F(s) \partial_s^{1-\alpha} u(s)) ds \right\|_{L^2(0,T;L^2(\Omega))}^2$$

$$\leq \frac{1}{\kappa_\alpha^2} \int_0^T \sum_{k=1}^\infty \lambda_k^{-1} |(\nabla \cdot (F(s) \partial_s^{1-\alpha} u(s)), e_k)|^2 ds$$

$$\leq \frac{1}{\kappa_\alpha^2} \int_0^T \|F(t) \partial_t^{1-\alpha} u(t)\|^2 dt$$

$$\leq \frac{1}{\kappa_\alpha^2} \|F\|_{L^\infty(0,T;L^2(\Omega))}^2 \|\partial_t^{1-\alpha} u\|_{L^2(0,T;L^2(\Omega))}^2.$$

Consequently, $\mathcal{D}(A^{\frac{1}{2}}) = H_0^1(\Omega)$ and the Young's inequality for the convolution yield that

$$\|\partial_t^{1-\alpha} u\|_{L^2(0,T;H_0^1(\Omega))} \leq M_\alpha \sqrt{\frac{T^{2\alpha-1}}{2\alpha-1}} \|u_0\|_{H_0^1(\Omega)} + \frac{2}{\alpha} M_\alpha \kappa_\alpha^{-\frac{1}{2}} T^{\frac{\alpha}{2}} \|f\|_{L^2(0,T;L^2(\Omega))}$$

$$+ \|F\|_{L^\infty(0,T;L^2(\Omega))} \|\partial_t^{1-\alpha} u\|_{L^2(0,T;L^2(\Omega))}.$$

On the other hand, according to (4.19) and Lemma 4.2, we immediately estimate

$$\|u(t)\|_{\mathcal{D}(A)} \leq M_\alpha \kappa_\alpha^{-\frac{1}{2}} t^{-\frac{\alpha}{2}} \|u_0\|_{H_0^1(\Omega)} + M_\alpha \kappa_\alpha^{-1} \int_0^t (t-s)^{-\alpha} \|f(s)\| ds$$

$$+ M_\alpha \kappa_\alpha^{-1} \|F\|_{W^{1,\infty}} \int_0^t (t-s)^{-\alpha} \|\partial_s^{1-\alpha} u(s)\|_{H_0^1(\Omega)} ds.$$

Using the Young's inequality for the convolution again, one can obtain that $u \in L^2(0, T; \mathcal{D}(A))$.

The proof of (ii). For $t \in [0, T]$, let $h > 0$ be such that $t + h \in [0, T]$. We use the case $\beta = 1, \gamma = \gamma_1 \in (0, \frac{1}{2\alpha})$ of (4.12) to derive

$$\|[S_\alpha(t+h) - S_\alpha(t)] v\|_{\mathcal{D}(A^{\gamma_1})} \leq \frac{M_\alpha \kappa_\alpha^{-\gamma_1}}{\alpha \gamma_1} (t^{-\alpha\gamma_1} - (t+h)^{-\alpha\gamma_1}) \|v\|, \quad \text{for } v \in L^2(\Omega).$$

$$(4.21)$$

In addition, it is easy to know that $\|[S_\alpha(t+h) - S_\alpha(t)] u_0\|_{\mathcal{D}(A^{\gamma_1})} \to 0$ as $h \to 0^+$ for $u_0 \in \mathcal{D}(A^{\gamma_1})$, and then according to the fundamental arguments, it follows that $u \in C([0, T]; \mathcal{D}(A^{\gamma_1}))$. Particularly, if $\gamma_1 = 0$, it also holds that $u \in C([0, T]; L^2(\Omega))$.

The proof of (iii). By the same way as we derived (4.16), one can get

$$\left\| \int_0^t P_\alpha(t-s)f(s)ds \right\|^2_{L^2(0,T;\mathcal{D}(A))} = \left\| \int_0^t AP_\alpha(t-s)f(s)ds \right\|^2_{L^2(0,T;L^2(\Omega))}$$

$$\leq \frac{1}{\kappa_\alpha^2}\|f\|^2_{L^2(0,T;L^2(\Omega))}$$

and

$$\left\| \int_0^t P_\alpha(t-s)\nabla \cdot (F(s)\partial_s^{1-\alpha}u(s))ds \right\|^2_{L^2(0,T;\mathcal{D}(A))}$$

$$\leq \frac{1}{\kappa_\alpha^2}\|\nabla \cdot (F(\cdot)\partial_t^{1-\alpha}u(\cdot))\|^2_{L^2(0,T;L^2(\Omega))}$$

$$\leq \frac{1}{\kappa_\alpha^2}\|F\|^2_{W^{1,\infty}}\|\partial_t^{1-\alpha}u\|^2_{L^2(0,T;H_0^1(\Omega))}.$$

Further, $P_\alpha(\cdot)u_0 \in L^2(0,T;\mathcal{D}(A))$ for $u_0 \in \mathcal{D}(A)$, then $\partial_t^{1-\alpha}u \in L^2(0,T;\mathcal{D}(A))$. $\square$

In fact, the decay estimates of u and $\partial_t^{1-\alpha}u$ can be obtained when we use the following condition instead of $f \in L^2(0,T;L^2(\Omega))$.

(Q) Assume that $f \in L^1(0,T;L^2(\Omega))$ satisfies $\|f(t)\| \leq M_f t^\vartheta$ for some $\vartheta \in (\frac{\alpha}{2}-1,0)$ and a.e. $t \in [0,T]$.

Remark 4.2. Let $\alpha \in (\frac{1}{2},1)$, $u_0 \in H_0^1(\Omega)$. Assume that $F \in W^{1,1}(0,T;L^2(\Omega)) \cap W^{1,\infty}(0,T;\Omega)$ and (Q) hold. Then we also have $u \in L^q(0,T;H^2(\Omega))$ for $q \in [1,\frac{2}{\alpha})$ and $\partial_t^{1-\alpha}u \in L^2(0,T;H_0^1(\Omega))$. Moreover, there exist $C_0 = C(\alpha, \|F\|_{W^{1,\infty}}, T)$ and $C_1 = C^*(\alpha, \|F\|_{W^{1,\infty}}, T)$ such that

$$\|\partial_t^{1-\alpha}u(t)\|_{\mathcal{D}(A^{\frac{1}{2}})} \leq C_0 t^{\alpha-1}, \quad \|u(t)\|_{\mathcal{D}(A)} \leq C_1 t^{-\frac{\alpha}{2}}.$$

Proof. Using Lemma 4.2 again, we can deduce from $u_0 \in H_0^1(\Omega)$ that

$$\|\partial_t^{1-\alpha}u(t)\|_{\mathcal{D}(A^{\frac{1}{2}})} \leq M_\alpha t^{\alpha-1}\|u_0\|_{H_0^1(\Omega)} + M_\alpha\kappa_\alpha^{-\frac{1}{2}}\int_0^t (t-s)^{\frac{\alpha}{2}-1}\|f(s)\|ds$$

$$+ M_\alpha\kappa_\alpha^{-\frac{1}{2}}\int_0^t (t-s)^{\frac{\alpha}{2}-1}\|\nabla \cdot (F(s)\partial_s^{1-\alpha}u(s))\|ds$$

$$\leq M_\alpha t^{\alpha-1}(\|u_0\|_{H_0^1(\Omega)} + M_f\kappa_\alpha^{-\frac{1}{2}}B(\frac{\alpha}{2},1+\vartheta)T^{1+\vartheta-\frac{\alpha}{2}})$$

$$+ M_\alpha\kappa_\alpha^{-\frac{1}{2}}\|F\|_{W^{1,\infty}}\int_0^t (t-s)^{\frac{\alpha}{2}-1}\|\partial_s^{1-\alpha}u(s)\|_{H_0^1(\Omega)}ds.$$

It follows from the generalized Gronwall's inequality (see e.g. [165, Lemma 7.1.1]) that there exists $C(\alpha, \|F\|_{W^{1,\infty}}, T)$ such that

$$\|\partial_t^{1-\alpha}u(t)\|_{\mathcal{D}(A^{\frac{1}{2}})} \leq aC(\alpha, \|F\|_{W^{1,\infty}}T)t^{\alpha-1},$$

where $a = \kappa_\alpha^{-\frac{1}{2}}\frac{M_\alpha}{\alpha}(\|u_0\|_{H_0^1(\Omega)} + M_f\kappa_\alpha^{-\frac{1}{2}}B(\frac{\alpha}{2},1+\vartheta)T^{1+\vartheta-\frac{\alpha}{2}}).$

On the other hand, in view of (4.19), we also obtain

$$\|u(t)\|_{\mathcal{D}(A)} \leq M_\alpha \kappa_\alpha^{-\frac{1}{2}} t^{-\frac{\alpha}{2}} \|u_0\|_{H_0^1(\Omega)} + M_\alpha \kappa_\alpha^{-1} \int_0^t (t-s)^{-\alpha} \|f(s)\| ds$$

$$+ M_\alpha \kappa_\alpha^{-1} \|F\|_{W^{1,\infty}} \int_0^t (t-s)^{-\alpha} \|\partial_s^{1-\alpha} u(s)\|_{H_0^1(\Omega)} ds$$

$$\leq M_\alpha t^{-\frac{\alpha}{2}} [\kappa_\alpha^{-\frac{1}{2}} \|u_0\|_{H_0^1(\Omega)} + M_f \kappa_\alpha^{-1} B(1-\alpha, 1+\vartheta) T^{1+\vartheta-\frac{\alpha}{2}}$$

$$+ C_0 T^{\frac{\alpha}{2}} \|F\|_{W^{1,\infty}} B(1-\alpha, \alpha)].$$

Therefore, it follows that $u \in L^q(0, T; H^2(\Omega))$ for $q \in [1, \frac{2}{\alpha})$. $\qquad\square$

We now verify under what conditions a mild solution may become classical for as long as the former exists.

Theorem 4.5. *Let $\alpha \in (\frac{1}{2}, 1)$, $u_0 \in \mathcal{D}(A)$. Assume that $f \in L^2(0, T; L^2(\Omega))$ and $F \in W^{1,1}(0, T; L^2(\Omega)) \cap W^{1,\infty}(0, T; \Omega)$. Then the mild solution u of the equation (4.1) obtained in Theorem 4.3 is also a classical solution.*

Proof. We differentiate (4.19) and use (4.17) to obtain that

$$\partial_t u(t) = S_\alpha'(t) u_0 + f(t) + \int_0^t S_\alpha'(t-s) f(s) ds - \nabla \cdot (F(t) \partial_t^{1-\alpha} u(t))$$

$$- \int_0^t S_\alpha'(t-s) \nabla \cdot (F(s) \partial_s^{1-\alpha} u(s)) ds. \tag{4.22}$$

Similar to the proof of (4.16), it follows from Theorem 4.4(i) and $u_0 \in \mathcal{D}(A)$ that $\partial_t u \in L^2(0, T; L^2(\Omega))$. Further, we can immediately from $-\Delta e_k = \lambda_k e_k$ and $S_\alpha'(t) v = -\kappa_\alpha A P_\alpha(t) v$ for $v \in L^2(\Omega)$ that

$$\partial_t u(t, x) - \kappa_\alpha \partial_t^{1-\alpha} \Delta u + \nabla \cdot (F \partial_t^{1-\alpha} u) = f(t, x), \quad (t, x) \in (0, T) \times \Omega.$$

This combined with Theorem 4.4 shows that the claim follows. $\qquad\square$

Remark 4.3. In Theorems 4.4 and 4.5, it needs to turn out that $\partial_t^{1-\alpha} u \in L^2(0, T; L^2(\Omega))$, so we can't get rid of the restriction on $\alpha \in (\frac{1}{2}, 1)$, which verifies the results in [217] from another perspective.

4.1.6 *Classical Solutions in Case of $\alpha \in (0, 1)$*

In this subsection, we study the existence and uniqueness of classical solutions in weighted Hölder continuous function spaces, in which the restriction on $\alpha \in (\frac{1}{2}, 1)$ can be removed. We introduce a working space for $0 < \theta < \eta < \min\{\frac{\alpha}{2}, 1-\alpha\}$,

$$\bar{X}_T = \{w \in C([0, T]; L^2(\Omega)) : w' \in \mathcal{F}^{\eta, \theta}((0, T]; L^2(\Omega))\},$$

whose norm is $\|w\|_{\bar{X}_T} = \|w\|_{C([0,T];L^2(\Omega))} + \|w'\|_{\mathcal{F}^{\eta,\theta}}$. Clearly, it is a Banach space. In view of the singular terms in integral equations and the space $\mathcal{F}^{\eta,\theta}((0,T];L^2(\Omega))$, our analysis makes heave use of the following inequality.

Lemma 4.5. *Let* $\mu \in (0,1)$, $\nu,\sigma \in [0,1)$ *and* $0 \le t_1 < t_2 \le T$. *The following inequality holds:*

$$\int_0^{t_1} (t_1 - s)^{\mu-1}(t_2 - s)^{\nu-1} s^{-\sigma} ds \le \left(\frac{2^\sigma \pi}{\sin(\mu\pi)} + \frac{2}{1-\sigma} \right) t_1^{-\sigma} t_2^\nu (t_2 - t_1)^{\mu-1}.$$

Proof. We split the integral into two parts:

$$\int_0^{t_1} (t_1 - s)^{\mu-1}(t_2 - s)^{\nu-1} s^{-\sigma} ds$$

$$= \int_0^{\frac{t_1}{2}} s^{\mu-1}(t_2 - t_1 + s)^{\nu-1}(t_1 - s)^{-\sigma} ds + \int_{\frac{t_1}{2}}^{t_1} s^{\mu-1}(t_2 - t_1 + s)^{\nu-1}(t_1 - s)^{-\sigma} ds$$

$$\le 2^\sigma t_1^{-\sigma} t_2^\nu \int_0^{\frac{t_1}{2}} s^{\mu-1}(t_2 - t_1 + s)^{-1} ds$$

$$+ 2t_1^{-1} t_2^\nu (t_2 - t_1)^{\mu-1} \int_{\frac{t_1}{2}}^{t_1} (t_2 - t_1)^{1-\mu} s^\mu (t_2 - t_1 + s)^{-1}(t_1 - s)^{-\sigma} ds.$$

According to

$$\int_0^\infty s^{\mu-1}(t_2 - t_1 + s)^{-1} ds = \frac{\pi}{\sin(\mu\pi)}(t_2 - t_1)^{\mu-1} \quad \text{and} \quad (t_2 - t_1)^{1-\mu} s^\mu \le t_2 - t_1 + s,$$

then the above integral is bounded by $\left(\frac{2^\sigma \pi}{\sin(\mu\pi)} + \frac{2}{1-\sigma} \right) t_1^{-\sigma} t_2^\nu (t_2 - t_1)^{\mu-1}$. $\qquad\square$

The following result shows that the solution of the integral equation (4.10) belongs to $\bar{X}_T$.

Lemma 4.6. *Let* $u_0 \in L^2(\Omega)$ *and* $f \in \mathcal{F}^{\eta,\theta}((0,T];L^2(\Omega))$. *Suppose that* $F \in W^{1,1}(0,T;L^2(\Omega)) \cap W^{1,\infty}(0,T;\Omega) \cap C^\theta([0,T];L^2(\Omega))$. *Then the equation* (4.10) *has a unique solution* $w \in \bar{X}_T$.

Proof. Consider the operator $\mathcal{H} : X_T \to X_T$ given as (4.11). Then it is well-defined. In fact, let $w \in X_T$, then $\mathcal{H}w \in C([0,T];L^2(\Omega))$. It suffices to prove $(\mathcal{H}w)' \in \mathcal{F}^{\eta,\theta}((0,T];L^2(\Omega))$. To do this, let us consider the operator

$$(\mathcal{H}w)'(t) = P_\alpha(t)u_0 + \int_0^t P_\alpha(t-s)f(s)ds - \int_0^t P_\alpha(t-s)\nabla \cdot (F(s)w'(s))ds$$

$$=: \mathcal{H}_1(t) + \mathcal{H}_2(t) + \mathcal{H}_3(t).$$

We only need to show $\mathcal{H}_i \in \mathcal{F}^{\eta,\theta}((0,T];L^2(\Omega))$ for $i = 1,2,3$. Noting that, from the definition of $P_\alpha(t)$, it is clear to know $\lim_{t\to 0} t^{1-\eta} P_\alpha(t)u_0$ exists for $\eta < \frac{\alpha}{2}$. Further, for $0 \le t_1 < t_2 \le T$, in view of (4.12), one can see

$$\|P_\alpha(t_2)u_0 - P_\alpha(t_1)u_0\| \le \frac{M_\alpha}{1-\alpha}\|u_0\|(t_1^{\alpha-1} - t_2^{\alpha-1}) \le \frac{M_\alpha\|u_0\|}{1-\alpha} \frac{(t_2 - t_1)^{1-\alpha}}{(t_1 t_2)^{1-\alpha}}.$$

$$(4.23)$$

It implies that

$$\frac{t_1^{1-\eta+\theta}\|\mathcal{H}_1(t_2)-\mathcal{H}_1(t_1)\|}{(t_2-t_1)^{\theta}} \leq \frac{M_\alpha\|u_0\|}{1-\alpha}t_1^{\alpha-\eta}\left(\frac{t_1}{t_2}\right)^{\theta}\left(\frac{t_2-t_1}{t_2}\right)^{1-\alpha-\theta}$$

$$\leq \frac{M_\alpha\|u_0\|}{1-\alpha}T^{\alpha-\eta}.$$

Therefore, we have

$$\lim_{t_2\to 0}\sup_{0\leq t_1<t_2}\frac{t_1^{1-\eta+\theta}\|\mathcal{H}_1(t_2)-\mathcal{H}_1(t_1)\|}{(t_2-t_1)^{\theta}}=0.$$

This shows that $\mathcal{H}_1\in\mathscr{J}^{\eta,\theta}((0,T];L^2(\Omega))$.

Under the assumption of f, one can see that

$$\|\mathcal{H}_2(t)\|\leq M_\alpha\int_0^t(t-s)^{\alpha-1}\|f(s)\|ds\leq M_\alpha B(\alpha,\eta)\|f\|_{\mathscr{F}^{\eta,\theta}}t^{\alpha+\eta-1},$$

this implies that $\lim_{t\to 0}t^{1-\eta}\|\mathcal{H}_2(t)\|$ exists. Further, we denote

$$\mathcal{H}_{21}(t):=\int_0^t P_\alpha(t-s)[f(s)-f(t)]ds,\quad \mathcal{H}_{22}(t):=\int_0^t P_\alpha(t-s)f(t)ds=Q(t)f(t).$$

Using (4.23) and Lemma 4.5, we have

$$\|\mathcal{H}_{21}(t_2)-\mathcal{H}_{21}(t_1)\|$$

$$\leq\left\|\int_0^{t_1}[P_\alpha(t_2-s)-P_\alpha(t_1-s)][f(s)-f(t_1)]ds\right\|$$

$$+\left\|\int_0^{t_1}P_\alpha(t_2-s)[f(t_1)-f(t_2)]ds\right\|+\left\|\int_{t_1}^{t_2}P_\alpha(t_2-s)[f(s)-f(t_2)]ds\right\|$$

$$\leq M_\alpha\|f\|_{\mathscr{F}^{\eta,\theta}}\left[\frac{1}{1-\alpha}(t_2-t_1)^{1-\alpha}\int_0^{t_1}(t_1-s)^{\alpha+\theta-1}(t_2-s)^{\alpha-1}s^{-(1-\eta+\theta)}ds\right.$$

$$+(t_2-t_1)^{\theta}\int_0^{t_1}(t_2-s)^{\alpha-1}s^{-(1-\eta+\theta)}ds+\left.\int_{t_1}^{t_2}(t_2-s)^{\alpha+\theta-1}s^{-(1-\eta+\theta)}ds\right]$$

$$\leq M_\alpha\|f\|_{\mathscr{F}^{\eta,\theta}}\left[\frac{C^*(\alpha)}{1-\alpha}t_1^{\eta-\theta-1}t_2^{\alpha}(t_2-t_1)^{\theta}\right.$$

$$\left.+B(\alpha,\eta-\theta)t_1^{\alpha+\eta-\theta-1}(t_2-t_1)^{\theta}+\frac{t_1^{\eta-\theta-1}}{\alpha+\theta}(t_2-t_1)^{\alpha+\theta}\right],$$

$$(4.24)$$

where $C^*(\alpha)=\frac{2^{1-\eta+\theta}\pi}{\sin((\alpha+\theta)\pi)}+\frac{2}{\eta-\theta}$, which implies that

$$\frac{t_1^{1-\eta+\theta}\|\mathcal{H}_{21}(t_2)-\mathcal{H}_{21}(t_1)\|}{(t_2-t_1)^{\theta}}\leq M_\alpha\|f\|_{\mathscr{F}^{\eta,\theta}}\left[\frac{C^*(\alpha)}{1-\alpha}t_2^{\alpha}+B(\alpha,\eta-\theta)t_1^{\alpha}+\frac{(t_2-t_1)^{\alpha}}{\alpha+\theta}\right]$$

$$\leq\bar{C}M_\alpha\|f\|_{\mathscr{F}^{\eta,\theta}}T^{\alpha},$$

where $\bar{C}$ is a constant depending on α,η,θ, and

$$\lim_{t_2\to 0}\sup_{0\leq t_1<t_2}\frac{t_1^{1-\eta+\theta}\|\mathcal{H}_{21}(t_2)-\mathcal{H}_{21}(t_1)\|}{(t_2-t_1)^{\theta}}=0.$$

On the other hand, since $\lim_{t\to 0} t^{1-\eta}Q(t)f(t) = 0$ is obvious, then a similar argument yields that

$$\|\mathcal{H}_{22}(t_2) - \mathcal{H}_{22}(t_1)\|$$

$$\leq \left\| \int_0^{t_1} [P_\alpha(t_2 - s) - P_\alpha(t_1 - s)]f(t_2)ds \right\|$$

$$+ \left\| \int_0^{t_1} P_\alpha(t_1 - s)[f(t_2) - f(t_1)]ds \right\| + \left\| \int_{t_1}^{t_2} P_\alpha(t_2 - s)f(t_2)ds \right\|$$

$$\leq M_\alpha \|f\|_{\mathcal{F}^{\eta,\theta}} \left[\frac{t_2^{\eta-1}}{1-\alpha} \int_0^{t_1} [(t_1 - s)^{\alpha-1} - (t_2 - s)^{\alpha-1}]ds \right. \tag{4.25}$$

$$\left. + t_1^{\eta-\theta-1}(t_2 - t_1)^\theta \int_0^{t_1} (t_1 - s)^{\alpha-1}ds + t_2^{\eta-1} \int_{t_1}^{t_2} (t_2 - s)^{\alpha-1}ds \right]$$

$$\leq M_\alpha \|f\|_{\mathcal{F}^{\eta,\theta}} \frac{t_1^{\eta-\theta-1}}{\alpha}(t_2 - t_1)^\theta \left[\frac{2-\alpha}{1-\alpha} t_1^\theta(t_2 - t_1)^{\alpha-\theta} + t_1^\alpha \right],$$

it also shows

$$\frac{t_1^{1-\eta+\theta}\|\mathcal{H}_{22}(t_2) - \mathcal{H}_{22}(t_1)\|}{(t_2 - t_1)^\theta} \leq \frac{M_\alpha}{\alpha}\|f\|_{\mathcal{F}^{\eta,\theta}} \left[\frac{2-\alpha}{1-\alpha} t_1^\theta(t_2 - t_1)^{\alpha-\theta} + t_1^\alpha \right]$$

$$\leq \frac{M_\alpha}{\alpha}\|f\|_{\mathcal{F}^{\eta,\theta}} \bar{\bar{C}} T^\alpha,$$

where $\bar{\bar{C}}$ is a constant depending on α, and then,

$$\lim_{t_2\to 0} \sup_{0\leq t_1 < t_2} \frac{t_1^{1-\eta+\theta}\|\mathcal{H}_{21}(t_2) - \mathcal{H}_{21}(t_1)\|}{(t_2 - t_1)^\theta} = 0.$$

Therefore $\mathcal{H}_2 \in \mathcal{F}^{\eta,\theta}((0,T]; L^2(\Omega))$.

For $\mathcal{H}_3(t)$, taking account of $F \in W^{1,\infty}(0,T;\Omega) \cap C^\theta((0,T]; L^2(\Omega))$, it holds that $Fw' \in \mathcal{F}^{\eta,\theta}((0,T]; L^2(\Omega))$. Indeed, we can clearly see that $\lim_{t\to 0} t^{1-\eta}F(t)w'(t)$ exists for the continuity of F and $w' \in \mathcal{F}^{\eta,\theta}((0,T]; L^2(\Omega))$. Further, from the Hölder continuity of F, one can know that

$$\|F(t_2)w'(t_2) - F(t_1)w'(t_1)\|$$

$$\leq \|F(t_2) - F(t_1)\|\|w'(t_1)\| + \|F(t_2)\|\|w'(t_2) - w'(t_1)\|$$

$$\leq L_F\|w'\|_{\mathcal{F}^{\eta,\theta}} t_1^{\eta-1}(t_2 - t_1)^\theta + \|F\|_{W^{1,\infty}}\|w'\|_{\mathcal{F}^{\eta,\theta}} t_1^{\eta-\theta-1}(t_2 - t_1)^\theta,$$

where $L_F = \sup_{0\leq t_1 < t_2 \leq T} \frac{\|F(t_2)-F(t_1)\|}{(t_2-t_1)^\theta}$, this also shows

$$\sup_{0\leq t_1 < t_2 \leq T} \frac{t_1^{1-\eta+\theta}\|F(t_2)w'(t_2) - F(t_1)w'(t_1)\|}{(t_2 - t_1)^\theta}$$

$$\leq L_F\|w'\|_{\mathcal{F}^{\eta,\theta}} T^\theta + \|F\|_{W^{1,\infty}}\|w'\|_{\mathcal{F}^{\eta,\theta}}$$

and

$$\lim_{t_2\to 0} \sup_{0\leq t_1 < t_2} \frac{t_1^{1-\eta+\theta}\|F(t_2)w'(t_2) - F(t_1)w'(t_1)\|}{(t_2 - t_1)^\theta} = 0.$$

As in the same way of the proof to $\mathcal{H}_2 \in \mathcal{F}^{\eta,\theta}((0,T];L^2(\Omega))$, it follows from (4.13) and (4.14) that $\mathcal{H}_3 \in \mathcal{F}^{\eta,\theta}((0,T];L^2(\Omega))$ provided $\theta < \frac{\alpha}{2}$. Consequently, combined with the above arguments, we have $(\mathcal{H}w)' \in \mathcal{F}^{\eta,\theta}((0,T];L^2(\Omega))$.

Using the Banach fixed point method and the continuation of the solution as we proved Lemma 4.3, one also obtains that the equation (4.10) has a unique solution in $\bar{X}_T$. $\qquad\square$

The next result shows that both the mild solution u of the equation (4.1) and $\partial_t^{1-\alpha}u$ belong to weighted Hölder continuous function spaces.

Theorem 4.6. *Let* $u_0 \in L^2(\Omega)$ *and* $f \in \mathcal{F}^{\eta,\theta}((0,T];L^2(\Omega))$. *Suppose that* $F \in W^{1,1}(0,T;L^2(\Omega)) \cap W^{1,\infty}(0,T;\Omega) \cap C^\theta([0,T];L^2(\Omega))$. *Then the equation* (4.1) *has a unique mild solution* $u \in \mathcal{F}^{\eta,\theta}((0,T];H_0^1(\Omega))$. *Moreover,* $\partial_t^{1-\alpha}u \in \mathcal{F}^{\eta,\theta}((0,T];H_0^1(\Omega))$.

Proof. Similar to the proof of Theorem 4.3, we know that the equation (4.1) has a unique mild solution $u : [0,T] \to L^2(\Omega))$ satisfying the integral equation (4.19). From Lemma 4.6, we also have that $\partial_t^{1-\alpha}u$ belongs to $\mathcal{F}^{\eta,\theta}((0,T];L^2(\Omega))$.

Next we would like to show $u \in \mathcal{F}^{\eta,\theta}((0,T];H_0^1(\Omega))$. In view of (4.20), one can see that

$$\|u(t)\|_{H_0^1(\Omega)} \leq M_\alpha \kappa_\alpha^{-\frac{1}{2}}\left[t^{-\frac{\alpha}{2}}\|u_0\| + \|f\|_{\mathcal{F}^{\eta,\theta}}\int_0^t (t-s)^{-\frac{\alpha}{2}}s^{\eta-1}ds\right]$$

$$+ M_\alpha \kappa_\alpha^{-1}\|F\|_{W^{1,\infty}}\|\partial_t^{1-\alpha}u\|_{\mathcal{F}^{\eta,\theta}}\int_0^t (t-s)^{-\alpha}s^{\eta-1}ds \qquad (4.26)$$

$$\leq M_\alpha \kappa_\alpha^{-\frac{1}{2}}\left[t^{-\frac{\alpha}{2}}\|u_0\| + \|f\|_{\mathcal{F}^{\eta,\theta}}B(1-\frac{\alpha}{2},\eta)t^{\eta-\frac{\alpha}{2}}\right]$$

$$+ M_\alpha \kappa_\alpha^{-1}t^{\eta-\alpha}\|F\|_{W^{1,\infty}}\|\partial_t^{1-\alpha}u\|_{\mathcal{F}^{\eta,\theta}}B(1-\alpha,\eta).$$

This implies that $\lim_{t\to 0} t^{1-\eta}u(t)$ exists. We denote

$$G_1(t) = S_\alpha(t)u_0, \quad G_2(t) = \int_0^t S_\alpha(t-s)f(s)ds,$$

$$G_3(t) = \int_0^t S_\alpha(t-s)\nabla \cdot (F(s)\partial_s^{1-\alpha}u(s))ds.$$

Using (4.21), we have, for $0 \leq t_1 < t_2 \leq T$,

$$\|G_1(t_2) - G_1(t_1)\|_{H_0^1(\Omega)} = \|(A^{\frac{1}{2}}S_\alpha(t_2) - A^{\frac{1}{2}}S_\alpha(t_1))u_0\|$$

$$\leq \frac{2M_\alpha \kappa_\alpha^{-\frac{1}{2}}}{\alpha}(t_1t_2)^{-\frac{\alpha}{2}}(t_2-t_1)^{\frac{\alpha}{2}}.$$

Similar to (4.24) and (4.25), we proceed to obtain that

$$\|G_2(t_2) - G_2(t_1)\|_{H_0^1(\Omega)}$$

$$\leq M_\alpha \kappa_\alpha^{-\frac{1}{2}}\|f\|_{\mathcal{F}^{\eta,\theta}}t_1^{\eta-\theta-1}(t_2-t_1)^\theta\left[\frac{2}{\alpha}C^*(1-\frac{\alpha}{2})t_2^{1-\frac{\alpha}{2}}\right.$$

$$+ B(1 - \frac{\alpha}{2}, \eta - \theta)t_1^{1-\frac{\alpha}{2}} + \frac{(t_2 - t_1)^{1-\frac{\alpha}{2}}}{1 - \frac{\alpha}{2} + \theta}\Bigg]$$

$$+ \frac{M_\alpha \kappa_\alpha^{-\frac{1}{2}}}{1 - \frac{\alpha}{2}}\|f\|_{\mathcal{F}^{\eta,\theta}}t_1^{\eta-\theta-1}(t_2 - t_1)^\theta\left[(1 + \frac{2}{\alpha})t_1^\theta(t_2 - t_1)^{1-\frac{\alpha}{2}-\theta} + t_1^{1-\frac{\alpha}{2}}\right].$$

For $G_3(t)$, we consider

$$\|G_3(t_2) - G_3(t_1)\|_{H_0^1(\Omega)}$$

$$= \left\| \int_0^{t_2} AS_\alpha(t_2 - s)A^{-\frac{1}{2}}\nabla \cdot (F(s)\partial_s^{1-\alpha}u(s))ds \right.$$

$$\left. - \int_0^{t_1} AS_\alpha(t_1 - s)A^{-\frac{1}{2}}\nabla \cdot (F(s)\partial_s^{1-\alpha}u(s))ds \right\|,$$

then

$$\|G_3(t_2) - G_3(t_1)\|_{H_0^1(\Omega)}$$

$$\leq M_\alpha \kappa_\alpha^{-1}\|F\partial_t^{1-\alpha}u\|_{\mathcal{F}^{\eta,\theta}}t_1^{\eta-\theta-1}(t_2 - t_1)^\theta$$

$$\times \left[\frac{1}{\alpha}C^*(1 - \alpha)t_2^{1-\alpha} + B(1 - \alpha, \eta - \theta)t_1^{1-\alpha} + \frac{(t_2 - t_1)^{1-\alpha}}{1 - \alpha + \theta}\right]$$

$$+ \frac{M_\alpha \kappa_\alpha^{-1}}{1 - \alpha}\|F\partial_t^{1-\alpha}u\|_{\mathcal{F}^{\eta,\theta}}t_1^{\eta-\theta-1}(t_2 - t_1)^\theta\left[(1 + \frac{1}{\alpha})t_1^\theta(t_2 - t_1)^{1-\alpha-\theta} + t_1^{1-\alpha}\right].$$

Consequently, combined with the above arguments, we have

$$\sup_{0 \leq t_1 < t_2 \leq T} \frac{t_1^{1-\eta+\theta}\|u(t_2) - u(t_1)\|_{H_0^1(\Omega)}}{(t_2 - t_1)^\theta} < \infty, \quad \text{for } \eta < \min\left\{1 - \alpha, \frac{\alpha}{2}\right\}$$

and $\lim_{t_2 \to 0} \sup_{0 \leq t_1 < t_2} \frac{t_1^{1-\eta+\theta}\|u(t_2)-u(t_1)\|_{H_0^1(\Omega)}}{(t_2-t_1)^\theta} = 0$, the desired result is obtained.

Next we show that $\partial_t^{1-\alpha}u \in \mathcal{F}^{\eta,\theta}((0,T]; H_0^1(\Omega))$. From Lemma 4.4, we know that

$$\partial_t^{1-\alpha}u(t) = P_\alpha(t)u_0 + \int_0^t P_\alpha(t - s)f(s)ds - \int_0^t P_\alpha(t - s)\nabla \cdot (F(s)\partial_s^{1-\alpha}u(s))ds.$$

To do this, we will estimate $\|\mathcal{H}_3(t)\|_{H_0^1(\Omega)}$. Let

$$\mathcal{H}_{31}(t) = \int_0^t AP_\alpha(t - s)A^{-\frac{1}{2}}\nabla \cdot [F(s)\partial_s^{1-\alpha}u(s) - F(t)\partial_t^{1-\alpha}u(t)]ds,$$

$$\mathcal{H}_{32}(t) = \int_0^t AP_\alpha(t - s)A^{-\frac{1}{2}}\nabla \cdot (F(t)\partial_t^{1-\alpha}u(t))ds.$$

According to Lemma 4.2 and $S_\alpha'(t)v = -\kappa_\alpha AP_\alpha(t)v$ for $v \in L^2(\Omega)$, one can see that

$$\|\mathcal{H}_3(t)\|_{H_0^1(\Omega)} \leq \|\mathcal{H}_{31}(t)\| + \|\mathcal{H}_{32}(t)\|$$

$$\leq M_\alpha \kappa_\alpha^{-1}\int_0^t (t - s)^{-1}\|F(s)\partial_s^{1-\alpha}u(s) - F(t)\partial_t^{1-\alpha}u(t)\|ds$$

$$+ \kappa_\alpha^{-1}\|(I - S_\alpha(t))A^{-\frac{1}{2}}\nabla \cdot (F(t)\partial_t^{1-\alpha}u(t))\|$$

$$\leq M_\alpha \kappa_\alpha^{-1} \|F\partial_t^{1-\alpha} u\|_{\mathcal{F}^{\eta,\theta}} \int_0^t (t-s)^{\theta-1} s^{-(1-\eta+\theta)} ds$$
$$+ \kappa_\alpha^{-1}(1+M_\alpha)\|F\partial_t^{1-\alpha} u\|_{\mathcal{F}^{\eta,\theta}} t^{\eta-1}$$
$$\leq \|F\partial_t^{1-\alpha} u\|_{\mathcal{F}^{\eta,\theta}} \kappa_\alpha^{-1}[M_\alpha B(\theta, \eta-\theta) + (1+M_\alpha)] t^{\eta-1},$$

then we immediately estimate

$$t^{1-\eta}\|\partial_t^{1-\alpha} u(t)\|_{H_0^1(\Omega)} \leq M_\alpha \kappa_\alpha^{-\frac{1}{2}}\left[t^{\frac{\alpha}{2}-\eta}\|u_0\| + t^{1-\eta}\int_0^t (t-s)^{\frac{\alpha}{2}-1}\|f(s)\|ds\right]$$
$$+ \|F\partial_t^{1-\alpha} u\|_{\mathcal{F}^{\eta,\theta}} \kappa_\alpha^{-1}[M_\alpha B(\theta, \eta-\theta) + (1+M_\alpha)]$$
$$\leq M_\alpha \kappa_\alpha^{-\frac{1}{2}}[t^{\frac{\alpha}{2}-\eta}\|u_0\| + \|f\|_{\mathcal{F}^{\eta,\theta}} B(\frac{\alpha}{2}, \eta)t^{\frac{\alpha}{2}}]$$
$$+ \|F\partial_t^{1-\alpha} u\|_{\mathcal{F}^{\eta,\theta}} \kappa_\alpha^{-1}[M_\alpha B(\theta, \eta-\theta) + (1+M_\alpha)].$$

This implies that $\lim_{t\to 0} t^{1-\eta}\partial_t^{1-\alpha} u(t)$ exists in $H_0^1(\Omega)$.

Similar to (4.24), (4.25) and the arguments on $G_3(t)$, we proceed to obtain that

$$\|\mathcal{H}_{31}(t_2) - \mathcal{H}_{31}(t_1)\|$$

$$\leq M_\alpha \kappa_\alpha^{-1}\|F\partial_t^{1-\alpha} u\|_{\mathcal{F}^{\eta,\theta}}\left[\frac{1}{\alpha}(t_2-t_1)\int_0^{t_1}(t_1-s)^{\theta-1}(t_2-s)^{-1}s^{-(1-\eta+\theta)}ds\right.$$

$$+ \left.\int_{t_1}^{t_2}(t_2-s)^{\theta-1}s^{-(1-\eta+\theta)}ds\right]$$

$$+ \left\|\int_0^{t_1} AP_\alpha(t_2-s)[F(t_1)\partial_s^{1-\alpha}u(t_1) - F(t_2)\partial_s^{1-\alpha}u(t_2)]ds\right\|$$

$$\leq M_\alpha\|F\partial_t^{1-\alpha} u\|_{\mathcal{F}^{\eta,\theta}} t_1^{\eta-\theta-1}(t_2-t_1)^\theta\left[\kappa_\alpha^{-1}\frac{C^*(0)}{\alpha} + 2\kappa_\alpha^{-1} + \frac{\kappa_\alpha^{-1}}{\theta}\right],$$

where we have used the following inequality

$$\left\|\int_0^{t_1} AP_\alpha(t_2-s)A^{-\frac{1}{2}}\nabla \cdot [F(t_1)\partial_s^{1-\alpha}u(t_1) - F(t_2)\partial_s^{1-\alpha}u(t_2)]ds\right\|$$
$$= \|\kappa_\alpha^{-1}[S_\alpha(t_2-t_1) - S_\alpha(t_2)]A^{-\frac{1}{2}}\nabla \cdot [F(t_1)\partial_t^{1-\alpha}u(t_1) - F(t_2)\partial_t^{1-\alpha}u(t_2)]\|$$
$$\leq 2M_\alpha\kappa_\alpha^{-1}\|F\partial_t^{1-\alpha} u\|_{\mathcal{F}^{\eta,\theta}} t_1^{\eta-\theta-1}(t_2-t_1)^\theta.$$

On the other hand, using $e^{-\rho} \leq \frac{1}{1+\rho}$ for $\rho > 0$, we have

$$|E_{\alpha,1}(-\lambda_k\kappa_\alpha t_2^\alpha) - E_{\alpha,1}(-\lambda_k\kappa_\alpha t_1^\alpha)|$$

$$= \left|\int_0^\infty \mathcal{M}_\alpha(\xi)e^{-\lambda_k\kappa_\alpha t_1^\alpha\xi}[e^{-\lambda_k\kappa_\alpha(t_2^\alpha-t_1^\alpha)\xi} - 1]d\xi\right|$$

$$= \left|\int_0^\infty \mathcal{M}_\alpha(\xi)\lambda_k^{\frac{\theta}{\alpha}}e^{-\lambda_k\kappa_\alpha t_1^\alpha\xi}\int_0^{t_2^\alpha-t_1^\alpha}\lambda_k^{1-\frac{\theta}{\alpha}}\kappa_\alpha\xi e^{-\lambda_k\kappa_\alpha\tau\xi}d\tau d\xi\right|$$

$$\leq \int_0^\infty \mathcal{M}_\alpha(\xi)\lambda_k^{\frac{\theta}{\alpha}}e^{-\lambda_k\kappa_\alpha t_1^\alpha\xi}\int_0^{t_2^\alpha-t_1^\alpha}\tau^{\frac{\theta}{\alpha}-1}(\kappa_\alpha\xi)^{\frac{\theta}{\alpha}}d\tau d\xi$$

$$\leq \frac{\alpha}{\theta}(t_2 - t_1)^\theta \int_0^\infty (\lambda_k \kappa_\alpha \xi)^{\frac{\theta}{\alpha}} M_\alpha(\xi) e^{-\lambda_k \kappa_\alpha t_1^\alpha \xi} d\xi$$

$$\leq \frac{\alpha}{\theta}(t_2 - t_1)^\theta t_1^{-\theta},$$

then the uniform convergence of integrals with variables shows that

$$\lim_{t_2 \to 0} \sup_{0 \leq t_1 < t_2} \int_0^\infty (\lambda_k \kappa_\alpha \xi)^{\frac{\theta}{\alpha}} M_\alpha(\xi) t_1^\theta e^{-\lambda_k \kappa_\alpha t_1^\alpha \xi} d\xi |(v, e_k)| = 0.$$

Therefore

$$\|[S_\alpha(t_2) - S_\alpha(t_1)]v\| \leq \frac{\alpha}{\theta}(t_2 - t_1)^\theta t_1^{-\theta}\|v\|$$

and

$$\lim_{t_2 \to 0} \sup_{0 \leq t_1 < t_2} \frac{t_1^{1-\eta+\theta}\|[S_\alpha(t_2) - S_\alpha(t_1)]v\|}{(t_2 - t_1)^\theta} = 0,$$

for $v \in L^2(\Omega)$. Thus

$$\|\mathcal{H}_{32}(t_2) - \mathcal{H}_{32}(t_1)\| \leq \kappa_\alpha^{-1}\|[I - S_\alpha(t_2)]A^{-\frac{1}{2}}\nabla \cdot [F(t_2)\partial_t^{1-\alpha}u(t_2) - F(t_1)\partial_t^{1-\alpha}u(t_1)]\|$$

$$+ \kappa_\alpha^{-1}\|[S_\alpha(t_2) - S_\alpha(t_1)]A^{-\frac{1}{2}}\nabla \cdot (F(t_1)\partial_t^{1-\alpha}u(t_1))\|$$

$$\leq \kappa_\alpha^{-1}\|F\partial_t^{1-\alpha}u\|_{\mathcal{F}^{\eta,\theta}} t_1^{\eta-\theta-1}(t_2 - t_1)^\theta [2M_\alpha + \frac{\alpha}{\theta}].$$

Further, using (4.14), we have

$$\|\mathcal{H}_1(t_2) - \mathcal{H}_1(t_1)\|_{H_0^1(\Omega)} \leq \frac{M_\alpha \kappa_\alpha^{-\frac{1}{2}}}{1 - \frac{\alpha}{2}}\|u_0\|(t_1^{\frac{\alpha}{2}-1} - t_2^{\frac{\alpha}{2}-1})$$

$$\leq \frac{M_\alpha \kappa_\alpha^{-\frac{1}{2}}}{1 - \frac{\alpha}{2}}\|u_0\|(t_1 t_2)^{\frac{\alpha}{2}-1}(t_2 - t_1)^{1-\frac{\alpha}{2}}.$$

Moreover, by using the similar way as in (4.24), (4.25) and the arguments on $G_3(t)$, one can also obtain that

$$\|\mathcal{H}_2(t_2) - \mathcal{H}_2(t_1)\|_{H_0^1(\Omega)}$$

$$\leq M_\alpha \kappa_\alpha^{-\frac{1}{2}}\|f\|_{\mathcal{F}^{\eta,\theta}} t_1^{\eta-\theta-1}(t_2 - t_1)^\theta \left[\frac{C^*(\frac{\alpha}{2})}{1-\frac{\alpha}{2}}t_2^{\frac{\alpha}{2}} + B(\frac{\alpha}{2}, \eta - \theta)t_1^{\frac{\alpha}{2}} + \frac{1}{\frac{\alpha}{2}+\theta}(t_2 - t_1)^{\frac{\alpha}{2}}\right]$$

$$+ \frac{2}{\alpha}M_\alpha \kappa_\alpha^{-\frac{1}{2}}\|f\|_{\mathcal{F}^{\eta,\theta}} t_1^{\eta-\theta-1}(t_2 - t_1)^\theta \left[\frac{2 - \frac{\alpha}{2}}{1-\frac{\alpha}{2}}t_1^\theta(t_2 - t_1)^{\frac{\alpha}{2}-\theta} + t_1^{\frac{\alpha}{2}}\right].$$

Consequently, combined with the above arguments, $\partial_t^{1-\alpha}u$ belongs to $\mathcal{F}^{\eta,\theta}((0,T];$ $H_0^1(\Omega))$. □

The regularity of mild solutions is also improved in the following results.

Remark 4.4. Let $u_0 \in L^2(\Omega)$ and $f \in \mathcal{F}^{\eta,\theta}((0,T]; L^2(\Omega))$. Suppose that $F \in W^{1,1}(0,T; L^2(\Omega)) \cap W^{1,\infty}(0,T;\Omega) \cap C^\theta((0,T]; H^1(\Omega))$. Then we have the following properties:

(i) if $u_0 \in L^2(\Omega)$, then $u \in \mathcal{F}^{\eta,\theta}((0,T]; H^2(\Omega))$;

(ii) if $u_0 \in H^1_0(\Omega)$, then $u \in L^\infty(0,T; H^1_0(\Omega))$ and $\partial_t^{1-\alpha} u \in \mathcal{F}^{\eta,\theta}((0,T]; H^2(\Omega))$.

Proof. The proof of (i). According to $F \in W^{1,\infty}(0,T;\Omega) \cap C^\theta((0,T]; H^1(\Omega))$ and $\partial_t^{1-\alpha} u \in \mathcal{F}^{\eta,\theta}((0,T]; H^1_0(\Omega))$, we have $F\partial_t^{1-\alpha} u \in \mathcal{F}^{\eta,\theta}((0,T]; H^1_0(\Omega))$. In view of (4.20), one can see that

$$\|u(t)\|_{H^2(\Omega)} \leq M_\alpha \kappa_\alpha^{-1}\Big[t^{-\alpha}\|u_0\| + \|f\|_{\mathcal{F}^{\eta,\theta}}\int_0^t (t-s)^{-\alpha}s^{\eta-1}ds$$

$$+ \|F\|_{W^{1,\infty}}\|\partial_t^{1-\alpha} u\|_{\mathcal{F}^{\eta,\theta}}\int_0^t (t-s)^{-\alpha}s^{\eta-1}ds\Big]$$

$$\leq M_\alpha \kappa_\alpha^{-1}\big[t^{-\alpha}\|u_0\| + t^{\eta-\alpha}\|f\|_{\mathcal{F}^{\eta,\theta}}B(1-\alpha,\eta)$$

$$+ t^{\eta-\alpha}\|F\|_{W^{1,\infty}}\|\partial_t^{1-\alpha} u\|_{\mathcal{F}^{\eta,\theta}}B(1-\alpha,\eta)\big].$$

This implies that $\lim_{t\to 0} t^{1-\eta}u(t)$ exists in $H^2(\Omega)$ provided $\eta \leq 1-\alpha$. Similar to the proof of Theorem 4.6, it is easy to show that

$$\sup_{0\leq t_1 < t_2 \leq T} \frac{t_1^{1-\eta+\theta}\|u(t_2) - u(t_1)\|_{H^2(\Omega)}}{(t_2-t_1)^\theta} < \infty$$

and

$$\lim_{t_2\to 0}\sup_{0\leq t_1 < t_2} \frac{t_1^{1-\eta+\theta}\|u(t_2) - u(t_1)\|_{H^2(\Omega)}}{(t_2-t_1)^\theta} = 0.$$

Thus $u \in \mathcal{F}^{\eta,\theta}((0,T]; H^2(\Omega))$.

The proof of (ii). If $u_0 \in H^1_0(\Omega)$, it follows from (4.26) that $u \in L^\infty(0,T; H^1_0(\Omega))$. To show $\partial_t^{1-\alpha} u \in \mathcal{F}^{\eta,\theta}((0,T]; H^2(\Omega))$, we consider

$$\|\partial_t^{1-\alpha} u(t)\|_{\mathcal{D}(A)} = \|AP_\alpha(t)u_0\| + \left\|\int_0^t AP_\alpha(t-s)f(s)ds\right\|$$

$$+ \left\|\int_0^t AP_\alpha(t-s)\nabla\cdot(F(s)\partial_s^{1-\alpha}u(s))ds\right\|.$$

In fact, we only need to show $AP_\alpha(t)u_0 \in \mathcal{F}^{\eta,\theta}((0,T]; L^2(\Omega))$. Since

$$\|AP_\alpha(t)u_0\| = \|A^{\frac{1}{2}}P_\alpha(t)A^{\frac{1}{2}}u_0\| \leq M_\alpha \kappa_\alpha^{-\frac{1}{2}}t^{\frac{\alpha}{2}-1}\|u_0\|_{H^1_0(\Omega)},$$

then $\lim_{t\to 0} t^{1-\eta}AP_\alpha(t)u_0$ exists for $\eta < \frac{\alpha}{2}$. Further, for $0 \leq t_1 < t_2 \leq T$, similar to (4.23), one can see that

$$\frac{t_1^{1-\eta+\theta}\|AP_\alpha(t_2)u_0 - AP_\alpha(t_1)u_0\|}{(t_2-t_1)^\theta} \leq \frac{M_\alpha \kappa_\alpha^{-\frac{1}{2}}}{1-\frac{\alpha}{2}}\|u_0\|_{H^1_0(\Omega)}t_1^{\frac{\alpha}{2}-\eta}\left(\frac{t_1}{t_2}\right)^\theta\left(\frac{t_2-t_1}{t_2}\right)^{1-\frac{\alpha}{2}-\theta}$$

$$\leq \frac{M_\alpha \kappa_\alpha^{-\frac{1}{2}}}{1-\frac{\alpha}{2}}\|u_0\|_{H^1_0(\Omega)}T^{\frac{\alpha}{2}-\eta}.$$

Hence

$$\lim_{t_2\to 0}\sup_{0\leq t_1 < t_2} \frac{t_1^{1-\eta+\theta}\|AP_\alpha(t_2)u_0 - AP_\alpha(t_1)u_0\|}{(t_2-t_1)^\theta} = 0,$$

and so $AP_\alpha(t)u_0 \in \mathcal{F}^{\eta,\theta}((0,T]; L^2(\Omega))$. By using the similar way as we dealt with $\|\mathcal{H}_3(t)\|_{H^1_0(\Omega)}$, $\mathcal{H}_{31}(t)$ and $\mathcal{H}_{32}(t)$, one can also deduce that $\partial_t^{1-\alpha} u \in \mathcal{F}^{\eta,\theta}((0,T]; H^2(\Omega))$. The desired result is obtained. $\square$

Finally, we also obtain a classical solution in a weighted Hölder continuous function space without the restriction on α.

Theorem 4.7. *Let $u_0 \in H_0^1(\Omega)$ and $f \in \mathcal{F}^{\eta,\theta}((0,T]; L^2(\Omega))$. Suppose that $F \in W^{1,1}(0,T; L^2(\Omega)) \cap W^{1,\infty}(0,T;\Omega) \cap C^\theta([0,T]; H^1(\Omega))$. Then the mild solution u of the equation (4.1) obtained in Theorem 4.6 is also a classical solution.*

Proof. From (4.22), we know that

$$\partial_t u(t) = -\kappa_\alpha A P_\alpha(t) u_0 + f(t) - \kappa_\alpha \int_0^t A P_\alpha(t-s) f(s) ds - \nabla \cdot (F(t) \partial_t^{1-\alpha} u(t))$$

$$+ \kappa_\alpha \int_0^t A P_\alpha(t-s) \nabla \cdot (F(s) \partial_s^{1-\alpha} u(s)) ds.$$

It follows from the proof of Remark 4.4 that $\partial_t u \in \mathcal{F}^{\eta,\theta}((0,T]; L^2(\Omega))$. Further, we can obtain immediately from $\partial_t^{1-\alpha} u$ that

$$\partial_t u(t,x) + \kappa_\alpha \partial_t^{1-\alpha} A u + \nabla \cdot (F \partial_t^{1-\alpha} u) = f(t,x), \quad (t,x) \in (0,T) \times \Omega.$$

Thus, the claim holds. $\qquad\square$

Remark 4.5. If we introduce another definition of mild solutions $u : [0,T] \to L^2(\Omega)$ satisfying the integral equation (4.19) instead of (4.9), the assumption $F \in W^{1,1}(0,T; L^2(\Omega)) \cap W^{1,\infty}(0,T;\Omega)$ in Subsection 4.1.5 can reduce to the assumption $F \in W^{1,\infty}(0,T;\Omega)$, and the assumption $F \in W^{1,1}(0,T; L^2(\Omega)) \cap W^{1,\infty}(0,T;\Omega) \cap C^\theta([0,T]; H^1(\Omega))$ in Subsection 4.1.6 can reduce to the assumption $W^{1,\infty}(0,T;\Omega) \cap C^\theta([0,T]; H^1(\Omega))$. Then the results on classical solutions would also hold.

4.2 Galerkin Approximation Method

4.2.1 Introduction

In this section, we study the existence, uniqueness and regularity of solutions to the following inhomogeneous, time-fractional Fokker-Planck initial-boundary value problem:

$$u_t(t,x) - \nabla \cdot (\partial_t^{1-\alpha} \kappa_\alpha \nabla u - \mathbf{F} \partial_t^{1-\alpha} u)(t,x) = g(t,x), \quad \text{for } (t,x) \in (0,T) \times \Omega, \tag{4.27a}$$

$$u(0,x) = u_0(x), \quad \text{for } x \in \Omega, \tag{4.27b}$$

$$u(t,x) = 0, \quad \text{for } x \in \partial\Omega \text{ and } 0 < t < T, \tag{4.27c}$$

where $\kappa_\alpha > 0$ is constant and Ω is an open bounded domain with C^2 boundary in $\mathbb{R}^d$ for some $d \geq 1$. In (4.27a), one has $0 < \alpha < 1$ and $\partial_t^{1-\alpha}$ is the standard Riemann–Liouville fractional derivative operator defined by $\partial_t^{1-\alpha} u = (J^\alpha u)_t$, where J^β denotes the Riemann–Liouville fractional integral operator of order β, viz.,

$$J^\beta u = \int_0^t \omega_\beta(t-s) u(s) ds, \quad \text{where } \omega_\beta(t) := \frac{t^{\beta-1}}{\Gamma(\beta)}, \quad \text{for } \beta > 0,$$

regularity hypotheses on $\mathbf{F}, g$ and u_0 will be imposed later.

The problem (4.27) was considered in [174, 218, 301]. We describe it as "general forcing" since $\mathbf{F} = \mathbf{F}(t, x)$; this is a more difficult problem than the special case where $\mathbf{F} = \mathbf{F}(x)$, which can be reduced to a problem already studied by several authors (see, e.g., [112, 240, 241, 272, 352]). More precisely, when the force $\mathbf{F}$ may depend on t as well as x, equation (4.27) cannot be rewritten in the form of the fractional evolution equation

$$J^{1-\alpha}(u_t) + Au = h(t, u, \nabla u, g, \mathbf{F}), \tag{4.28}$$

in which the first term is a Caputo fractional derivative, the operator $A = -\kappa_\alpha \Delta$, and the function h does not depend explicitly on $\partial_t^{1-\alpha} u$.

The regularity of the solution to the Cauchy problem for (4.28) was studied in [112]; there is a fundamental solution of that problem was constructed and investigated for a more general evolution equation where the operator A in (4.28) is a uniformly elliptic operator with variable coefficients that acts on the spatial variables. The Cauchy problem was also considered in [272] where $h = h(t, u, g, \mathbf{F})$ lies in a space of weighted Hölder continuous functions, and in [352] for the case where A is almost sectorial. existence and uniqueness of a solution to the initial-boundary value problem where (4.27a) is replaced by (4.28) is shown in [240, 241].

The section is organized as follows. Subsection 4.2.2 introduces our basic notation and the definitions of mild and classical solutions of (4.27). Various technical properties of fractional integral operators that will be used in our analysis are provided in Subsection 4.2.3. In Subsection 4.2.4, we introduce the Galerkin approximation of the solution of (4.27) and prove existence and uniqueness of approximate solutions. Properties of the mild and classical solutions are derived in Subsections 4.2.5 and 4.2.6, respectively. Finally, in Subsection 4.2.7, we provide estimates of the time derivatives of the classical solution in $L^2(\Omega)$ and $H^2(\Omega)$, needed for the error analysis of numerical methods for solving (4.27), see, e.g., [174, 218, 301].

4.2.2   *Notation and Definitions*

Throughout the section, we often suppress the spatial variables and write v or $v(t)$ instead of $v(t, \cdot)$ for various functions v. We also use the notation v' for the time derivative.

Let $\| \cdot \|$ denote the $L^2(\Omega)$ norm defined by $\|v\|^2 = \langle v, v \rangle$, where $\langle \cdot, \cdot \rangle$ is the $L^2(\Omega)$ inner product. Let $\| \cdot \|_{H^r(\Omega)}$ and $| \cdot |_{H^r(\Omega)}$ be the standard Sobolev norm and seminorm on the Hilbert space of functions whose rth-order derivatives lie in $L^2(\Omega)$. We borrow some standard notation from parabolic partial differential equations, e.g., $C([0, T]; L^2(\Omega))$.

Assume throughout the section that the forcing function $\mathbf{F} = (F_1, \ldots, F_d)^T \in W^{1,\infty}((0, T) \times \Omega)$ and that its divergence $\nabla \cdot \mathbf{F}$ is continuous on $[0, T] \times \overline{\Omega}$. Then $\mathbf{F}$ is continuous on $[0, T] \times \overline{\Omega}$ and we set

$$\|\mathbf{F}\|_\infty := \max_{1 \leq i \leq d} \max_{(t,x) \in [0,T] \times \overline{\Omega}} |F_i(t, x)|$$

and

$$\|\mathbf{F}\|_{1,\infty} := \|\mathbf{F}\|_\infty + \max_{(x,t)\in[0,T]\times\overline{\Omega}} |\nabla \cdot \mathbf{F}(t,x)|.$$

Stronger assumptions on the regularity of $\mathbf{F}$ will be made in some subsections.

We use C to denote a constant that depends on the data $\Omega, \kappa_\alpha, \mathbf{F}$ and T of the problem (4.27) but is independent of any dimension of finite-dimensional spaces to be used in our Galerkin approximations. Here the unsubscripted constants C are generic and can take different values in different places throughout the section.

We now recall the definitions of some Banach spaces from [118, p.301].

Definition 4.3. Let X be a real Banach space with norm $|\cdot|$. The space $C([0,T];X)$ comprises all continuous functions $v : [0,T] \to X$ with

$$\|v\|_{C([0,T];X)} := \max_{0\leq t\leq T} |v(t)|.$$

Let $p \in [1,\infty]$. The space $L^p(0,T;X)$ comprises all measurable functions $v : [0,T] \to X$ for which

$$\|v\|_{L^p(0,T;X)} := \begin{cases} \left(\int_0^T |v(t)|^p\,dt\right)^{1/p} < \infty & \text{when } 1 \leq p < \infty, \\ \text{esssup}_{0\leq t\leq T} |v(t)| < \infty & \text{when } p = \infty. \end{cases}$$

The space $W^{1,p}(0,T;X)$ comprises all measurable functions $v : [0,T] \to X$ for which

$$\|v\|_{W^{1,p}(0,T;X)} := \|v\|_{L^p(0,T;X)} + \|v'\|_{L^p(0,T;X)} \text{ is finite.}$$

Definition 4.4. Given a Banach space X with a norm (or seminorm) $|\cdot|$, define $L^2_\alpha(0,T;X)$ to be the space of functions $v : [0,T] \to X$ for which the following norm (or seminorm) is finite:

$$\|v\|_{L^2_\alpha(0,T;X)} := \max_{0\leq t\leq T} \left[J^\alpha(|v|^2)(t)\right]^{1/2}$$

$$= \max_{0\leq t\leq T} \left[\frac{1}{\Gamma(\alpha)} \int_0^t (t-s)^{\alpha-1}|v(s)|^2\,ds\right]^{1/2}.$$

For any Banach space X, clearly $L^2_\alpha(0,T;X) \subset L^2(0,T;X)$ for $0 < \alpha < 1$ and $\|\cdot\|_{L^2_1(0,T;X)} = \|\cdot\|_{L^2(0,T;X)}$ if we formally put $\alpha = 1$ in Definition 4.4. For brevity, when $X = L^2(\Omega)$ we write

$$\|v\|_{L^2_\alpha} = \|v\|_{L^2_\alpha(0,T;L^2(\Omega))} \quad \text{and} \quad \|v\|_{L^2} = \|v\|_{L^2(0,T;L^2(\Omega))}.$$

The Mittag-Leffler function $E_\alpha(z)$ that is used in the fractional Gronwall's inequality of Lemma 4.7 is defined by

$$E_\alpha(z) := \sum_{k=0}^\infty \frac{z^k}{\Gamma(k\alpha + 1)},$$

for $z \in \mathbb{C}$.

We now introduce the definitions of mild solutions and classical solutions to problem (4.27). Set

$$G(t) := u_0 + \int_0^t g(s)\, ds, \quad \text{for } 0 \le t \le T.$$

Definition 4.5 (Mild solutions). A mild solution of the problem (4.27) is a function $u \in L^2(0,T; L^2(\Omega))$ such that $J^\alpha u \in L^2(0,T; H^2(\Omega) \cap H_0^1(\Omega))$ and u satisfies

$$u - \kappa_\alpha \Delta(J^\alpha u) + \nabla \cdot (\mathbf{F} J^\alpha u) - \nabla \cdot \left(\int_0^t \mathbf{F}'(s) J^\alpha u(s)\, ds \right) = G(t), \quad \text{a.e. on } (0,T) \times \Omega.$$

$$(4.29)$$

Definition 4.6 (Classical solutions). A classical solution of the problem (4.27) is a function u belonging to the space $C([0,T]; L^2(\Omega)) \cap L^\infty(0,T; H_0^1(\Omega)) \cap L^2(0,T; H^2(\Omega))$ such that

$$u' \in L^2(0,T; L^2(\Omega)) \quad \text{and} \quad \partial_t^{1-\alpha} u \in L^2(0,T; H^2(\Omega)),$$

with u satisfying (4.27a) a.e. on $(0,T) \times \Omega$, and (4.27b) a.e. on Ω.

4.2.3   Technical Preliminaries

This subsection provides some properties of fractional integrals that will be needed in our analysis.

Lemma 4.7. *[371, Corollary 2] Let $\beta > 0$. Assume that a and b are non-negative and non-decreasing functions on the interval $[0,T]$, with $a \in L^1(0,T)$ and $b \in C[0,T]$. If $y \in L^1(0,T)$ satisfies*

$$0 \le y(t) \le a(t) + b(t) \int_0^t \omega_\beta(t-s) y(s)\, ds, \quad \text{for } 0 \le t \le T,$$

then

$$y(t) \le a(t) E_\beta\big(b(t) t^\beta\big), \quad \text{for } 0 \le t \le T.$$

The following lemmas will be used several times in our analysis.

Lemma 4.8. *[174, Lemma 2.3] If $\alpha \in (1/2, 1)$ and $v(x, \cdot) \in L^2(0,T)$ for each $x \in \Omega$, then for $t \in [0,T]$,*

$$J^\alpha \langle J^\alpha v, v \rangle(t) \ge \frac{1}{2} \| J^\alpha v(t) \|^2 \quad \text{and} \quad \int_0^t \langle J^\alpha v, v \rangle(s)\, ds \ge \frac{1}{2} J^{1-\alpha}(\| J^\alpha v \|^2)(t).$$

Lemma 4.9. *[273, Lemma 3.1 (ii)] If $\alpha \in (0,1)$ and $v(x, \cdot) \in L^2(0,T)$ for each $x \in \Omega$, then for $t \in [0,T]$,*

$$\int_0^t \langle J^\alpha v, v \rangle(s)\, ds \ge \cos(\alpha\pi/2) \int_0^t \| J^{\alpha/2} v \|^2(s)\, ds.$$

Lemma 4.10. *[174, Lemma 2.1] Let $\beta \in (0,1)$. If $\phi(\cdot,t) \in L^2(\Omega)$ for $t \in [0,T]$, then*

$$\|J^\beta \phi(t)\|^2 \leq \omega_{\beta+1}(t) \, J^\beta(\|\phi\|^2)(t), \quad \text{for } 0 \leq t \leq T.$$

Proof. As the proof is short, we give it here for completeness. The Cauchy–Schwarz's inequality yields

$$\begin{aligned}
\|J^\beta \phi(t)\|^2 &= \int_\Omega \left[\int_0^t \omega_\beta(t-s)\phi(x,s)ds \right]^2 dx \\
&\leq \int_\Omega \left[\int_0^t \omega_\beta(t-s)\,ds \right]\left[\int_0^t \omega_\beta(t-s)\phi^2(x,s)\,ds \right] dx \\
&= \omega_{\beta+1}(t) \int_0^t \omega_\beta(t-s) \int_\Omega \phi^2(x,s)\,dx\,ds \\
&= \omega_{\beta+1}(t) \, J^\beta(\|\phi\|^2)(t).
\end{aligned}$$

$\square$

Lemma 4.11. *For any $t > 0$ and $\beta > 0$,*

$$\|J^\beta \phi(t)\| \leq \frac{t^\beta}{\Gamma(\beta+1)}\|\phi\|_{L^\infty(0,t;L^2)}, \quad \text{for all } \phi \in L^\infty(0,t;L^2).$$

If $\beta > 1/2$, then

$$\|J^\beta \phi(t)\| \leq \frac{t^{\beta-1/2}}{\Gamma(\beta)\sqrt{2\beta-1}}\|\phi\|_{L^2(0,t;L^2)}, \quad \text{for all } \phi \in L^2(0,t;L^2).$$

Proof. The Minkowski's integral inequality gives

$$\begin{aligned}
\|J^\beta \phi(t)\| &= \left[\int_\Omega \left(\int_0^t \omega_\beta(t-s)\phi(s)\,ds \right)^2 dx \right]^{1/2} \\
&\leq \int_0^t \omega_\beta(t-s)\|\phi(s)\|\,ds \\
&\leq \|\phi\|_{L^\infty(0,t;L^2)} \int_0^t \omega_\beta(t-s)\,ds \\
&= \frac{t^\beta}{\Gamma(\beta+1)}\|\phi\|_{L^\infty(0,t;L^2)}.
\end{aligned}$$

(4.30)

To prove the second inequality, apply the Hölder's inequality to (4.30) to obtain

$$\begin{aligned}
\|J^\beta \phi(t)\| &\leq \int_0^t \omega_\beta(t-s)\|\phi(s)\|\,ds \\
&\leq \left(\int_0^t \omega_\beta^2(t-s)\,ds \right)^{1/2} \|\phi\|_{L^2(0,t;L^2)} \\
&= \frac{t^{\beta-1/2}}{\sqrt{2\beta-1}\,\Gamma(\beta)}\|\phi\|_{L^2(0,t;L^2)},
\end{aligned}$$

for any $\beta > 1/2$, which completes the proof of this lemma. $\square$

Lemma 4.12. *[259, Theorem A.1] For $t > 0$,*

$$\int_0^t \langle \partial_s^{1-\alpha} v, v \rangle \, ds \geq \rho_\alpha t^{\alpha-1} \int_0^t \|v(s)\|^2 \, ds,$$

where

$$\rho_\alpha = \pi^{1-\alpha} \frac{(1-\alpha)^{1-\alpha}}{(2-\alpha)^{2-\alpha}} \sin(\tfrac{1}{2}\pi\alpha).$$

The following estimate involving the force $\mathbf{F}$ is used several times in our analysis.

Lemma 4.13. *If $\phi : [0, T] \to H^1(\Omega)$, then*

$$\left\| \nabla \cdot \left(\mathbf{F}(t)\phi(t) \right) \right\| \leq \|\mathbf{F}\|_{1,\infty} \|\phi(t)\|_{H^1(\Omega)}, \quad \text{for } 0 \leq t \leq T.$$

Proof. The vector field identity $\nabla \cdot (\mathbf{F}\phi) = (\nabla \cdot \mathbf{F})\phi + \mathbf{F} \cdot (\nabla \phi)$ implies that

$$\left\| \nabla \cdot \left(\mathbf{F}(t)\phi(t) \right) \right\|^2 \leq \left(\|\nabla \cdot \mathbf{F}(t)\|_{L^\infty(\Omega)}^2 + \|\mathbf{F}(t)\|_{L^\infty(\Omega,\mathbb{R}^d)}^2 \right) \left(\|\phi(t)\|^2 + \|\nabla\phi(t)\|^2 \right)$$

$$\leq \left(\|\nabla \cdot \mathbf{F}(t)\|_{L^\infty(\Omega)} + \|\mathbf{F}(t)\|_{L^\infty(\Omega,\mathbb{R}^d)} \right)^2 \|\phi(t)\|_{H^1(\Omega)}^2$$

$$\leq \left(\|\mathbf{F}\|_{1,\infty} \|\phi(t)\|_{H^1(\Omega)} \right)^2,$$

which gives the desired estimate. $\qquad\square$

We now recall a fundamental compactness result that will be used several times in the proofs of our main results.

Lemma 4.14 (Aubin–Lions–Simon). *Let $B_0 \subset B_1 \subset B_2$ be three Banach spaces. Assume that the embedding of B_1 in B_2 is continuous and that the embedding of B_0 in B_1 is compact. Let p and r satisfy $1 \leq p, r \leq +\infty$. For $T > 0$, define the Banach space*

$$E_{p,r} := \{ v \in L^p(0,T; B_0) : \partial_t v \in L^r(0,T; B_2) \}$$

with norm

$$\|v\|_{E_{p,r}} := \|v\|_{L^p(0,T;B_0)} + \|v'\|_{L^r(0,T;B_2)}.$$

Then,

- *the embedding $E_{p,r}$ in $L^p(0,T; B_1)$ is compact when $p < +\infty$, and*
- *the embedding $E_{p,r}$ in $C([0,T]; B_1)$ is compact when $p = +\infty$ and $r > 1$.*

Proof. See, e.g., [54, Theorem II.5.16]. $\qquad\square$

4.2.4　Galerkin Approximation of Solution

In this subsection we prove existence and uniqueness of a finite-dimensional Galerkin approximation of the solution of (4.27). This is a standard classical tool for deriving existence and regularity results for parabolic initial-boundary value problems, see, e.g., [118, Section 7.1.2].

Let $\{w_k\}_{k=1}^{\infty}$ be a complete set of eigenfunctions for the operator $-\Delta$ in $H_0^1(\Omega)$, with $\{w_k\}$ an orthonormal basis of $L^2(\Omega)$ and an orthogonal basis of $H_0^1(\Omega)$, see [118, Section 6.5.1]. For each positive integer m, set $W_m = \mathrm{span}\{w_1, w_2, \ldots, w_m\}$ and consider $u_m : [0, T] \to W_m$ given by

$$u_m(t) := \sum_{k=1}^{m} d_m^k(t) w_k(x).$$

Let Π_m be the orthogonal projector from $L^2(\Omega)$ onto W_m defined by: for each $v \in L^2(\Omega)$, one has

$$\Pi_m v \in W_m \quad \text{and} \quad \langle \Pi_m v, w \rangle = \langle v, w \rangle, \quad \text{for all } w \in W_m.$$

The projections of the source term and initial data are denoted by

$$g_m(t) := \Pi_m g(t) \quad \text{and} \quad u_{0m} := \Pi_m u_0.$$

We aim to choose the functions d_m^k so that for $k = 1, 2, \ldots, m$ and $t \in (0, T]$ one has

$$u_m' - \kappa_\alpha \partial_t^{1-\alpha} \Delta u_m + \Pi_m \big(\nabla \cdot (\mathbf{F}(t) \partial_t^{1-\alpha} u_m)\big) = g_m(t), \tag{4.31a}$$

$$d_m^k(0) = \langle u_0, w_k \rangle. \tag{4.31b}$$

Existence and uniqueness of a solution to (4.31) are guaranteed by the following lemma.

Lemma 4.15. *[218, Theorem 3.1] Let $F \in W^{1,\infty}(0, T; L^\infty(\Omega))$ and $g \in L^1(0, T; L^2(\Omega))$. Then for each positive integer m, the system of equations (4.31) has a solution $\{d_m^k\}_{k=1}^m$ with $u_m : [0, T] \to H^2(\Omega) \cap H_0^1(\Omega)$ absolutely continuous. This solution is unique among the space of absolutely continuous functions mapping $[0, T]$ to $H_0^1(\Omega)$.*

Proof. Our argument is mainly based on the proof of [218, Theorem 3.1], but we fill a gap in that argument by verifying that u_m is absolutely continuous. Define the linear operator $B_m(t) : W_m \to W_m$ by

$$\langle B_m(t)v, w \rangle := -\kappa_\alpha \langle \Delta v, w \rangle + \langle \Pi_m (\nabla \cdot (\mathbf{F}(t, \cdot)v)), w \rangle, \quad \text{for all } v, w \in W_m,$$

and rewrite (4.31a) as

$$u_m'(t) + B_m(t) \partial_t^{1-\alpha} u_m(t) = g_m(t).$$

Formally integrating this equation in time we obtain the Volterra integral equation [218, p.1768]:

$$u_m(t) + \int_{s=0}^{t} K_m(t, s) u_m(s) \, ds = G_m(t), \quad \text{for } 0 \le t \le T, \tag{4.32}$$

where

$$K_m(t,s) = B_m(t)w_\alpha(t-s) - \int_s^t B_m'(\tau)w_\alpha(\tau - s)\,d\tau$$

and

$$G_m(t) := u_{0m} + \int_0^t g_m(s)\,ds.$$

It is shown in [218] that (4.32) has a unique solution $u_m \in C([0,T]; H_0^1(\Omega))$.

Now, $g \in L^1(0,T; L^2(\Omega))$ implies that $g_m \in L^1(0,T; L^2(\Omega))$, and it follows that $G_m : [0,T] \to L^2(\Omega)$ is absolutely continuous. Furthermore, Theorem 2.5 of [101] implies (using the continuity of u_m) that $t \mapsto \int_{s=0}^t K_m(t,s)u_m(s)\,ds$ is absolutely continuous. Hence, (4.32) shows that $u_m : [0,T] \to L^2(\Omega)$ is absolutely continuous.

We are now able to differentiate (4.32) (to differentiate the integral term, imitate the calculation in the proof of [101, Lemma 2.12]), obtaining

$$u_m'(t) + \int_{s=0}^t B_m(t)w_\alpha(t-s)u_m'(s)\,ds = g_m(t), \quad \text{for almost all } t \in [0,T].$$

The absolute continuity of $u_m(t)$ implies that $\partial_t^{1-\alpha}u_m(t)$ exists for almost all $t \in [0,T]$ by [101, Lemma 2.12]. Hence from the above equation, u_m satisfies (4.31a). From (4.32), one sees immediately that u_m satisfies (4.31b), so we have demonstrated the existence of a solution to (4.31).

To see that this solution of (4.31) is unique among the space of absolutely continuous functions, one can use the proof of [218, Theorem 3.1] since the absolute continuity of the solution is now known a priori. $\qquad\square$

4.2.5 *Existence and Uniqueness of Mild Solution*

In this subsection, we assume that $\alpha \in (0,1)$, $\mathbf{F} \in W^{1,\infty}((0,T) \times \Omega)$ and that the initial data $u_0 \in L^2(\Omega)$.

4.2.5.1 *A priori estimates*

In order to prove a priori estimates, we consider the integrated form of equation (4.31a):

$$u_m(t) - \kappa_\alpha J^\alpha \Delta u_m(t) + \int_0^t \Pi_m\big(\nabla \cdot (\mathbf{F}(s)\partial_s^{1-\alpha}u_m(s))\big)\,ds = G_m(t), \qquad (4.33)$$

where $G_m(t) = \Pi_m G(t)$ as in (4.32).

Let C_P denote the Poincaré constant for Ω, viz., $\|v\|^2 \le C_P\|\nabla v\|^2$ for $v \in H_0^1(\Omega)$.

Lemma 4.16. *Let m be a positive integer. Let $u_m(t)$ be the absolutely continuous solution of (4.31a) that is guaranteed by Lemma 4.15. Then for any $t \in [0,T]$ one has*

$$\cos(\alpha\pi/2) \int_0^t \|J^{\alpha/2}u_m(s)\|^2\,ds + \kappa_\alpha \int_0^t \|J^\alpha u_m(s)\|_{H^1(\Omega)}^2\,ds \le C_1 \int_0^t \|G_m(s)\|^2\,ds$$

$$(4.34)$$

and

$$\int_0^t \|u_m(s)\|^2 ds \le C_3 \int_0^t \|G_m(s)\|^2\, ds, \tag{4.35}$$

where

$$C_1 := \frac{1 + C_P}{2} \left[1 + \frac{C_2\omega_{1+\alpha/2}^2(t)}{\cos(\alpha\pi/2)} E_{\alpha/2}\left(\frac{C_2\omega_{1+\alpha/2}(t)}{\cos(\alpha\pi/2)}t^{\frac{\alpha}{2}}\right)\right],$$

$$C_2 := 2\left(1 + \frac{\|\mathbf{F}\|_\infty^2}{\kappa_\alpha}\right) + \frac{T^2\|\mathbf{F}'\|_\infty^2}{\kappa_\alpha},$$

$$C_3 := 2 + \frac{C_1}{\kappa_\alpha}\left(4\|\mathbf{F}\|_{1,\infty}^2 + 2T^2\|\mathbf{F}'\|_{1,\infty}^2\right).$$

Proof. Taking the inner product of both sides of (4.33) with $J^\alpha u_m(t) \in W_m$, then integrating by parts with respect to x, we obtain

$$\langle J^\alpha u_m(t), u_m(t)\rangle + \kappa_\alpha \|J^\alpha \nabla u_m(t)\|^2$$
$$= \left\langle \int_0^t \mathbf{F}(s)\partial_s^{1-\alpha} u_m(s)ds, J^\alpha \nabla u_m(t)\right\rangle + \langle G_m(t), J^\alpha u_m(t)\rangle$$
$$\le \frac{\kappa_\alpha}{2}\|J^\alpha \nabla u_m(t)\|^2 + \frac{1}{2\kappa_\alpha}\left\|\int_0^t \mathbf{F}(s)\partial_s^{1-\alpha} u_m(s)\, ds\right\|^2 \tag{4.36}$$
$$+ \frac{1}{4}\|G_m(t)\|^2 + \|J^\alpha u_m(t)\|^2.$$

Integrating by parts with respect to the time variable, and using the Minkowski's integral inequality and the Hölder's inequality, we have

$$\left\|\int_0^t \mathbf{F}(s)\partial_s^{1-\alpha} u_m(s)\, ds\right\|^2$$
$$= \left\|\mathbf{F}(t)J^\alpha u_m(t) - \int_0^t \mathbf{F}'(s)J^\alpha u_m(s)\, ds\right\|^2$$
$$\le 2\|\mathbf{F}\|_\infty^2 \|J^\alpha u_m(t)\|^2 + 2\|\mathbf{F}'\|_\infty^2 \left(\int_0^t \|J^\alpha u_m(s)\|\, ds\right)^2 \tag{4.37}$$
$$\le 2\|\mathbf{F}\|_\infty^2 \|J^\alpha u_m(t)\|^2 + 2t\|\mathbf{F}'\|_\infty^2 \int_0^t \|J^\alpha u_m(s)\|^2\, ds.$$

It follows from (4.36) and (4.37) that

$$\langle J^\alpha u_m(t), u_m(t)\rangle + \frac{\kappa_\alpha}{2}\|J^\alpha \nabla u_m(t)\|^2$$
$$\le \frac{1}{4}\|G_m(t)\|^2 + \left(1 + \frac{\|\mathbf{F}\|_\infty^2}{\kappa_\alpha}\right)\|J^\alpha u_m(t)\|^2 + \frac{t\|\mathbf{F}'\|_\infty^2}{\kappa_\alpha}\int_0^t \|J^\alpha u_m(s)\|^2\, ds.$$

Integrating in time and invoking Lemma 4.9, we deduce that

$$\cos(\alpha\pi/2)\int_0^t \|J^{\alpha/2}u_m(s)\|^2\,ds + \kappa_\alpha \int_0^t \|J^\alpha \nabla u_m(s)\|^2\,ds$$

$$\leq \frac{1}{2}\int_0^t \|G_m(s)\|^2\,ds + 2\left(1+\frac{\|\mathbf{F}\|_\infty^2}{\kappa_\alpha}\right)\int_0^t \|J^\alpha u_m(s)\|^2\,ds$$

$$+\frac{2\|\mathbf{F}'\|_\infty^2}{\kappa_\alpha}\int_0^t s\int_0^s \|J^\alpha u_m(\tau)\|^2 d\tau\,ds$$

$$\leq \frac{1}{2}\int_0^t \|G_m(s)\|^2\,ds + C_2\int_0^t \|J^\alpha u_m(s)\|^2\,ds. \tag{4.38}$$

But Lemma 4.10 gives us

$$\|J^\alpha u_m(s)\|^2 = \|J^{\alpha/2}(J^{\alpha/2}u_m)(s)\|^2 \leq \omega_{1+\alpha/2}(s)J^{\alpha/2}(\|J^{\alpha/2}u_m\|^2)(s). \tag{4.39}$$

Thus, setting $\psi_m(t) := J^1(\|J^{\alpha/2}u_m\|^2)(t)$, we deduce from (4.38) that

$$\psi_m(t) \leq \frac{1}{2\cos(\alpha\pi/2)}\int_0^t \|G_m(s)\|^2\,ds + \frac{C_2\omega_{1+\alpha/2}(t)}{\cos(\alpha\pi/2)}J^{1+\alpha/2}(\|J^{\alpha/2}u_m\|^2)(t)$$

$$= \frac{1}{2\cos(\alpha\pi/2)}\int_0^t \|G_m(s)\|^2\,ds + \frac{C_2\omega_{1+\alpha/2}(t)}{\cos(\alpha\pi/2)}J^{\alpha/2}\psi_m(t).$$

Applying Lemma 4.7, one obtains

$$\psi_m(t) \leq E_{\alpha/2}\left(\frac{C_2\omega_{1+\alpha/2}(t)}{\cos(\alpha\pi/2)}t^{\frac{\alpha}{2}}\right)\frac{1}{2\cos(\alpha\pi/2)}\int_0^t \|G_m(s)\|^2\,ds, \quad \text{for } 0 \leq t \leq T. \tag{4.40}$$

This inequality and (4.39) together yield

$$\int_0^t \|J^\alpha u_m(s)\|^2\,ds$$

$$\leq \omega_{1+\alpha/2}(t)J^{\alpha/2}\psi_m(t)$$

$$\leq \frac{\omega_{1+\alpha/2}(t)}{2\cos(\alpha\pi/2)}\int_0^t \omega_{\alpha/2}(t-s)E_{\alpha/2}\left(\frac{C_2\omega_{1+\alpha/2}(s)}{\cos(\alpha\pi/2)}s^{\frac{\alpha}{2}}\right)\int_0^s \|G_m(z)\|^2\,dz\,ds$$

$$\leq \frac{\omega_{1+\alpha/2}^2(t)}{2\cos(\alpha\pi/2)}E_{\alpha/2}\left(\frac{C_2\omega_{1+\alpha/2}(t)}{\cos(\alpha\pi/2)}t^{\frac{\alpha}{2}}\right)\int_0^t \|G_m(s)\|^2\,ds.$$

Now (4.34) follows immediately on recalling (4.38)–(4.40) and the Poincaré's inequality.

In a similar fashion, we take the inner product of both sides of (4.33) with $u_m(t) \in W_m$ and then integrate by parts with respect to x, to obtain

$$\|u_m(t)\|^2 + \kappa_\alpha\langle J^\alpha \nabla u_m(t), \nabla u_m(t)\rangle$$

$$= -\left\langle\int_0^t \nabla\cdot(\mathbf{F}(s)\partial_s^{1-\alpha}u_m(s))\,ds, u_m(t)\right\rangle + \langle G_m(t), u_m(t)\rangle \tag{4.41}$$

$$\leq \|G_m(t)\|^2 + \frac{1}{2}\|u_m(t)\|^2 + \left\|\int_0^t \nabla\cdot(\mathbf{F}(s)\partial_s^{1-\alpha}u_m(s))\,ds\right\|^2.$$

Using Lemma 4.13 and the same arguments as in the proof of (4.37), we also have

$$\left\| \int_0^t \nabla \cdot (\mathbf{F}(s) \partial_s^{1-\alpha} u_m(s)) \, ds \right\|^2 \leq 2 \|\mathbf{F}\|_{1,\infty}^2 \|J^\alpha u_m(t)\|_{H^1(\Omega)}^2$$

$$+ 2t \|\mathbf{F}'\|_{1,\infty}^2 \int_0^t \|J^\alpha u_m(s)\|_{H^1(\Omega)}^2 \, ds. \tag{4.42}$$

This estimate and (4.41) together imply

$$\frac{1}{2} \|u_m(t)\|^2 + \kappa_\alpha \langle J^\alpha \nabla u_m(t), \nabla u_m(t) \rangle \leq \|G_m(t)\|^2 + 2 \|\mathbf{F}\|_{1,\infty}^2 \|J^\alpha u_m(t)\|_{H^1(\Omega)}^2$$

$$+ 2t \|\mathbf{F}'\|_{1,\infty}^2 \int_0^t \|J^\alpha u_m(s)\|_{H^1(\Omega)}^2 \, ds.$$

Integrating in time, we get

$$\int_0^t \|u_m(s)\|^2 \, ds \leq 2 \int_0^t \|G_m(s)\|^2 \, ds$$

$$+ \left(4 \|\mathbf{F}\|_{1,\infty}^2 + 2t^2 \|\mathbf{F}'\|_{1,\infty}^2 \right) \int_0^t \|J^\alpha u_m(s)\|_{H^1(\Omega)}^2 \, ds.$$

Now apply the inequality (4.34) to complete the proof. □

Lemma 4.17. *Let m be a positive integer, and let $u_m(t)$ be the absolutely continuous solution of (4.31a) that is guaranteed by Lemma 4.15. Then, for any $t \in [0, T]$,*

$$\cos(\alpha\pi/2) \int_0^t \|J^{\alpha/2} \nabla u_m(s)\|^2 \, ds + \kappa_\alpha \int_0^t \|J^\alpha \Delta u_m(s)\|^2 \, ds \leq C_4 \int_0^t \|G_m(s)\|^2 \, ds \tag{4.43}$$

and

$$\|J^1 u_m(t)\|_{H^1(\Omega)}^2 \leq C_5 \int_0^t \|G_m(s)\|^2 \, ds, \tag{4.44}$$

where

$$C_4 := \frac{2}{\kappa_\alpha} + \frac{2C_1}{\kappa_\alpha^2} \left(2 \|\mathbf{F}\|_{1,\infty}^2 + T^2 \|\mathbf{F}'\|_{1,\infty}^2 \right)$$

and

$$C_5 := \frac{C_4 T^{1-\alpha} (1 + C_P)}{(1 - \alpha) \cos(\alpha\pi/2) \Gamma(1 - \alpha/2)^2}.$$

Proof. Taking the inner product of both sides of (4.33) with $-J^\alpha \Delta u_m(t) \in W_m$ and then integrating by parts with respect to x, we obtain

$$\langle J^\alpha \nabla u_m(t), \nabla u_m(t) \rangle + \kappa_\alpha \|J^\alpha \Delta u_m(t)\|^2$$

$$= \left\langle \int_0^t \nabla \cdot (\mathbf{F}(s) \partial_s^{1-\alpha} u_m(s)) \, ds, J^\alpha \Delta u_m(t) \right\rangle - \langle G_m(t), J^\alpha \Delta u_m(t) \rangle$$

$$\leq \frac{\kappa_\alpha}{2} \|J^\alpha \Delta u_m(t)\|^2 + \frac{1}{\kappa_\alpha} \|G_m(t)\|^2 + \frac{1}{\kappa_\alpha} \left\| \int_0^t \nabla \cdot (\mathbf{F}(s) \partial_s^{1-\alpha} u_m(s)) \, ds \right\|^2.$$

This inequality and (4.42) together imply

$$\langle J^\alpha \nabla u_m(t), \nabla u_m(t)\rangle + \frac{\kappa_\alpha}{2}\|J^\alpha \Delta u_m(t)\|^2$$

$$\leq \frac{1}{\kappa_\alpha}\|G_m(t)\|^2 + \frac{2}{\kappa_\alpha}\|\mathbf{F}\|_{1,\infty}^2 \|J^\alpha u_m(t)\|_{H^1(\Omega)}^2$$

$$+ \frac{2t}{\kappa_\alpha}\|\mathbf{F}'\|_{1,\infty}^2 \int_0^t \|J^\alpha u_m(s)\|_{H^1(\Omega)}^2\, ds.$$

Integrating in time and invoking Lemma 4.9, we deduce that

$$2\cos(\alpha\pi/2)J^1(\|J^{\alpha/2}\nabla u_m\|^2)(t) + \kappa_\alpha \int_0^t \|J^\alpha \Delta u_m(s)\|^2\, ds$$

$$\leq \frac{2}{\kappa_\alpha}\int_0^t \|G_m(s)\|^2\, ds + \frac{2}{\kappa_\alpha}(2\|\mathbf{F}\|_{1,\infty}^2 + t^2\|\mathbf{F}'\|_{1,\infty}^2)\int_0^t \|J^\alpha u_m(s)\|_{H^1(\Omega)}^2\, ds,$$

which, after applying inequality (4.34) of Lemma 4.16, completes the proof of (4.43).

Applying (4.30) with $\phi = J^{\alpha/2}u_m$ and $\beta = 1 - \alpha/2$ gives

$$\|J^1 u_m(t)\|_{H^1(\Omega)} = \|J^{1-\alpha/2}J^{\alpha/2}u_m(t)\|_{H^1(\Omega)}$$

$$\leq J^{1-\alpha/2}(\|J^{\alpha/2}u_m\|_{H^1(\Omega)})(t) = (\omega_{1-\alpha/2} * z)(t),$$

where $z(t) = \|J^{\alpha/2}u_m(t)\|_{H^1(\Omega)}$. Using the Young's convolution inequality we get

$$\|J^1 u_m(t)\|_{H^1(\Omega)}^2 \leq \|\omega_{1-\alpha/2} * z\|_{L^\infty(0,t)}^2 \leq \|\omega_{1-\alpha/2}\|_{L^2(0,t)}^2\|z\|_{L^2(0,t)}^2$$

$$= \frac{t^{1-\alpha}}{(1-\alpha)\Gamma(1-\alpha/2)^2}\int_0^t \|J^{\alpha/2}u_m(s)\|_{H^1(\Omega)}^2\, ds.$$

The inequality (4.44) now follows immediately from (4.43). $\qquad\square$

4.2.5.2   *The mild solution*

Our assumption that Ω has a C^2 boundary ensures that if $v \in H_0^1(\Omega)$ satisfies $\Delta v \in L^2(\Omega)$, then $v \in H^2(\Omega)$. Moreover, there is a regularity constant C_R, depending only on Ω, such that

$$\|v\|_{H^2(\Omega)} \leq C_R\|\Delta v\|, \quad \text{for } v \in H_0^1(\Omega). \tag{4.45}$$

Our next result requires a strengthening of the regularity hypothesis on $\mathbf{F}$.

Theorem 4.8. *Assume that* $u_0 \in L^2(\Omega)$, $\mathbf{F} \in W^{2,\infty}((0,T)\times\Omega)$ *and* $g \in L^2(0,T;L^2 (\Omega))$. *Then there exists a unique mild solution* u *of* (4.27) *(in the sense of Definition 4.5) such that*

$$\|u\|_{L^2(0,T;L^2)}^2 + \|J^\alpha u\|_{L^2(0,T;H^2)}^2 \leq (C_4 + \kappa_\alpha^{-1}C_4C_R)\|G\|_{L^2}^2. \tag{4.46}$$

Proof. In order to prove the existence of a mild solution, we first prove the convergence of the approximate solutions u_m, and then find the limit of equation (4.33) as m tends to infinity.

Note first that $\|G_m(s)\| \leq \|G(s)\|$ because

$$|\langle G_m(s), w\rangle| = |\langle G(s), \Pi_m w\rangle| \leq \|G(s)\|\|\Pi_m w\| \leq \|G(s)\|\|w\|, \quad \text{for all } w \in L_2(\Omega).$$

Hence Lemma 4.16 shows that the sequence $\{\partial_t(J^1 u_m)\}_{m=1}^\infty = \{u_m\}_{m=1}^\infty$ is bounded in $L^2(0,T;L^2(\Omega))$, and Lemma 4.17 shows that the sequence $\{J^1 u_m\}_{m=1}^\infty$ is bounded in $L^\infty(0,T;H_0^1(\Omega))$. Applying Lemma 4.14 with $B_0 = H_0^1(\Omega)$, $B_1 = B_2 = L^2(\Omega)$, $p = +\infty$ and $r = 2$, it follows that there exists a subsequence of $\{J^1 u_m\}_{m=1}^\infty$, again denoted by $\{J^1 u_m\}_{m=1}^\infty$, and a $v \in C([0,T];L^2(\Omega))$, such that

$$J^1 u_m \to v \text{ strongly in } C([0,T];L^2(\Omega)). \tag{4.47}$$

Furthermore, from the above bounds on $\{J^1 u_m\}_{m=1}^\infty$ and well-known results [54, Theorem II.2.7] for weak and weak-$\star$ compactness, by choosing sub-subsequences we get

$$J^1 u_m \to v \text{ weak-}\star \text{ in } L^\infty(0,T;H_0^1(\Omega)) \text{ and}$$
$$u_m = \partial_t(J^1 u_m) \to \partial_t v \text{ weakly in } L^2(0,T;L^2(\Omega)). \tag{4.48}$$

By letting $u := \partial_t v \in L^2(0,T;L^2(\Omega))$, we have $v = J^1 u$. It remains to prove that $J^\alpha u_m$ converges weakly to $J^\alpha u$ in $L^2(0,T;H^2(\Omega))$. Applying Lemma 4.11 with $\phi = J^1 u_m$ and $\beta = \alpha$, for any $t \in [0,T]$ we deduce that

$$\|J^{1+\alpha} u_m(t)\|_{H^1(\Omega)} \le \frac{t^\alpha}{\Gamma(\alpha+1)} \|J^1 u_m\|_{L^\infty(0,t;H^1(\Omega))}.$$

This inequality, together with Lemma 4.17, implies that the sequence $\{J^{1+\alpha} u_m\}_{m=1}^\infty$ is bounded in $L^\infty(0,T;H_0^1(\Omega))$.

Also, Lemma 4.16 shows that the sequence $\{\partial_t(J^{1+\alpha} u_m)\}_{m=1}^\infty = \{J^\alpha u_m\}_{m=1}^\infty$ is bounded in $L^2(0,T;H_0^1(\Omega))$. It now follows from Lemma 4.14, again with $B_0 = H_0^1(\Omega)$, $B_1 = B_2 = L^2(\Omega)$, $p = +\infty$ and $r = 2$, that there exists a subsequence of $\{J^{1+\alpha} u_m\}_{m=1}^\infty$ (still denoted by $\{J^{1+\alpha} u_m\}_{m=1}^\infty$) and $\bar{u} \in C([0,T];L^2(\Omega))$ such that

$$J^{1+\alpha} u_m \to \bar{u} \text{ strongly in } C([0,T];L^2(\Omega)). \tag{4.49}$$

Furthermore, from the upper bound (4.43) of $\{\partial_t(J^{1+\alpha} u_m)\}_{m=1}^\infty$ in $L^2(0,T;H^2(\Omega))$, by choosing a subsequence one gets

$$J^\alpha u_m = \partial_t(J^{1+\alpha} u_m) \to \partial_t \bar{u} \text{ weakly in } L^2(0,T;H^2(\Omega)). \tag{4.50}$$

On the other hand, by applying Lemma 4.11 with $\phi = J^1(u_m - u)$ and $\beta = \alpha$, we deduce that for any $t \in [0,T]$,

$$\|J^{1+\alpha}(u_m - u)(t)\|_{L^2(\Omega)} \le \frac{t^\alpha}{\Gamma(\alpha+1)} \|J^1(u_m - u)\|_{L^\infty(0,t;L^2(\Omega))}.$$

Hence, (4.47) implies that $\lim_{m\to\infty} \|J^{1+\alpha}(u_m - u)\|_{L^\infty(0,T;L^2(\Omega))} = 0$. Recalling (4.49), we have $\bar{u} = J^{1+\alpha} u$. By choosing subsequences, we obtain

$$J^{1+\alpha} u_m \to J^{1+\alpha} u \text{ strongly in } C([0,T];L^2(\Omega)), \tag{4.51}$$
$$J^{1+\alpha} u_m \to J^{1+\alpha} u \text{ weak-}\star \text{ in } L^\infty(0,T;H_0^1(\Omega)), \tag{4.52}$$
$$J^\alpha u_m \to J^\alpha u \text{ weakly in } L^2(0,T;H^2(\Omega)), \tag{4.53}$$

where we used the boundedness of $\{J^{1+\alpha} u_m\}_{m=1}^\infty$ in $L^\infty(0,T;H_0^1(\Omega))$ that was already mentioned, and (4.50).

Multiplying both sides of (4.33) by a test function $\xi \in C_c^\infty((0,T) \times \Omega)$, integrating over $(0,T) \times \Omega$ and noting that Π_m is a self-adjoint operator on $L^2(\Omega)$ gives

$$
\begin{aligned}
&\langle u_m, \xi \rangle_{L^2(0,T;L^2)} - \kappa_\alpha \langle J^\alpha \Delta u_m, \xi \rangle_{L^2(0,T;L^2)} + \langle h_m, \Pi_m \xi \rangle_{L^2(0,T;L^2)} \\
&= \langle G_m, \xi \rangle_{L^2(0,T;L^2)},
\end{aligned}
\tag{4.54}
$$

where $h_m(t) := \int_0^t \nabla \cdot (\mathbf{F}(s) \partial_s^{1-\alpha} u_m(s)) \, ds$. Using (4.48) and (4.53), as $m \to \infty$ one has

$$
\begin{aligned}
\langle u_m, \xi \rangle_{L^2(0,T;L^2)} &\to \langle u, \xi \rangle_{L^2(0,T;L^2)}, \\
\langle G_m, \xi \rangle_{L^2(0,T;L^2)} &\to \langle G, \xi \rangle_{L^2(0,T;L^2)}, \\
\langle J^\alpha \Delta u_m, \xi \rangle_{L^2(0,T;L^2)} &\to \langle J^\alpha \Delta u, \xi \rangle_{L^2(0,T;L^2)}.
\end{aligned}
\tag{4.55}
$$

To find the limit of the most complicated term $\langle h_m, \Pi_m \xi \rangle_{L^2(0,T;L^2)}$ in (4.54), we first integrate by parts twice with respect to the time variable:

$$
\begin{aligned}
h_m(t) &= \int_0^t \nabla \cdot (\mathbf{F}(s) \partial_s^{1-\alpha} u_m(s)) \, ds \\
&= \nabla \cdot (\mathbf{F}(t) J^\alpha u_m(t)) - \int_0^t \nabla \cdot (\mathbf{F}'(s) J^\alpha u_m(s)) \, ds \\
&= \nabla \cdot (\mathbf{F}(t) J^\alpha u_m(t)) - \nabla \cdot (\mathbf{F}'(t) J^{1+\alpha} u_m(t)) \\
&\quad + \int_0^t \nabla \cdot (\mathbf{F}''(s) J^{1+\alpha} u_m(s)) \, ds.
\end{aligned}
\tag{4.56}
$$

It now follows from the boundedness of $\{J^\alpha u_m\}_{m=1}^\infty$ in $L^2(0,T;H^2(\Omega))$, and $\{J^{1+\alpha} u_m\}_{m=1}^\infty$ in $L^\infty(0,T;H_0^1(\Omega))$, that $\{h_m\}_{m=1}^\infty$ is bounded in $L^2(0,T;L^2(\Omega))$. Hence,

$$
\lim_{m\to\infty} \langle h_m, \Pi_m \xi - \xi \rangle_{L^2(0,T;L^2)} = 0.
\tag{4.57}
$$

On the other hand, by using (4.56) and integration by parts with respect to x, we have

$$
\begin{aligned}
\langle h_m, \xi \rangle_{L^2(0,T;L^2)} &= \langle \nabla \cdot (\mathbf{F} J^\alpha u_m), \xi \rangle_{L^2(0,T;L^2)} - \langle \nabla \cdot (\mathbf{F}' J^{1+\alpha} u_m), \xi \rangle_{L^2(0,T;L^2)} \\
&\quad - \int_0^T \int_\Omega \left(\int_0^t \mathbf{F}''(s) J^{1+\alpha} u_m(s) \, ds \right) \cdot \nabla \xi(t) \, dx \, dt.
\end{aligned}
$$

Combining this identity with (4.51)–(4.53) gives

$$
\begin{aligned}
\lim_{m\to\infty} \langle h_m, \xi \rangle_{L^2(0,T;L^2)} &= \langle \nabla \cdot (\mathbf{F} J^\alpha u), \xi \rangle_{L^2(0,T;L^2)} - \langle \nabla \cdot (\mathbf{F}' J^{1+\alpha} u), \xi \rangle_{L^2(0,T;L^2)} \\
&\quad - \int_0^T \int_\Omega \left(\int_0^t \mathbf{F}''(s) J^{1+\alpha} u(s) \, ds \right) \cdot \nabla \xi(t) \, dx \, dt \\
&= \langle \nabla \cdot (\mathbf{F} J^\alpha u), \xi \rangle_{L^2(0,T;L^2)} \\
&\quad - \int_0^T \int_\Omega \nabla \cdot \left(\int_0^t \mathbf{F}'(s) J^\alpha u(s) \, ds \right) \xi(t) \, dx \, dt.
\end{aligned}
$$

Now invoking (4.57) yields

$$\lim_{m\to\infty} \langle h_m, \Pi_m \xi \rangle_{L^2(0,T;L^2)} = \langle \nabla \cdot (\mathbf{F} J^\alpha u), \xi \rangle_{L^2(0,T;L^2)}$$

$$- \int_0^T \int_\Omega \nabla \cdot \left(\int_0^t \mathbf{F}'(s) J^\alpha u(s) \, ds \right) \xi(t) \, dx \, dt. \tag{4.58}$$

Let $m \to \infty$ in (4.54). Using (4.55) and (4.58), we deduce that for any $\xi \in C_c^\infty((0,T) \times \Omega)$ one has

$$\left\langle u - \kappa_\alpha J^\alpha \Delta u + \nabla \cdot (\mathbf{F} J^\alpha u) - \nabla \cdot \left(\int_0^t \mathbf{F}'(s) J^\alpha u(s) \, ds \right), \xi \right\rangle_{L^2(0,T;L^2)}$$

$$= \langle g, \xi \rangle_{L^2(0,T;L^2)}.$$

Since $C_c^\infty((0,T) \times \Omega)$ is dense in $L^2((0,T) \times \Omega)$, the above equation also holds true for any test function $\xi \in L^2((0,T) \times \Omega)$. Hence, u satisfies (4.29) a.e. on $(0,T) \times \Omega$.

The weak convergence of u_m described in (4.48), and [54, Corollary II.2.8] with (4.35) together yield $\|u\|_{L^2(0,T;L^2)}^2 \le C_3 \|G\|_{L^2(0,T;L^2)}^2$. Similarly, (4.53) and (4.43) imply that $\kappa_\alpha \|J^\alpha \Delta u\|_{L^2(0,T;H^2)}^2 \le C_4 \|G\|_{L^2(0,T;L^2)}^2$. Thus, (4.46) is proved.

The uniqueness of the solution u follows from linearity and (4.46), because if $u_0 = 0$ and $g = 0$, then $G = 0$ and hence $u = 0$. $\qquad\square$

4.2.6 Existence and Uniqueness of Classical Solution

In the rest of this section, we assume that

$$\frac{1}{2} < \alpha < 1. \tag{4.59}$$

The assumption (4.59) is not overly restrictive because (4.27) is usually considered as a variant of the case $\alpha = 1$. We cannot avoid this restriction on α in Subsections 4.2.6 and 4.2.7 since our analysis makes heavy use of $\partial_t^{1-\alpha} u$, and for typical solutions u of (4.27), it will turn out that $\|\partial_t^{1-\alpha} u\|_{L^2(0,T;L^2(\Omega))} < \infty$ only for $1/2 < \alpha < 1$. To see this heuristically, assume that $u(t,x) = \phi(x) + v(t,x)$, where v vanishes as $t \to 0$ so $\phi(x)$ is the dominant component near $t = 0$; then $\partial_t^{1-\alpha} u(t,x) = \phi(x) \omega_\alpha(t) + \partial_t^{1-\alpha} v(t,x) \approx \phi(x) \omega_\alpha(t)$ near $t = 0$, and $\int_0^T \omega_\alpha^2(t) \, dt$ is finite only if $\alpha > 1/2$.

4.2.6.1 A priori estimates

Since $\partial_t^{1-\alpha} 1 = \omega_\alpha(t)$, we can rewrite (4.31a) in terms of

$$v_m(t) := u_m(t) - u_{0m}$$

as

$$v_m' - \kappa_\alpha \partial_t^{1-\alpha} \Delta v_m + \Pi_m(\nabla \cdot (\mathbf{F}(t) \partial_t^{1-\alpha} v_m))$$

$$= \Pi_m g(t) - \omega_\alpha(t) [\Pi_m(\nabla \cdot (\mathbf{F}(t) u_{0m}) - \kappa_\alpha \Delta u_{0m}]. \tag{4.60}$$

We will require the following bound for u_{0m}.

Lemma 4.18. *If $u_0 \in H^2(\Omega)$, then $\|u_{0m}\|_{H^2(\Omega)} \le C_R \|u_0\|_{H^2(\Omega)}$ for all m, where the constant C_R was defined in (4.45).*

Proof. Write $d_{0m}^k = d_m^k(0) = \langle u_0, w_k \rangle$ for $1 \le k \le m$, and let $\lambda_k > 0$ denote the kth Dirichlet eigenvalue of the Laplacian so that $-\Delta w_k = \lambda_k w_k$ for all k. In this way,

$$u_{0m}(x) = \sum_{k=1}^{m} d_{0m}^k w_k(x) \quad \text{and} \quad -\Delta u_{0m}(x) = \sum_{k=1}^{m} d_{0m}^k \lambda_k w_k(x).$$

If $u_0 \in H^2(\Omega)$, then $\Delta u_0 \in L_2(\Omega)$. So using the Parseval's identity,

$$\|\Delta u_{0m}\|^2 = \sum_{k=1}^{m} |\langle \Delta u_{0m}, w_k \rangle|^2 = \sum_{k=1}^{m} \lambda_k^2 |\langle u_0, w_k \rangle|^2$$

$$\le \sum_{k=1}^{\infty} \lambda_k^2 |\langle u_0, w_k \rangle|^2 = \sum_{k=1}^{\infty} |\langle u_0, \Delta w_k \rangle|^2 = \|\Delta u_0\|^2.$$

Thus, $\|u_{0m}\|_{H^2(\Omega)} \le C_{\mathrm{R}} \|\Delta u_{0m}\| \le C_{\mathrm{R}} \|\Delta u_0\| \le C_{\mathrm{R}} \|u_0\|_{H^2(\Omega)}$. $\qquad\square$

We now prove upper bounds for $\|\partial_t^{1-\alpha} v_m(t)\|$ and $J^\alpha(\|\partial_t^{1-\alpha} \nabla v_m\|^2)(t)$ for any $t \in [0, T]$. The argument used in the following lemma is based on the proof of [174, Theorem 3.1].

Lemma 4.19. *Let m be a positive integer. Let $v_m(t)$ be the absolutely continuous solution of (4.60) that is guaranteed by Lemma 4.15. Then, for almost all $t \in [0, T]$,*

$$\|\partial_t^{1-\alpha} v_m(t)\|^2 \le E_{2\alpha-1}\big(C_7 t^{2\alpha-1}\big)\left(C_6 \|u_0\|_{H^2(\Omega)}^2 + \int_0^t \|g(s)\|^2\, ds\right) \qquad (4.61)$$

and

$$J^\alpha\big(\|\partial_t^{1-\alpha} \nabla v_m\|^2\big)(t) \le \frac{1 + C_7 E_{2\alpha-1}\big(C_7 t^{2\alpha-1}\big) \omega_{2\alpha}(t)}{\kappa_\alpha}$$
$$\times \left(C_6 \|u_0\|_{H^2(\Omega)}^2 + \int_0^t \|g(s)\|^2\, ds\right), \qquad (4.62)$$

where

$$C_6 := C_{\mathrm{R}}^2 \big(\kappa_\alpha + \|\mathbf{F}\|_{1,\infty}\big)^2$$

and

$$C_7 := \frac{\Gamma(2\alpha - 1)}{\Gamma(\alpha)^2}\left(1 + \frac{T^{2\alpha-1}}{(2\alpha - 1)\Gamma(\alpha)^2} + \frac{\|\mathbf{F}\|_\infty^2 \Gamma(\alpha) T^{1-\alpha}}{\kappa_\alpha \Gamma(2\alpha - 1)}\right).$$

Proof. For notational convenience, set $z_m(t) := \partial_t^{1-\alpha} v_m(t) \in W_m$. Taking the inner product of both sides of (4.60) with $z_m(t)$ and integrating by parts with respect to x, we obtain

$$\langle v_m', z_m \rangle + \kappa_\alpha \|\nabla z_m\|^2 = \langle g(t), z_m \rangle + \langle \mathbf{F}(t) z_m, \nabla z_m \rangle$$
$$- \langle \nabla \cdot (\mathbf{F}(t) u_{0m}) - \kappa_\alpha \Delta u_{0m}, z_m \rangle \omega_\alpha(t). \qquad (4.63)$$

By the Cauchy-Schwarz's inequality and the Young's inequality, one has

$$|\langle \mathbf{F}(t) z_m, \nabla z_m \rangle| \le \|\mathbf{F}\|_\infty \|z_m\| \|\nabla z_m\| \le \frac{\kappa_\alpha}{2} \|\nabla z_m\|^2 + \frac{\|\mathbf{F}\|_\infty^2}{2\kappa_\alpha} \|z_m\|^2$$

and, using Lemma 4.13,

$$\left|\langle \nabla \cdot (\mathbf{F}(t)u_{0m}) - \kappa_\alpha \Delta u_{0m}, z_m \rangle\right| \leq \left[\|\mathbf{F}\|_{1,\infty}\|u_{0m}\|_{H^1(\Omega)} + \kappa_\alpha\|\Delta u_{0m}\|\right]\|z_m\|.$$

Substituting these bounds into (4.63) and then applying Lemma 4.18, we obtain

$$\langle v_m', z_m \rangle + \frac{\kappa_\alpha}{2}\|\nabla z_m\|^2 \leq \|g(t)\|\,\|z_m\| + \frac{\|\mathbf{F}\|_\infty^2}{2\kappa_\alpha}\|z_m\|^2 + \sqrt{C_6}\,\|u_0\|_{H^2(\Omega)}\|z_m\|\,\omega_\alpha(t).$$

$$(4.64)$$

But $v_m(0) = 0$, so $z_m = \partial_t^{1-\alpha}v_m = {}^C\partial_t^{1-\alpha}v_m = J^\alpha(v_m')$, thus $\langle v_m', z_m \rangle = \langle v_m', J^\alpha(v_m')\rangle$. Applying J^α to both sides of (4.64) and invoking Lemma 4.8 to handle the first term, we get

$$\|z_m\|^2 + \kappa_\alpha J^\alpha\left(\|\nabla z_m\|^2\right)$$

$$\leq 2J^\alpha\left(\|g\|\|z_m\|\right) + \frac{\|\mathbf{F}\|_\infty^2}{\kappa_\alpha}J^\alpha\left(\|z_m\|^2\right) \qquad (4.65)$$

$$+ 2\sqrt{C_6}\,\|u_0\|_{H^2(\Omega)}J^\alpha(\|z_m\|\,\omega_\alpha), \quad \text{for } 0 < t \leq T.$$

By the Cauchy-Schwarz's inequality and the Young's inequality,

$$2J^\alpha(\|g\|\|z_m\|)(t) = 2\int_{s=0}^t \frac{(t-s)^{\alpha-1}}{\Gamma(\alpha)}\|g(s)\|\|z_m(s)\|\,ds$$

$$\leq 2\left(\int_{s=0}^t \|g(s)\|^2\,ds\right)^{1/2}\left(\int_{s=0}^t \frac{(t-s)^{2\alpha-2}}{\Gamma(\alpha)^2}\|z_m(s)\|^2\,ds\right)^{1/2}$$

$$\leq \int_{s=0}^t \|g(s)\|^2\,ds + \frac{\Gamma(2\alpha-1)}{\Gamma(\alpha)^2}\int_{s=0}^t \omega_{2\alpha-1}(t-s)\|z_m(s)\|^2\,ds,$$

and if $0 \leq s < t$, then $(t-s)^{\alpha-1} = (t-s)^{1-\alpha}(t-s)^{2\alpha-2} \leq t^{1-\alpha}(t-s)^{2\alpha-2}$, so

$$\frac{\|\mathbf{F}\|_\infty^2}{\kappa_\alpha}J^\alpha(\|z_m\|^2)(t) \leq \frac{\|\mathbf{F}\|_\infty^2 t^{1-\alpha}}{\kappa_\alpha\Gamma(\alpha)}\int_{s=0}^t (t-s)^{2\alpha-2}\|z_m(s)\|^2\,ds.$$

For the final term in (4.65), we have

$$2\sqrt{C_6}\,\|u_0\|_{H^2(\Omega)}J^\alpha(\|z_m\|\,\omega_\alpha) \leq C_6\|u_0\|_{H^2(\Omega)}^2 + \left[J^\alpha(\|z_m\|\,\omega_\alpha)\right]^2$$

with

$$\left[J^\alpha(\|z_m\|\,\omega_\alpha)(t)\right]^2 = \left(\int_0^t \omega_\alpha(t-s)\|z_m(s)\|\,\omega_\alpha(s)\,ds\right)^2$$

$$\leq \left(\int_0^t \omega_\alpha(s)^2\,ds\right)\left(\int_0^t \omega_\alpha(t-s)^2\|z_m(s)\|^2\,ds\right)$$

$$= \frac{\Gamma(2\alpha-1)t^{2\alpha-1}}{(2\alpha-1)\Gamma(\alpha)^4}\int_0^t \omega_{2\alpha-1}(t-s)\|z_m(s)\|^2\,ds.$$

Hence, (4.65) yields

$$\|z_m(t)\|^2 + \kappa_\alpha J^\alpha\left(\|\nabla z_m\|^2\right)(t)$$

$$\leq C_6\|u_0\|_{H^2(\Omega)}^2 + \int_0^t \|g(s)\|^2\,ds \qquad (4.66)$$

$$+ C_7\int_0^t \omega_{2\alpha-1}(t-s)\|z_m(s)\|^2\,ds, \quad \text{for } 0 < t \leq T.$$

Discard the κ_α term and then apply the fractional Gronwall's inequality (Lemma 4.7) to get (4.61). Finally, after substituting the bound (4.61) into the right-hand side of (4.66), it is straightforward to deduce (4.62). $\qquad\square$

The next corollary follows easily from Lemma 4.19.

Corollary 4.1.
$$\max_{0 \le t \le T} \|\partial_t^{1-\alpha} v_m(t)\|^2 \le C_8 \big[\|u_0\|_{H^2(\Omega)}^2 + \|g\|_{L^2}^2\big],$$
$$\|\partial_t^{1-\alpha} \nabla v_m\|_{L_\alpha^2}^2 \le C_9 \big[\|u_0\|_{H^2(\Omega)}^2 + \|g\|_{L^2}^2\big].$$
Here, for $i = 8, 9$, the constants $C_i = C_i(\alpha, T, \kappa_\alpha, \|\boldsymbol{F}\|_{1,\infty})$ blow up as $\alpha \to (1/2)^+$ but are bounded as $\alpha \to 1^-$.

Corollary 4.1 implies an $L^2(\Omega)$ bound on $u_m(t)$, which we give in Corollary 4.2.

Corollary 4.2. *With C_8 as in Corollary 4.1, one has*
$$\|v_m(t)\|^2 \le C_8 \omega_{2-\alpha}^2(t) \big[\|u_0\|_{H^2(\Omega)}^2 + \|g\|_{L^2}^2\big], \quad \text{for} \quad 0 \le t \le T \tag{4.67a}$$
and
$$\|v_m\|_{L_\alpha^2}^2 \le \frac{C_8 T^{2-\alpha}}{\Gamma(2-\alpha)} \big[\|u_0\|_{H^2(\Omega)}^2 + \|g\|_{L^2}^2\big]. \tag{4.67b}$$

Proof. As $u_m(t)$ is absolutely continuous, by using [101, Theorem 2.2], we have $v_m(t) = (J^1 u_m')(t) = J^{1-\alpha}(J^\alpha u_m')(t)$. Thus [101, Theorem 2.22] can be invoked, which yields
$$v_m(t) = J^{1-\alpha} \partial_t^{1-\alpha}(u_m(t) - u_m(0)), \quad \text{for almost all } t.$$
Set $z_m(t) = \partial_t^{1-\alpha} v_m(t)$, so $v_m(t) = J^{1-\alpha} z_m(t)$. Now Lemma 4.10 and Corollary 4.1 give
$$\|v_m(t)\|^2 = \|J^{1-\alpha} z_m(t)\|^2$$
$$\le \omega_{2-\alpha}(t) \int_0^t \omega_{1-\alpha}(t-s) \|z_m(s)\|^2 \, ds$$
$$\le \omega_{2-\alpha}(t) C_8 \big[\|u_0\|_{H^2(\Omega)}^2 + \|g\|_{L^2(0,T;L^2)}^2\big] \int_0^t \omega_{1-\alpha}(t-s) \, ds$$
$$= C_8 \big[\|u_0\|_{H^2(\Omega)}^2 + \|g\|_{L^2(0,T;L^2)}^2\big] \omega_{2-\alpha}^2(t).$$
As $u_m(t)$ is continuous, the inequality (4.67a) is valid for all t.

Next, using (4.67a) and the semigroup property $\omega_\alpha * \omega_\beta = \omega_{\alpha+\beta}$, we get
$$\|v_m\|_{L_\alpha^2}^2 = \max_{0 \le t \le T} \int_{s=0}^t \omega_\alpha(t-s) \|v_m(s)\|^2 \, ds$$
$$\le C_8 \big[\|u_0\|_{H^2(\Omega)}^2 + \|g\|_{L^2(0,T;L^2)}^2\big] \max_{0 \le t \le T} \int_{s=0}^t \omega_\alpha(t-s) \omega_{2-\alpha}^2(s) \, ds$$
$$\le C_8 \big[\|u_0\|_{H^2(\Omega)}^2 + \|g\|_{L^2(0,T;L^2)}^2\big] \Big(\max_{0 \le t \le T} \omega_{2-\alpha}(t)\Big) \Big(\max_{0 \le t \le T} \omega_2(t)\Big),$$
which gives (4.67b). $\qquad\square$

In the next lemma, we also provide upper bounds for $\{v_m\}_{m=1}^\infty$ in $W^{1,2}(0,T;L^2) \cap L^2(0,T;H^2)$ and $\{\partial_t^{1-\alpha}\Delta v_m\}_{m=1}^\infty$ in $L^2(0,T;L^2)$. Recall that the constant $\rho_\alpha > 0$ was defined in Lemma 4.12.

Lemma 4.20. *Let m be a positive integer. Let $v_m(t)$ be the absolutely continuous solution of (4.60) that is guaranteed by Lemma 4.15. Then for almost all $t \in [0,T]$, one has*

$$\|\nabla v_m(t)\|^2 + \kappa_\alpha \rho_\alpha t^{\alpha-1} \int_0^t \|\Delta v_m\|^2 \, ds \leq \frac{t^{1-\alpha}}{\kappa_\alpha \rho_\alpha} [(C_{10} + C_{11})\|u_0\|_{H^2(\Omega)}^2 + C_{10}\|g\|_{L^2}^2],$$

(4.68)

$$\int_0^t \|v_m'\|^2 \, ds \leq (C_{10} + C_{11})C_R^2\|u_0\|_{H^2(\Omega)}^2 + C_{10}\|g\|_{L^2}^2$$

(4.69)

and

$$\int_0^t \|\partial_s^{1-\alpha}\Delta v_m(s)\|^2 \, ds \leq \frac{1}{\kappa_\alpha^2} [(C_{10} + C_{11})\|u_0\|_{H^2(\Omega)}^2 + C_{10}\|g\|_{L^2}^2],$$

(4.70)

where

$$C_{10} := 3[1 + \|\mathbf{F}\|_{1,\infty}^2(C_8 T + C_9\Gamma(\alpha)T^{1-\alpha})]$$

and

$$C_{11} := \frac{6C_R^2(\kappa_\alpha^2 + \|\mathbf{F}\|_{1,\infty}^2)}{(2\alpha - 1)\Gamma(\alpha)^2} T^{2\alpha-1}.$$

Proof. Take the inner product of both sides of (4.60) with $-\Delta v_m \in W_m$ and integrate by parts with respect to x to get

$$\frac{1}{2}\frac{d}{dt}\|\nabla v_m\|^2 + \kappa_\alpha\langle\partial_t^{1-\alpha}\Delta v_m, \Delta v_m\rangle = \langle\nabla \cdot (\mathbf{F}(t)\partial_t^{1-\alpha}v_m), \Delta v_m\rangle - \langle g(t), \Delta v_m\rangle$$
$$- \langle\kappa_\alpha\Delta u_{0m} - \nabla \cdot (\mathbf{F}(t)u_{0m}), \Delta v_m\rangle\omega_\alpha(t).$$

Integrating in time and noting that, by Lemma 4.12,

$$\rho_\alpha t^{\alpha-1} \int_0^t \|\Delta v_m\|^2 \, ds \leq \int_0^t \langle\partial_s^{1-\alpha}\Delta v_m, \Delta v_m\rangle \, ds,$$

we obtain

$$\frac{1}{2}\|\nabla v_m(t)\|^2 + \kappa_\alpha\rho_\alpha t^{\alpha-1} \int_0^t \|\Delta v_m\|^2 \, ds$$
$$\leq 3\epsilon \int_0^t \|\Delta v_m\|^2 \, ds + \frac{1}{4\epsilon} \int_0^t \left[\|\nabla \cdot (\mathbf{F}(s)\partial_s^{1-\alpha}v_m)\|^2 + \|g(s)\|^2\right.$$
$$\left.+ \|\kappa_\alpha\Delta u_{0m} - \nabla \cdot (\mathbf{F}(s)u_{0m})\|^2\omega_\alpha(s)^2\right] ds,$$

with a free parameter $\epsilon > 0$. Choosing $\epsilon = \kappa_\alpha\rho_\alpha t^{\alpha-1}/6$ and recalling Lemma 4.13 yields

$$\|\nabla v_m(t)\|^2 + \kappa_\alpha\rho_\alpha t^{\alpha-1} \int_0^t \|\Delta v_m\|^2 \, ds$$

$$\leq \frac{3}{\kappa_\alpha \rho_\alpha t^{\alpha-1}} \int_0^t \left(\|g(s)\|^2 + 2\kappa_\alpha^2 \|\Delta u_{0m}\|^2 \omega_\alpha(s)^2 \right) ds$$

$$+ \frac{3\|\mathbf{F}\|_{1,\infty}^2}{\kappa_\alpha \rho_\alpha t^{\alpha-1}} \int_0^t \left(\|\partial_s^{1-\alpha} v_m\|_{H^1(\Omega)}^2 + 2\|u_{0m}\|_{H^1(\Omega)}^2 \omega_\alpha(s)^2 \right) ds$$

$$\leq \frac{t^{1-\alpha}}{\kappa_\alpha \rho_\alpha} \left(C_{11} \|u_0\|_{H^2(\Omega)}^2 + 3\|g\|_{L^2}^2 + 3\|\mathbf{F}\|_{1,\infty}^2 \int_0^t \|\partial_s^{1-\alpha} v_m\|_{H^1(\Omega)}^2 \, ds \right),$$

by Lemma 4.18. Invoking Corollary 4.1, we have

$$\int_0^t \|\partial_s^{1-\alpha} v_m\|_{H^1(\Omega)}^2 \leq \int_0^t \left(\|\partial_s^{1-\alpha} v_m\|^2 + \frac{\omega_\alpha(t-s)}{\omega_\alpha(t)} \|\partial_s^{1-\alpha} \nabla v_m\|^2 \right) ds \tag{4.71}$$

$$\leq \left(C_8 \, t + C_9 \Gamma(\alpha) t^{1-\alpha} \right) \left[\|u_0\|_{H^2(\Omega)}^2 + \|g\|_{L^2}^2 \right],$$

and the bound (4.68) follows.

In a similar fashion, we next take the inner product of both sides of (4.60) with $v_m' \in W_m$ and integrate by parts with respect to x to obtain

$$\|v_m'(t)\|^2 + \kappa_\alpha \langle J^\alpha \nabla v_m', \nabla v_m' \rangle = - \langle \nabla \cdot (\mathbf{F}(t) \partial_t^{1-\alpha} v_m), v_m' \rangle$$

$$+ \langle g(t), v_m' \rangle - \langle \nabla \cdot (\mathbf{F}(t) u_{0m}) - \kappa_\alpha \Delta u_{0m}, v_m' \rangle \omega_\alpha(t)$$

$$\leq 3\epsilon \|v_m'(t)\|^2 + \frac{1}{4\epsilon} \left(\|g(t)\|^2 + \|\nabla \cdot (\mathbf{F}(t) \partial_t^{1-\alpha} v_m)\|^2 \right.$$

$$\left. + \|\nabla \cdot (\mathbf{F}(t) u_{0m}) - \kappa_\alpha \Delta u_{0m}\|^2 \omega_\alpha^2(t) \right).$$

Choosing $\epsilon = 1/6$ and invoking Lemma 4.13 gives

$$\|v_m'\|^2 + 2\kappa_\alpha \langle J^\alpha \nabla v_m', \nabla v_m' \rangle \leq 3\|g(t)\|^2 + 3\|\mathbf{F}\|_{1,\infty}^2 \|\partial_t^{1-\alpha} v_m\|_{H^1(\Omega)}^2$$

$$+ 6(\kappa_\alpha^2 + \|\mathbf{F}\|_{1,\infty}^2) \|u_{0m}\|_{H^2(\Omega)}^2 \omega_\alpha^2(t).$$

Integrating both sides of the inequality in time and invoking Lemma 4.8, we deduce that

$$\int_0^t \|v_m'\|^2 \, ds \leq 3\|g\|_{L^2(0,T;L^2)}^2 + 3\|\mathbf{F}\|_{1,\infty}^2 \|\partial_t^{1-\alpha} v_m\|_{L^2(0,t;H^1(\Omega))}^2 + C_{11} \|u_{0m}\|_{H^2(\Omega)}^2. \tag{4.72}$$

The second result (4.69) now follows from (4.71), (4.72) and Lemma 4.18. Using similar arguments, take the inner product of both sides of (4.60) with $-\partial_t^{1-\alpha} \Delta v_m \in W_m$, integrate by parts with respect to x and note that $\partial_t^{1-\alpha} \Delta v_m = J^\alpha \Delta v_m'$ to obtain

$$\langle \nabla v_m', J^\alpha \nabla v_m' \rangle + \kappa_\alpha \|\partial_t^{1-\alpha} \Delta v_m\|^2$$

$$= \langle \nabla \cdot (\mathbf{F}(t) \partial_t^{1-\alpha} v_m), \partial_t^{1-\alpha} \Delta v_m \rangle - \langle g(t), \partial_t^{1-\alpha} \Delta v_m \rangle$$

$$- \langle \kappa_\alpha \Delta u_{0m} - \nabla \cdot (\mathbf{F}(t) u_{0m}), \partial_t^{1-\alpha} \Delta v_m \rangle \omega_\alpha(t)$$

$$\leq \frac{\kappa_\alpha}{2} \|\partial_t^{1-\alpha} \Delta v_m\|^2 + \frac{3}{2\kappa_\alpha} \left(\|\nabla \cdot (\mathbf{F}(t) \partial_t^{1-\alpha} v_m)\|^2 + \|g(t)\|^2 \right.$$

$$\left. + \|\kappa_\alpha \Delta u_{0m} - \nabla \cdot (\mathbf{F}(t) u_{0m})\|^2 \omega_\alpha^2(t) \right).$$

Now integrate in time, invoking Lemma 4.8 and using (4.71), to deduce that

$$\int_0^t \|\partial_s^{1-\alpha} \Delta v_m(s)\|^2 \, ds$$

$$\leq \frac{3}{\kappa_\alpha^2} \int_0^t \left(\|\nabla \cdot (\mathbf{F}(s) \partial_s^{1-\alpha} v_m)\|^2 + \|g(s)\|^2 \right.$$

$$\left. + \|\kappa_\alpha \Delta u_{0m} - \nabla \cdot (\mathbf{F}(s) u_{0m})\|^2 \omega_\alpha^2(s) \right) ds$$

$$\leq \frac{3}{\kappa_\alpha^2} \int_0^t \left(\|g(s)\|^2 + 2\kappa_\alpha^2 \|\Delta u_{0m}\|^2 \omega_\alpha^2(s) \right) ds$$

$$+ \frac{3\|\mathbf{F}\|_{1,\infty}^2}{\kappa_\alpha^2} \int_0^t \left(\|\partial_s^{1-\alpha} v_m\|_{H^1(\Omega)}^2 + 2\|u_{0m}\|_{H^1(\Omega)}^2 \omega_\alpha^2(s) \right) ds,$$

and (4.70) now follows by (4.71) and Lemma 4.18, which completes the proof of the lemma. $\qquad\square$

Inequality (4.71) may also be derived (with a different constant factor) by applying (4.45) to (4.70).

We remark that the function $\alpha \mapsto \rho_\alpha$ is monotone increasing for $\alpha \in (0,1)$, with $\rho_\alpha \to 1$ as $\alpha \to 1$. Thus, $\rho_{1/2} < \rho_\alpha < 1$ for $1/2 < \alpha < 1$, with $\rho_{1/2} = \sqrt{2\pi/27} = 0.48240\ldots$.

4.2.6.2 *The classical solution*

In the sequel, by using the method of compactness, we show that there is a subsequence of $\{v_m\}_{m=1}^\infty$ such that the sum of its limit and the initial data satisfies the equation (4.27) almost everywhere.

Theorem 4.9. *Assume that $\alpha \in (1/2, 1)$, $u_0 \in H^2(\Omega) \cap H_0^1(\Omega)$, $\mathbf{F} \in W^{1,\infty}((0,T) \times \Omega)$ and $g \in L^2(0,T; L^2)$. Then there exists a unique classical solution of (4.27), in the sense of Definition 4.6, such that*

$$\sup_{0 \leq t \leq T} \|u(t)\|_{H^1(\Omega)}^2 + \|u'\|_{L^2}^2 + \|\partial_t^{1-\alpha} u\|_{L^2(0,T;H^2)}^2 \leq C_{12} \left[\|u_0\|_{H^2(\Omega)}^2 + \|g\|_{L^2}^2 \right], \quad (4.73)$$

where

$$C_{12} := (C_{10} + C_{11}) \left(1 + C_R^2 + \frac{T^{1-\alpha}}{\kappa_\alpha \rho_\alpha} + \frac{1}{\kappa_\alpha^2} \right).$$

Proof. From Lemma 4.20 we obtain

$$\sup_{0 \leq t \leq T} \|v_m(t)\|_{H^1(\Omega)}^2 + \|v_m'\|_{L^2}^2 + \|\partial_t^{1-\alpha} v_m\|_{L^2(0,T;H^2)}^2 \leq C_{12} \left[\|u_0\|_{H^2(\Omega)}^2 + \|g\|_{L^2}^2 \right],$$

$$(4.74)$$

which shows that the sequence $\{v_m\}_{m=1}^\infty$ is bounded in $L^\infty(0,T; H^1) \cap L^2(0,T; H^2 \cap H_0^1)$ and that the sequence $\{v_m'\}_{m=1}^\infty$ is bounded in $L^2(0,T; L^2)$. Since the embeddings $H^2 \hookrightarrow H^1 \hookrightarrow L^2$ are compact, it follows from Lemma 4.14 that there exists a subsequence of $\{v_m\}_{m=1}^\infty$ (still denoted by $\{v_m\}_{m=1}^\infty$) such that

$$v_m \to v \text{ strongly in } C([0,T]; L^2) \cap L^2(0,T; H^1). \quad (4.75)$$

Furthermore, from the upper bounds of $\{v_m\}_{m=1}^{\infty}$ we have

$$v_m \to v \text{ weakly in } L^{\infty}(0,T;H^1) \cap L^2(0,T;H^2 \cap H_0^1)$$
$$\text{and } v_m' \to v' \text{ weakly in } L^2(0,T;L^2). \tag{4.76}$$

By virtue of Lemma 4.11, the strong convergence in (4.75) implies that

$$J^{\alpha} v_m \to J^{\alpha} v \text{ strongly in } C([0,T];L^2) \cap L^2(0,T;H^1).$$

This, together with Corollary 4.1 and (4.70), yields

$$\partial_t J^{\alpha} v_m \to \partial_t J^{\alpha} v \text{ weakly in } L^{\infty}(0,T;L^2) \cap L^2(0,T;H^2). \tag{4.77}$$

Multiplying both sides of (4.60) by a test function $\xi \in L^2(0,T;L^2)$, integrating over $(0,T) \times \Omega$ and noting that Π_m is a self-adjoint operator on $L^2(\Omega)$, we deduce that

$$\langle v_m', \xi \rangle_{L^2(0,T;L^2)} - \kappa_\alpha \langle \partial_t^{1-\alpha} \Delta v_m, \xi \rangle_{L^2(0,T;L^2)} + \langle \nabla \cdot (\mathbf{F} \partial_t^{1-\alpha} v_m), \Pi_m \xi \rangle_{L^2(0,T;L^2)}$$
$$= \langle g, \Pi_m \xi \rangle_{L^2(0,T;L^2)} - \langle \omega_\alpha [\nabla \cdot (\mathbf{F} u_{0m}) - \kappa_\alpha \Delta u_{0m}], \Pi_m \xi \rangle_{L^2(0,T;L^2)}.$$

Now let $m \to \infty$ in this equation and recall (4.76) and (4.77). We get

$$\langle u', \xi \rangle_{L^2(0,T;L^2)} - \kappa_\alpha \langle \partial_t^{1-\alpha} \Delta u, \xi \rangle_{L^2(0,T;L^2)} + \langle \nabla \cdot (\mathbf{F} \partial_t^{1-\alpha} u), \xi \rangle_{L^2(0,T;L^2)}$$
$$= \langle g, \xi \rangle_{L^2(0,T;L^2)}, \tag{4.78}$$

for all $\xi \in L^2(0,T;L^2)$, where $u := v + u_0$. From (4.75)–(4.77), we have

$$u' \in L^2(0,T;L^2) \quad \text{and} \quad \partial_t^{1-\alpha} u \in L^2(0,T;H^2).$$

Hence, it follows from (4.78) that u satisfies (4.27) a.e. in $(0,T) \times \Omega$.

Taking the limit as $m \to \infty$ in (4.74), we obtain (4.73). The uniqueness of the solution u follows from (4.73), which completes the proof of the theorem. $\qquad \square$

Remark 4.6. It follows from the uniqueness in Theorems 4.8 and 4.9 that the mild solution will become the classical solution when $\alpha \in (1/2,1)$ and $u_0 \in H^2(\Omega) \cap H_0^1(\Omega)$. Furthermore, the continuous dependence of both the mild and classical solutions on the initial data u_0 follows from (4.46) and (4.73).

4.2.7 *Regularity of Classical Solution*

Recall that $1/2 < \alpha < 1$ and that in general $C = C(\Omega, \kappa_\alpha, \mathbf{F}, T)$. From Lemma 4.22 onwards, we allow $C = C(\Omega, \kappa_\alpha, \mathbf{F}, T, q)$, where q appears in the statements of our results below.

From Theorem 4.9, for almost every $(t,x) \in (0,T) \times \Omega$ the solution $u(t,x)$ satisfies (4.27). Using the identity $\partial_t^{1-\alpha} u = (J^{\alpha} u)'(t) = (J^{\alpha} u')(t) + u(0)\omega_\alpha(t)$, we rewrite (4.27) as

$$u' - \nabla \cdot (\kappa_\alpha \nabla J^{\alpha} u' - \mathbf{F} J^{\alpha} u') = g(t) + \nabla \cdot [\kappa_\alpha \nabla u_0 - \mathbf{F}(t) u_0] \omega_\alpha(t). \tag{4.79}$$

From this equation and the fact that $J^\alpha \phi(0) = 0$ for any function $\phi \in L^2(0, T; L^2)$, we deduce that $u'(t) = O(t^{\alpha-1})$ when t is close to 0. By letting $z(t, x) := tu'(t, x)$, we have $z(0) = 0$. The regularity of z is examined in the following lemma.

Lemma 4.21. *Assume that $\int_0^T \|tg'(t)\|^2 \, dt$ is finite. Then, the function z defined above satisfies*

$$
\sup_{0 \leq t \leq T} \|z(t)\|_{H^1(\Omega)}^2 + \|z'\|_{L^2}^2 + \|\partial_t^{1-\alpha} z\|_{L^2(0,T;H^2)}^2
$$
$$
\leq C_{13} \big(\|u_0\|_{H^2(\Omega)}^2 + \|g\|_{L^2}^2 + \|g_1\|_{L^2}^2 \big),
\tag{4.80}
$$

for some constant C_{13}.

Proof. For any $t > 0$, multiplying both sides of (4.79) by t and using the elementary identity

$$
\begin{aligned}
t(J^\alpha u')(t) &= (J^\alpha z)(t) + \alpha(J^{\alpha+1} u')(t) \\
&= (J^\alpha z)(t) + \alpha\big((J^\alpha u)(t) - u_0 \omega_{\alpha+1}(t)\big) \\
&= (J^\alpha z)(t) + \alpha(J^\alpha u)(t) - u_0 t \omega_\alpha(t),
\end{aligned}
\tag{4.81}
$$

we obtain a differential equation for z:

$$
z - \nabla \cdot \big(J^\alpha \kappa_\alpha \nabla z - \mathbf{F} J^\alpha z \big) = tg(t) + \alpha \nabla \cdot \big(\kappa_\alpha \nabla J^\alpha u - \mathbf{F} J^\alpha u \big).
$$

Differentiating both sides of this equation with respect to t and noting that $J^\alpha z' = \partial_t^{1-\alpha} z$, we have

$$
z' - \nabla \cdot (\partial_t^{1-\alpha} \kappa_\alpha \nabla z - \mathbf{F} \partial_t^{1-\alpha} z) = \bar{G}(t, x),
\tag{4.82}
$$

where

$$
\bar{G} := g + tg' + \alpha \nabla \cdot \big(\kappa_\alpha \nabla \partial_t^{1-\alpha} u - \mathbf{F}'(t) J^\alpha u - \mathbf{F}(t) \partial_t^{1-\alpha} u \big).
$$

Applying Lemma 4.13 and letting $g_1(t) = tg'(t)$, we find that

$$
\begin{aligned}
\|\bar{G}\|_{L^2}^2 \leq 4 \Big(&\|g\|_{L^2}^2 + \|g_1\|_{L^2}^2 + \alpha^2 (\kappa_\alpha + \|\mathbf{F}\|_{1,\infty})^2 \|\partial_t^{1-\alpha} u\|_{L^2(0,T;H^2)}^2 \\
&+ \alpha^2 \|\mathbf{F}'\|_{1,\infty}^2 \|J^\alpha u\|_{L^2(0,T;H^1)}^2 \Big),
\end{aligned}
$$

with Lemma 4.16 and Theorem 4.9 implying that

$$
\|\bar{G}\|_{L^2}^2 \leq C \big(\|u_0\|_{H^2(\Omega)}^2 + \|g\|_{L^2}^2 + \|g_1\|_{L^2}^2 \big), \quad \text{for some constant } C.
$$

Thus, applying Theorem 4.9 to equation (4.82) with initial data $z(0) = 0$, we deduce the bound (4.80). $\qquad \square$

From Theorem 4.9, for almost every $(t, x) \in (0, T) \times \Omega$, we have the identity

$$
u_t - \nabla \cdot (\partial_t^{1-\alpha} \kappa_\alpha \nabla u)) = f,
\tag{4.83}
$$

where $f := g - \nabla \cdot (\mathbf{F} \partial_t^{1-\alpha} u) \in L^2(0, T; H^1)$. The regularity of solutions to problem (4.83) subject to the initial condition $u_0 \in H^2(\Omega) \cap H_0^1(\Omega)$ was studied in [258]. In order to apply [258, Theorem 5.7], we need at least an upper bound for $\int_0^t s\|f'(s)\|ds$ which is proved in the following lemma. Here and subsequently, notation such as f' and $f^{(j)}$ indicates time derivatives, and we denote higher-order fractional derivatives by $\partial_t^{j-\alpha}u := \partial_t^{j-1}\partial_t^{1-\alpha}u = (J^\alpha u)^{(j)}$ for $j \in \{1, 2, \ldots\}$ and $0 < \alpha < 1$.

Lemma 4.22. *Let u be the solution of (4.27) and $f := g - \nabla \cdot (F\partial_t^{1-\alpha}u)$. Then, for $q \in \{0, 1, 2, \ldots\}$, $F \in W^{q,\infty}(0, T; L^2(\Omega))$, and for any $t \in (0, T]$, there is a constant $C = C(\Omega, \kappa_\alpha, F, q)$ such that*

$$\int_0^t s^{2q}\|f^{(q)}(s)\|^2\,ds \leq C\left(\|u_0\|_{H^2(\Omega)}^2 + \sum_{j=0}^q \int_0^t s^{2j}\|g^{(j)}(s)\|^2\,ds\right). \qquad (4.84)$$

Proof. Inequality (4.84) holds for $q = 0$ by virtue of (4.73) (with t playing the role of T) because

$$\|f(t)\|^2 \leq C\big(\|g(t)\|^2 + \|\partial_t^{1-\alpha}u\|_{H^1(\Omega)}\big)^2.$$

For the case $q = 1$, we note first that

$$t^2\|f'(t)\|^2 = t^2\big\|g'(t) - \nabla \cdot \big(F'(t)\partial_t^{1-\alpha}u + F(t)\partial_t^{2-\alpha}u\big)\big\|^2$$
$$\leq Ct^2\big(\|g'(t)\|^2 + \|\partial_t^{1-\alpha}u\|_{H^1(\Omega)}^2 + \|\partial_t^{2-\alpha}u\|_{H^1(\Omega)}^2\big).$$

By (4.81), we have $(J^\alpha z)(t) = t(J^\alpha u')(t) - \alpha(J^{\alpha+1}u')(t)$, and differentiating with respect to t gives

$$\partial_t^{1-\alpha}z = t(J^\alpha u')'(t) - (\alpha - 1)(J^\alpha u')(t) = t\partial_t^{2-\alpha}u - (\alpha - 1)\partial_t^{1-\alpha}u,$$

where we used the identities $(J^\alpha u')(t) = \partial_t^{1-\alpha}u - u_0\omega_\alpha(t)$ and $(\alpha - 1)\omega_\alpha(t) = \frac{\alpha-1}{\Gamma(\alpha)}t^{\alpha-1}$. Thus,

$$t\partial_t^{2-\alpha}u = \partial_t^{1-\alpha}z + (\alpha - 1)\partial_t^{1-\alpha}u. \qquad (4.85)$$

Hence, by Theorem 4.9 and Lemma 4.21 (with t again playing the role of T),

$$\int_0^t s^2\|\partial_s^{2-\alpha}u\|_{H^2(\Omega)}^2\,ds \leq C\left(\|u_0\|_{H^2(\Omega)}^2 + \int_0^t \big[\|g(s)\|^2 + s^2\|g'(s)\|^2\big]\,ds\right), \qquad (4.86)$$

implying that the desired inequality (4.84) holds for $q = 1$.

Multiply both sides of (4.85) by t and then differentiate with respect to t, obtaining

$$t^2\partial_t^{3-\alpha}u = \partial_t^{1-\alpha}z + t\partial_t^{2-\alpha}z + (\alpha - 1)\partial_t^{1-\alpha}u + (\alpha - 3)t\,\partial_t^{2-\alpha}u. \qquad (4.87)$$

Since z satisfies (4.82) — an equation similar to (4.27a) but with a different source $\bar{G}$ and with $z(0) = 0$ — we get an estimate for z corresponding to (4.86):

$$\int_0^t s^2\|\partial_s^{2-\alpha}z\|_{H^2(\Omega)}^2\,ds \leq C\int_0^t \big[\|\bar{G}(s)\|^2 + s^2\|\bar{G}'(s)\|^2\big]\,ds.$$

This inequality, together with (4.73), (4.80), (4.86) and (4.87), yields

$$\int_0^t s^4\|\partial_t^{3-\alpha}u\|_{H^2(\Omega)}^2\,ds \leq C\left(\|u_0\|_{H^2(\Omega)}^2 + \int_0^t \big[\|g(s)\|^2\right.$$
$$\left. + s^2\|g'(s)\|^2 + s^4\|g''(s)\|^2\big]\,ds\right), \qquad (4.88)$$

which implies the desired inequality (4.84) for $q = 2$.

The general case follows by iterating the arguments above, cf. [260]. $\qquad\square$

We can now prove regularity estimates for the classical solution u.

Theorem 4.10. *Let $g_j(t) := t^j g^{(j)}(t)$ for $j = 1, 2, \ldots$. For $q \in \{1, 2, \ldots\}$, $\mathbf{F} \in W^{q,\infty}(0, T; L^2(\Omega))$ and for any $t \in (0, T]$,*

$$t^q \|\Delta u^{(q)}(t)\| \le Ct^{-(\alpha-1/2)} \left(\|u_0\|_{H^2(\Omega)} + \sum_{j=0}^{q+1} \|g_j\|_{L^2} \right)$$

and

$$t^q \|u^{(q)}(t)\| \le Ct^{1/2} \left(\|u_0\|_{H^2(\Omega)} + \sum_{j=0}^{q} \|g_j\|_{L^2} \right).$$

Proof. By (4.83), it follows from [258, Theorem 4.4] with $r = 2$ and $\nu = \alpha$, and from [258, Theorem 5.6] with $r = 0$, $\mu = 2$ and $\nu = \alpha$, that

$$t^q \|\Delta u^{(q)}(t)\| \le C \left(\|u_0\|_{H^2(\Omega)} + t^{-\alpha} \sum_{j=0}^{q+1} \int_0^t s^j \|f^{(j)}(s)\| \, ds \right).$$

Similarly, from [258, Theorem 4.4] with $r = 2$ and $\nu = \alpha$, and from [258, Theorem 5.4] with $r = \mu = 0$, and

$$t^q \|u^{(q)}(t)\| \le C \left(t^\alpha \|u_0\|_{H^2(\Omega)} + \sum_{j=0}^{q} \int_0^t s^j \|f^{(j)}(s)\| \, ds \right).$$

The theorem follows by Lemma 4.22 since

$$\int_0^t s^j \|f^{(j)}(s)\| \, ds \le t^{1/2} \left(\int_0^t s^{2j} \|f^{(j)}(s)\|^2 \, ds \right)^{1/2}.$$

$\square$

Corollary 4.3. *Let $\eta > 1/2$. If*
$$\|g^{(j)}(t)\| \le Mt^{\eta-1-j} \quad \text{for} \quad 0 \le j \le q+1, \ \mathbf{F} \in W^{q,\infty}(0, T; L^2(\Omega)) \quad \text{and} \quad t \in (0, T],$$
then

$$t^q \|\Delta u^{(q)}(t)\| \le C \left(t^{-(\alpha-1/2)} \|u_0\|_{H^2(\Omega)} + Mt^{\eta-\alpha} \right)$$

and

$$t^q \|u^{(q)}(t)\| \le C \left(t^{1/2} \|u_0\|_{H^2(\Omega)} + Mt^\eta \right).$$

Proof. The assumption on g ensures that $\|g_j\| \le Mt^{\eta-1/2}$. $\square$

The alternative and longer analysis in [260, Theorems 6.2 and 6.3] shows that these bounds can be improved to
$$t^q \|\Delta u^{(q)}(t)\| \le C \left(\|u_0\|_{H^2(\Omega)} + Mt^{\eta-\alpha} \right)$$
and
$$t^q \|u^{(q)}(t)\| \le C \left(t^\alpha \|u_0\|_{H^2(\Omega)} + Mt^\eta \right),$$
for any $\alpha \in (0, 1)$ and $\eta > 0$.

4.3 Notes and Remarks

The contents in Section 4.1 are taken from Peng and Zhou [294]. The results of Section 4.2 are due to Le, McLean and Stynes [219].

Chapter 5

Fractional Schrödinger Equations

In this chapter, we firstly study the linear fractional Schrödinger equation on a Hilbert space, with a fractional time derivative of order $0 < \alpha < 1$, and a self-adjoint generator A. Using the spectral theorem we prove existence and uniqueness of strong solutions, and we show that the solutions are governed by an operator solution family $\{U_\alpha(t)\}_{t \geq 0}$. Moreover, we prove that the solution family $U_\alpha(t)$ converges strongly to the family of unitary operators e^{-itA}, as α approaches to 1. Next, we apply the tools of harmonic analysis to study the Cauchy problem for time-fractional Schrödinger equation. Some fundamental properties of two solution operators are estimated. The existence and a sharp decay estimate for solutions of the given problem in two different spaces are addressed. Finally, we study fractional Schrödinger equations with potential and optimal controls. Existence, uniqueness, local stability and attractivity, and data continuous dependence of mild solutions are also presented respectively. Further, we study the optimal control problems for the controlled fractional Schrödinger equations with potential. Existence and uniqueness of optimal pairs for the standard Lagrange problem are presented.

5.1 Linear Equations on Hilbert Space

5.1.1 *Introduction*

The Schrödinger equation is the basic equation of quantum mechanics, it describes the evolution in time of a quantum system. More recently, Laskin has introduced the fractional Schrödinger equation, as a result of extending the Feynman path integral, the resulting equation is a fundamental equation in fractional quantum mechanics [213–215]. Furthermore, Laskin [213] states that "the fractional Schrödinger equation provides us with a general point of view on the relationship between the statistical properties of the quantum mechanical path and the structure of the fundamental equations of quantum mechanics". Naber [274] introduced and examined some properties of the time-fractional Schrödinger equation,

$$\begin{cases} {}_0^C D_t^\alpha u(t) = (-i)^\alpha A u(t), \\ u(0) = u_0, \end{cases} \tag{5.1}$$

in which $(-i)^{\alpha} = e^{-i\alpha\frac{\pi}{2}}$. It was shown in [274] that the above equation (5.1) is equivalent to the usual Schrödinger equation with a time dependent Hamiltonian. On the other hand, it was point out that the so-called quantum comb model [35, 38, 180, 181], leads to a time-fractional Schrödinger equation with $\alpha = \frac{1}{2}$. The equation (5.1) describes non-Markovian evolution in Quantum Mechanics. As a result this system has memory. Different aspects of the time-fractional Schrödinger equation have already been studied. Particular solutions were sought in [38, 106, 274] and numerical analysis performed in [128]. Nevertheless, to the best of our knowledge, there are no results in the literature which show in full generality the existence and uniqueness of solutions to the abstract Schrödinger equation on a Hilbert space.

The purpose of this section is to consider the abstract fractional evolution equation (5.1) on a Hilbert space $\mathcal{H}$, in which A is a positive self adjoint operator on $\mathcal{H}$, and $\frac{\partial^{\alpha} u}{\partial t^{\alpha}}$ is the Caputo fractional derivative of order $\alpha \in (0,1)$. We show that A generates a family of bounded operators $\{U_{\alpha}(t)\}_{t \geq 0}$ which are defined by the functional calculus of A via the Mittag-Leffler function when evaluated at A. Moreover if u_0 belongs to the domain of A then we show that $u(t) = U_{\alpha}(t)u_0$ is the unique strong solution of problem (5.1). We also study the problem of the continuous dependence on α for $U_{\alpha}(t)$, and we show that

$$\lim_{\alpha \to 1^-} U_{\alpha}(t) = e^{-itA},$$

where e^{-itA} is the unitary group whose infinitesimal generator corresponds to the self adjoint operator A. Thus, we recover in the limit as $\alpha \to 1$ the classical Theorem of Stone.

In Subsection 5.1.2, we introduce the notations and give the definition of strong solution to the fractional Schrödinger equation. Moreover, we formulate and prove some technical but very crucial lemma. The main result about existence and uniqueness of solution is shown in Subsection 5.1.3. The properties of the solution operator are formulated and proven in Subsection 5.1.4. We give an example in Subsection 5.1.5.

5.1.2 *Preliminaries*

We recall the definition of the Riemann-Liouville fractional integral by the convolution product,

$$J^{\alpha}f(t) = \frac{1}{\Gamma(\alpha)} \int_0^t (t-s)^{\alpha-1} f(s)ds,$$

for a given locally integrable function f defined on the half line $\mathbb{R}_+ = [0, \infty)$ and taking values on a Banach space X. Henceforth we use the notation,

$$J^{\alpha}f(t) = (g_{\alpha} * f)(t), \tag{5.2}$$

where $g_{\alpha}(t) = \frac{t^{\alpha-1}}{\Gamma(\alpha)}$, for $\alpha > 0, t > 0$. Then the following property holds: $J^{\alpha+\beta}f = J^{\alpha}J^{\beta}f$, for $\alpha, \beta > 0$, f is suitable enough. Assume that $u \in C([0, \infty); X)$ and that the convolution $g_{1-\alpha} * u$ belongs to $C^1((0, \infty); X)$. Then the Caputo fractional derivative of order $\alpha \in (0, 1)$, can be interpreted as

$$\,_{0}^{C}D_t^\alpha u(t) = \frac{d}{dt}(g_{1-\alpha} * u)(t) - u(0)g_{1-\alpha}(t). \tag{5.3}$$

Henceforth we shall denote the Caputo derivative either by $\,_{0}^{C}D_t^\alpha u(t)$ or $\frac{\partial^\alpha u}{\partial t^\alpha}(t)$, indistinctly.

Let X be a Banach space, and suppose that $u_0 \in X$ and $\omega \in \mathbb{C}$. If $0 < \alpha < 1$, then the equation

$$\,_{0}^{C}D_t^\alpha u(t) = \omega u(t), \quad u(0) = u_0 \tag{5.4}$$

has a unique solution given by

$$u(t) = u_0 E_\alpha(\omega t^\alpha),$$

see [41, 146, 197]. Moreover, the uniqueness of the solution of (5.4) follows by the uniqueness theorem for the Laplace transform.

Let A be a densely defined self-adjoint operator on a Hilbert space $\mathcal{H}$, and let $0 < \alpha < 1$. For a given $u_0 \in \mathcal{H}$, we study the following equation of fractional order α,

$$\begin{cases} \,_{0}^{C}D_t^\alpha u(t) = (-i)^\alpha Au(t), & t > 0, \\ u(0) = u_0. \end{cases} \tag{5.5}$$

We first introduce the notion of strong solution for the abstract fractional Cauchy problem (5.5).

Definition 5.1. Let $0 < \alpha < 1$. Assume that $u_0 \in D(A)$. A function u is called a strong solution of (5.5) if $u \in C(\mathbb{R}_+, D(A))$ and $g_{1-\alpha} * u$ belongs to $C^1((0, \infty); \mathcal{H})$, and (5.5) holds for all $t > 0$.

We will show that the strong solution of (5.5) is determined by the functional calculus for a self-adjoint operator when it is applied to the Mittag-Leffler function. Moreover, the following lemma will give us the necessary bounds we need in the proof of the qualitative properties of the solution operator.

In order to prove the next lemma we recall from [147] that the Mittag-Leffler function has the following representation for $\alpha \in (0, 1]$,

$$E_\alpha(z) = \int_0^\infty K_\alpha(r, z)dr + \frac{1}{\alpha}e^{z^{1/\alpha}}, \quad |\arg(z)| < \pi/\alpha \quad \text{and} \quad z \neq 0, \tag{5.6}$$

in which

$$K_\alpha(r, z) = -\frac{e^{-r^{1/\alpha}} z \sin(\pi\alpha)}{\pi\alpha(r^2 - 2rz\cos(\pi\alpha) + z^2)}. \tag{5.7}$$

Lemma 5.1. (a) *Let* $\alpha_0 \leq \alpha < 1/2$, *in which* $\alpha_0 > 0$. *Then there is a positive constant* $M(\alpha)$ *such that for all* $t \geq 0$,

$$\sup_{\omega \geq 0} |E_\alpha((-it)^\alpha \omega)| \leq M(\alpha). \tag{5.8}$$

(b) *There is $M > 0$ such that for all $t \geq 0$ and all $\alpha \in [1/2, 1)$,*

$$\sup_{w \geq 0} |E_\alpha((-it)^\alpha w)| \leq M. \tag{5.9}$$

Proof. First we show (5.8) that is $\alpha_0 \leq \alpha < 1/2$. We notice that it suffices to prove assertion (5.8) for $t = 1$. Indeed, let us assume that (5.8) holds for $t = 1$, then for any $t > 0$, we have

$$|E_\alpha((-it)^\alpha w)| = |E_\alpha((-i)^\alpha t^\alpha w)| \leq M(\alpha).$$

To begin we assume $w \geq 1/\alpha_0$. Next we recall that $(-i)^\alpha = e^{-i\alpha\pi/2}$. Then we proceed to estimate $|K_\alpha(r, (-i)^\alpha w)|$ for arbitrary $w \geq 1/\alpha_0$. Thus,

$$|K_\alpha(r, (-i)^\alpha w)| \leq \frac{B}{\pi\alpha} \frac{e^{-r^{1/\alpha} w}}{|(r^2 - 2r(a - ib)wA + w^2(A - iB)|}, \tag{5.10}$$

where $A = \cos(\pi\alpha)$, $B = \sin(\pi\alpha)$, $a = \cos(\alpha\pi/2)$, $b = \sin(\alpha\pi/2)$ and these quantities are all positive since $0 < \alpha < 1/2$. Next we set $u(r)$ and $v(r)$ the real and imaginary parts respectively of the denominator on the right hand side of (5.10), that is

$$u(r) = r^2 - 2rawA + w^2A \text{ and } v(r) = 2rbwA - w^2B.$$

Hence,

$$|K_\alpha(r, (-i)^\alpha w)| \leq \frac{B}{\pi\alpha} \frac{e^{-r^{1/\alpha} w}}{|(r^2 - 2r(a - ib)wA + w^2(A - iB)|} \leq \frac{Be^{-r^{1/\alpha} w}}{\pi\alpha |u(r)|}. \tag{5.11}$$

On the other hand the quadratic $u(r) = r^2 - 2rawA + w^2A$ is positive for all real r and its minimum equals to $w^2A(1 - a^2A) > 0$ since $a^2A < 1$, and $A > 0$. But then,

$$u(r) = r^2 - 2rawA + w^2A \geq w^2A(1 - a^2A) > 0.$$

Hence the right side of (5.11) turns out to be less than or equals to

$$\frac{Be^{-r^{1/\alpha} w}}{\pi\alpha w^2 A(1 - a^2A)}. \tag{5.12}$$

Therefore, from (5.11) and (5.12) follows that

$$|K_\alpha(r, (-i)^\alpha w)| \leq \frac{e^{-r^{1/\alpha}}}{\pi A(1 - a^2A)},$$

since $0 < \alpha_0 \leq \alpha < 1/2$ and $w \geq 1/\alpha_0$. Furthermore

$$|K_\alpha(r, (-i)^\alpha w)| \leq \begin{cases} \dfrac{e^{-r}}{\pi A(1 - a^2A)} & \text{for } r > 1, \\ \dfrac{1}{\pi A(1 - a^2A)} & \text{for } r \leq 1. \end{cases} \tag{5.13}$$

Therefore, from (5.13) we obtain that the integral

$$\int_0^\infty K_\alpha(r, (-i)^\alpha w) dr$$

is bounded independently of $\omega \geq \frac{1}{\alpha_0}$. But then, it follows from the integral representation (5.6) that there is a bound $M_1(\alpha)$ such that

$$\sup_{\omega \geq 1/\alpha_0} |E_\alpha((-i)^\alpha \omega)| \leq M_1(\alpha). \qquad (5.14)$$

Now if $\omega \leq 1/\alpha_0$, then we have that

$$|(-i)^\alpha \omega| \leq \frac{1}{\alpha_0}.$$

Thus, from the very definition of the Mittag-Leffler function, we obtain that

$$|E_\alpha((-i)^\alpha \omega)| \leq \sum_{k=0}^\infty \frac{\left(\frac{1}{\alpha_0}\right)^k}{\Gamma(\alpha k + 1)} = E_\alpha(1/\alpha_0).$$

Moreover, by the Stirling's formula

$$\Gamma(x) = \sqrt{2\pi} x^{x-\frac{1}{2}} e^{-x + \frac{\theta}{12x}},$$

where $\theta \in [0, 1]$ and for x large enough, we have

$$\frac{\left(\frac{1}{\alpha_0}\right)^k}{\Gamma(\alpha k + 1)} \leq \frac{e^2}{\sqrt{2\pi}} \left(\frac{e}{\alpha_0^{\alpha_0 + 1} k^{\alpha_0}}\right)^k.$$

Hence, by the Lebesgue's theorem, we obtain that the map $[\alpha_0, 1] \ni \alpha \mapsto E_\alpha(1/\alpha_0)$ is continuous. Therefore, there exists $M(\alpha_0)$ such that

$$\sup_{\omega \leq 1/\alpha_0} |E_\alpha((-i)^\alpha \omega)| \leq \sup_{\alpha \in [\alpha_0, 1/2]} E_\alpha(1/\alpha_0) = M(\alpha_0). \qquad (5.15)$$

Now, the proof of assertion (5.8) follows from (5.14) together with (5.15).

Next we show (5.9). First we assume that $\omega \geq 2$ under the condition $1/2 \leq \alpha < 1$ from the hypothesis. Again it suffices to prove assertion (5.9) for $t = 1$. We notice that $A \leq 0$, and B, a, and b are all positive. Thus $|v(r)| = |2rb\omega A - B\omega^2| = -2rb\omega A + B\omega^2 \geq B\omega^2 > 0$. Hence,

$$|K_\alpha(r, (-i)^\alpha \omega)| \leq \frac{e^{-r^{1/\alpha}} B\omega}{\pi\alpha |v(r)|} \leq \frac{e^{-r^{1/\alpha}}}{\pi\alpha\omega} \leq \frac{e^{-r^{1/\alpha}}}{\pi}.$$

Furthermore,

$$|K_\alpha(r, (-i)^\alpha \omega)| \leq \begin{cases} \dfrac{e^{-r}}{\pi} & \text{for } r > 1, \\ \dfrac{1}{\pi} & \text{for } r \leq 1. \end{cases}$$

Hence, reasoning as in the proof of (5.8) we obtain that there is a positive constant M which in this case does not depend on the value of $\alpha \in [1/2, 1)$, so that

$$\sup_{\omega \geq 2} |E_\alpha((-i)^\alpha \omega)| \leq M.$$

Now by an application of the same argument as in (5.15), we can show that there is a $M_2 > 0$ independent of $\alpha \in [1/2, 1)$ such that

$$\sup_{\omega \leq 2} |E_\alpha((-i)^\alpha \omega)| \leq M_2.$$

Thus the proof of (5.9) now follows from these last two inequalities. $\qquad \square$

5.1.3 Existence of Dynamics

In this part of our section, we state and prove our principal assertion.

Theorem 5.1. *Let $\mathcal{H}$ be a Hilbert space and let A be a positive self-adjoint operator on $\mathcal{H}$. Then there exists a unique strong solution to the problem*

$$\begin{cases} {}^{C}_{0}D^{\alpha}_{t}u(t) = (-i)^{\alpha}Au(t), & t > 0, \\ u(0) = u_0, & u_0 \in D(A). \end{cases} \tag{5.16}$$

Moreover, there is a measure space (Ω, μ), a measurable function a on Ω and a unitary map $W : L^2(\Omega) \to \mathcal{H}$ such that the unique solution of the problem (5.16) has the following representation

$$u(t) = W(E_{\alpha}((-it)^{\alpha}\, a(\cdot\,))W^{-1}u_0.$$

Proof. Let us recall that because of the spectral theorem for a self-adjoint operator $A : D(A) \subset \mathcal{H} \to \mathcal{H}$, there exists a measure space (Ω, μ), and a Borel measurable function a and a unitary map $W : L^2(\Omega, \mu) \to \mathcal{H}$ such that the following diagram commutes

$$
\begin{array}{ccc}
L^2(\Omega, \mu) & \xrightarrow{\ M_a\ } & L^2(\Omega, \mu) \\
{\scriptstyle W}\big\downarrow & & \big\uparrow{\scriptstyle W^{-1}} \\
\mathcal{H} & \xrightarrow[\ A\ ]{} & \mathcal{H}
\end{array}
$$

for each $f \in L^2(\Omega, \mu)$ such that $Wf \in D(A)$. Moreover, if $f \in L^2(\Omega, \mu)$ is given, then $Wf \in D(A)$ if and only if $M_a f \in L^2(\Omega, \mu)$, see e.g. [309, 327], where $M_a f(x) = a(x)f(x)$.

Thus, the spectral theorem ensures us that there exists a unitary map W from $L^2(\Omega)$ onto $\mathcal{H}$ such that

$$W^{-1}AW\varphi(\xi) = a(\xi)\varphi(\xi), \quad \xi \in \Omega. \tag{5.17}$$

Now the proof of the theorem falls naturally into two parts.
  Uniqueness.
  Let us assume that u is a strong solution to the problem (5.16). We define $v(t, \xi) = (W^{-1}u(t))(\xi)$. Then it follows from (5.17) that

$$(W^{-1}Au(t))(\xi) = W^{-1}AWv(t)(\xi) = a(\xi)v(t, \xi). \tag{5.18}$$

Let us observe that $g_{1-\alpha} * v \in C^1((0, \infty); L^2(\Omega))$. Indeed, we shall show that

$$\frac{d}{dt}g_{1-\alpha} * v = W^{-1}\left(\frac{d}{dt}g_{1-\alpha} * u\right).$$

Next we set $\Theta(t) = g_{1-\alpha} * v$. Then using the fact that W is an isometry, we get that

$$\left\| \frac{\Theta(t+h) - \Theta(t)}{h} - W^{-1}(\frac{d}{dt}g_{1-\alpha} * u) \right\|_{L^2(\Omega)}$$

$$= \left\| W^{-1}\left(\frac{g_{1-\alpha} * u(t+h) - g_{1-\alpha} * u(t)}{h} - \frac{d}{dt}g_{1-\alpha} * u(t) \right) \right\|_{L^2(\Omega)}$$

$$= \left\| \frac{g_{1-\alpha} * u(t+h) - g_{1-\alpha} * u(t)}{h} - \frac{d}{dt}g_{1-\alpha} * u(t) \right\|_{\mathcal{H}} \to 0, \text{ as } h \to 0,$$

where the convergence follows from the assumptions on the function u.

Moreover, we can easily check that derivative Θ' is a continuous function, thus we obtain that $g_{1-\alpha} * v \in C^1((0, \infty); L^2(\Omega))$. Furthermore, by the continuity of W, we obtain

$$W^{-1}\left(\frac{d}{dt}g_{1-\alpha} * u \right) = \frac{d}{dt}g_{1-\alpha} * W^{-1}u.$$

Thus, from the definition of the Caputo derivative, denote $v_0 = W^{-1}u_0$, we obtain that

$$W^{-1}\frac{\partial^\alpha u}{\partial t^\alpha}(t) = W^{-1}\left(\frac{d}{dt}g_{1-\alpha} * u - u_0 g_{1-\alpha} \right)$$

$$= \frac{d}{dt}g_{1-\alpha} * W^{-1}u - v_0 g_{1-\alpha}$$

$$= \frac{\partial^\alpha v(t, \cdot)}{\partial t^\alpha}.$$

Now if we apply W^{-1} to both sides of equation (5.16), then we obtain the following equation on $L^2(\Omega)$,

$$\begin{cases} \dfrac{\partial^\alpha v(t, \cdot)}{\partial t^\alpha} = (-i)^\alpha a(\cdot)v(t, \cdot), & t > 0 \\ v(0, \cdot) = v_0, \end{cases} \tag{5.19}$$

where $u_0 \in D(A)$. Now, it follows that the unique solution of the above fractional differential equation (5.19) is given by

$$v(t, \xi) = E_\alpha((-it)^\alpha a(\xi))v_0.$$

Since $(W^{-1}u(t))(\xi) = E_\alpha((-it)^\alpha a(\xi))v_0$, we get that u is given by

$$u(t) = W(E_\alpha((-it)^\alpha a(\cdot))W^{-1}u_0), \quad u_0 \in D(A). \tag{5.20}$$

This finishes with the proof of the uniqueness property.

Existence.

Next, we shall show that $u(t)$ given by formula (5.20) is indeed a strong solution to the initial value problem (5.16). First of all, we prove that $u \in C(\mathbb{R}_+, D(A))$. We need to show that $u(t) \in D(A)$, for all $t \geq 0$. For this purpose let us recall that

$$Ah = W(a(\cdot)(W^{-1}h)(\cdot)), \quad \text{for } h \in D(A).$$

Thus, by the spectral theorem, we know that $h \in D(A)$ if and only if $a(\cdot)(W^{-1}h)(\cdot) \in L^2(\Omega)$, see [309, 327]. Hence, $u_0 \in D(A)$ if and only if $a(\xi)(W^{-1}u_0)(\xi)$ belongs to $L^2(\Omega)$. But then, from the fact that $\xi \mapsto E_\alpha((-it)^\alpha a(\xi))$ is bounded by Lemma 5.1, it follows that the function

$$a(\xi)(W^{-1}u(t))(\xi) = E_\alpha((-it)^\alpha a(\xi))a(\xi)(W^{-1}u_0)(\xi), \quad \xi \in \Omega$$

is in $L^2(\Omega)$ for all $t \geq 0$ and effectively we get that $u(t) \in D(A)$. Moreover, since the mapping $t \mapsto E_\alpha((-it)^\alpha a(\xi))$ is continuous, the map u is continuous. Indeed, let us take $t_0, t \in \mathbb{R}_+$, then we have

$$\left\| W\left((E_\alpha((-i(t+t_0))^\alpha a(\cdot)) - E_\alpha((-it_0)^\alpha a(\cdot)))W^{-1} \right) u_0 \right\|_{\mathcal{H}}$$
$$= \left\| \left(E_\alpha((-i(t+t_0))^\alpha a(\cdot)) - E_\alpha((-it_0)^\alpha a(\cdot)) \right) W^{-1} u_0 \right\|_{L^2(\Omega)}.$$

Since $(E_\alpha((-i(t+t_0))^\alpha a(\xi)) - E_\alpha((-it_0)^\alpha a(\xi)))$ is bounded by Lemma 5.1. Hence, there exists M_α such that

$$\left| (E_\alpha((-i(t+t_0))^\alpha a(\xi)) - E_\alpha((-it_0)^\alpha a(\xi)))(W^{-1}u_0)(\xi) \right| \leq M_\alpha |(W^{-1}u_0)(\xi)|.$$

Thus, by an application of the Lebesgue's dominated convergence theorem the proof of the continuity of the function u defined in (5.20) is finished.

Next, we prove that the map

$$\Phi(t) = \frac{1}{\Gamma(1-\alpha)} \int_0^t (t-s)^{-\alpha} u(s) ds$$

belongs to $C^1((0,\infty); \mathcal{H})$. For this purpose we consider the following mapping

$$\phi(t) = \frac{1}{\Gamma(1-\alpha)} \int_0^t (t-s)^{-\alpha} E_\alpha((-is)^\alpha a(\xi)) ds.$$

Once more by the definition of the Caputo derivative we get

$$\phi'(t) = (-i)^\alpha a(\xi) E_\alpha((-it)^\alpha a(\xi)) + \frac{1}{\Gamma(1-\alpha)} \frac{1}{t^\alpha}.$$

Now, we shall show that

$$\Phi'(t) = W \phi'(t) W^{-1} u_0. \tag{5.21}$$

Let us notice that

$$\lim_{h \to 0} \left| \left(\frac{\phi(t+h) - \phi(t)}{h} - \phi'(t) \right) W^{-1} u_0 \right| = 0.$$

Moreover, by the mean value theorem and Lemma 5.1, we have

$$\left| \left(\frac{\phi(t+h) - \phi(t)}{h} - \phi'(t) \right) W^{-1} u_0 \right| = \left| \left(\frac{1}{h} \int_t^{t+h} \phi'(s) ds - \phi'(t) \right) W^{-1} u_0 \right|$$
$$\leq C(c_\alpha(t) + |a(\xi)|) |W^{-1} u_0|,$$

for some constants C and $c_\alpha(t)$ independent on h. Since $u_0 \in D(A)$, we have $a(\xi)W^{-1}u_0, W^{-1}u_0$ belongs to $L^2(\Omega)$. Moreover,

$$\left\| \frac{\Phi(t+h) - \Phi(t)}{h} - W\phi'(t)W^{-1}u_0 \right\|_{\mathcal{H}}$$
$$= \left\| W\left(\frac{\phi(t+h) - \phi(t)}{h} - \phi'(t) \right) W^{-1} u_0 \right\|_{\mathcal{H}}. \tag{5.22}$$

Since W is unitary, it follows that

$$\left\| W \left(\frac{\phi(t+h) - \phi(t)}{h} - \phi'(t) \right) W^{-1} u_0 \right\|_{\mathcal{H}}$$

$$= \left\| \left(\frac{\phi(t+h) - \phi(t)}{h} - \phi'(t) \right) W^{-1} u_0 \right\|_{L^2(\Omega)}.$$

(5.23)

Therefore from (5.22) and (5.23), we obtain

$$\left\| \frac{\Phi(t+h) - \Phi(t)}{h} - W\phi'(t)W^{-1} u_0 \right\|_{\mathcal{H}}$$

$$= \left\| \left(\frac{\phi(t+h) - \phi(t)}{h} - \phi'(t) \right) W^{-1} u_0 \right\|_{L^2(\Omega)}.$$

Hence by the Lebesgue's dominated convergence theorem, we have

$$\left\| \left(\frac{\phi(t+h) - \phi(t)}{h} - \phi'(t) \right) W^{-1} u_0 \right\|_{L^2(\Omega)} \to 0, \quad \text{as } h \to 0.$$

Thus, the proof of (5.21) is complete and hence we have the differentiability of the function Φ. Furthermore, arguing as above, we get that $\Phi' \in C((0, \infty); \mathcal{H})$.

It remains to prove that the function u defined in (5.20) satisfies equation (5.16). In order to show this last claim we compute the Caputo derivative of the u. Thus,

$$\begin{aligned} {}^{C}_{0}D_t^\alpha u(t) &= \Phi'(t) - \frac{1}{\Gamma(1-\alpha)} \frac{u_0}{t^\alpha} \\ &= W(-i)^\alpha a(\xi) E_\alpha((-it)^\alpha a(\xi)) W^{-1} u_0 \\ &= W(-i)^\alpha a(\xi) W^{-1} W E_\alpha((-it)^\alpha a(\xi)) W^{-1} u_0 \\ &= (-i)^\alpha A u(t), \end{aligned}$$

and the whole proof of Theorem 5.1 is now finished. $\qquad\square$

Remark 5.1. Let A be a self-adjoint operator. Then we shall denote by $U_\alpha(t)$ the corresponding solution operator family given by Theorem 5.1. To be more explicit

$$U_\alpha(t)\phi = W(E_\alpha((-it)^\alpha a(\cdot))W^{-1})\phi, \quad \phi \in \mathcal{H}, \ t \geq 0.$$

5.1.4 *Properties of the Solution Operator*

In this subsection, we study the properties of the solution operator U_α.

Proposition 5.1. *The family $\{U_\alpha(t)\}_{t \geq 0}$ satisfies*

(i) *$U_\alpha(t)$ is strongly continuous for $t \geq 0$ and $U_\alpha(0) = I$;*
(ii) *$U_\alpha(t)(D(A)) \subseteq D(A)$ and $AU_\alpha(t)x = U_\alpha(t)Ax$, for all $x \in D(A)$, $t \geq 0$.*

Proof. (i) This follows from the proof of Theorem 5.1.

(ii) Using similar consideration as in the proof of Theorem 5.1 we get that

$$U_\alpha(t)(D(A)) \subseteq D(A).$$

Next, the commutation property $[U_\alpha(t), A] = 0$ on $D(A)$ follows from the fact that

$$A = W M_{a(\cdot)} W^{-1}, \quad U_\alpha(t) = W M_{E_\alpha((-it)^\alpha a(\cdot))} W^{-1}, \ t \geq 0.$$

Thus

$$A U_\alpha(t) \phi = U_\alpha(t) A \phi, \quad \text{for all } \phi \in D(A), \ t \geq 0.$$

$\square$

Next, we state some further properties of the solution operator U_α.

Proposition 5.2. *Let* $\alpha \subset (0, 1)$. *Then the solution operator enjoys the following properties.*

(i) $U_\alpha(t)^* = W E_\alpha((it)^\alpha a(\cdot)) W^{-1}, \quad t > 0.$

(ii) $U_\alpha(t) U_\alpha(t)^* = U_\alpha^*(t) U_\alpha(t) = W |E_\alpha((it)^\alpha a(\cdot))|^2 W^{-1}, \ t > 0.$

(iii) *Let* e^{-itA} *be the unitary group generated by the self-adjoint operator* A. *Then*

$$\lim_{\alpha \to 1^-} U_\alpha(t) \phi = e^{-itA} \phi, \quad \text{for every } \phi \in \mathcal{H}, \quad \text{and } t \geq 0.$$

Remark 5.2. In the paper of Dong and Xu [106], it has been pointed out that the quantity $\|U_\alpha(t) u_0\|$ is not conserved during the evolution.

Proof. (i) Let us take $\phi, \psi \in \mathcal{H}$. Using the fact that W is a unitary operator, we get

$$\begin{aligned}
(U_\alpha(t)\psi, \phi)_{\mathcal{H}} &= (W E_\alpha((-it)^\alpha a(\cdot)) W^{-1}\psi, \phi)_{\mathcal{H}} \\
&= (E_\alpha((-it)^\alpha a(\cdot)) W^{-1}\psi, W^{-1}\phi)_{L^2(\Omega)} \\
&= \int_\Omega E_\alpha((-it)^\alpha a(x)) W^{-1}\psi(x) \overline{W^{-1}\phi(x)} dx \\
&= \int_\Omega W^{-1}\psi(x) \overline{E_\alpha((it)^\alpha a(x)) W^{-1}\phi(x)} dx \\
&= (W^{-1}\psi, E_\alpha((it)^\alpha a(\cdot)) W^{-1}\phi)_{L^2(\Omega)} \\
&= (\psi, W E_\alpha((it)^\alpha a(\cdot)) W^{-1}\phi)_{\mathcal{H}}.
\end{aligned}$$

Hence, we obtain

$$U_\alpha(t)^* = W E_\alpha((it)^\alpha a(\cdot)) W^{-1}.$$

The proof of (ii) follows from the very definition of $U_\alpha(t)$. Next, we show (iii). We will prove that

$$\lim_{\alpha \to 1^-} \|U_\alpha(t)\phi - e^{-itA}\phi\|_{\mathcal{H}} = 0, \quad t \geq 0, \quad \phi \in \mathcal{H}. \tag{5.24}$$

Since, W is an isometry and

$$e^{-itA} = W e^{-ita(\xi)} W^{-1},$$

we have that

$$\|U_\alpha(t)\phi - e^{-itA}\phi\|_{\mathcal{H}}^2 = \int_\Omega \left| E_\alpha((-it)^\alpha a(\xi)) - e^{-ita(\xi)} \right|^2 \left|(W^{-1}\phi)(\xi)\right|^2 d\xi. \quad (5.25)$$

According to Lemma 5.1 part (b), for each $t > 0$ the function $|E_\alpha((-it)^\alpha a(\xi))|$ is bounded independently of $\xi \in \Omega$ and $\alpha \in [1/2, 1)$. But then, there is M such that for all $\alpha \in [1/2, 1)$, and $\xi \in \Omega$

$$|E_\alpha((-it)^\alpha a(\xi)) - e^{-ita(\xi)}| \le M.$$

Hence

$$|E_\alpha((-it)^\alpha a(\xi)) - e^{-ita(\xi)}||(W^{-1}\phi)(\xi)| \le M|(W^{-1}\phi)(\xi)|.$$

Moreover

$$\lim_{\alpha \to 1^-} \left(E_\alpha((-it)^\alpha a(\xi)) - e^{-ita(\xi)} \right) = 0$$

and $W^{-1}\phi \in L^2(\Omega)$. Then the dominated convergence theorem applies to (5.25) when $\alpha \to 1^-$, and thus the proof of (iii) is finished. □

5.1.5 *An Example*

We consider $A = -\Delta$, the Laplacian operator on $L^2(\mathbb{R}^n)$. Then by the spectral theorem, we have

$$Au := \mathcal{F}^{-1}(|\xi|^2 \mathcal{F})u, \quad \text{for } u \in D(A) := \mathcal{S}(\mathbb{R}^n).$$

Next we find the strong solution of the following fractional Schrödinger equation. Suppose that $0 < \alpha < 1$ and consider the initial value problem

$$\begin{cases} \dfrac{\partial^\alpha u}{\partial t^\alpha}(t, x) = (-i)^\alpha(-\Delta)u(t, x), & t > 0, x \in \mathbb{R}^n, \\ u(0, \cdot) = g(\cdot) \in C_0^\infty(\mathbb{R}^n). \end{cases} \quad (5.26)$$

We will show that the strong solution of (5.26) is defined by a convolution kernel which is given by the Fourier transform in the distributional sense of the Mittag-Leffler function. To prove this claim we first recall some basic facts. We denote by $\mathcal{S}(\mathbb{R}^n)$ and by $\mathcal{S}'(\mathbb{R}^n)$ the Schwartz space and the space of tempered distributions respectively. Let φ be a function of $\mathcal{S}(\mathbb{R}^n)$. Then we recall that the action of the dilation operator on φ is defined as $\varphi_\lambda(x) = \varphi(\lambda x)$, $\lambda \in \mathbb{R}$, $x \in \mathbb{R}^n$. Furthermore the action on the Fourier transform $\mathcal{F}$ is

$$(\mathcal{F}\varphi)_\lambda = \frac{1}{\lambda^n}\mathcal{F}\varphi_{1/\lambda} \quad \text{and} \quad \mathcal{F}\varphi_\lambda = \frac{1}{\lambda^n}(\mathcal{F}\varphi)_{1/\lambda}, \quad \lambda > 0. \quad (5.27)$$

If u is a distribution then we recall that $\langle u_\lambda, \varphi \rangle = \frac{1}{\lambda^n}\langle u, \varphi_{1/\lambda} \rangle$, and the same identities as (5.27) are also verified.

Next we set

$$e(\xi) = E_\alpha((-i)^\alpha|\xi|^2), \quad \xi \in \mathbb{R}^n.$$

Thus,

$$e_{t^{\alpha/2}}(\xi) = E_\alpha((-it)^\alpha |\xi|^2).$$

We see that the hypotheses of Theorem 5.1 are satisfied. Hence the strong solution of (5.26) is given by

$$u(t, x) = \mathcal{F}^{-1}(e_{t^{\alpha/2}}(\mathcal{F}g))(x). \tag{5.28}$$

We notice that the function $\xi \mapsto e_{t^{\alpha/2}}(\xi)$ is bounded for each $t \geq 0$ by Lemma 5.1. Thus $e_{t^{\alpha/2}}$ defines a tempered distribution by integration. But then, $\mathcal{F}(e_{t^{\alpha/2}})$ is also a tempered distribution. Now if $u \in \mathcal{S}'(\mathbb{R}^n)$ and $\varphi \in \mathcal{S}(\mathbb{R}^n)$ then we have that $u * \varphi \in C^\infty(\mathbb{R}^n) \cap \mathcal{S}'(\mathbb{R}^n)$ and $(\mathcal{F}u)(\mathcal{F}\varphi) = \mathcal{F}(u * \varphi)$, see [230]. And then, we have that

$$\mathcal{F}^{-1}(e_{t^{\alpha/2}}(\mathcal{F}g)) = (\mathcal{F}^{-1}(e_{t^{\alpha/2}}) * g)$$

as tempered distributions. Moreover

$$\mathcal{F}^{-1}(e_{t^{\alpha/2}}) = \frac{1}{t^{n\alpha/2}} (\mathcal{F}^{-1}e)_{1/t^{\alpha/2}}.$$

Hence taking into account the above considerations we can represent the solution of (5.26) as

$$u(t, x) = \frac{1}{t^{n\alpha/2}} ((\mathcal{F}^{-1}e)_{1/t^{\alpha/2}} * g)(x).$$

Since $e_{t^{\alpha/2}} \in C^\infty \cap L^\infty$, we have that $e_{t^{\alpha/2}}(\mathcal{F}g) \in \mathcal{S}$. Thus, from formula (5.28), we get that the function $x \mapsto u(t, x)$ belongs to the Schwartz space for each $t \geq 0$.

Using Proposition 5.2 and the above considerations, we close the section with the following observation.

Proposition 5.3. *Let $\phi \in \mathcal{S}(\mathbb{R}^n)$, then*

$$\frac{1}{t^{n\alpha/2}} ((\mathcal{F}^{-1}e)_{1/t^{\alpha/2}} * \phi) \xrightarrow[\alpha \to 1]{L^2} \frac{1}{(4\pi i t)^{\frac{n}{2}}} \int_{\mathbb{R}^n} e^{i\frac{|\cdot - y|^2}{4t}} \phi(y) dy.$$

5.2 Nonlinear Schrödinger Equation

5.2.1 *Introduction*

The nonlinear Schrödinger equation received a great deal of attention due to its extensive applications in quantum mechanics, optics, seismology and plasma physics [179, 286]. For more details, see the monographs by Bourgain [53], Cazenave [65] and the references therein.

In this section, we consider the following nonlinear time-fractional Schrödinger equation

$$\begin{cases} i\partial_t^\alpha u(t, x) + \Delta u(t, x) + \lambda |u(t, x)|^p u(t, x) = 0, & t \in \mathbb{R}^+, \ x \in \mathbb{R}^N, \\ u(0, x) = u_0(x), & x \in \mathbb{R}^N, \end{cases} \tag{5.29}$$

where ∂_t^α denotes the Caputo fractional derivative of order $\alpha \in (0, 1)$, $\lambda \in \mathbb{C}$, $p \geq 0$ and some assumptions on p will be specified later.

Let $v : [0, \infty) \times \mathbb{R}^N \to \mathbb{R}^N$. The Caputo fractional derivative of order α for the function v and is defined by

$$\partial_t^\alpha v(t, x) = \frac{1}{\Gamma(1 - \alpha)} \int_0^t (t - s)^{-\alpha} \frac{\partial}{\partial s} v(s, x) ds, \quad t > 0.$$

In this section, we make use of the tools of harmonic analysis to study the existence and decay estimate of solutions for (5.29). The section is organized as follows. In Subsection 5.2.2, we give the concept of solutions for problem (5.29) and discuss some properties of the operators appearing in the solution via the tools of harmonic analysis. Subsection 5.2.3 is concerned with the global existence of solutions for problem (5.29). It is also shown that the solution continuously depends on the initial value and has a sharp decay estimate. Finally, we switch our discussion to Sobolev spaces.

5.2.2 Solution Operators

In this subsection, we give an integral formula which is formally equivalent to (5.29).

We denote by $L^q(\mathbb{R}^N)$ ($q > 0$) the space of q-integral functions with the norm $\| \cdot \|_q$. $\mathcal{L}$ denotes the Laplace transform. $\mathcal{F}$ and $\mathcal{F}^{-1}$ represent the Fourier transform and its inverse respectively. Let $H_q^s := \mathcal{F}^{-1}[(1 + |\xi|^2)^{\frac{s}{2}} \mathcal{F}(L^q(\mathbb{R}^N))]$ and $\dot{H}_q^s := \mathcal{F}^{-1}[|\xi|^s \mathcal{F}(L^q(\mathbb{R}^N))]$ denote the Sobolev space and the homogeneous Sobolev space equipped with the norms $\|f\|_{H_q^s} := \|\mathcal{F}^{-1}[(1 + |\xi|^2)^{\frac{s}{2}} \mathcal{F}(f)]\|_q$ and $\|f\|_{\dot{H}_q^s} := \|\mathcal{F}^{-1}[|\xi|^s \mathcal{F}(f)]\|_q$. In the sequel, $*$ denotes the convolution.

Now we study the formulation of mild solutions to equation (5.29). Let us recall the Wright-type function

$$\mathcal{M}_\alpha(\theta) = \sum_{k=0}^\infty \frac{(-\theta)^k}{k! \Gamma(1 - \alpha(1 + k))}$$

and Mittag-Leffler functions

$$E_{\alpha,1}(t) = \sum_{k=0}^\infty \frac{t^k}{\Gamma(\alpha k + 1)}, \quad E_{\alpha,\alpha}(t) = \sum_{k=0}^\infty \frac{t^k}{\Gamma(\alpha k + \alpha)}.$$

To achieve our purpose, we transform the Cauchy problem (5.29) into an integral equation in the following lemma.

Lemma 5.2. *If u satisfies problem* (5.29), *then we have*

$$u(t, x) = \mathcal{S}_\alpha(t) u_0(x) + i\lambda \int_0^t (t - \tau)^{\alpha - 1} \mathcal{P}_\alpha(t - \tau) |u(\tau, x)|^p u(\tau, x) d\tau, \qquad (5.30)$$

where the solution operators $\mathcal{S}_\alpha(t)$ and $\mathcal{P}_\alpha(t)$ are given by

$$\mathcal{S}_\alpha(t)\varphi(x) = \left((4\pi i t^\alpha)^{-\frac{N}{2}} \int_0^\infty \theta^{-\frac{N}{2}} \mathcal{M}_\alpha(\theta) \exp(\frac{i|\cdot|^2}{4\theta t^\alpha}) d\theta * \varphi \right)(x),$$

$$\mathcal{P}_\alpha(t)\varphi(x) = \left((4\pi i t^\alpha)^{-\frac{N}{2}} \int_0^\infty \alpha \theta^{1-\frac{N}{2}} \mathcal{M}_\alpha(\theta) \exp(\frac{i|\cdot|^2}{4\theta t^\alpha}) d\theta * \varphi \right)(x).$$

Proof. Applying the space Fourier transform to (5.29), we get

$$\begin{cases} i\partial_t^\alpha \mathcal{F}(u)(t,\xi) - |\xi|^2 \mathcal{F}(u)(t,\xi) + \lambda \mathcal{F}(|u|^P u)(t,\xi) = 0, \quad t > 0, \\ \mathcal{F}(u)(0,\xi) = \mathcal{F}(u_0), \end{cases} \tag{5.31}$$

which, on taking temporal Laplace transform, leads to the equation

$$\mathcal{L}[\mathcal{F}(u)](s,\xi) = \frac{s^{\alpha-1}}{s^\alpha + i|\xi|^2} \mathcal{L}[\mathcal{F}(u_0)] + \frac{i\lambda}{s^\alpha + i|\xi|^2} \mathcal{L}[\mathcal{F}(|u|^P u)](s,\xi). \tag{5.32}$$

Taking inverse Laplace transform of (5.32) together with the convolution theorem and the facts that $\mathcal{L}[E_{\alpha,1}(-zt^\alpha)](s) = \frac{s^{\alpha-1}}{s^\alpha+z}$ and $\mathcal{L}[t^{\alpha-1}E_{\alpha,\alpha}(-zt^\alpha)](s) = \frac{1}{s^\alpha+z}$ for $z \in \mathbb{C}$, we obtain

$$\mathcal{F}(u)(t,\xi) = E_{\alpha,1}(-i|\xi|^2 t^\alpha)\mathcal{F}(u_0)$$
$$+ i\lambda \int_0^t (t-\tau)^{\alpha-1} E_{\alpha,\alpha}(-i|\xi|^2(t-\tau)^\alpha)\mathcal{F}(|u|^P u)(\tau,\xi)d\tau.$$

Applying the inverse Fourier transform to the above equation yields

$$u(t,x) = \mathcal{F}^{-1}[E_{\alpha,1}(-i|\xi|^2 t^\alpha)\mathcal{F}(u_0)]$$
$$+ i\lambda \int_0^t (t-\tau)^{\alpha-1} \mathcal{F}^{-1}[E_{\alpha,\alpha}(-i|\xi|^2(t-\tau)^\alpha)\mathcal{F}(|u|^P u)(\tau,\xi)]d\tau.$$

According to the convolution theorem for Fourier transform, we obtain

$$\mathcal{F}^{-1}[E_{\alpha,1}(-i|\xi|^2 t^\alpha)\mathcal{F}(u_0)] = \{\mathcal{F}^{-1}[E_{\alpha,1}(-i|\xi|^2 t^\alpha)]\} * u_0(x), \tag{5.33}$$

$$\mathcal{F}^{-1}[E_{\alpha,\alpha}(-i|\xi|^2 t^\alpha)\mathcal{F}(|u|^P u)(\tau,\xi)]$$
$$= \{\mathcal{F}^{-1}[E_{\alpha,\alpha}(-i|\xi|^2 t^\alpha)]\} * [|u(t,x)|^P u(t,x)]. \tag{5.34}$$

Using $E_{\alpha,1}(-z) = \int_0^\infty M_\alpha(\theta)\exp(-z\theta)d\theta$ for $z \in \mathbb{C}$ and the Fubini's theorem, (5.33) takes the form

$$\mathcal{F}^{-1}[E_{\alpha,1}(-i|\xi|^2 t^\alpha)] = \left(\frac{1}{2\pi}\right)^N \int_{\mathbb{R}^N} \exp(ix\cdot\xi) \int_0^\infty M_\alpha(\theta)\exp(-\theta it^\alpha|\xi|^2)d\theta d\xi$$
$$= \left(\frac{1}{2\pi}\right)^N \int_0^\infty M_\alpha(\theta)d\theta \int_{\mathbb{R}^N} \exp(ix\cdot\xi - \theta it^\alpha|\xi|^2)d\xi$$
$$= (4\pi it^\alpha)^{-\frac{N}{2}} \int_0^\infty \theta^{-\frac{N}{2}} M_\alpha(\theta)\exp\left(\frac{i|x|^2}{4\theta t^\alpha}\right)d\theta.$$

Notice that $E_{\alpha,\alpha}(-z) = \int_0^\infty \alpha\theta M_\alpha(\theta)\exp(-z\theta)d\theta$ for $z \in \mathbb{C}$. Using a similar argument for (5.34) yields

$$\mathcal{F}^{-1}[E_{\alpha,\alpha}(-i|\xi|^2 t^\alpha)] = (4\pi it^\alpha)^{-\frac{N}{2}} \int_0^\infty \alpha\theta^{1-\frac{N}{2}} M_\alpha(\theta)\exp\left(\frac{i|x|^2}{4\theta t^\alpha}\right)d\theta.$$

Consequently, we have

$$u(t,x) = \mathcal{S}_\alpha(t)u_0(x) + i\lambda \int_0^t (t-\tau)^{\alpha-1} \mathcal{P}_\alpha(t-\tau)|u(\tau,x)|^P u(\tau,x)d\tau.$$

$\square$

Now we introduce the concept of mild solutions with the aid of the above lemma.

Definition 5.2. For $T > 0$, by a mild solution of equation (5.29) on $[0, T]$ corresponding to an initial value u_0, we mean that u satisfies integral equation (5.30).

Lemma 5.3. *[65] For $t \neq 0$, let the operator $\mathscr{T}(t)$ be defined by*

$$\mathscr{T}(t)\varphi(x) = (4\pi i t)^{-\frac{N}{2}} \exp\left(\frac{i|\cdot|^2}{4t}\right) * \varphi,$$

for all $\varphi \in \mathcal{S}(\mathbb{R}^N)$. If $q \in [2, \infty]$ and $t \neq 0$, then $\mathscr{T}(t)$ maps $L^{q'}(\mathbb{R}^N)$ continuously to $L^q(\mathbb{R}^N)$ and that

$$\|\mathscr{T}(t)\varphi\|_q \leq (4\pi t)^{-N(\frac{1}{2}-\frac{1}{q})}\|\varphi\|_{q'}, \quad \text{for } \varphi \in L^{q'}(\mathbb{R}^N).$$

Before proceeding further, we state the fundamental estimates for the solution operators $\mathcal{S}_\alpha(t)$ and $\mathcal{P}_\alpha(t)$, which will play an important role in proving the existence results.

Lemma 5.4. *If $q \in [2, \infty]$ satisfies $N(\frac{1}{2} - \frac{1}{q}) < 1$ and $t \neq 0$, then $\mathcal{S}_\alpha(t)$ and $\mathcal{P}_\alpha(t)$ map $L^{q'}(\mathbb{R}^N)$ continuously to $L^q(\mathbb{R}^N)$ and there exists a constant C_1 such that*

$$\|\mathcal{S}_\alpha(t)\varphi\|_q \leq C_1 t^{-\alpha N(\frac{1}{2}-\frac{1}{q})}\|\varphi\|_{q'} \quad \text{and} \quad \|\mathcal{P}_\alpha(t)\varphi\|_q \leq C_1 t^{-\alpha N(\frac{1}{2}-\frac{1}{q})}\|\varphi\|_{q'}.$$

Proof. By the definitions of $\mathcal{S}_\alpha(t)$ and $\mathcal{P}_\alpha(t)$, it follows that

$$\mathcal{S}_\alpha(t)\varphi(x) = \int_0^\infty M_\alpha(\theta)\mathscr{T}(\theta t^\alpha)\varphi(x)d\theta,$$

$$\mathcal{P}_\alpha(t)\varphi(x) = \int_0^\infty \alpha\theta M_\alpha(\theta)\mathscr{T}(\theta t^\alpha)\varphi(x)d\theta.$$

From Lemma 5.3, we have

$$\|\mathcal{S}_\alpha(t)\varphi\|_q \leq \int_0^\infty M_\alpha(\theta)\|\mathscr{T}(\theta t^\alpha)\varphi\|_q d\theta$$

$$\leq (4\pi t^\alpha)^{-N(\frac{1}{2}-\frac{1}{q})}\int_0^\infty \theta^{-N(\frac{1}{2}-\frac{1}{q})}M_\alpha(\theta)d\theta \cdot \|\varphi\|_{q'}$$

$$\leq C_1 t^{-\alpha N(\frac{1}{2}-\frac{1}{q})}\|\varphi\|_{q'}$$

and

$$\|\mathcal{P}_\alpha(t)\varphi\|_q \leq \int_0^\infty \alpha\theta M_\alpha(\theta)\|\mathscr{T}(\theta t^\alpha)\varphi\|_q d\theta$$

$$\leq \alpha(4\pi t^\alpha)^{-N(\frac{1}{2}-\frac{1}{q})}\int_0^\infty \theta^{1-N(\frac{1}{2}-\frac{1}{q})}M_\alpha(\theta)d\theta \cdot \|\varphi\|_{q'}$$

$$\leq C_1 t^{-\alpha N(\frac{1}{2}-\frac{1}{q})}\|\varphi\|_{q'},$$

where

$$C_1 = \max\left\{(4\pi)^{-N(\frac{1}{2}-\frac{1}{q})}\frac{\Gamma(1-N(\frac{1}{2}-\frac{1}{q}))}{\Gamma(1-\alpha N(\frac{1}{2}-\frac{1}{q}))}, \alpha(4\pi)^{-N(\frac{1}{2}-\frac{1}{q})}\frac{\Gamma(2-N(\frac{1}{2}-\frac{1}{q}))}{\Gamma(1+\alpha-\alpha N(\frac{1}{2}-\frac{1}{q}))}\right\}.$$

$\square$

Lemma 5.5. *If $q \in [2, \infty]$ satisfies $N(\frac{1}{2} - \frac{1}{q}) < 1$ and $t \neq 0$, then there exists a constant C_1 such that*

$$\|S_\alpha(t)\varphi\|_{H_q^s} \leq C_1 t^{-\alpha N(\frac{1}{2} - \frac{1}{q})} \|\varphi\|_{H_{q'}^s},$$

and

$$\|P_\alpha(t)\varphi\|_{H_q^s} \leq C_1 t^{-\alpha N(\frac{1}{2} - \frac{1}{q})} \|\varphi\|_{H_{q'}^s}.$$

Proof. Fix $t \neq 0$ and let $w \in \mathcal{S}(\mathbb{R}^N)$. Then

$$\begin{aligned}
\mathcal{F}^{-1}[w\mathcal{F}(S_\alpha(t)\varphi)] &= \mathcal{F}^{-1}[wE_{\alpha,1}(-i|\xi|^2 t^\alpha)\mathcal{F}(\varphi)] \\
&= \mathcal{F}^{-1}\{E_{\alpha,1}(-i|\xi|^2 t^\alpha)\mathcal{F}\mathcal{F}^{-1}[w\mathcal{F}(\varphi)]\} \\
&= S_\alpha(t)(\mathcal{F}^{-1}[w\mathcal{F}(\varphi)]).
\end{aligned}$$

In view of Lemma 5.4, we get

$$\|\mathcal{F}^{-1}[w\mathcal{F}(S_\alpha(t)\varphi)]\|_q \leq C_1 t^{-\alpha N(\frac{1}{2} - \frac{1}{q})} \|\mathcal{F}^{-1}[w\mathcal{F}(\varphi)]\|_{q'}, \quad \text{for } 2 \leq q \leq \infty.$$

Choosing $w = (1 + |\xi|^2)^{\frac{s}{2}}$ or $w = |\xi|^s$, the results follows immediately from the definitions of the various Sobolev norms.

On the other hand, we can proceed analogously with the operator $P_\alpha(t)$ to obtain the desired results.   □

Remark 5.3. Note that the estimates given in the above lemma can be derived for the norms of $\dot{H}_q^s(\mathbb{R}^N)$ and $\dot{H}_{q'}^s(\mathbb{R}^N)$.

5.2.3 *Existence*

This subsection is concerned with the existence of solutions for equation (5.29). In the sequel, we need the assumption that p satisfies

$$p_0 < p < \frac{4}{N-2}, \quad \text{if } N > 2 \ (p_0 < p < \infty \ \text{if } N \leq 2), \tag{5.35}$$

where p_0 is the positive root of the equation $Nx^2 + (N-2)x - 4 = 0$.

Let

$$\gamma = \frac{4 - p(N-2)}{2p(p+2)}.$$

Denote by X_p the space consisting of all Bochner measurable functions $u : (0, +\infty) \to L^{p+2}(\mathbb{R}^N)$ such that

$$\sup_{t>0} t^{\alpha\gamma} \|u(t)\|_{p+2} < \infty.$$

Obviously X_p is a complete metric space equipped with the norm

$$\|u\|_{X_p} = \sup_{t>0} t^{\alpha\gamma} \|u(t)\|_{p+2}.$$

Let $\eta \in \left[0, \frac{1}{\alpha} - \gamma(p+1)\right)$. For convenience, we set

$$B_\alpha = B\left(\alpha\left[1 - \frac{Np}{2(p+2)}\right], 1 - \alpha\gamma(p+1)\right),$$

$$B_\eta = B\left(\alpha\left[1 - \frac{Np}{2(p+2)}\right], 1 - \alpha\left[\gamma(p+1) + \eta\right]\right).$$

Notice that $B_\alpha \leq B_\eta$.

Theorem 5.2. *Let p satisfy (5.35). If there exists $\varepsilon > 0$ such that*

$$\|\mathcal{S}_\alpha(t)u_0\|_{X_p} \leq \varepsilon, \tag{5.36}$$

then the equation (5.29) admits a unique solution $u(t,x) \in X_p$ satisfying $\|u\|_{X_p} \leq 2\varepsilon$.

Proof. Set

$$\Omega_\varepsilon = \{u \in X_p : \|u\|_{X_p} \leq 2\varepsilon\}.$$

It is easy to see that Ω_ε is a complete metric space when equipped with the metric

$$d(u,v) = \sup_{t>0} t^{\alpha\gamma}\|u(t) - v(t)\|_{p+2} = \|u - v\|_{X_p}.$$

For $u, v \in \Omega_\varepsilon$, it follows from the estimate

$$\left||u|^p u - |v|^p v\right| \leq (1+p)\left(|u|^p + |v|^p\right)|u - v|$$

and the Hölder's inequality that

$$\left\||u|^p u - |v|^p v\right\|_{\frac{p+2}{p+1}} \leq C\|u - v\|_{p+2}\left(\|u\|_{p+2}^p + \|v\|_{p+2}^p\right). \tag{5.37}$$

Let an operator Θ be defined by

$$\Theta u(t,x) = \mathcal{S}_\alpha(t)u_0(x) + i\lambda \int_0^t (t-\tau)^{\alpha-1}\mathcal{P}_\alpha(t-\tau)|u(\tau,x)|^p u(\tau,x)d\tau.$$

From Lemma 5.4, we obtain

$$\|\Theta u(t)\|_{p+2} \leq \|\mathcal{S}_\alpha(t)u_0\|_{p+2} + |\lambda| \int_0^t (t-\tau)^{\alpha-1}\|\mathcal{P}_\alpha(t-\tau)|u(\tau)|^p u(\tau)\|_{p+2}d\tau$$

$$\leq \|\mathcal{S}_\alpha(t)u_0\|_{p+2} + |\lambda|C_1 \int_0^t (t-\tau)^{\alpha[1-\frac{Np}{2(p+2)}]-1}\||u(\tau)|^p u(\tau)\|_{\frac{p+2}{p+1}}d\tau$$

$$\leq \|\mathcal{S}_\alpha(t)u_0\|_{p+2} + |\lambda|CC_1 \int_0^t (t-\tau)^{\alpha[1-\frac{Np}{2(p+2)}]-1}\|u(\tau)\|_{p+2}^{p+1}d\tau.$$

Setting

$$M(t) = \sup_{0 \leq \tau \leq t} \tau^{\alpha\gamma}\|u(\tau)\|_{p+2},$$

we conclude that

$$t^{\alpha\gamma}\|\Theta u(t)\|_{p+2}$$

$$\leq t^{\alpha\gamma}\|S_\alpha(t)u_0\|_{p+2}$$

$$+ |\lambda|CC_1 t^{\alpha\gamma}(M(t))^{p+1}\int_0^t (t-\tau)^{\alpha[1-\frac{Np}{2(p+2)}]-1}\tau^{-\alpha\gamma(p+1)}d\tau \tag{5.38}$$

$$\leq \varepsilon + |\lambda|CC_1 t^{\alpha[1-\gamma p-\frac{Np}{2(p+2)}]}(M(t))^{p+1}\int_0^1 (1-\tau)^{\alpha[1-\frac{Np}{2(p+2)}]-1}\tau^{-\alpha\gamma(p+1)}d\tau.$$

On the other hand, as $\gamma = \frac{4-p(N-2)}{2p(p+2)}$ and p satisfies (5.35), we get

$$\alpha\gamma(p+1) < 1, \quad \frac{Np}{2(p+2)} < 1, \quad \gamma p + \frac{Np}{2(p+2)} = 1. \tag{5.39}$$

Thus it follows that

$$\sup_{t>0} t^{\alpha\gamma}\|\Theta u(t)\|_{p+2} \leq \varepsilon + |\lambda|B_\alpha CC_1 \sup_{t>0}(M(t))^{p+1},$$

that is,

$$\|\Theta u(t)\|_{X_p} \leq \varepsilon + |\lambda|B_\alpha CC_1 \|u\|_{X_p}^{p+1}.$$

Then we choose $\varepsilon < (\frac{1}{2^{p+2}|\lambda|B_\alpha CC_1})^{\frac{1}{p}}$ such that $\varepsilon + |\lambda|B_\alpha CC_1(2\varepsilon)^{p+1} \leq 2\varepsilon$. This implies that $\|\Theta u(t)\|_{X_p} \leq 2\varepsilon$ for $u \in \Omega_\varepsilon$, that is, the operator Θ maps Ω_ε into itself.

Following the method of derivation of (5.38), we further obtain from Lemma 5.4 and (5.37) that

$$t^{\alpha\gamma}\|\Theta u(t) - \Theta v(t)\|_{p+2}$$

$$\leq \lambda C_1 t^{\alpha\gamma}\int_0^t (t-\tau)^{\alpha[1-\frac{Np}{2(p+2)}]-1}\||u(\tau)|^p u(\tau) - |v(\tau)|^p v(\tau)\|_{\frac{p+2}{p+1}}d\tau$$

$$\leq |\lambda|CC_1 t^{\alpha\gamma}\int_0^t (t-\tau)^{\alpha[1-\frac{Np}{2(p+2)}]-1}\|u(\tau)-v(\tau)\|_{p+2}\big(\|u(\tau)\|_{p+2}^p + \|v(\tau)\|_{p+2}^p\big)d\tau$$

$$\leq 2|\lambda|CC_1(2\varepsilon)^p d(u,v)t^{\alpha\gamma}\int_0^t (t-\tau)^{\alpha[1-\frac{Np}{2(p+2)}]-1}\tau^{-\alpha\gamma(p+1)}d\tau$$

$$= 2|\lambda|B_\alpha CC_1(2\varepsilon)^p d(u,v).$$

With the choice of ε, we have

$$\|\Theta u - \Theta v\|_{X_p} \leq 2|\lambda|B_\alpha CC_1(2\varepsilon)^p d(u,v) < \frac{1}{2}d(u,v),$$

which shows that Θ is a strict contraction on Ω_ε. Thus Θ has a fixed point u, which is the unique solution of (5.29). $\square$

Now we prove that mild solutions of (5.29) depend continuously on the initial value u_0 and have a sharp decay estimate.

Theorem 5.3. *Let u_0 and u_0' satisfy (5.36). Assume that u and u' are mild solutions of the problem (5.29) with the initial values u_0 and u_0', respectively. Then*

$$\|u - u'\|_{X_p} \leq 2\|S_\alpha(t)(u_0 - u_0')\|_{X_p}.$$

Moreover, if $\sup_{t>0} t^{\alpha\gamma}(1+t)^{\alpha\eta}\|S_\alpha(t)(u_0 - u_0')\|_{p+2} < \infty$ for $\alpha[\gamma(p+1)+\eta] < 1$, $\eta > 0$, then

$$\|u(t) - u'(t)\|_{p+2} \leq Ct^{-\alpha\gamma}(1+t)^{-\alpha\eta}.$$

Proof. Since u and u' are mild solutions of the equation (5.29) with the initial values u_0 and u_0', we can argue as in the proof of Theorem 5.2 to deduce that

$$t^{\alpha\gamma}\|u(t) - u'(t)\|_{p+2}$$

$$\leq t^{\alpha\gamma}\|\mathcal{S}_\alpha(t)(u_0 - u_0')\|_{p+2}$$

$$+ |\lambda|C_1 t^{\alpha\gamma} \int_0^t (t-\tau)^{\alpha[1-\frac{Np}{2(p+2)}]-1} \||u(\tau)|^p u(\tau) - |u'(\tau)|^p u'(\tau)\|_{\frac{p+2}{p+1}} d\tau$$

$$\leq t^{\alpha\gamma}\|\mathcal{S}_\alpha(t)(u_0 - u_0')\|_{p+2} + 2|\lambda|B_\eta CC_1(2\varepsilon)^p \sup_{t>0} t^{\alpha\gamma}\|u(t) - u'(t)\|_{p+2}.$$

Further, the choice of ε ensures that $2|\lambda|B_\eta CC_1(2\varepsilon)^p < \frac{1}{2}$. Thus

$$\sup_{t>0} t^{\alpha\gamma}\|u(t) - u'(t)\|_{p+2} \leq 2\sup_{t>0} t^{\alpha\gamma}\|\mathcal{S}_\alpha(t)(u_0 - u_0')\|_{p+2},$$

that is, $\|u - u'\|_{X_p} \leq 2\|\mathcal{S}_\alpha(t)(u_0 - u_0')\|_{X_p}$.

On the other hand, we find that

$$(1+t)^{\alpha\eta}(t-\tau)^{\alpha[1-\frac{Np}{2(p+2)}]-1}\|u(\tau) - u'(\tau)\|_{p+2}(\|u(\tau)\|_{p+2}^p + \|u'(\tau)\|_{p+2}^p)$$

$$\leq 2(2\varepsilon)^p (\frac{1+t}{1+\tau})^{\alpha\eta}(t-\tau)^{\alpha[1-\frac{Np}{2(p+2)}]-1}\tau^{-\alpha\gamma(p+1)}$$

$$\times \sup_{0\leq\tau\leq t} \{\tau^{\alpha\gamma}(1+\tau)^{\alpha\eta}\|u(\tau) - u'(\tau)\|_{p+2}\}$$

$$\leq 2(2\varepsilon)^p t^{\alpha\eta}(t-\tau)^{\alpha[1-\frac{Np}{2(p+2)}]-1}\tau^{-\alpha\gamma(p+1)-\alpha\eta}$$

$$\times \sup_{0\leq\tau\leq t} \{\tau^{\alpha\gamma}(1+\tau)^{\alpha\eta}\|u(\tau) - u'(\tau)\|_{p+2}\},$$

where we have used the inequality $(\frac{1+t}{1+\tau})^{\alpha\eta} \leq (\frac{t}{\tau})^{\alpha\eta}$. Therefore, we have

$$t^{\alpha\gamma}(1+t)^{\alpha\eta}\|u(t) - u'(t)\|_{p+2}$$

$$\leq t^{\alpha\gamma}(1+t)^{\alpha\eta}\|\mathcal{S}_\alpha(t)(u_0 - u_0')\|_{p+2}$$

$$+ 2|\lambda|CC_1(2\varepsilon)^p t^{\alpha(\eta+\gamma)} \sup_{0\leq\tau\leq t} \{\tau^{\alpha\gamma}(1+\tau)^{\alpha\eta}\|u(\tau) - u'(\tau)\|_{p+2}\}$$

$$\times \int_0^t (t-\tau)^{\alpha[1-\frac{Np}{2(p+2)}]-1}\tau^{-\alpha[\gamma(p+1)+\eta]}d\tau.$$

In view of $\alpha[\gamma(p+1)+\eta] < 1$ and (5.39), one can obtain

$$\sup_{t>0}\{t^{\alpha\gamma}(1+t)^{\alpha\eta}\|u(t) - u'(t)\|_{p+2}\}$$

$$\leq \sup_{t>0}\{t^{\alpha\gamma}(1+t)^{\alpha\eta}\|\mathcal{S}_\alpha(t)(u_0 - u_0')\|_{p+2}\}$$

$$+ 2|\lambda|B_\eta CC_1(2\varepsilon)^p \sup_{t>0}\{\tau^{\alpha\gamma}(1+\tau)^{\alpha\eta}\|u(\tau) - u'(\tau)\|_{p+2}\}.$$

Using $2|\lambda|B_\eta CC_1(2\varepsilon)^p < 1$ together with $\sup_{t>0} t^{\alpha\gamma}(1+t)^{\alpha\eta}\|\mathcal{S}_\alpha(t)(u_0 - u_0')\|_{p+2} < \infty$ leads to the estimate

$$\sup_{t>0}\{t^{\alpha\gamma}(1+t)^{\alpha\eta}\|u(t) - u'(t)\|_{p+2}\} \leq C_2,$$

which implies that

$$\|u(t) - u'(t)\|_{p+2} \leq C_2 t^{-\alpha\gamma}(1+t)^{-\alpha\eta}.$$

$\square$

Now we discuss the analogue results in Sobolev spaces. To achieve the purpose, we firstly need the estimate for the nonlinear operator $|u|^p u$ on $\dot{H}_q^s(\mathbb{R}^N)$. Let $q = \frac{N(p+2)}{N+ps}$. Then we have the following estimate:

$$\||u|^p u - |v|^p v\|_{\dot{H}_{q'}^s} \leq C\|u - v\|_{\dot{H}_q^s}\left(\|u\|_{\dot{H}_q^s}^p + \|v\|_{\dot{H}_q^s}^p\right), \tag{5.40}$$

where s satisfies

$$0 \leq s < \min\left\{p, \frac{N}{q}\right\}, \quad \text{if } p \notin 2\mathbb{N}^+; \quad 0 \leq s < \frac{N}{q}, \quad \text{if } p \in 2\mathbb{N}^+.$$

Set

$$\gamma = \frac{1}{2}\left(\frac{2}{p} - \frac{N}{q} + s\right),$$

and denote by X_q the space consisting of all Bochner measurable functions $u :$ $(0, +\infty) \to \dot{H}_q^s(\mathbb{R}^N)$ such that

$$\sup_{t>0} t^{\alpha\gamma}\|u(t)\|_{\dot{H}_q^s} < \infty.$$

It is obvious that X_q is a complete metric space endowed with the norm

$$\|u\|_{X_q} = \sup_{t>0} t^{\alpha\gamma}\|u(t)\|_{\dot{H}_q^s}.$$

Secondly, we need to impose a second restriction on s, that is,

$$s \in \Delta_p := \left(\frac{N}{2} - \frac{p+2}{p}, \frac{N}{2} - \frac{p+2}{p(p+1)}\right). \tag{5.41}$$

Therefore the values of q and γ together with (5.41) ensure that

$$\alpha\gamma(p+1) < 1, \quad \frac{Np}{2(p+2)} < 1, \quad \gamma p + \frac{N}{2}\left(1 - \frac{2}{q}\right) = 1.$$

Let us set

$$\Lambda_p = \begin{cases} \{0\} \cap \Delta_p, & \text{if } 0 < p < 1, \\ [0, p) \cap \Delta_p, & \text{if } p \geq 1 \text{ and } p \notin 2\mathbb{N}^+, \\ [0, \infty) \cap \Delta_p, & \text{if } p \geq 1 \text{ and } p \in 2\mathbb{N}^+. \end{cases}$$

Note that p is called to be admissible if the set Λ_p is nonempty.

Since the proofs of the following results are similar to that of Theorems 5.2 and 5.3, we just state the results.

Theorem 5.4. *Let p be admissible and $s \in \Lambda_p$. If there exists $\varepsilon > 0$ such that*

$$\|\mathcal{S}_\alpha(t)u_0\|_{X_q} \leq \varepsilon, \tag{5.42}$$

then the problem (5.29) admits a unique solution $u(t,x) \in X_q$ satisfying $\|u\|_{X_q} \leq 2\varepsilon$.

Theorem 5.5. *Let u_0 and u_0' satisfy (5.42). Assume that u and u' are mild solutions of the equation (5.29) with the initial values u_0 and u_0', respectively. Then*

$$\|u - u'\|_{X_q} \leq 2\|\mathcal{S}_\alpha(t)(u_0 - u_0')\|_{X_q}.$$

Moreover, if $\sup_{t>0} t^{\alpha\gamma}(1+t)^{\alpha\eta}\|\mathcal{S}_\alpha(t)(u_0 - u_0')\|_{\dot{H}_q^s} < \infty$ for $\alpha[\gamma(p+1) + \eta] < 1$, $\eta > 0$, then

$$\|u(t) - u'(t)\|_{\dot{H}_q^s} \leq Ct^{-\alpha\gamma}(1+t)^{-\alpha\eta}.$$

5.3   Notes and Remarks

The results in Section 5.1 due to Górka, Prado and Trujillo [152]. The contents in Section 5.2 are taken from Peng, Zhou and Ahmad [299].

Bibliography

[1] E. Affili, E. Valdinoci, Decay estimates for evolution equations with classical and fractional time-derivatives. J. Differential Equations **266**(7), 4027-4060(2019)

[2] K. Abe, Y. Giga, Analyticity of the Stokes semigroup in spaces of bounded functions. Acta Math. **211**(1), 1-46(2013)

[3] M. Abramowitz, I.A. Stegun, *Handbook of Mathematical Functions with Formulas, Graphs, and Mathematical Tables* in Applied mathematics series (Washington, D.C., 1967)

[4] R.A. Adams, *Sobolev Spaces* (Academic Press, London, 1975)

[5] P. Agarwal, J.J. Nieto, M.J. Luo, Extended Riemann-Liouville type fractional derivative operator with applications. Open Math. **15**(1), 1667-1681(2017)

[6] R.P. Agarwal, D. Baleanu, J.J. Nieto, D.F.M. Torres, Y. Zhou, A survey on fuzzy fractional differential and optimal control nonlocal evolution equations. J. Comput. Appl. Math. **339**, 3-29(2018)

[7] O.P. Agrawal, Fractional variational calculus in terms of Riesz fractional derivatives. J. Phy. A: Math. Theoretical **40**, 6287-6303(2007)

[8] O.P. Agrawal, Fractional variational calculus and the transversality conditions. J. Phys. A **39**, 10375-10384(2006)

[9] B. Ahmad, S.K. Ntouyas, A. Alsaedi, Existence of solutions for fractional differential equations with nonlocal and average type integral boundary conditions. J. Appl. Math. Comput. **53**, 129-145(2017)

[10] A.A. Alikhanov, A priori estimates for solutions of boundary value problems for fractional-order equations. Differ. Equ. **46**(5), 660-666(2010)

[11] M. Allen, L. Caffarelli, A. Vasseur, A parabolic problem with a fractional time derivative. Arch. Ration. Mech. Anal. **221**(2), 603-630(2016)

[12] M.F. de Almeida, L.C.F. Ferreira, On the Navier-Stokes equations in the half-space with initial and boundary rough data in Morrey spaces. J. Differential Equations. **254**, 1548-1570(2013)

[13] M.F. de Alemida, J.C.P. Precioso, Existence and symmetries of solutions in Besov-Morrey spaces for a semilinear heat-wave type equation. J. Math. Anal. Appl. **432**, 338-355(2015)

[14] E. Alvarez, C.G. Gal, V. Keyantuo, M. Warma, Well-posedness results for a class of semi-linear super-diffusive equations. Nonlinear Anal. **181**, 24-61(2019)

[15] H. Amann, *Linear and Quasilinear Parabolic Problems* (Birkhäuser, Berlin, 1995)

[16] C. Amrouche, A. Rejaiba, L^p-theory for Stokes and Navier-Stokes equations with Navier boundary condition. J. Differential Equations **256**, 1515-1547(2014)

[17] G.A. Anastassiou, *Fractional Differentiation Inequalities* (Springer, New York, 2009)

[18] B. de Andrade, A. Viana, Abstract Volterra integrodifferential equations with applications to parabolic models with memory. Math. Ann. **369**, 1131-1175(2017)

[19] B. de Andrade, A.N. Carvalho, P.M. Carvalho-Neto, P. Marin-Rubio, Semilinear fractional differential equations: global solutions, critical nonlinearities and comparison results. Topol. Methods Nonlinear Anal. **45**, 439-467(2015)

[20] D. Araya, C. Lizama, Almost automorphic mild solutions to fractional differential equations. Nonlinear Anal. **69**, 3692-3705(2008)

[21] W. Arendt and R. Mazzeo. Friedlander's eigenvalue inequalities and the Dirichlet-to-Neumann semigroup. Commun. Pure Appl. Anal. **11**(6), 2201-2212(2012)

[22] W. Arendt and A.F.M. ter Elst. The Dirichlet-to-Neumann operator on rough domains. J. Differential Equations. **251**(8), 2100-2124(2011)

[23] W. Arendt, A.F.M. ter Elst, J.B. Kennedy and M. Sauter, The Dirichlet-to-Neumann operator via hidden compactness. J. Funct. Anal. **266**(3), 1757-17862014

[24] W. Arendt, C. Batty, M. Hieber, F. Neubrander, Vector-valued Laplace Transforms and Cauchy Problems. Israel J. Math. **59**(3), 327-352(1987)

[25] W. Arendt, A.F.M. ter Elst and M. Warma, Fractional powers of sectorial operators via the Dirichlet-to-Neumann operator. Comm. Partial Differential Equations **43**(1), 1-24(2018)

[26] W. Arendt, A.F.M. ter Elst, Gaussian estimates for second order elliptic operators with boundary conditions. J. Operator Theory **38**, 87-130(1997)

[27] W. Arendt, M. Warma, The Laplacian with Robin boundary conditions on arbitrary domains. Potential Anal. **19**, 341-363(2003)

[28] W. Arendt, M. Warma, Dirichlet and Neumann boundary conditions: What is in between?. J. Evol. Equ. **3**, 119-135(2003)

[29] M.J. Arrieta, A.N. Carvalho, Abstract parabolic problems with critical nonlinearities and applications to Navier-Stokes and heat equations. Trans. Amer. Math. Soc. **352**, 285-310(1999)

[30] A. Ashyralyev, F. Tetikoglu, A note on fractional spaces generated by the positive operator with periodic conditions and applications. Bound. Value Probl. **2015**, 17(2015)

[31] T.M. Atanacković, S. Pilipović, B. Stanković, D. Zorica, *Fractional Calculus with Applications in Mechanics* (Wiley-ISTE, New York, 2014)

[32] N.T. Bao, H.L. Nguyen, A.V. Van, H.T. Nguyen, Y. Zhou, Existence and regularity of inverse problem for the nonlinear fractional Rayleigh-Stokes equations. Math. Methods Appl. Sci. **44**, 2532-2558(2021)

[33] V. Barbu, Optimal control of Navier-Stokes equations with periodic inputs. Nonlinear Anal. **31**(1), 15-31(1998)

[34] E. Barkai, R. Metzler, J. Klafter, From continuous time random walks to the fractional Fokker-Planck equation. Physical Review E **61**, 132-138(2000)

[35] E. Baskin, A. Iomin, Superdiffusion on a Comb Structure. Phys. Rev. Lett. **93**, 120603(2004)

[36] L. Baudouin, O. Kavian, J.P. Puel, Regularity for a Schrödinger equation with singular potentials and application to bilinear optimal control. J. Diff. Eq. **216**, 188-222(2005)

[37] L. Baudouin, J. Salomon, Constructive solution of a bilinear optimal control problem for a Schrödinger equation. Systems Control Lett. **57**, 453-464(2008)

[38] S. Bayin, Time fractional Schrödinger equation: Fox's H-functions and the effective potential. J. Math. Phys. **54**, 012103(2013)

[39] E. Bazhlekova, B. Jin, R. Lazarov, Z. Zhou, An analysis of the Rayleigh-Stokes problem for a generalized second-grade fluid. Numer. Math. **131**(1), 1-31(2015)

[40] E. Bazhlekova, Subordination principle for a class of fractional order differential equations. Mathematics **2**, 412-427(2015)

[41] E. Bazhlekova, *Fractional Evolution Equations in Banach Spaces* (Eindhoven University of Technology, Netherlands, 2001)

[42] G.S. Beavers and D. D. Joseph, Boundary conditions at a naturally permeable wall. J. Fluid Mech. **30**(1), 197-207(2006)

[43] M. Ben-Artzi, J.P. Croisille, D. Fishelov, *Navier-Stokes Equations in Planar Domains* (World Scientific, Singapore, 2013)

[44] D.A. Benson, S.W. Wheatcraft, M.M. Meerschaert, The fractional-order governing equation of Lévy motion. Water Resour. Res. **36**, 1413-1423(2000)

[45] J. Bergh, J. Löfström, *Interpolation Spaces: An Introduction* (Springer-Verlag, Berlin, 1976)

[46] B. Berkowitz, J. Klafter, R. Metzler, H. Scher, Physical pictures of transport in heterogeneous media: Advection-dispersion, random-walk, and fractional derivative formulations. Water Res. Res. **38**, 9-1-9-12(2002)

[47] U. Biccari, M. Warma and E. Zuazua, Local elliptic regularity for the Dirichlet fractional Laplacian. Adv. Nonlinear Stud. **17**(4), 837-839(2017)

[48] U. Biccari, M. Warma and E. Zuazua, Local elliptic regularity for the Dirichlet fractional Laplacian. Adv. Nonlinear Stud. **17**(2), 387-409(2017)

[49] K. Bogdan, K. Burdzy and Z.-Q. Chen, Censored stable processes. Probab. Theory Related Fields **127**(1), 89-152(2003)

[50] A. V. Bobylev, C. Cercignani, The inverse Laplace transform of some analytic functions with an application to the eternal solutions of the Boltzmann equation. Appl. Math. Lett. **15**, 807-813(2002)

[51] V. Bögelein, F. Duzaar, L. Schätzler, C. Scheven, Existence for evolutionary problems with linear growth by stability methods. J. Differential Equations **266**(11), 7709-7748(2019)

[52] V. Bögelein, F. Duzaar, P. Marcellini, C. Scheven, Doubly nonlinear equations of porous medium type. Arch. Ration. Mech. Anal. **229**(2), 503-545(2018)

[53] J. Bourgain, *Global Solutions of Nonlinear Schrödinger Equations* (American Mathematical Society, America, 1999)

[54] F. Boyer and P. Fabrie, *Mathematical Tools for the Study of the Incompressible Navier-Stokes Equations and Related Models* (Springer Science and Business Media, Berlin, 2012)

[55] H. Brezis, *Functional Analysis, Sobolev Spaces and Partial Differential Equations* (Springer Science & Business Media, Berlin, 2010)

[56] V.S. Buslaev, C. Sulem, On asymptotic stability of solitary waves for nonlinear Schrödinger equations. Annales de l'Institut Henri Poincare. Nonlinear Anal. **20**, 419-475(2003)

[57] L.A. Caffarelli and L. Silvestre, An extension problem related to the fractional Laplacian. Comm. Partial Differential Equations **32**(7-9), 1245-1260(2007)

[58] L.A. Caffarelli and P.R. Stinga, Fractional elliptic equations, Caccioppoli estimates and regularity. Nonlinear Anal. **33**(3), 767-807(2016)

[59] M. Cannone, *Ondelettes, Paraproduits et Navier-Stokes* (Diderot Editeur, Paris, 1995)

[60] M. Cannone, A generalization of a theorem by Kato on Navier-Stokes equations. Rev. Mat. Iberoamericana **13**, 515-541(1997)

[61] S. Carillo, Existence, uniqueness and exponential decay: An evolution problem in heat conduction with memory. Quart. Appl. Math. **69**(4), 635-649(2011)

[62] Y. Cao, J.X. Yin, C.P. Wang, Cauchy problems of semilinear pseudo-parabolic equations. J. Differential Equations **46**, 4568-4590(2009)

[63] C.M. Carracedo, M.S. Alix, *The Theory of Fractional Powers of Operators, North-Holland Mathematics Studies* (Elsevier, San Diego, 2001)

[64] P.M. de Carvalho-Neto, P. Gabriela, Mild solutions to the time fractional Navier-Stokes equations in $\mathbb{R}^N$. J. Differential Equations **259**, 2948-2980(2015)

[65] T. Cazenave, *Semilinear Schrödinger Equations* (American Mathematical Society, New York, 2003)

[66] T. Cazenave, P.L. Lions, Orbital stability of standing waves for some nonlinear Schrödinger equations. Comm. Math. Phys. **85**, 549-561(1982)

[67] T. Cazenave, Stable solutions of the logarithmic Schrödinger equation. Nonlinear Anal. **7**, 1127-1140(1983)

[68] T. Cazenave, B. Weissler, The cauchy problem for the critical nonlinear Schrödinger equation in H^s. Nonlinear Anal. **14**(10), 807-836(1990)

[69] A. Çeşmelioglu and B. Rivière, Existence of a weak solution for the fully coupled Navier-Stokes/Darcy-transport problem. J. Diff. Eq. **252**, 4138-4175(2012)

[70] A. Çeşmelioglu, V. Girault and B. Riviere, Time-dependent coupling of Navier-Stokes and Darcy flows. ESAIM Math. Model. Numer. Anal. **47**, 539-554(2013)

[71] J.Y. Chemin, I. Gallagher, Large, global solutions to the Navier-Stokes equations, slowly varying in one direction. Trans. Amer. Math. Soc. **362**, 2859-2873(2010)

[72] J.Y. Chemin, I. Gallagher, M. Paicu, Global regularity for some classes of large solutions to the Navier-Stokes equations. Ann. of Math. **173**, 983-1012(2011)

[73] C.M. Chen, F. Liu, V. Anh, A Fourier method and an extrapolation technique for Stokes' first problem for a heated generalized second grade fluid with fractional derivative. J. Comput. Appl. Math. **223**(2), 777-789(2009)

[74] C.M. Chen, F. Liu, K. Burrage, Y. Chen, Numerical methods of the variable-order Rayleigh-Stokes problem for a heated generalized second grade fluid with fractional derivative. IMA J. Appl. Math. **78**(5), 924-944(2013)

[75] J. Chen, F. Liu, V. Anh, Analytical solution for the time-fractional telegraph equation by the method of separating variables. J. Math. Anal. Appl. **338**, 1364-1377(2008)

[76] Y. Chen, X. Zhao and Y. Huang, Mortar element method for the time dependent coupling of Stokes and Darcy flows. J. Sci. Comput. **80**(2), 1310-1329(2019)

[77] Y. Chen, H. Gao, M. Garrido-Atienza, B. Schmalfuss, Pathwise solutions of SPDEs driven by Hölder-continuous integrators with exponent larger than $1/2$ and random dynamical systems. Discrete Contin. Dyn. Syst. **34**, 79-98(2014)

[78] P. Chidyagwai and B. Rivière, On the solution of the coupled Navier-Stokes and Darcy equations. Comput. Methods Appl. Mech. Eng. **198**, 3806-3820(2009)

[79] R. Chill, V. Keyantuo, M. Warma, Generation of cosine families on $L^p(0,1)$ by elliptic operators with Robin boundary conditions. 113-130, *Functional Analysis and Evolution Equations* (Birkhäuser, Basel, 2008)

[80] H.J. Choe, Boundary regularity of suitable weak solution for the Navier-Stokes equations. J. Funct. Anal. **268**, 2171-2187(2015)

[81] B. Claus, M. Warma, Realization of the fractional laplacian with nonlocal exterior conditions via forms method. J. Evol. Equ. **20**, 1597-1631(2020)

[82] Ph. Clément, G. Gripenberg, S.O. Londen, Regularity properties of solutions of fractional evolution equations. *Evolution equations and their applications in physical and life sciences* (Dekker, New York, 2001)

[83] Ph. Clément, S.O. Londen, G. Simonett, Quasilinear evolutionary equations and continuous interpolation spaces. J. Differential Equations. **196**, 418-447(2004)

[84] Ph. Clément, R. Zacher, Global smooth solutions to a fourth-order quasilinear fractional evolution equation. 131-146, *Functional Analysis and Evolution Equations* (Birkhauser, Basel, 2008)

[85] P. Constantin, C. Foias, *Navier-Stokes Equations* (University of Chicago Press, Chicago, 1988)

[86] J.M. Coron and P. Lissy, Local null controllability of the three-dimensional Navier-Stokes system with a distributed control having two vanishing components. Invent. Math. **198**(3), 833-880(2014)

[87] J.M. Coron and S. Guerrero, Local null controllability of the two-dimensional Navier-Stokes system in the torus with a control force having a vanishing component. J. Math. Pures Appl. **92**(5), 528-545(2009)

[88] J.M. Coron and A.V. Fursikov, Global exact controllability of the 2D Navier-Stokes equations on a manifold without boundary. Russ. J. Math. Phys. **4**(4), 429-448(1996)

[89] R. Courant, D. Hilbert, *Methods of Mathematical Physics: Partial Differential Equations* (John Wiley and Sons, New Jersey, 2008)

[90] S. Cuccagna, Stabilization of solutions to nonlinear Schrödinger equations. Comm. Pure Appl. Math. **54**, 1110-1145(2001)

[91] E. Cuesta, Asymptotic behaviour of the solutions of fractional integro-differential equations and some time discretizations. Discrete Contin. Dyn. Syst. (SUPPL), 277-285(2007)

[92] E. Cuesta, Some advances on image processing by means of fractional calculus. 265-271, *Nonlinear Science and complexity* (Springer, Dordrecht, 2011)

[93] R. Danchin, Global existence in critical spaces for compressible Navier-Stokes equations. Invent. Math. **141**, 579-614(2000)

[94] D. Daners, Robin boundary value problems on arbitrary domains. Trans. Amer. Math. Soc. **352**, 4207-4236(2000)

[95] D.T. Dang, E. Nane, D.M. Nguyen, N.H. Tuan, Continuity of solutions of a class of fractional equations. Potential Anal. **49**, 423-478(2018)

[96] E.B. Davies, *Heat Kernels and Spectral Theory* (Cambridge University Press, Cambridge, 1989)

[97] M. Dehghan, A computational study of the one-dimensional parabolic equation subject to nonclassical boundary specifications. Numer. Methods Partial Differential Equations **22**(1), 220-257(2006)

[98] M. Dehghan, M. Abbaszadeh, A finite element method for the numerical solution of Rayleigh-Stokes problem for a heated generalized second grade fluid with fractional derivatives. Engin. Comput. **33**, 587-605(2017)

[99] G. Desch, S.O. Londen, Evolutionary equations driven by fractional Brownian motion. Stoch. Partial Differ. Equ. Anal. Comput. **1**, 424-454(2013)

[100] E. Di Nezza, G. Palatucci, and E. Valdinoci, Hitchhiker's guide to the fractional Sobolev spaces. Bull. Sci. Math. **136**(5), 521-573(2012)

[101] K. Diethelm, *The Analysis of Fractional Differential Equations: An Application-oriented Exposition Using Differential Operators of Caputo Type* (Springer Science and Business Media, Berlin, 2010)

[102] H.F. Di, Y.D. Shang, X.M. Peng, Blow-up phenomena for a pseudo-parabolic equation with variable exponents. Appl. Math. Lett. **64**, 67-73(2017)

[103] X.L. Ding, J.J. Nieto, Analytical solutions for multi-term time-space fractional partial differential equations with nonlocal damping terms. Fract. Calc. Appl. Anal. **21**, 312-335(2018)

[104] S. Dipierro, X. Ros-Oton, E. Valdinoci, Nonlocal problems with Neumann boundary conditions, Rev. Mat. Iberoam. **33**(2), 377-416(2017)

[105] H. Dong, D. Kim, L_p-estimates for time fractional parabolic equations with coefficients measurable in time. Adv. Math. **345**, 289-345(2019)

[106] J. Dong, M. Xu, Space-time fractional Schrödinger equation with time-independent potentials. J. Math. Anal. Appl. **344**, 1005-1017(2008)

[107] F. Duzaar, J. Habermann, Partial regularity for parabolic systems with non-standard growth. J. Evol. Equ. **12**(1), 203-244(2012)

[108] P.G. Drazin, N. Riley, *The Navier-Stokes Equations: A Classification of Flows and Exact Solutions* (Cambridge University Press, Cambridge, 2006)

[109] S. Dubois, Mild solutions to the Navier-Stokes equations and energy equalities. J. Differential Equations **189**(1), 99-147(2003)

[110] J. Duchon, R. Robert, Inertial energy dissipation for weak solutions of incompressible Euler and Navier-Stokes equations. Nonlinearity **13**(1), 249-255(2000)

[111] R. Eid, S.I. Muslih, D. Baleanu, E. Rabei, On fractional Schrödinger equation in α-dimensional fractional space. Nonlinear Anal. Real World Appl. 10(2009), 1299-1304.

[112] S.D. Eidelman, A.N. Kochubei, Cauchy problem for fractional diffusion equations. J. Differential Equations **199**, 211-255(2004)

[113] M. El-Shahed, A. Salem, On the generalized Navier-Stokes equations. Appl. Math. Comput. **156**(1), 287-293(2004)

[114] M.M. El-Borai, Some probability densities and fundamental solutions of fractional evolution equations. Chaos Solitons Fractals **14**, 433-440(2002)

[115] O. El-Mennaoui, V. Keyantuo, Trace theorems for holomorphic semigroups and the second order Cauchy problem. Proc. Amer. Math. Soc. **124**, 1445-1458(1996)

[116] K.J. Engel, R. Nagel, *One-parameter Semigroups for Linear Evolution Equations* (Springer Verlag, Berlin, 2000)

[117] N. Engheta, On fractional calculus and fractional multipoles in electromagnetism. IEEE Trans. Antennas and Propagation **44**, 554-566(1996)

[118] L.C. Evans, *Partial Differential Equation* (American Mathematical Society, Providence, Rhode Island, 1997)

[119] L.C. Evans, *Partial Differential Equations*, (American Mathematical Society, Providence, Rhode Island, 2nd Ed. 2010)

[120] E. Feireisl, A. Novotný, H. Petzeltová, On the existence of globally defined weak solutions to the Navier-Stokes equations. J. Math. Fluid Mech. **3**(4), 358-392(2001)

[121] L.C.F. Ferreira, E.J. Villamizar-Roa, Self-similar solutions, uniqueness and long-time asymptotic behavior for semilinear heat equations. Differential Integral Equations **19**(12), 1349-1370(2006)

[122] C. Fetecau, C. Fetecau, The Rayleigh-Stokes problem for heated second grade fluids. Internat. J. Non-Linear Mech. **37**(6), 1011-1015(2002)

[123] C. Fetecau, The Rayleigh-Stokes problem for an edge in an Oldroyd-B fluid. C. R. Math. Acad. Sci. Paris **335**(11), 979-984(2002)

[124] C. Fetecau, J. Zierep, The Rayleigh-Stokes-problem for a Maxwell fluid. Z. Angew. Math. Phys. **54**(6), 1086-1093(2003)

[125] C. Fetecau, M. Jamil, C. Fetecau, D. Vieru, The Rayleigh-Stokes problem for an edge in a generalized Oldroyd-B fluid. Z. Angew. Math. Phys. **60**(5), 921-933(2009)

[126] G. Fibich, Singular solutions of the subcritical nonlinear Schrödinger equation. Phys. D **240**, 1119-1122(2011)

[127] A. Floer, A. Weinstein, Nonspreading wave packets for the cubic Schrödinger equation with a bounded potential. J. Funct. Anal. **69**, 397-408(1986)

[128] N.J. Ford, M.M. Rodrigues, N. Vieira, A numerical method for the fractional Schrödinger type equation of spatial dimension two. Fract. Calc. Appl. Anal. **16**(2), 454-468(2013)

[129] D. Fujiwara, Concrete characterization of the domains of fractional powers of some elliptic differential operators of the second order. Proc. Japan Academy **43**(2), 82-86(1967)

[130] M. Fukushima, Y. Oshima, and M. Takeda, *Dirichlet Forms and Symmetric Markov Processes* (Walter de Gruyter & Co., Berlin, 2011)

[131] C.G. Gal, On the strong-to-strong interaction case for doubly nonlocal Cahn-Hilliard equations. Discrete Contin. Dyn. Syst. **37**(1), 131-167(2017)

[132] C.G. Gal and M. Warma, Fractional in time semilinear parabolic equations and applications. HAL Id: hal-01578788(2017)

[133] G.P. Galdi, *An Introduction to the Mathematical Theory of the Navier-Stokes Equations: Steady-State Problems* (Springer Science and Business Media, Berlin, 2011)

[134] Z.Z. Ganji, D.D. Ganji, A. Ganji, M. Rostamian, Analytical solution of time-fractional Navier-Stokes equation in polar coordinate by homotopy perturbation method. Numer. Methods Partial Differential Equations **26**, 117-124(2010)

[135] T. Ghosh, M. Salo, G. Uhlmann, The Calderón problem for the fractional Schrödinger equation. Anal. PDE. **13**(2), 455-475(2020)

[136] D. Gilbarg, David, N.S. Trudinger, *Elliptic Partial Differential Equations of Second Order* (Springer Verlag, Berlin, 2001)

[137] Y. Giga and T. Namba, Well-posedness of Hamilton-Jacobi equations with Caputo's time fractional derivative. Comm. Partial Differential Equations **42**(7), 1088-1120(2017)

[138] Y. Giga, T. Miyakawa, Navier-Stokes flow in $\mathbb{R}^3$ with measures as initial vorticity and Morrey spaces. Comm. Partial Differential Equations **14**, 577-618(1989)

[139] Y. Giga, T. Miyakawa, Solutions in L_r of the Navier-Stokes initial value problem. Arch. Ration. Mech. Anal. **89**(3), 267-281(1985)

[140] Y. Giga, S. Yoshikazu, Abstract L^p estimates for the Cauchy problem with applications to the Navier-Stokes equations in exterior domains. J. Funct. Anal. **102**, 72-94(1991)

[141] J. Ginibre and G. Velo, The global Cauchy problem for nonlinear Klein-Gordon equation. Math. Z **189**, 487-505(1985)

[142] V. Girault and B. Rivière, DG approximation of coupled Navier-Stokes and Darcy equations by Beaver-Joseph-Saffman interface condition. SIAM J. Numer. Anal. **47**, 2052-2089(2009)

[143] W.G. Glöckle, T.F. Nonnenmacher, A fractional calculus approach to self-similar protein dynamics. Biophys J. **68**(1), 46-53(1995)

[144] F. Golse and L. Saint-Raymond, The Navier-Stokes limit of the Boltzmann equation for bounded collision kernels. Invent. Math. **155**(1), 81-161(2004)

[145] J.A. Goldstein, *Semigroups of Linear Operators and Applications* (Oxford University Press, New York, 1985)

[146] R. Gorenflo, A. Iskenderov, Y. Luchko, Mapping between solutions of fractional diffusion-wave equations. Fract. Calc. Appl. Anal. **3**(1), 75-86(2000)

[147] R. Gorenflo, J. Loutchko, Y. Luchko, Computation of the Mittag-Leffler function $E_{\alpha,\beta}(z)$ and its derivatives. Frac. Calc. Appl. Anal. **5**(4), 491-518(2002)

[148] R. Gorenflo, F. Mainardi, On Mittag-Leffler-type functions in fractional evolution processes. J. Comput. Appl. Math. **118**, 283-299(2000)

[149] R. Gorenflo, Y. Luchko, F. Mainardi, Analytical properties and applications of the Wright function. Fract. Calc. Appl. Anal. **2**, 383-414(1999)

[150] Y. V. Gorbatenko. Existence and uniqueness of mild solutions of second order semilinear differential equations in Banach space. Methods Funct. Anal. Topology **17**(1), 1-9(2011)

[151] R. Gorenflo, A. A. Kilbas, F. Mainardi, and S. V. Rogosin. *Mittag-Leffler Functions, Related Topics and Applications* (Springer, Berlin, 2014)

[152] P. Górka, H. Prado and J. Trujillo, The time fractional Schrödinger equation on Hilbert space. Integral Equations Operator Theory **87**, 1-14(2017)

[153] G. Grubb, Fractional Laplacians on domains, a development of Hörmander's theory of μ-transmission pseudo-differential operators. Adv. Math. **268**, 478-528(2015)

[154] P. Guerrero, J.L. López, J.J. Nieto, Global H^1 solvability of the 3D logarithmic Schrödinger equation. Nonlinear Anal. Real World Appl. **11**, 79-87(2010)

[155] V.S. Guliyev, M.N. Omarova, M.A. Ragusa, Characterizations for the genuine Calderon-Zygmund operators and commutators on generalized Orlicz-Morrey spaces. Adv Nonlinear Anal. **12**(1), (2023)

[156] B.L. Guo, Y.P. Wu, Orbital stability of solitary waves for the nonlinear derivative Schrödinger equation. J. Differential Equations **123**, 35-55(1995)

[157] B.H. Guswanto, T. Suzuki, Existence and uniqueness of mild solutions for fractional semilinear differential equations. Electron. J. Differential Equations **2015**, 1-16(2015)

[158] H. Hajaiej, X. Yu, Z. Zhai, Fractional Gagliardo-Nirenberg and Hardy inequalities under Lorentz norms. J. Math. Anal. Appl. **396**, 569-577(2012)

[159] B. Han, Refinable functions and cascade algorithms in weighted spaces with Hölder continuous masks. SIAM J. Math. Anal. **40**, 70-102(2008)

[160] J.W. He, L. Peng, Approximate controllability for a class of fractional stochastic wave equations. Comput. Math. Appl. **78**, 1463-1476(2019)

[161] J.W. He, Y. Zhou, On a backward problem for nonlinear time fractional wave equations. Proc. Roy. Soc. Edinburgh Sect. A. **152**(6), 1589-1612(2022)

[162] J.W. He, Y. Zhou, Local/global existence analysis of fractional wave equations with exponential nonlinearity. Bull. Sci. Math. **189**, 103357(2023)

[163] J.W. He, Y. Zhou, L. Peng, On well-posedness of semilinear Rayleigh-Stokes problem with fractional derivative on $\mathbb{R}^N$, Adv. Nonlinear Anal. **11**, 580-597(2022)

[164] H. Heck, H. Kim, H. Kozono, Weak solutions of the stationary Navier-Stokes equations for a viscous incompressible fluid past an obstacle. Math. Ann. **356**, 653-681(2013)

[165] D. Henry, *Geometric Theory of Semilinear Parabolic Equations* (Springer-Verlag, Berlin-New York, 1981)

[166] B.I. Henry, T.A.M. Langlands, P. Straka, Fractional Fokker-Planck equations for subdiffusion with space- and time-dependent forces, Phys. Rev. Lett. **105**, 170602(2010)

[167] R. Herrmann, *Fractional Calculus: An Introduction for Physicists* (World Scientific, Singapore, 2011)

[168] M. Hieber, Integrated semigroups and differential operators on L^p-spaces. Math. Ann. **291**, 1-16(1991)

[169] M. Hieber, O. Sawada, The Navier-Stokes equations in $\mathbb{R}^n$ with linearly growing initial data. Arch. Ration. Mech. Anal. **175**(2), 269-285(2005)

[170] M. Hieber, Y. Shibata, The Fujita-Kato approach to the Navier-Stokes equations in the rotational framework. Math. Z. **265**(2), 481-491(2010)

[171] R. Hilfer, *Applications of Fractional Calculus in Physics* (World Scientific, Singapore, 2000)

[172] H. Hirata, C. Miao, Space-time estimates of linear flow and application to some nonlinear integro-differential equations corresponding to fractional-order time derivative. Adv. Differential Equations **7**(2), 217-236(2002)

[173] J. Huang, G. Wang, J. Xiong, A maximum principle for partial information backward stochastic control problems with applications. SIAM J. Control Optim. **48**(4), 2106-2117(2009)

[174] C. Huang, K.N. Le, M. Stynes, A new analysis of a numerical method for the time-fractional Fokker-Planck equation with general forcing. IMA J. Numer. Anal.**40**(2), 1217-1240(2020)

[175] N.Q. Hung, J.L. Vazquez, Porous medium equation with nonlocal pressure in a bounded domain. Comm. Partial Differential Equations **43**(10), 1502-1539(2018)

[176] V. Hutson, J.S. Pym, M.J. Cloud, *Applications of Functional Analysis and Operator Theory* (Elsevier,Amsterdam, 2005)

[177] L.N. Huynh, Y. Zhou, D. O'Regan, N.H. Tuan, Fractional Landweber method for an initial inverse problem for time-fractional wave equations. Appl. Anal. **100**(4), 860-878(2021)

[178] S. Ibrahim, S. Keraani, Global small solutions for the Navier-Stokes-Maxwell system. SIAM J. Math. Anal. **43**(5), 2275-2295(2011)

[179] M. Inc, A.I. Aliyu, A. Yusuf and D. Balean, Dark-bright optical solitary waves and modulation instability analysis with $(2+1)$-dimensional cubic-quintic nonlinear Schrödinger equation. Waves Random Complex Media **29**(3), 393-402(2019)

[180] A. Iomin, Fractional-time quantum dynamics. Phys. Rev. E (3) **80**(2), 022103(2009)

[181] A. Iomin, Fractional-time Schrödinger equation: fractional dynamics on a comb. Chaos Solitons Fractals **44**(4-5), 348-352(2011)

[182] T. Iwabuchi, R. Takada, Global solutions for the Navier-Stokes equations in the rotational framework. Math. Ann. **357**, 727-741(2013)

[183] T. Iwabuchi, R. Takada, Global well-posedness and ill-posedness for the Navier-Stokes equations with the Coriolis force in function spaces of Besov type. J. Funct. Anal. **267**(5), 1321-1337(2014)

[184] H. Jia, V. Šverák, Local-in-space estimates near initial time for weak solutions of the Navier-Stokes equations and forward self-similar solutions. Invent. Math. **196**(1), 233-265(2014)

[185] A. Jüngel, Global weak solutions to compressible Navier-Stokes equations for quantum fluids. SIAM J. Math. Anal. **42**(3), 1025-1045(2010)

[186] T. Kato, *Perturbation Theory for Linear Operators* (Springer-Verlag, Berlin Heidelberg, 1995)

[187] O.D. Kellogg, *Foundations of Potential Theory* (Springer-Verlag, Berlin-New York 1967)

[188] J. Kemppainen, J. Siljander, V. Vergara, R. Zacher, Decay estimates for time-fractional and other non-local in time subdiffusion equations in $\mathbb{R}^d$. Math. Ann. **366**(3), 941-979(2016)

[189] V. Keyantuo and C. Lizama, Mild well-posedness of abstract differential equations. 371-387, *Functional Analysis and Evolution Equations* (Birkhäuser, Basel, 2008)

[190] V. Keyantuo, C. Lizama, M. Warma, Asymptotic behavior of fractional-order semilinear evolution equations. Differential Integral Equations **26**, 757-780(2013)

[191] V. Keyantuo, C. Lizama, M. Warma, Spectral criteria for solvability of boundary value problems and positivity of solutions of time-fractional differential equations. Abstr. Appl. Anal. **11**, 614328(2013)

[192] D.Q. Khai, Well-posedness for the Navier-Stokes equations with datum in the Sobolev spaces. Acta Math. Vietnam. **42**(3), 431-443(2017)

[193] D.Q. Khai, V.T.T. Duong, On the initial value problem for the Navier-Stokes equations with the initial datum in the Sobolev spaces. arXiv preprint arXiv:1603.04219(2016)

[194] D.Q. Khai, N.M. Tri, Well-posedness for the Navier-Stokes equations with data in homogeneous Sobolev-Lorentz spaces. Nonlinear Anal. **149**, 130-145(2017)

[195] D.Q. Khai, N.M. Tri, Well-posedness for the Navier-Stokes equations with datum in Sobolev-Fourier-Lorentz spaces. J. Math. Anal. Appl. **437**, 754-781(2016)

[196] M. Khan, The Rayleigh-Stokes problem for an edge in a viscoelastic fluid with a fractional derivative model. Nonlinear Anal. Real World Appl. **10**(5), 3190-3195(2009)

[197] A.A. Kilbas, H.M. Srivastava, J.J. Trujillo, *Theory and Applications of Fractional Differential Equations* (Elsevier Science B.V., Amsterdam, 2006)

[198] I. Kim, K.H. Kim and S. Lim, An $L_q(L_p)$-theory for the time fractional evolution equations with variable coefficients. Adv. Math. **306**, 123-176(2017)

[199] V. Kiryakova, *Generalized Fractional Calculus and Applications* (CRC press, Florida, 1993)

[200] E. Klann, R. Ramlau, Regularization by fractional filter methods and data smoothing. Inverse Problems **24**(2), 025018(2008)

[201] R. Kutner, Coherent spatio-temporal coupling in fractional wanderings. Renewed approach to continuous-time Lévy flights. 1-14, *Anomalous Diffusion From Basics to Applications* (Springer, Berlin, Heidelberg, 1999).

[202] G. Koch, N. Nadirashvili, G.A. Seregin, V. Šverák, Liouville theorems for the Navier-Stokes equations and applications. Acta Math. **203**, 83-105(2009)

[203] H. Koch, D. Tataru, Well-posedness for the Navier-Stokes equations. Adv. Math. **157**(1), 22-35(2001)

[204] A. N. Kochubei, Distributed order calculus and equation of ultraslow diffusion. J. Math. Anal. Appl. **340**(1), 252-281(2008)

[205] M. Kostić, *Generalized Semigroups and Cosine Functions* (Matematicki Institut SANU, Belgrade, 2011)

[206] H. Kozono, L^1-solutions of the Navier-Stokes equations in exterior domains. Math. Ann. **312**, 319-340(1998)

[207] H. Kozono, M. Yamazaki, The stability of small stationary solutions in Morrey spaces of the Navier-Stokes equation. Indiana Univ. Math. J. **44**(4), 1307-1336(1995)

[208] H. Kozono, S. Shimizu, Stability of stationary solutions to the Navier-Stokes equations in the Besov space. Math. Nachr. **296**(5), 1964-1982(2023)

[209] P. Kumar, D.N. Pandey, D. Bahuguna, Approximations of solutions to a fractional differential equation with a deviating argument. Differential Equations Dynam. Systems **22**(4), 333-352(2014)

[210] D. Lan, Regularity and stability analysis for semilinear generalized Rayleigh-Stokes equations. Evol. Equ. Control The. **11**(1), 259-282(2022)

[211] T. Langlands, B. Henry, S. Wearne, Fractional cable equation models for anomalous electrodiffusion in nerve cells: infinite domain solutions. J. Math. Biol. **59**(6), 761-808(2009)

[212] T. Langlands, B. Henry, S. Wearne, Fractional cable equation models for anomalous electrodiffusion in nerve cells: finite domain solutions. SIAM J. Appl. Math. **71**(4), 1168-1203(2011)

[213] N. Laskin, Fractional Schrödinger equation. Phys. Rev. E **66**, 056108(2002)

[214] N. Laskin, Factional quantum mechanics. Phys. Rev. E **62**, 3135-3145(2000)

[215] N. Laskin, Fractional quantum mechanics and Levy path integrals. Phys. Lett. A **268**, 298-305(2000)

[216] P.D. Lax, *Functional analysis* (Wiley Interscience, New York, 2002)

[217] K.N. Le, W. McLean, M. Stynes, Existence, uniqueness and regularity of the solution of the time-fractional Fokker-Planck equation with general forcing. Commun. Pure Appl. Anal. **18**(5), 2789-2811(2019)

[218] K.N. Le, W. McLean, K. Mustapha, Numerical solution of the time-fractional Fokker-Planck equation with general forcing. SIAM J. Numer. Anal. **54**, 1763-1784(2016)

[219] K.N. Le, W. McLean, M. Stynes, Existence, uniqueness and regularity of the solution of the time-fractional Fokker-Planck equation with general forcing. Commun. Pure Appl. Anal. **18**(5), 2765-2787(2019)

[220] Z. Lei, F.H. Lin, Global mild solutions of Navier-Stokes equations. Comm. Pure Appl. Math. **64**(9), 1297-1304(2011)

[221] P.G. Lemarié-Rieusset, *Recent Developments in the Navier-Stokes Problem* (Chapman CRC Press, 2002)

[222] P.G. Lemarié-Rieusset, The Navier-Stokes equations in the critical Morrey-Campanato space. Rev. Mat. Iberoamericana **23**(3), 897-930(2007)

[223] J. Leray, Sur le mouvement d'un liquide visqueux emplissant l'espace. Acta Math. **63**, 193-248(1934)

[224] L. Li, G.L. Liu, Some compactness criteria for weak solutions of time fractional PDEs. SIAM J. Math. Anal. **50**(4), 3963-3995(2018)

[225] L. Li, J.G. Liu, L. Wang, Cauchy problems for Keller-Segel type time-space fractional diffusion equation. J. Differential Equations **265**(3), 1044-1096(2018)

[226] Y.N. Li, H.R. Sun, Integrated fractional resolvent operator function and fractional abstract Cauchy problem. Abstr. Appl. Anal. **2014**, 1-9(2014)

[227] Y.N. Li, H.R. Sun, Regularity of mild solutions to fractional Cauchy problems with Riemann-Liouville fractional derivative. Electron. J. Differential Equations **2014**(184), 1-13(2014)

[228] Z. Li, Y. Liu, M. Yamamoto, Initial-boundary value problems for multi-term time-fractional diffusion equations with positive constant coefficients. Appl. Math. Comput. **257**, 381-397(2015)

[229] W. Lian, J. Wang, R. Xu, Global existence and blow up of solutions for pseudo-parabolic equation with singular potential. J. Differential Equations **269**(6), 4914-4959(2020)

[230] F. Linares, G. Ponce, *Introduction to Nonlinear Dispersive Equations* (Springer, Berlin, 2009)

[231] J.L. Lions, Sur l'existence de solutions des équations de Navier-Stokes. Rend. Sem. Mat. Univ. Padova **30**, 16-23(1960)

[232] F. Liu, V. Anh, I. Turner, Numerical solution of the space fractional Fokker-Planck equation, J. Computat. Appl. Math. **166**, 209-219(2004)

[233] S. Liu, X. Wu, X.F. Zhou, W. Jiang, Asymptotical stability of Riemann-Liouville fractional nonlinear systems. Nonlinear Dynam. **86**, 65-71 (2016)

[234] W. Liu, M. Rockner, da L.J. Silva, Quasi-linear (stochastic) partial differential equations with time-fractional derivatives. SIAM J. Math. Anal. **50**, 2588-2607(2018)

[235] Y. Liu, W.S. Jiang, F.L. Huang, Asymptotic behaviour of solutions to some pseudo-parabolic equations. Appl. Math. Lett. **25**, 111-114(2012)

[236] W.J. Liu, J.Y. Yu, A note on blow-up of solution for a class of semilinear pseudo-parabolic equations. J. Funct. Anal. **274**, 1276-1283(2018)

[237] Z. Liu, X. Li, Approximate controllability of fractional evolution systems with Riemann-Liouville fractional derivatives. SIAM J. Control Optim. **53**, 1920-1933(2015)

[238] C. Lizama, Regularized solutions for abstract Volterra equations. J. Math. Anal. Appl. **243**, 278-292(2000)

[239] N.H. Luc, N.H. Tuan, Y. Zhou, Regularity of the solution for a final value problem for the Rayleigh-Stokes equation. Math. Methods Appl. Sci. **42**(10), 3481-3495(2019)

[240] Y. Luchko, Some uniqueness and existence results for the initial-boundary value problems for the generalized time-fractional diffusion equation. Comput. Math. Appl. **59**, 1766-1772(2010)

[241] Y. Luchko, Maximum principle for the generalized time-fractional diffusion equation. J. Math. Anal. Appl. **351**, 218-223(2009)

[242] G. Lukaszewicz, P. Kalita, *Navier-Stokes Equations: An Introduction with Applications* (Springer, Berlin, 2016)

[243] J.A.T. Machado, Discrete-time fractional-order controllers. Fract. Calc. Appl. Anal. **4**(1), 47-66(2001)

[244] J.T. Machado, V. Kiryakova, F. Mainardi, Recent history of fractional calculus. Commun. Nonlinear Sci. Numer. Simul. **16**, 1140-1153(2011)

[245] R.L. Magin, *Fractional Calculus in Bioengineering* (Begell House Publishers Inc., Redding, 2006)

[246] P. Magal, S. Ruan, On integrated semigroups and age structured models in L^p-spaces. Differential Integral Equations **20**, 197-239(2007)

[247] P. Magal, S. Ruan, *Center Manifolds for Semilinear Equations with Non-dense Domain and Applications to Hopf Bifurcation in Age Structured Models* (Amer. Math. Soc., 2009)

[248] A. Mahmood, S. Parveen, A. Ara, N.A. Khan, Exact analytic solutions for the unsteady flow of a non-Newtonian fluid between two cylinders with fractional derivative model. Commun. Nonlinear Sci. Numer. Simul. **14**(8), 3309-3319(2009)

[249] N. Masmoudi, T.K. Wong, Local-in-time existence and uniqueness of solutions to the Prandtl equations by energy methods. Comm. Pure Appl. Math. **68**, 1683-1741(2015)

[250] F. Mainardi, Fractional relaxation-oscillation and fractional diffusion-wave phenomena. Chaos Solitons Fractals **7**(9), 1461-1477(1996)

[251] F. Mainardi, *Fractional Calculus and Waves in Linear Viscoelasticity, An Introduction to Mathematical Models* (Imperial College Press, London, 2010)

[252] F. Mainardi, *Fractional Calculus: Some Basic Problems in Continuum and Statistical Mechanics* (Springer Vienna, 1997)

[253] F. Mainardi, R. Gorenflo, On Mittag-Leffler-type functions in fractional evolution processes. J. Comput. Appl. Math. **118**(1-2), 283-299(2000)

[254] F. Mainardi, P. Paradisi, R. Gorenflo, Probability distributions generated by fractional diffusion equations. arXiv preprint arXiv **0704**, 0320(2007)

[255] F. Mainardi, P. Paradisi, A model of diffusive waves in viscoelasticity based on fractional calculus. Proceedings of the 36th IEEE Conference on Decision and Control **5**, 4961-4966(1997)

[256] A.J. Majda, A.L. Bertozzi, Vorticity and Incompressible Flow. Cambridge texts in applied mathematics. Appl. Mech. Rev. **55**(4), B77-B78(2002)

[257] W. McLean, W.C. McLean, *Strongly Elliptic Systems and Boundary Integral Equations* (Cambridge university press, Cambridge, 2000)

[258] W. McLean, Regularity of solutions to a time-fractional diffusion equation. ANZIAM J. **52**, 123-138(2010)

[259] W. McLean, Fast summation by interval clustering for an evolution equation with memory. SIAM J. Sci. Comput. **34**, 3039-3056(2012)

[260] W. McLean, K. Mustapha, R. Ali, O. Knio, Well-posedness and regularity of time-fractional advection-diffusion-reaction equations. Frac. Calc. Appl. Anal. **22**(4), 918-944(2019)

[261] W. McLean, K. Mustapha, R. Ali, O. Knio, Regularity theory for time-fractional advection-diffusion-reaction equations. Comput. Math. Appl. **79**(4), 947-961(2020)

[262] L. Mehrdad, M. Dehghan, The use of Chebyshev cardinal functions for the solution of a partial differential equation with an unknown time-dependent coefficient subject to an extra measurement. J. Comput. Appl. Math. **235**(3), 669-678(2010)

[263] R. Metzler, J. Klafter, The random walks guide to anomalous diffusion: A fractional dynamics approach. Phys. Rep. **339**(1), 1-77(2000)

[264] R. Metzler, E. Barkai, J. Klafter, Deriving fractional Fokker-Planck equations from a generalised master equation. Phys. Rev. Lett. **46**, 431-436(1999)

[265] R. Metzler, E. Barkai, J. Klafter, Anomalous diffusion and relaxation close to thermal equilibrium: A fractional Fokker-Planck equation approach. Physical Review Lett. **82**, 3563-3567(1999)

[266] A. Mikelic, A global existence result for the equations describing unsaturated flow in porous media with dynamic capillary pressure. J. Diff. Eq. **248**, 1561-1577(2010)

[267] R. Mikulevicius, H. Pragarauskas, N. Sonnadara, On the Cauchy-Dirichlet problem in the half space for parabolic SPDEs in weighted Hölder spaces. Acta Appl. Math. **97**, 129-149(2007)

[268] K.S. Miller, B. Ross, *An Introduction to the Fractional Calculus and Fractional Differential Equations* (John Wiley & Sons, Inc., New York, 1993)

[269] H. Miura, Remark on uniqueness of mild solutions to the Navier-Stokes equations. J. Funct. Anal. **218**(1), 110-129(2005)

[270] S. Momani, O. Zaid, Analytical solution of a time-fractional Navier-Stokes equation by Adomian decomposition method. Appl. Math. Comput. **177**, 488-494(2006)

[271] S. Morigi, L. Reichel, F. Sgallari, Fractional Tikhonov regularization with a nonlinear penalty term. E J. Comput. Appl. Math. **324**, 142-154(2017)

[272] J. Mu, B. Ahmad, S. Huang, Existence and regularity of solutions to time-fractional diffusion equations. Comput. Math. Appl. **73**(6), 985-996(2017)

[273] K. Mustapha, D. Schötzau, Well-posedness of hp-version discontinuous Galerkin methods for fractional diffusion wave equations. IMA J. Numer. Anal. **34**, 1426-1446(2014)

[274] M. Naber, Time fractional Schrödinger equation. J. Math. Phys. **45**(8), 3339-3352(2004)

[275] S. Nadeem, S. Asghar, T. Hayat, M. Hussain, The Rayleigh Stokes problem for rectangular pipe in Maxwell and second grade fluid. Meccanica **43**(5), 495-504(2008)

[276] M.T. Nair, *Linear Operator Equation: Approximation and Regularization* (World Scientific, 2009)

[277] F. Neubrander, Integrated semigroups and their applications to the abstract Cauchy problem. Pacific J. Math. **135**, 111-155(1988)

[278] H.L. Nguyen, H.T. Nguyen, Y. Zhou, Regularity of the solution for a final value problem for the Rayleigh-Stokes equation. Math. Methods Appl. Sci. **42**, 3481-3495(2019)

[279] H.T. Nguyen, L.D. Long, V.T. Nguyen, T. Tran, On a final value problem for the time-fractional diffusion equation with inhomogeneous source. Inverse Probl. Sci. Eng. **25**, 1367-1395(2017)

[280] H.T. Nguyen, Y. Zhou, N.T. Tran, H.C. Nguyen, Initial inverse problem for the nonlinear fractional Rayleigh-Stokes equation with random discrete data. Commun. Nonlinear Sci. Numer. Simul. **78**, 104873(2019)

[281] R.R. Nigmatullin, The realization of the generalized transfer equation in a medium with fractal geometry. Phys. Stat. Sol. B **133**, 425-430(1986)

[282] R.H. Nochetto, E. Otárola, A.J. Salgado, A PDE approach to space-time fractional wave problems. SIAM J. Numer. Anal. **54**, 848-873(2016)

[283] S.K. Ntouyas, Boundary value problems for nonlinear fractional differential equations and inclusions with nonlocal and fractional integral boundary conditions. Opuscula Math. **33**, 117-138(2013)

[284] Y. Oka, E. Zhanpeisov, Existence of solutions to fractional semilinear parabolic equations in Besov-Morrey spaces. (2023) arXiv:2301.04263

[285] E. Orsingher, L.Beghin, Time-fractional telegraph equations and telegraph process with Brownian time. Probab. Theory Related Fields **128**(1), 141-160(2004)

[286] M.S. Osman, J.A.T. Machado, D. Baleanu, On nonautonomous complex wave solutions described by the coupled Schrödinger-Boussinesq equation with variable-coefficients. Optical and Quantum Elec. **50**(2), 1-11(2018)

[287] J. Otero, A uniqueness theorem for a Robin boundary value problem of physical geodesy. Quart. Appl. Math. **56**, 245-257(1998)

[288] E.M. Ouhabaz, *Analysis of Heat Equations on Domains* (Princeton University Press, Princeton, NJ, 2005)

[289] V. Pandey, S. Holm, Linking the fractional derivative and the Lomnitz creep law to non-Newtonian time-varying viscosity. Phys. Rev. E **94**(3), 032606(2016)

[290] M.D. Paola, A. Pirrotta, A. Valenza, Visco-elastic behavior through fractional calculus: an easier method for best fitting experimental results. Mech. Mater. **43**, 799-806(2011)

[291] A. Pazy, *Semigroup of Linear Operators and Applications to Partial Differential Equations* (Springer-Verlag, New York, 1983)

[292] H. Pecher, L^p-Abschätzungen und klassische Lösungen für nichtlineare Wellengleichungen. Manuscripta Math. **20**(3), 227-244(1977)

[293] L. Peng, A. Debbouche, Y. Zhou, Existence and approximations of solutions for time-fractional Navier-stokes equations. Math. Methods Appl. Sci. **41**, 8973-8984(2018)

[294] L. Peng, Y. Zhou, The existence of mild and classical solutions for time fractional Fokker-Planck equations. Monatsh. Math. **199**, 377-410(2022)

[295] L. Peng, Y. Zhou, The analysis of approximate controllability for distributed order fractional diffusion problems. Appl. Math. Optim. **86**(2), 22(2022)

[296] L. Peng, Y. Zhou, The well-posdness results of solutions in Besov-Morrey spaces for fractional Rayleigh-Stokes equations. Qual. Theory Dyn. Syst. **23**, 43(2024)

[297] L. Peng, Y. Zhou, Characterization of solutions in Besov spaces for fractional Rayleigh-Stokes equations. Commun. Nonlinear Sci. Numer. Simul. to appress.

[298] L. Peng, Y. Zhou, B. Ahmad, A. Alsaedi, The Cauchy problem for fractional Navier-Stokes equations in Sobolev spaces. Chaos Solitons Fractals **102**, 218-228(2017)

[299] L. Peng, Y. Zhou, B. Ahmad, The well-posedness for fractional nonlinear Schrödinger equations. Comput. Math. Appl. **77**(7), 1998-2005(2019)

[300] Li. Peng, Y. Zhou and J.W. He, The well-posedness analysis of distributed order fractional diffusion problems on $\mathbb{R}^N$. Monatsh. Math. **198**, 445-463 (2022)

[301] L. Pinto, E. Sousa, Numerical solution of a time-space fractional Fokker-Planck equation with variable force field and diffusion. Commun. Nonlinear Sci. Numer. Simul. **50**, 211-228(2017)

[302] F. Planchon, Global strong solutions in Sobolev or Lebesgue spaces to the incompressible Navier-Stokes equations in $\mathbb{R}^3$. Ann. Inst. H. Poincare Anal. Non Lineaire **13**, 319-336(1996)

[303] I. Podlubny, *Fractional Differential Equations* (Academic Press, San Diego, 1999)

[304] I. Podlubny, Fractional-order systems and $PI^\lambda D^\mu$-controllers. IEEE Trans. Automat. Control **44**(1), 208-214(1999)

[305] R. Ponce, Hölder continuous solutions for fractional differential equations and maximal regularity. J. Differential Equations **255**, 3284-3304(2013)

[306] J.C. Pozo, V. Vergara, A non-local in time telegraph equation. Nonlinear Anal. **193**, 111411(2020)

[307] J. Prüss, *Evolutionary Integral Equations and Applications* (Birkhäuser Verlag, Basel, 1993)

[308] G. Raugel, G.R. Sell, Navier-Stokes equations on thin 3D domains, I, Global attractors and global regularity of solutions. J. Amer. Math. Soc. **6**, 503-568(1993)

[309] M. Reed, B. Simon, *Methods of Modern Mathematical Physics I: Functional Analysis. Revised and enlarged edition* (Elsevier, Amsterdam,1980)

[310] J.C. Robinson, J.L. Rodrigo, W. Sadowski, *The Three-Dimensional Navier-Stokes Equations: Classical Theory* (Cambridge University Press, London, 2016)

[311] M. Sack, M. Struwe, Scattering for a critical nonlinear wave equation in two space dimensions. Math. Ann. **365**, 969-985(2016)

[312] P.G. Saffman, On the boundary condition at the surface of porous medium. Stud. Appl. Math. **50**(2), 93-101(1971)

[313] K. Sakamoto, M. Yamamoto, Initial value/boundary value problems for fractional diffusion-wave equations and applications to some inverse problems. J. Math. Anal. Appl. **382**(1), 426-447(2011)

[314] F. Salah, Z.A. Aziz, D.L.C. Ching, New exact solution for Rayleigh-Stokes problem of Maxwell fluid in a porous medium and rotating frame. Results Phys. **1**(1), 9-12(2011)

[315] T.O. Salim, Solution of fractional order Rayleigh-Stokes equations. Adv. Theor. Appl. Mech. **1**(5), 241-254(2008)

[316] S.G. Samko, A.A. Kilbas, O.I. Marichev, *Fractional Integrals and Derivatives. Theory and Applications* (Gordon and Breach Science, 1987)

[317] L. Saint-Raymond, From the BGK model to the Navier-stokes equations. Ann. Sci. Ecole Norm. Sup. (4) **36**(2), 271-317(2003)

[318] E. Scalas, R. Gorenflo, F. Mainardi, Fractional calculus and continuous-time finance. Phys. A **284**(1-4), 376-384(2000)

[319] G. Seregin, *Lecture Notes on Regularity Theory for the Navier-Stokes Equations* (World Scientific, Singapore, 2014)

[320] R. Servadei and E. Valdinoci, On the spectrum of two different fractional operators. Proc. Roy. Soc. Edinburgh Sect. A **144**(4), 831-855(2014)

[321] F. Shen, W. Tan, Y. Zhao, T. Masuoka, The Rayleigh-Stokes problem for a heated generalized second grade fluid with fractional derivative model. Nonlinear Anal. Real World Appl. **7**(5), 1072-1080(2006)

[322] Y.L. Shi, L. Li, Z.H. Shen, Boundedness of p-adic Singular integrals and multilinear commutator on Morrey-Herz spaces. Journal of Function Spaces, **2023**, 9965919(2023)

[323] R.E. Showalter, The final value problem for evolution equations. J. Math. Anal. Appl. **47**(3), 563-572(1974)

[324] D. Stan, F. del Teso, J.L. Vázquez, Existence of weak solutions for a general porous medium equation with nonlocal pressure. Arch. Ration. Mech. Anal. **233**(1), 451-496(2019)

[325] M. Subaşi, An optimal control problem governed by the potential of a linear Schrödinger equation. Appl. Math. Comput. **131**, 95-106(2002)

[326] P.L. Sulem, C. Sulem, *The Nonlinear Schrödinger Equation: Self-focusing and Wave Collapse* (Springer Science & Business Media, Berlin, 2007)

[327] M. Taylor, *Partial Differential Equations II: Qualitative Studies of Linear Equations* (Springer Science & Business Media, Berlin, 2013)

[328] M. Taylor, Analysis on Morrey spaces and applications to Navier-Stokes equation. Comm. Partial Differential Equations. **17**, 1407-1456(1992)

[329] R. Temam, *Navier-Stokes Equations and Nonlinear Functional Analysis* (Society for Industrial and Applied Mathematics, 1995)

[330] R. Temam, *Navier-Stokes Equations: Theory and Numerical Analysis* (North-Holland, Amsterdam, 1977)

[331] E. Topp, M. Yangari, Existence and uniqueness for parabolic problems with Caputo time derivative. J. Differential Equations, **262**(12), 6018-6046(2017)

[332] T.P. Tsai, Asymptotic dynamics of nonlinear Schrödinger equations with many bound states. J. Differential Equations **192**, 225-282(1995)

[333] N.A. Triet, L.V.C. Hoan, N.H. Luc, N.H. Tuan, N.V. Thinh, Identification of source term for the Rayleigh-Stokes problem with Gaussian random noise. Math. Methods Appl. Sci. **41**(14), 5593-5601(2018)

[334] N.H. Tuan, V.V. Au, A.T. Nguyen, Mild solutions to a time-fractional Cauchy problem with nonlocal nonlinearity in Besov spaces. Arch. Math. **118**(3), 305-314(2022)

[335] P.T. Tuan, T.D. Ke, N.N. Thang, Final value problem for Rayleigh-Stokes type equations involving weak-valued nonlinearities. Fract. Calc. Appl. Anal. **26**(2), 694-717(2023)

[336] N.H. Tuan, T. Caraballo, T.B. Ngoc and Y. Zhou, Existence and regularity results for terminal value problem for nonlinear fractional wave equations. Nonlinearity, **34**(3), 1448(2021)

[337] N.H. Tuan, T.B. Ngoc, Y. Zhou and D. O'Regan, On existence and regularity of a terminal value problem for the time fractional diffusion equation. Inverse Problems **36**(5), 055011(2020)

[338] N.H. Tuan, N.D. Phuong, T.N. Thach, New well-posedness results for stochastic delay Rayleigh-Stokes equations. Discrete Contin. Dyn. Syst. Ser. B **28**(1), 347-358(2022)

[339] N.H. Tuan, Y. Zhou, T.N. Thach, N.H. Can, Initial inverse problem for the nonlinear fractional Rayleigh-Stokes equation with random discrete data. Commun. Nonlinear Sci. Numer. Simul. **78**, 104873(2019)

[340] W. Varnhorn, *The Stokes Equations* (Akademie Verlag, Berlin, 1994)

[341] A.F. Vasseur, C. Yu, Existence of global weak solutions for 3D degenerate compressible Navier-Stokes equations. Invent. math. **206**(3), 935-974(2016)

[342] H. Vivian, J. Pym, M. Cloud, *Applications of Functional Analysis and Operator Theory* (Elsevier, Amsterdam, 2005)

[343] W. Von Wahl, *The Equations of Navier-Stokes and Abstract Parabolic Equations* (Braunschweig, Vieweg, 1985)

[344] B.X. Wang, Z.H. Huo, C.C. Hao, *Harmonic Analysis Method for Nonlinear Evolution Equations* (World Scientific, Singapore, 2011)

[345] J. Wang, T. Wei, Y. Zhou, Optimal error bound and simplified Tikhonov regularization method for a backward problem for the time-fractional diffusion equation. J. Comput. Appl. Math. **279**, 277-292 (2015)

[346] J.N. Wang, A. Alsaedi, B. Ahmad, Y. Zhou, Well-posedness and blow-up results for nonlinear fractional Rayleigh-Stokes problem. Adv. Nonlinear Anal. **11**, 1579-1597 (2022)

[347] J.N. Wang, Y. Zhou, J.W. IIe, Existence and regularization of solutions for nonlinear fractional Rayleigh-Stokes problem with final condition. Math. Methods Appl. Sci. **44**, 13493-13508 (2021)

[348] J.N. Wang, Y Zhou, A. Alsaedi, B. Ahmad, Well-posedness and regularity of fractional Rayleigh-Stokes problems. Z. Angew. Math. Phys. **73**(4), 161(2022)

[349] T. Wei, J.G. Wang, A modified quasi-boundary value method for an inverse source problem of the time-fractional diffusion equation. Appl. Numer. Math. **78**, 95-111(2014)

[350] Y. Wang, Global existence and blow up of solutions for the inhomogeneous nonlinear Schrödinger equation in $\mathbb{R}^2$. J. Math. Anal. Appl. **338**, 1008-1019(2008)

[351] Y. Wang, S. Li and Z. Si, A second order in time incremental pressure correction finite element method for the Navier-Stokes/Darcy problem. ESAIM Math. Model. Numer. Anal. **52**(4), 1477-1500(2018)

[352] R.N. Wang, D.H. Chen, T.J. Xiao, Abstract fractional Cauchy problems with almost sectorial operators. J. Differential Equations **252**, 202-235(2012)

[353] J.G. Wang, Y.B. Zhou, T. Wei, A posteriori regularization parameter choice rule for the quasiboundary value method for the backward time-fractional diffusion problem. Appl. Math. Lett. **26**(7), 741-747(2013)

[354] J.R. Wang, Y. Zhou, A class of fractional evolution equations and optimal controls. Nonlinear Anal. Real World Appl. **12**, 262-272(2011)

[355] J.R. Wang, Y. Zhou, W. Wei, Fractional Schrödinger equations with potential and optimal controls. Nonlinear Anal. Real World Appl. **13**, 2755-2766(2012)

[356] M. Warma, A fractional Dirichlet-to-Neumann operator on bounded Lipschitz domains. Commun. Pure Appl. Anal. **14**(5), 2043-2067(2015)

[357] M. Warma, The fractional relative capacity and the fractional Laplacian with Neumann and Robin boundary conditions on open sets. Potential Anal. **42**(2), 499-547(2015)

[358] M. Warma, The fractional Neumann and Robin type boundary conditions for the regional fractional p-Laplacian. NoDEA Nonlinear Differential Equations Appl. **23**(1), 1-46(2016)

[359] M. Warma, The Robin and Wentzell-Robin Laplacians on Lipschitz domains. Semigroup Forum **73**, 10-30(2006)

[360] M. Warma, Approximate controllability from the exterior of space-time fractional diffusive equations. SIAM J. Control Optim. **57**(3), 2037-2063(2019)

[361] M. Warma, *The Laplacian with General Robin Boundary Conditions* (Ph.D. Dissertation, University of Ulm (Germany), 2002)

[362] F.B. Weissler, The Navier-Stokes initial value problem in L^p. Arch. Ration. Mech. Anal. **74**, 219-230(1980)

[363] G. M. Wright, The generalized Bessel function of order greater than one. Q. J. Math. **11**, 36-48(1940)

[364] X.X. Xi, M. Hou, X.F. Zhou, Y. Wen. Approximate controllability for mild solution of time-fractional Navier-Stokes equations with delay. Z. Angew. Math. Phys. **72**(3), 113(2021)

[365] R.Z. Xu, J. Su, Global existence and finite time blow-up for a class of semilinear pseudo-parabolic equations. J. Funct. Anal. **264**, 2732-2763(2013)

[366] C. Xue, J. Nie, Exact solutions of Rayleigh-Stokes problem for heated generalized Maxwell fluid in a porous half-space. Math. Probl. Eng. **2008**(2008)

[367] C. Xue, J. Nie, Exact solutions of the Rayleigh-Stokes problem for a heated generalized second grade fluid in a porous half-space. Appl. Math. Modelling **33**(1), 524-531(2009)

[368] A. Yagi, *Abstract Parabolic Evolution Equations and Their Applications* (Springer Science & Business Media, Berlin, 2009)

[369] X. Yang. X. Jiang, Numerical algorithm for two dimensional fractional Stokes' first problem for a heated generalized second grade fluid with smooth and non-smooth solution. Comput. Math. Appl. **78**(5), 1562-1571(2019)

[370] M. Yamazaki, The Navier-Stokes equations in the weak-L^n space with time-dependent external force. Math. Ann. **317**(4), 635-675(2000)

[371] H. Ye, J. Gao, Y. Ding, A generalized Gronwall inequality and its application to a fractional differential equation. J. Math. Anal. Appl. **328**(2), 1075-1081(2007)

[372] H. Yetişkin, M. Subaşi, On the optimal control problem for Schrödinger equation with complex potential. Appl. Math. Comput. **216**, 1896-1902(2010)

[373] B. Yildiz, M. Subaşi, On the optimal control problem for linear Schrödinger equation. Appl. Math. Comput. **121**, 373-381(2001)

[374] K. Yosida, *Functional Analysis, Sixth edition* (Springer-Verlag, Berlin, 1999)

[375] R. Zacher, A De Giorgi-Nash type theorem for time fractional diffusion equations. Math. Ann. **356**, 99-146(2013)

[376] R. Zacher, Maximal regularity of type L_p for abstract parabolic Volterra equations. J. Evol. Equ. **5**, 79-103(2005)

[377] R. Zacher, A De Giorgi-Nash type theorem for time fractional diffusion equations. Math. Ann. **356**, 99-146(2013)

[378] A.M. Zaky, An improved tau method for the multi-dimensional fractional Rayleigh-Stokes problem for a heated generalized second grade fluid. Comput. Math. Appl. **75**(7), 2243-2258(2018)

[379] G.M. Zaslavsky, Chaos, fractional kinetics, and anomalous transport. Phys. Rep. **371**, 461-580(2002)

[380] G.M. Zaslavsky, Fractional kinetic equation for Hamiltonian chaos. Phys. D **76**, 110-122(1994)

[381] E. Zeidler, *Nonlinear Functional Analysis and Its Application II/A* (Springer-Verlag, Berlin, 1990)

[382] K. Zhang, On Shinbrot's conjecture for the Navier-Stokes equations. Proc. R. Soc. Lond. Ser. A Math. Phys. Eng. Sci. **440**(1910), 537-540(1993)

[383] Q.G. Zhang, H.R. Sun, The blow-up and global existence of solutions of Cauchy problems for a time fractional diffusion equation. Topol. Methods Nonlinear Anal. **46**(1), 69-92(2015)

[384] M. Zheng, F. Liu, I. Turner, et al., A novel high order space-time spectral method for the time fractional Fokker-Planck equation, SIAM J. Sci. Comput. **37**, A701-A724(2015)

[385] G.H. Zhou, Z.B. Guo, Boundary feedback stabilization for an unstable time fractional reaction diffusion equation. SIAM J. Control Optim. **56**, 75-10(2018)

[386] Y. Zhou, *Basic Theory of Fractional Differential Equations* (World Scientific, Singapore, 2014)

[387] Y. Zhou, *Fractional Evolution Equations and Inclusions: Analysisand Control* (Academic Press, London, 2016)

[388] Y. Zhou, Existence and uniqueness of solutions for a system of fractional differential equations. Frac. Calc. Appl. Anal. **12**(2), 195-204(2009)

[389] Y. Zhou, Attractivity for fractional evolution equations with almost sectorial operators. Frac. Calc. Appl. Anal. **21**(3), 786-800(2018)

[390] Y. Zhou, Attractivity for fractional differential equations. Appl. Math. Lett. **75**, 1-6(2018)

[391] Y. Zhou, B. Ahmad, A. Alsaedi, Existence of nonoscillatory solutions for fractional neutral differential equations. Appl. Math. Lett. **72**, 70-74(2017)

[392] Y. Zhou, J.W. He, Well-posedness and regularity for fractional damped wave equations. Monatsh. Math. **194**(2), 425-458(2021)

[393] Y. Zhou, J.W. He, New results on controllability of fractional evolution systems with order $\alpha \in (1, 2)$, Evol. Equ. Control Theory **10**, 491-509(2021)

[394] Y. Zhou, J.W. He, A. Alsaedi, B. Ahmad, The well-posedness for semilinear time fractional wave equations on $\mathbb{R}^n$. Elec. Res. Arch. **30**(8), 2981-3003(2022)

[395] Y. Zhou, J.W. He, B. Ahmad, N.H. Tuan, Existence and regularity results of a backward problem for fractional diffusion equations. Math. Methods Appl. Sci. **42**, 6775-6790(2019)

[396] Y. Zhou, F. Jiao, Existence of mild solutions for fractional neutral evolution equations. Comput. Math. Appl. **59**, 1063-1077(2010)

[397] Y. Zhou, F. Jiao, Nonlocal Cauchy problem for fractional evolution equations. Nonlinear Anal. Real World Appl. **11**, 4465-4475(2010)

[398] Y. Zhou, F. Jiao, J. Li, Existence and uniqueness for fractional neutral differential equations with infinite delay. Nonlinear Anal. **71**(7/8), 3249-3256(2009)

[399] Y. Zhou, F. Jiao, J. Li, Existence and uniqueness for p-type fractional neutral differential equations. Nonlinear Anal. **71**(7/8), 2724-2733(2009)

[400] Y. Zhou, F. Jiao, J. Pečarić, Abstract Cauchy problem for fractional functional differential equations. Topol. Methods Nonlinear Anal. **42**, 119-136(2013)

[401] Y. Zhou, J. Manimaran, L. Shangerganesh, A. Debbouche, Weakness and Mittag-Leffler stability of solutions for time-fractional keller-segel models. Int. J. Nonlinear Sci. Numer. Simul. **19**(7/8), 753-761(2018)

[402] Y. Zhou, L. Peng, On the time-fractional Navier-Stokes equations. Comput. Math. Appl. **73**(6), 874-891(2017)

[403] Y. Zhou, L. Peng, Weak solutions of the time-fractional Navier-Stokes equations and optimal control. Comput. Math. Appl. **73**(6), 1016-1027(2017)

[404] Y. Zhou, L. Peng, Y.Q. Huang, Existence and Hölder continuity of solutions for time-fractional Navier-Stokes equations. Math. Methods Appl. Sci. **41**, 7830-7838(2018)

[405] Y. Zhou L. Peng, Y.Q. Huang, Duhamel's formula for time-fractional Schrödinger equations. Math. Methods Appl. Sci. **41**, 8345-8349(2018)

[406] Y. Zhou, L. Peng, B. Ahmad and A. Alsaedi, Energy methods for the time-fractional Navier-Stokes equations. Chaos Solitons Fractals **102**, 78-85(2017)

[407] Y. Zhou, L. Shangerganesh, J. Manimaran, A. Debbouche, A class of time-fractional reaction-diffusion equation with nonlocal boundary condition. Math. Methods Appl. Sci. **41**, 2987-2999(2018)

[408] Y. Zhou, S. Suganya, B. Ahmad, Approximate controllability of impulsive fractional integro-differential equation with state-dependent delay in Hilbert spaces. IMA J. Math. Control Inform. **36**, 603-622(2019)

[409] Y. Zhou, J.N. Wang, The nonlinear Rayleigh-Stokes problem with Riemann-Liouville fractional derivative. Math. Methods Appl. Sci. **44**(3), 2431-2438(2021)

[410] Y. Zhou, V. Vijayakumar, R. Murugesu, Controllability for fractional evolution inclusions without compactness. Evol. Equ. Control Theory **4**, 507-524(2015)

[411] Y. Zhou, V. Vijayakumar, C. Ravichandran, R. Murugesu, Controllability results for fractional order neutral functional differential inclusions with infinite delay. Fixed Point Theory **18**, 773-798(2017)

Index

Printed in the United States
by Baker & Taylor Publisher Services

Printed in the United States
by Baker & Taylor Publisher Services